The AutoCAD® Tutor

for

Engineering Graphics

Release 14

by Alan J. Kalameja

The AutoCAD® Tutor

for

Engineering Graphics

Release 14

by Alan J. Kalameja

Press

I(T)P® International Thomson Publishing

Albany • Bonn • Boston • Cincinnati • Detroit • London • Madrid
Melbourne • Mexico City • New York • Pacific Grove • Paris • San Francisco
Singapore • Tokyo • Toronto • Washington

NOTICE TO THE READER

Trademarks

AutoCAD® and the AutoCAD® logo are registered trademarks of Autodesk, Inc.
Windows is a trademark of the Microsoft Corporation.
All other product names are acknowledged as trademarks of their respective owners.

Cover: Background image Copyright © 1996 PhotoDisc, Inc.

COPYRIGHT © 1998
Delmar Publishers Inc.
Autodesk Press imprint
an International Thomson Publishing Company

The ITP logo is a trademark under license
Printed in the United States of America

For more information, contact:

Autodesk Press
3 Columbia Circle, Box 15-015
Albany, New York 12212-5015

International Thomson Editores
Campos Eliseos 385, Piso 7
Colonia Polanco
11560 Mexico D. F. Mexico

International Thomson Publishing Europe
Berkshire House 168-173
High Holborn
London, WC1V7AA
United Kingdom

International Thomson Publishing GmbH
Königswinterer Strasse 418
53227 Bonn Germany

Thomas Nelson Australia
102 Dodds Street
South Melbourne, 3205
Victoria, Australia

International Thomson Publishing France
Tour Maine-Montparnesse
33, Avenue du Maine
75755 Paris Cedex 15, France

Nelson Canada
1120 Birchmont Road
Scarborough, Ontario
Canada, M1K 5G4

International Thomson Publishing - Japan
Hirakawacho Kyowa Building, 3F
2-2-1 Hirakaw-cho Chiyoda-ku
Tokyo 102 Japan

International Thomson Publishing Southern Africa
Building 18, Constantia Park
240 Old Pretoria Road
P.O. Box 2459
Halfway House, 1685 South Africa

International Thomson Publishing Asia
221 Henderson Road
S#05 -10 Henderson Building
Singapore 0315

4 5 6 7 8 9 10 XXX 03 02 01 00 99

Library of Congress Cataloging-in-Publication Data

Kalameja, Alan J.
    The AutoCAD tutor for engineering graphics release 14 / by Alan J. Kalameja.
        p. cm.
    Includes index.
    ISBN: 0-7668-0131-4
    1. Engineering graphics. 2. AutoCAD (Computer file). I. Title

T385.K3452 1998
604.2'0285'5369–dc21   97-25353
                                                          CIP

Contents

Unit 2 Object Construction and Manipulation 2–1

Unit 3   Geometric Constructions 3–1

Unit 6   Analyzing 2D Drawings 6–1

Appendix A Bonus Commands A–1

Index

Preface

Engineering graphics has been around for a long time as a means of defining an object graphically before it is constructed and used by consumers. Previously, the process for producing the drawing involved drawing aids such as pencils, ink pens, triangles, t-squares, etc. to place the idea on paper before making changes and producing blue-line prints for distribution. The ability to produce these drawings on a computer is quite new, however the principles and basics of engineering drawing remain the same.

This text uses engineering drawing basics to produce drawings using AutoCAD Release 14 and a series of tutorial exercises that follow each unit. Following the tutorials, extra problems are provided to enhance your skills in constructing engineering drawings. A brief description of each unit follows:

Unit 1 - AutoCAD Fundamentals

This first unit is provided to introduce you to such fundamental AutoCAD concepts as basic drawing; screen elements; the use of function keys; starting a drawing file using the Wizard; setting drawing units and limits; opening up an existing drawing file; the importance organizing a drawing through layers; using object snaps and the AutoSnap feature; magnifying the drawing using the ZOOM command; using the PAN command; productive uses of realtime zooms and pans; understanding absolute; relative; and polar coordinates; using the Direct Distance mode; drawing lines; erasing lines; using the DDVIEW dialog box; and saving drawings. Two layer tutorials and one drawing tutorial follow and additional problems are provided at the end of the unit.

Unit 2 - Object Construction and Manipulation

This unit provides a brief explanation of all AutoCAD drawing and editing commands. The use of blocks is also discussed. The topic of grips and how they are used to enhance drawing accuracy and productivity is presented. A series of tutorials follow used to complement the topics covered in the unit. Additional problems related to the tutorial exercises follow.

Unit 3 - Geometric Constructions

This unit discusses how AutoCAD commands and Object Snap options may be used for geometric construction purposes. Two tutorial exercises follow along with additional problems at the end of the unit.

Unit 4 - Shape Description/Multiview Projection

Shape description and multi-view drawing using AutoCAD are the focus of this unit. The basics of shape description are discussed along with proper use of linetypes, fillets and rounds, chamfers, and runouts. One tutorial outlines the steps used for creating a multi-view drawing using AutoCAD. Another tutorial approaches multi-view drawing through the use of .XYZ filters and tracking techniques. A series of sketching exercises along with additional drawing problems are provided at the end of the unit.

Unit 5 - Dimensioning Techniques

Dimensioning techniques using AutoCAD is the topic of this unit. Basic dimensioning rules are discussed before concentrating on all AutoCAD dimensioning commands. A thorough discussion of the use of the Dimension Styles dialog box is included in this unit. Three tutorials follow along with additional dimensioning problems.

Unit 6 - Analyzing 2-D Drawings

This unit provides information on analyzing drawings for accuracy purposes. The AREA, ID, LIST, DIST, and DDMODIFY commands and how they are used on various objects for analysis purposes are discussed in detail. A series of tutorial exercises follow to allow users to test their accuracy of drawing. Numerous problems along with questions on each problem are provided at the end of this unit to further test a user's drawing accuracy.

Unit 7 - Region Modeling Techniques

This unit provides instruction on using the BOUNDARY and REGION commands to create a region; this can be thought of as an alternate means of constructing and analyzing objects. Drawing features such as holes and slots are formed through the use of the Boolean operations of union, subtraction, and intersection. Mass property calculations are then performed on a region which yield area, perimeter, and centroid information. A tutorial exercise and various drawing problems are provided at the end of this unit.

Unit 8 - Section Views

Section views are described in this unit including full, half, assembly, aligned, offset, broken, revolved, removed, and isometric sections. Hatching techniques through the use of the BHATCH command in AutoCAD are also discussed. The advantages of associative hatching will also be discussed. Three tutorial exercises follow along with additional problems dealing with the topic of section views.

Unit 9 - Auxiliary Views

Producing auxiliary views using AutoCAD is discussed in this unit. Items discussed include rotating the snap at an angle to project lines of sight perpendicular to a surface to be used for the preparation of the auxiliary view. One tutorial exercise and additional problems follow in this unit.

Unit 10 - Isometric Drawings

This unit discusses constructing isometric drawings with particular emphasis on using the Snap-Style option in AutoCAD used to create an isometric grid. Methods of toggling between right, top, and left isometric modes will be explained. In addition to isometric basics, creating circles and angles in isometric are also discussed. Two tutorial exercises follow along with additional problems at the end of this unit.

Unit 11 - Solid Modeling Fundamentals

This unit begins with a comparison between isometric, extruded, wireframe, surfaced, and solid model drawings. The unit continues with a detailed discussion of the use of the User Coordinate System and how it is positioned to construct objects in 3D. The display of 3D images through the VPOINT and DDVPOINT commands is discussed. Creating various solid primitives such as cones and cylinders is discussed in addition to the ability to construct complex solid objects through the use of the Boolean operations of union, subtraction, and intersection. The unit continues on to a discussion of extruding and rotating operations for creating solid models in addition to filleting and chamfering solid models. Analyzing solid models is also covered in this unit. Once the solid model is created, the SOLVIEW command is used to layout 2-D views of the model while the SOLDRAW command is used to draw the 2-D views. Because of the importance of this design paradigm, six tutorial exercises follow along with additional problems at the end of this unit.

Online Companion

This new edition contains a special Internet companion piece. The Online Companion™ is your link to AutoCAD on the Internet. We've compiled supporting resources with links to a variety of sites. Not only can you find out about training and education, industry sites, and the online community, we also point to valuable archives compiled for AutoCAD users from various Web sites. In addition, there is information of special interest to users of *The AutoCAD Tutor Engineering Graphics*. These include updates, information about the author, and a page where you can send us your comments. You can find the Online Companion at:

http://www.autodeskpress.com/onlinecompanion.html

When you reach the Online Companion page, click on the title *The AutoCAD Tutor for Engineering Graphics*.

Acknowledgments

I wish to thank the staff at Autodesk Press for their assistance with this document, especially Sandy Clark, Peg Gantz, Mary Beth Vought, and Jennifer Gaines. Thanks also go out to John Shanley of Phoenix Creative Graphics for his part in the desktop publishing aspects of this document. I would also like to thank Don Baer of Trident Technical College for performing the technical edit on the entire manuscript.

Special thanks go out to Barbara Savins who assisted with the problems at the end of the Shape Description/Multi-View Projection Unit.

Finally, I dedicate this book in memory of my father, Edward J. Kalameja who influenced my appreciation in the discipline and applications of design and engineering drawing techniques.

Alan J. Kalameja

The publisher and author would like to thank and acknowledge the many professionals who reviewed the manuscript to help us publish this AutoCAD Release 14 text. A special acknowledgment is due the following instructors who reviewed the chapters in detail:

Roger Burgess
Greenville Technical College

Mike Carmen
Oklahoma State University – Okmulgee

Jerry B. Davis
Abraham Baldwin Agricultural College

Steve Luft
College of Technology

Augusto Op den Bosch
Georgia Institute of Technology

Mr. Porras
North Valley Occupational Center, CA

Special thanks for his thorough technical edit of the material are extended to:

Keith Donald Baer
Trident Technical College

About the Author

Alan J. Kalameja is the Department Head of Computer-Aided Design at Trident Technical College located in Charleston, South Carolina. He has been at the College for over 16 years and has been using AutoCAD since 1984. He directs the Authorized AutoCAD Training Center at Trident, which is charged with providing industry training to local and regional companies. Currently, he is an Education Training Specialist with Autodesk, is a member of the AutoCAD Certification Exam Board, and has authored *The AutoCAD Release 12 Certification Exam Preparation Manual* and *The AutoCAD Release 13 Certification Exam Preparation Manual* both by Delmar Publishers/Autodesk Press.

Conventions

All tutorials in this publication use the following conventions in the instructions:

Whenever you are told to enter text, the text appears in **boldface** type. This may take the form of entering an AutoCAD command or entering such information as absolute, relative or polar coordinates. You must follow these and all text inputs by pressing the Return or Enter key to execute the input. For example, to draw a line using the LINE command from coordinate value (3,1) to coordinate value (8,2), the sequence would look line the following:

Command: **LINE**
From point: **3,1**
To point: **8,2**
To point: *(Press* ENTER *to exit this command)*

Instructions in this tutorial are designed to enter all commands, options, coordinates, etc., from the keyboard. You may enter the same commands by selecting them from the screen menu, digitizing tablet, pull-down menu area, or floating toolbar.

Instructions for selecting objects are in italic type. When instructed to select an object, move the pickbox on the object to be selected and press the pick button on the mouse or digitizing puck.

If you enter the wrong command for a particular step, you may cancel the command by pressing the ESC key. This key is located in the upper left hand corner of any standard keyboard.

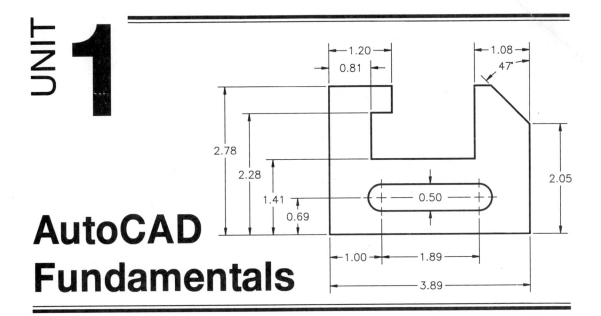

AutoCAD Fundamentals

Welcome to the world of Computer-Aided Design (CAD). CAD is used as a tool to produce all types of engineering graphics drawings whether they be two-dimensional or three-dimensional or whether they be architectural, electrical, or mechanical in application. Drafting and design has always been compared to a language such as English or German; however, the design process and the ability to capture a design technically on paper was and still is considered the language of industry. Due to the evolution of the computer, the design has shifted from paper to the video display, although paper output through a plotter is still considered in some applications an absolute must. The same tools that were available for the manual production of drawings have changed considerably using the computer and, in some cases, the results may appear unnoticeable therefore casting doubt on the justification of the CAD terminal versus the drawing board. However, if ever there was an instrument that could have been used to eliminate the drudgery and tedious nature of the manual drawing board, the CAD terminal has found its place in the modern design office. When drawings were first laid out manually, if the drawing was not properly centered for appearance pur-

poses, a new calculation was performed, lines were erased (if they were drawn lightly), and new lines were constructed at the new location. Although it may be argued that basic manual practices such as centering drawings should always be done, with CAD, most manual practices are considered unnecessary. In CAD, if a drawing appears off-center, simply exercise a popular CAD command to move the drawing into the proper position. Once dimensions are added and the drawing once again appears off-center, move it again and again to achieve the proper appearance. This is but one of hundreds of time-savers used to justify the existence of a computer in the design process; but it just does not stop here for 2D drawings.

The ability to use the computer to model an object in three dimensions has always been considered a form of art when performed on the manual drawing board. Not any more. When growing up as children and when asked to draw a house, we all tended to create a picture complete with receding lines and depth because this is what our eyes actually saw. Then to our surprise, this ability was taken away as we were told to look at the picture as consisting of a series of primary views and draw them flat. Now with 3D graphics available on most CAD systems, an individual not only is able to construct in three dimensions, but is also able to construct a prototype called a solid model to be used for analysis.

Use the text in the following pages and units to get a better idea of the world of 2D and 3D design and how it has been changed using the tool called the computer.

The AutoCAD Drawing Screen—Typical Installation

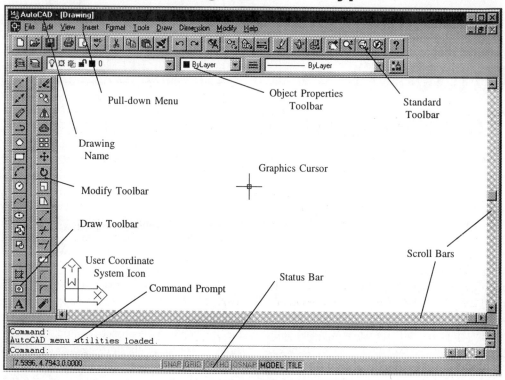

Figure 1–1

When performing a typical installation, and when first launching AutoCAD, the AutoCAD Release 14 screen is illustrated in Figure 1–1. Included in this screen is the pull-down menu area used to locate various commands under such categories as FILE, EDIT, VIEW, and INSERT (to name a few). Directly beneath the pull-down menu area is the Standard toolbar containing such commands as New and Open in addition to displaying numerous flyouts supporting the ZOOM, OBJECT SNAPS, and AERIAL VIEW commands. The Object Properties toolbar (below the Standard toolbar) is designed to manipu-

late layers, linetypes, and color. Two additional toolbars appear on the left side of Figure 1–1. The Draw toolbar holds such commands as LINE, CIRCLE, ARC, and BHATCH. The Modify toolbar contains FILLET, CHAMFER, ERASE, and ARRAY to name just a few. The User Coordinate System icon is displayed in the lower corner of the drawing editor to alert you to the current coordinate system in use. The "W" in the icon indicates that you are in the World coordinate system. At the lower part of the display screen is the Command prompt area which prompts you for input depending on the command currently in progress. At the very bottom is the Status area. Use this to control the Coordinate, Snap, and Grid displays in addition to toggling Ortho mode on or off. These modes are activated by double-clicking on the button to turn the mode on or off. Scroll bar areas are present just below and to the right of the drawing editor screen. These may be used to pan to a different screen location especially when the screen has been magnified using one of the ZOOM command options. Picking of points on the screen or selecting objects to edit is performed by the graphics cursor.

The AutoCAD Drawing Screen—Full Installation

If a Full Installation is performed, the screen in Figure 1–2 is displayed. One key addition to this screen is the appearance of three additional loating toolbars; the Bonus Standard Toolbar, the Bonus Text Toolbar, and the Bonus Layer Toolbar. Commands that give you the ability to trim or extend to a block, perform a multiple stretch, and create a revision cloud just to name a few are found in the Bonus Standard Toolbar. Commands that give you the ability to perform global searches and replacements on text, place text along an arc, and explode text into individual objects just to name a few are found in the Bonus Text Toolbar. The Bonus Layer Toolbar allows you to turn layers on or off, freeze or thaw layers, and lock or unlock layers by picking on an object belonging to the layer. A majority of these

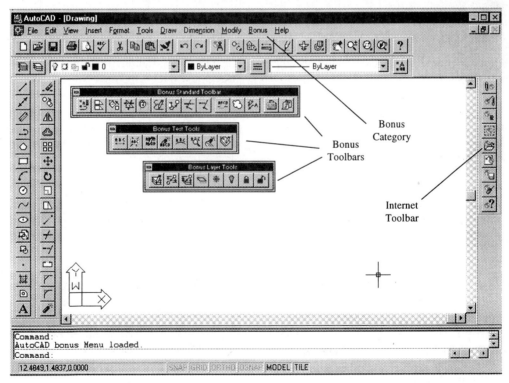

Figure 1–2

Bonus commands are explained and illustrated in Appendix A. Another addition to the screen above is the Bonus category that has been added to the pull-down menu area. If the floating Bonus toolbars are not displayed, Bonus commands may be picked from the pull-down menu in the Bonus area. To assist with the posting of drawings over the Internet, an Internet toolbar is displayed on the right side of the screen.

The features displayed above are only present if performing a Full Installation. If the Typical Installation was performed, it is possible to activate the above items by performing a Custom Installation. This enables you to pick the items to install individually. Click on Bonus Tools and Internet Tools when performing a Custom Installation.

Using Keyboard Function Keys

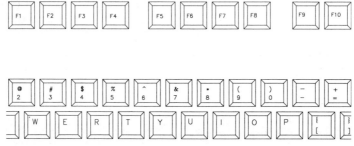

Figure 1–3A

Once in a drawing file, you have additional aids to control such settings as Grid and Snap. Illustrated in Figure 1–3A is a typical keyboard complete with function keys above the alphanumeric keys. These function keys are labeled F1 through F12; only keys F1 through F10 will be discussed here. Software companies commonly program certain functions into these keys to assist you in their application and AutoCAD is no exception. Most keys act as switches which turn functions on or off (see Figure 1–3B). When pressed, the F1 key takes you from the graphics display screen to the Windows Online Help capabilities. Pressing F2 takes you to the text screen consisting of a series of previously prompt sequences. This may be helpful to view the previous command sequence in text form. Pressing F3 activates the Osnap Settings dialog box used to set certain Osnap modes. This will be discussed later in this unit. Use F4 to toggle tablet mode on or off. This mode is only activated when the digitizing tablet has been calibrated for the purpose of tracing a draw-

ing into the computer. Pressing F5 scrolls you through the three supported isoplane modes used to construct isometric drawings. The F6 key toggles the coordinate display located in the lower left corner of the status line on or off. When off, the coordinates update when an area of the screen is picked

Function Keys Definitions	
F1	= Windows Online Help
F2	= Toggle Text/Graphics Screen
F3	= Osnap Settings Dialog Box
F4	= Toggle Tablet Mode On/Off
F5	= Toggle Isoplane Modes
F6	= Toggle Coordinates On/Off
F7	= Toggle Grid Mode On/Off
F8	= Toggle Ortho Mode On/Off
F9	= Toggle Snap Mode On/Off
F10	= Toggle Status Bar On/Off

Figure 1–3B

with a mouse or digitizer puck. When on, the coordinates dynamically change with the current position of the cursor. The F7 key turns the display of Grid on or off. The actual grid spacing is set by the GRID command and not by this function key. Orthogonal mode is toggled on or off using the F8 key. Use this key to force objects such as lines to be drawn horizontally or vertically. Use F9 to toggle Snap mode on or off. The SNAP command sets the current Snap value. Use F10 to toggle the Status bar located at the bottom of the display screen on or off. The Status bar illustrated in Figure 1–4 holds most functions controlled by the function keys. The buttons located in the status bar are dynamic in that they also control the same modes controlled by the function keys. For example, double-clicking on the GRID button turns the Grid on; double-clicking again on the GRID button turns the Grid off. This type of operation can control the coordinate display (F6), Snap (F9), Grid (F7), and Osnap (F3). The Paper Space environment may be entered by double-clicking on the TILE button. Once inside of Paper Space, floating Model Space mode may be entered by double-clicking on the Model button; this changes to Paper. Double-clicking on Paper returns to the Paper Space environment.

6.0000,3.0000,0.0000   SNAP GRID ORTHO OSNAP MODEL TILE

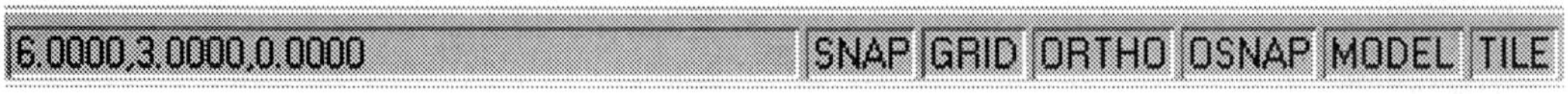

Figure 1–4

Methods of Choosing Commands

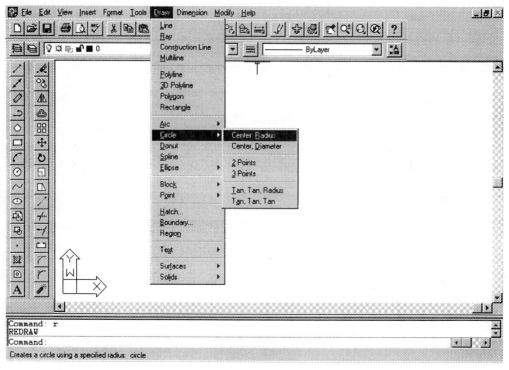

Figure 1–5

There are four ways to enter or select commands for constructing or editing drawings. In Figure 1–5, the pull-down menu is becoming very popular with many users. The pull-down menu allows you to pick certain areas that hold most AutoCAD commands. The optional screen menu is illustrated in Figure 1–6. To activate it, choose the PREFERENCES command located under the Tools pull-down menu area, click on the Display Tab, and place a check in the edit box to display the screen menu. Once activated, the screen menu displays along the right side of the display. Commands may also be entered in directly from the keyboard. This practice is popular for users familiar with the commands.

Illustrated in Figure 1–7 is an example of the fourth method for selecting commands, namely through a toolbar. The toolbar illustrated in Figure 1–6 holds all Osnap modes. Numerous toolbars may be found by selecting "View" in the pull-down menu area followed by "Toolbars...."

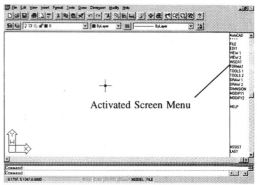

Activated Screen Menu

Figure 1–7   **Figure 1–6**

Activating toolbars

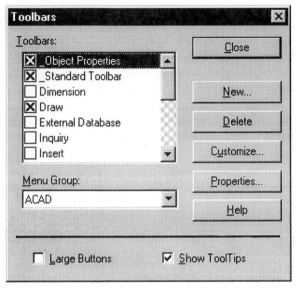

Figure 1-8B

Numerous toolbars are available to assist you in picking other types of commands. Clicking on Toolbars.... located in the View area of the pull-down menu illustrated in Figure 1-8A displays the Toolbars dialog box illustrated in Figure 1-8B. By default, four toolbars are already loaded or active in all drawings; they are Object Properties, Standard Toolbar, Draw, and Modify. To make another toolbar active, click in the empty box next to the name of the toolbar. This will display the toolbar and allow you to preview it before making it a part of the display screen. If this toolbar is correct, click on the Close button. This will close out the Toolbars dialog box but leave the selected toolbar. Illustrated in Figure 1-8C is the Dimension toolbar. This toolbar displayed after placing a check in the box next to Dimension in Figure 1-8B.

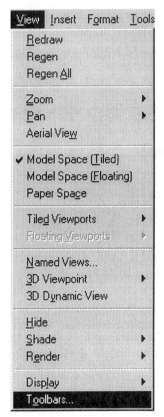

Figure 1-8A

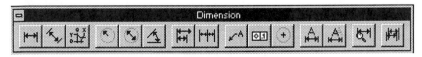

Figure 1-8C

Icon Menus

Icon menus display graphical pictures representing commands or command options. The icon menu in Figure 1–9 shows numerous hatch patterns. This icon menu is displayed after selecting the "Pattern..." option of the BHATCH command. In the past, operators had to know by name the hatch pattern to use such as "Plasti", Insul, or "ANSI32." Once operators entered the hatch pattern from the keyboard, they had no idea in some cases what the hatch pattern would look like until they performed the entire HATCH command (see Figure 1–10A). Sometimes the hatch pattern proved unsuitable for the particular application. Icon menus solve this di-

lemma. The graphic lets you preview what the pattern will look like before you place it in a drawing.

To select a desired pattern, move the selector arrow to the pattern and pick anywhere inside the pattern. A box will be placed outlining the pattern similar to the illustration in Figure 1–10B. If the wrong pattern is selected, select another pattern and that pattern will be highlighted within the box.

Icon menus allow for quick selection of items for use in a drawing because of their graphical nature.

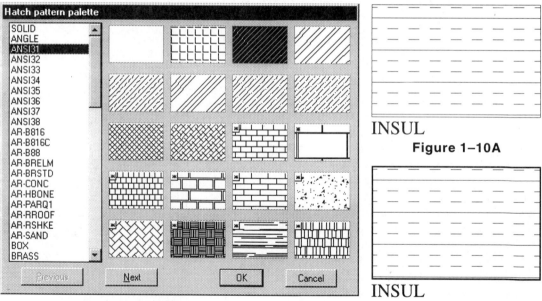

Figure 1–9

INSUL

Figure 1–10A

INSUL

Figure 1–10B

Commonly Used Command Aliases

To assist with the entry of AutoCAD commands from the keyboard, certain commands have been shortened and are considered "Aliases." The list in Figure 1–11 shows the command aliases commonly used throughout this text. Once you are comfortable with the keyboard, using command aliases provide a very fast and efficient method of activating AutoCAD commands.

A,	*ARC		INT,	*INTERSECT
AL,	*ALIGN		L,	*LINE
AR,	*ARRAY		LA,	*LAYER
BO,	*BOUNDARY		LE,	*LEADER
BR,	*BREAK		LEN,	*LENGTHEN
C,	*CIRCLE		LI,	*LIST
CH,	*DDCHPROP		LT,	*LINETYPE
CHA,	*CHAMFER		LTS,	*LTSCALE
CO,	*COPY		M,	*MOVE
D,	*DDIM		MA,	*MATCHPROP
DAL,	*DIMALIGNED		ME,	*MEASURE
DAN,	*DIMANGULAR		MI,	*MIRROR
DBA,	*DIMBASELINE		ML,	*MLINE
DCE,	*DIMCENTER		MO,	*DDMODIFY
DCO,	*DIMCONTINUE		MS,	*MSPACE
DDI,	*DIMDIAMETER		MT,	*MTEXT
DED,	*DIMEDIT		O,	*OFFSET
DI,	*DIST		OS,	*DDOSNAP
DIV,	*DIVIDE		P,	*PAN
DLI,	*DIMLINEAR		PE,	*PEDIT
DO,	*DONUT		PL,	*PLINE
DOR,	*DIMORDINATE		PO,	*POINT
DOV,	*DIMOVERRIDE		POL,	*POLYGON
DRA,	*DIMRADIUS		PR,	*PREFERENCES
DST,	*DIMSTYLE		PS,	*PSPACE
E,	*ERASE		R,	*REDRAW
ED,	*DDEDIT		RE,	*REGEN
EL,	*ELLIPSE		REG,	*REGION
EX,	*EXTEND		REV,	*REVOLVE
EXT,	*EXTRUDE		RO,	*ROTATE
F,	*FILLET		S,	*STRETCH
GR,	*DDGRIPS		SC,	*SCALE
H,	*BHATCH		SEC,	*SECTION
HE,	*HATCHEDIT		SHA,	*SHADE
HI,	*HIDE		SL,	*SLICE
I,	*DDINSERT		SP,	*SPELL
IN,	*INTERFERE		SPL,	*SPLINE

Figure 1–11

ST,	*STYLE	UN,	*DDUNITS
SU,	*SUBTRACT	UNI,	*UNION
T,	*DTEXT	V,	*DDVIEW
TO,	*TOOLBAR	VP,	*DDVPOINT
TOL,	*TOLERANCE	W,	*WBLOCK
TOR,	*TORUS	WE,	*WEDGE
TR,	*TRIM	X,	*EXPLODE
UC,	*DDUCS	XL,	*XLINE
UCP,	*DDUCSP	Z,	*ZOOM

Figure 1–11

Other Command Aliases

The list of command aliases on the previous page is by no means complete. Listed in Figure 1–12 are other command aliases supplied with AutoCAD. These commands do not nec-essarily apply to this text; however they may be used for other applications such as structured query language applications, external references, and image management.

3F,	*3DFACE	INS,	*INSERTOBJ
AAD,	*ASEADMIN	MV,	*MVIEW
AEX,	*ASEEXPORT	PA,	*PASTESPEC
ALI,	*ASELINKS	PRE,	*PREVIEW
ASQ,	*ASESQLED	PU,	*PURGE
ARO,	*ASEROWS	Q,	*QUIT
ASE,	*ASESELECT	RA	*REDRAWALL
AT,	*DDATTDEF	REA,	*REGENALL
ATE,	*DDATTE	REN,	*DDRENAME
B,	*BMAKE	RM,	*DDRMODES
-CH,	*CHANGE	RPR,	*RPREF
COL,	*DDCOLOR	RRE,	*RENDER
DR,	*DRAWORDER	SCR,	*SCRIPT
DV,	*DVIEW	SE,	*DDSELECT
EXIT,	*QUIT	SET,	*SETVAR
EXP,	*EXPORT	SO,	*SOLID
G,	*GROUP	SPE,	*SPLINEDIT
IAD,	*IMAGEADJUST	TA,	*TABLET
IAT,	*IMAGEATTACH	TH,	*THICKNESS
ICL,	*IMAGECLIP	XA,	*XATTACH
IIM,	*IMAGE	XC,	*XCLIP
IM,	*IMPORT	XR,	*XREF

Figure 1–12

Using the Preferences Dialog Box

Picking the Tools area of the pull-down menu area
in Figure 1–13 exposes the PREFERENCES command located at the very bottom of the pull-down
menu. Pick this command to make various changes
to the operating environment of AutoCAD. Seven
categories or Tabs are available to allow you to make
AutoCAD changes including to how Files are
pointed to, the Performance of the system, the Com-
patibility of the system, making General changes
to the operation of the software, making changes to
the Display, Point, and Printer, and saving changes
under a unique name or Profile.

The Display and Pointer areas will be discussed in
detail in the following pages. Clicking on the Display Tab activates the dialog box illustrated in Figure 1–13. This tab controls whether the AutoCAD
screen menu is displayed or not on the screen and
if scroll bars are displayed on the screen or not.
Parameters that deal with text include the number
of lines that make up the command line and the
number of text lines that make up the text window
of the system.

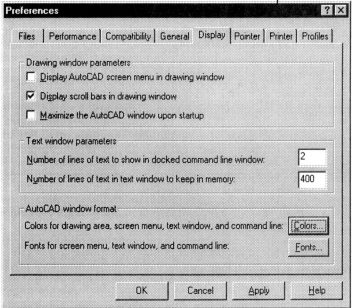

Figure 1–13

Changing the Color of the Display

Clicking on the Colors button of the Display Tab activates the AutoCAD Window Colors dialog box shown in Figure 1–14. Use this to make changes to the colors that make up the display screen. For example, to change the Graphics window background from the default color of White to a new background of Black, check to see that the proper area is displayed in the Window Element edit box and select the color to change. Another way of making changes is to pick on the icon under Graphics Window to change to a different Window Element. For example, actually pick on the icon at "A" in Figure 1–14. This will change the Window Element to Menu Background; or pick on the icon at "B" to change the Window Element to Text Window Text Background. This provides an easier way to manipulate AutoCAD screen colors.

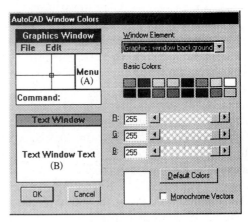

Figure 1–14

Changing the Size of the Cursor

Clicking on the Pointer Tab of the main Preferences dialog box activates the dialog box shown in Figure 1–15. In addition to a listing of commonly used input devices such as mouse and digitizer devices, notice the last area of the dialog box. Here, the size of the graphic cursor may be changed in size. By default, the cursor is displayed at 5 percent of the size of the display screen. To make it larger, increase the screen percentage. To make the cursor fill the total screen as in past versions of AutoCAD, set the Percentage of screen size to a value of 100 percent.

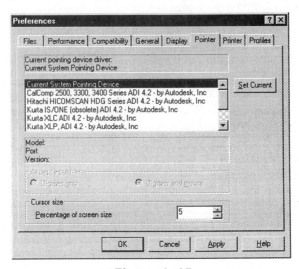

Figure 1–15

Starting a New Drawing

To begin a new drawing file, select the NEW command with the button shown in the following command sequence; or click on the NEW command located in the File pull-down menu area in Figure 1–16. The NEW command can also be entered in at the command prompt similar to the following command sequence:

 Command:**NEW**

Entering the NEW command displays the Create New Drawing dialog box illustrated in Figure 1–17. Four options are available in this dialog box: Using a Wizard, Using a Template, Starting from Scratch, and Instructions on each of the options.

The Wizard button has two options associated with it: Performing a Quick Setup or an Advanced Setup. Both of these options will be discussed in greater detail in the following pages.

Clicking on the button entitled "Use a Template" displays an edit box containing various files that have an extension of .DWT (see Figure 1–18). These stand for template files and are designed to conform to various standard drawing sheet sizes. Associated with each template file is a corresponding title block that is displayed in the Preview area. You may scroll through the various template files and get a glimpse of the title block tied to the template file.

Clicking on the button entitled "Start from Scratch" begins the AutoCAD drawing file with such default settings as a sheet of paper 12 units by 9 units and four-place decimal precision.

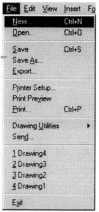

Figure 1–16

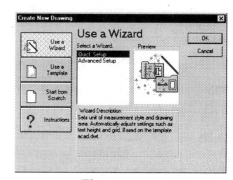

Figure 1–17

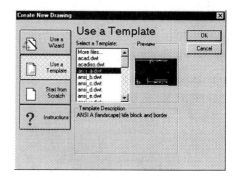

Figure 1–18

Using the Quick Wizard

When clicking on the "Use a Wizard" button in addition to the Quick setup displays the dialog box shown in Figure 1–19 which consists of two tabs identified by Step numbers and categories. Step 1 is shown in Figure 1–19 and is used to graphically control the units of the drawing. Five units of measure are available through this Wizard, namely Decimal, Engineering, Architectural, Fractional, and Scientific. Clicking on the radio button displays a sample of what the units will look like in the drawing. When completed with setting the units of the drawing, click on the Next >> button for the next dialog box in the Quick Setup sequence.

Clicking Next >> activates Step 2 of the quick drawing setup. This tab deals with the area of the drawing which is given as a Width and a Length value (see Figure 1–20). Setting this is comparable to using the LIMITS command which will be discussed later in this unit. As the width and length of the drawing area are entered, the Sample Area image updates to reflect the changes in the values. At times, operators mistakenly substitute the width value for the length and vice versa. The image will allow you to preview what the limits or drawing area will look like. If the drawing area is incorrect, changes can be made and the Sample Area image will update to the latest changes in the area. At any point, you may elect to return to the units area by clicking on the Step 1 Units tab to change the units of the drawing. When completed with the setting of the drawing units and area, click on the Done button which will return to the drawing editor where the units and area will be updated to reflect the changes made in the Quick Setup Wizard.

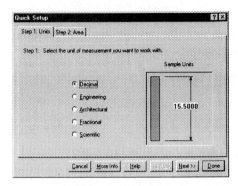

Figure 1–19

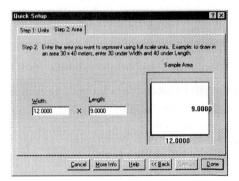

Figure 1–20

Using the Advanced Wizard

When activating the Quick Wizard, the operator is guided through two tabs to make changes in the units and area of the drawing. Clicking on Advanced Setup in the Create New Drawing dialog box in Figure 1–17 displays the Advanced Setup dialog box illustrated in Figure 1–21. The Advanced Wizard contains seven tabs to make changes to the initial drawing setup. The first tab deals with the drawing units of the drawing. This tab is almost identical to the dialog box found in the Quick Wizard; it is used to allow you to choose among Decimal, Engineering, Architectural, Fractional, and Scientific units. When changing the units, they will preview in the Sample Units image. An additional control for units in the Advanced Wizard allows you to change the precision of the main units. In Figure 1–21, the precision for the decimal units is four decimal places. Changing the precision will update the Sample Units image in Figure 1–21.

The second tab, Step 2 of the Advanced Setup Wizard, deals with how angles will be measured (see Figure 1–22). Five methods of angle measurements include the default of Decimal Degrees followed by Degrees/Minutes/Seconds, Grads, Radians, and Surveyors. A precision box for the measurement of angles allows you to change to different angular precision values. As with the Quick Setup Wizard, making changes to the measurement of angles and angular precision will be displayed in the Sample Angle image area.

The third tab, Step 3 of the Advanced Setup dialog box, displays how angles will be measured in the drawing. By default, angles are measured starting with East for an angle of 0, North for an angle of 90 degrees, West for 180 degrees, and South for 270 degrees. Also by default, these angles are measured counterclockwise. You have the option of changing the direction of zero and to make these changes update the Angle Zero Direction image illustrated in Figure 1–23.

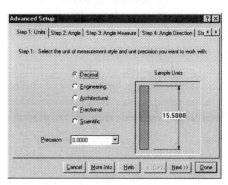

Figure 1–21

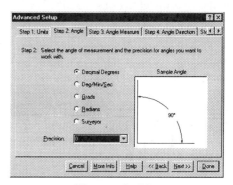

Figure 1–22

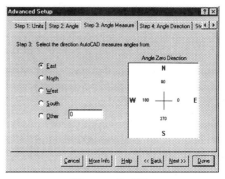

Figure 1–23

The fourth tab, Step 4 of the Advanced Setup dialog box, deals with Angle Direction. As noted in Figure 1–23, all angles are by default measured in the counterclockwise direction. Use the dialog box illustrated in Figure 1–24 to change from counterclockwise measurement of angles to clockwise angular measurements. Throughout the duration of this book, all examples dealing with angular measurements keep the default setting of counterclockwise angular measurement.

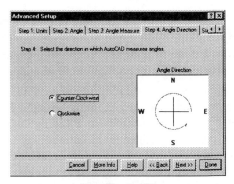

Figure 1–24

The fifth tab, Step 5 of the Advanced Setup dialog box, is identical to Step 2 of the Quick Wizard setup; use this tab to change the drawing area (see Figure 1–25). Enter values for the Width and Length in the edit boxes provided and notice the Sample Area image update to the new drawing area. This has the same effect as using the LIMITS command which will be discussed later in this unit.

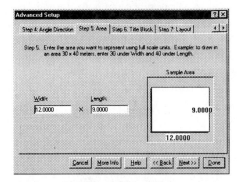

Figure 1–25

The sixth tab, Step 6 of the Advanced Setup dialog box, is designed to insert a title block into the current drawing. Once a title block is picked from the Title Block Description, it is displayed in the Sample Title Block image illustrated in Figure 1–26. The following standards are supported in the list of title blocks in the Title Block Description box: ANSI, ARCh/Eng, DIN, ISO, and JIS.

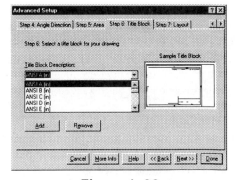

Figure 1–26

The last tab, Step 7 of the Advanced Setup dialog box, is designed to enable the paper space environment. See Figure 1–27. Paper space is a layout environment that holds title block and other drawing annotation information. Another important feature of paper space is the ability of the drawing to be plotted out at a scale factor of 1:1 even if it was drawn full size inside of model space, the default drawing environment. The paper space environment will be introduced in Unit 11, Solid Modeling Fundamentals.

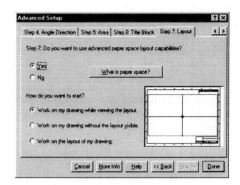

Figure 1–27

Opening an Existing Drawing

The OPEN command is used to edit a drawing that already exists or has already been created. Select this command from the "File" area of the pull-down menu area. When this command is selected, a dialog box appears similar to Figure 1–28. Listed in the edit box area are all files that match the type at the bottom of the dialog box. Since the file type is .dwg, listed are all valid drawing files supported by AutoCAD. To choose a different subdirectory, use the standard Windows95 file management techniques by clicking in the file box at "A." This will display all subdirectories associated with the drive. Clicking on the subdirectory will display any drawing files contained in it if they exist.

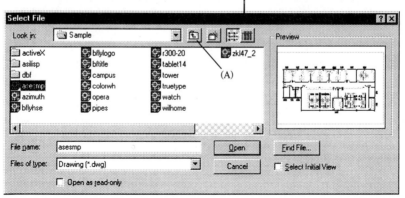

Figure 1–28

When first entering AutoCAD, the Start Up dialog box in Figure 1–29 displays. In addition to using the Quick and Advanced Wizards, Using a Template, or Starting a drawing from Scratch, an Open Drawing button allows you to open up an existing drawing. The "Select a File:" edit box displays the existing drawing files in the current subdirectory. To perform a search for drawing files in a different subdirectory, choose the "More files…" area. As you pick drawing files, valid Release 13 and Release 14 files will display in the Preview area of the dialog box. Drawing files created in versions prior to Release 13 must be opened in Release 14 and saved before they preview.

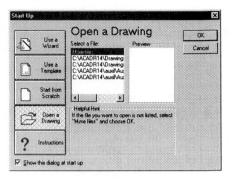

Figure 1–29

A view is a previously saved portion of the drawing display screen. Illustrated in Figure 1–30 are typical orthographic view names. Instead of opening a drawing file where the entire drawing is displayed, you have the option of opening up the drawing file and automatically going to one of the views. This is considered good practice; rather than bring up the entire drawing and then zoom into a specific area of the screen, this zoom operation is bypassed by going directly to the named view. If a drawing file has been previously saved containing views, the "Select Initial View" box is checked. When the drawing file loads, the dialog box in Figure 1–30 displays on the screen alerting you to select the desired view.

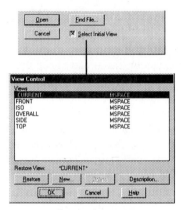

Figure 1–30

If the "Open as read-only" mode is checked, a drawing file is displayed on the screen. Objects may be drawn or edited as with all drawing files. However, none of the changes may be saved because the drawing may only be viewed; this is the purpose of read only mode. If any changes were made and a save is attempted, the alert box illustrated in Figure 1–31 informs you the current file is write protected and no changes may be permanently made.

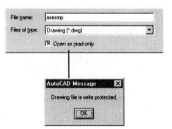

Figure 1–31

Using the Browse Button

Clicking on the Find File button of the Select File dialog box displays the Browse/Search dialog box illustrated in Figure 1–32. Use it to preview groups of drawings; select the desired file to open it as the current drawing file. Clicking on the Search Tab activates a dialog box that enables you to enter drive information, file type information, and dates and times of the file in order to perform a search.

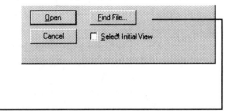

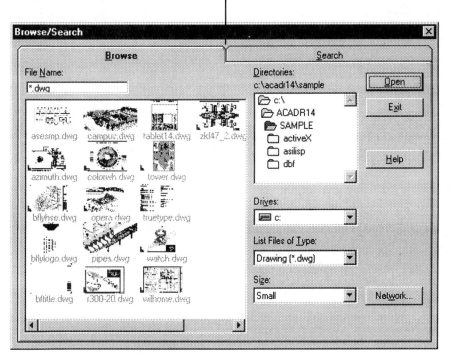

Figure 1–32

Using the Units Control Dialog Box

The Units Control dialog box is available to interactively set the units of a drawing. This dialog box is selected from the pull-down menu under "Format." Selecting "Units…" activates the dialog box illustrated in Figure 1-33.

By default, decimal units are set. The number of decimal places past the zero is also set to four. The following systems of units are available: Scientific, Decimal, Engineering, Architectural, and Fractional. Scientific units are displayed in exponential format. Engineering units are displayed in feet and decimal inches. Architectural units are displayed in feet and fractional inches. Fractional units are displayed in fractional inches.

The following methods of measuring angles are supported in the Units Control dialog box: Decimal Degrees, Degrees/Minutes/Seconds, Grads, Radians, and Surveyor's Units. Accuracy of decimal degree angles may be set between zero and eight places.

Selecting "Direction…" in the Units Control dialog box displays another dialog called Direction Control. Use this dialog box to control the direction of angle zero in addition to changing the way angles are measured from counterclockwise to clockwise. Angles are measured by default in the counterclockwise direction.

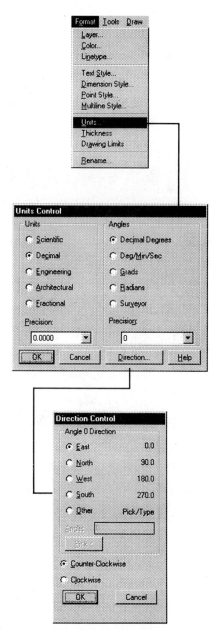

Figure 1–33

Using the LIMITS Command

By default, the size of the drawing display screen in a new drawing file measures 12 units in the X direction and 9 units in the Y direction. As this size may be ideal for small objects, larger drawings require a more drawing screen area. Use the LIMITS command for increasing the size of the drawing area. Select this command from the "Format" area of the pull-down menu (see Figure 1–34A); this command may also be entered directly at the command prompt. Illustrated in Figure 1–34B is a section view drawing. This drawing fits in a screen size of 24 units in the X direction and 18 units in the Y direction. Follow the prompt sequence below for changing the limits of a drawing.

Command: **LIMITS**
ON/OFF/<Lower left corner>:
   <0.0000,0.0000>: *(Press* ENTER *to accept this value)*
Upper right corner <12.0000,9.0000>:**24,18**

Before continuing, perform a ZOOM-All to change the size of the display screen to reflect the changes in the limits of the drawing. This command can be found in the View pull-down menu area. It is also found on the Zoom Toolbar, which is activated through the View pull-down menu area, clicking on Toolbar, and placing a check in the box next to Zoom.

 Command: **ZOOM**
All/Center/Dynamic/Extents/Left/Previous/
   Vmax/Window/<Scale(X/XP)>:**ALL**

Figure 1–34A

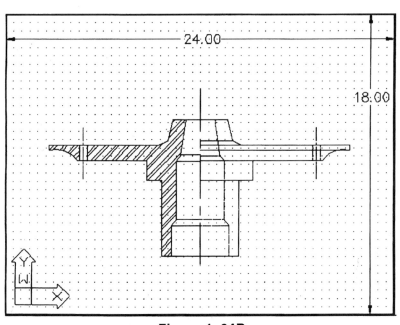

Figure 1–34B

Calculating the Limits of the Drawing

Before attempting any drawing, the limits or sheet size of the drawing must first be calculated. This is to ensure the entire drawing plots out. To properly calculate the limits of the drawing, two items are first needed; namely the scale of the drawing and the size paper the drawing will be plotted out on. Most operators who use CAD systems in industry already have some idea as to the scale of the drawing and if it will fit on a certain size sheet of paper.

For the purposes of this example, the following steps will illustrate the creation of drawing limits based on a "D"-size paper and at a scale of 1/4" = 1'0." This scale is commonly used to construct residential house plans. As the following steps use this scale and sheet, the method works for calculating limits at any scale and on any sheet of paper.

Step #1

Activate the Print/Plot Configuration dialog box in Figure 1–35 by using the PRINT command in the File pull-down menu area. Be sure to examine the current plot device located in the Device and Default Information area of the dialog box. For the purposes of this example, the HP Draftmaster I plotter will be used as the output device.

Step #2

Click on the Size button of the Paper Size and Orientation area of the main Plot dialog box. This will display the Paper Size dialog box shown in Figure 1–36. A "D"-size sheet of paper normally measures 36" wide and 24" high. However, since this plotting device is a pen plotter, room must be allotted for the pen carriage in addition to the roll bars that hold the paper. This is the reason the paper measures 33" wide and 21" high. You must use the values in the Paper Size dialog box to calculate the limits; otherwise, if the "D" size sheet of paper measuring 36" x 24" is used, the top and right borders of the paper will not plot out. This of course is not the case if you use preprinted border and title blocks.

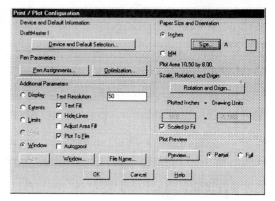

Figure 1–35

Figure 1–36

Size	Width	Height	Size	Width	Height
A	10.50	8.00	USER:		
B	16.00	10.00			
C	21.00	16.00	USER1:		
D	33.00	21.00			
E	43.00	33.00	USER2:		
F	40.00	28.00			
A4	11.20	7.80	USER3:		
A3	15.60	10.70			
A2	22.40	15.60	USER4:		
A1	32.20	22.40			
MAX	44.70	35.31			

Step #3

Determine the multiplication factor based on the drawing scale. In computer-aided design, most drawings are performed at full size. Since the drawing is at full size, the limits of the drawing must hold a border and title block designed to match the full drawing size. The multiplication factor for the scale 1/4" = 1'0" is 48; this is found by dividing 1' by 1/4." Use this multiplier to increase the size of the paper enabling an operator to produce a full size drawing on the computer. This is vastly different from manual drawing practices where an object had to be reduced 1/48th its normal size in order to fit on the "D"-size paper at a scale of 1/4" = 1'0".

Step #4

Using the multiplication factor of 48, increase the size of the "D"-size sheet of paper:

Paper Size from the Plot dialog box	33	21
Multiplication Factor	48	48
	1,584	1,008

Convert the previous values to feet:

1,584 /12 = 132'   1,008 /12 = 84'

The limits of the drawing become 132' wide and 84' high.

Step #5

Use the Units Control dialog box to set the current units to Architectural (Decimal mode does not accept feet as valid units). Next, use the LIMITS command and enter the lower left corner as 0,0 and the upper right corner as 132',84'. Once the limits have been set, use the ZOOM command and the ALL option to regenerate the entire screen and display the new sheet of paper. To view the active drawing area, use the RECTANG command to construct a rectangle around the drawing using 0,0 as the lower left corner and 132',84' as the upper right corner. Construct the drawing inside of the rectangular area.

Step #6

Once the drawing is constructed, dimensioned, etc., and is ready for plotting, use the PRINT command to activate the Print/Plot Configuration dialog box shown in Figure 1–37. If the Scale, Rotation, and Origin are of the main PLOT dialog box, enter 1/4 for Plotted Inches and 1' for Drawing Units. Notice these values convert to 0.25 = 12. This means for every 1' distance on the drawing screen, the plotting pen moves 1/4."

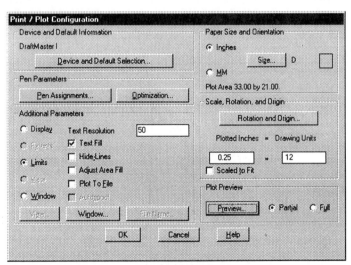

Figure 1–37

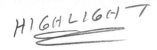

Limit Settings for Typical Sheet Sizes

SCALE	SCALE FACTOR	ANSI "A" 11"x 8.5"	ANSI "B" 17"x11"	ANSI "C" 22"x17"	ANSI "D" 34"x22"	ANSI "E" 44"x34"
DECIMAL SCALES						
.125=1	8	88,68	136,88	176,136	272,176	352,272
.25=1	4	44,34	68,44	88,68	136,88	176,136
.50=1	2	22,17	34,22	44,34	68,44	88,68
1.00=1	1	11,8.5	17,11	22,17	34,22	44,34
2.00=1	.50	5.5,4.25	8.5,5.5	11,8.5	17,11	22,17

SCALE	SCALE FACTOR	ANSI "A" 11"x 8.5"	ANSI "B" 17"x11"	ANSI "C" 22"x17	ANSI "D" 34"x22"	ANSI "E" 44"x34"
ARCHITECTURAL SCALES						
1/8"=1'-0"	96	88',68'	136',88'	176',136'	272',176'	352',272'
1/4"=1'-0"	48	44',34'	68',44'	88',68'	136',88'	176',136'
1/2"=1'-0"	24	22',17'	34',22'	44',34'	68',44'	88',68'
3/4"=1'-0"	16	14.7',11.3'	22.7',14.7'	29.3',22.7'	45.3',29.3'	58.7',45.3'
1"=1'-0"	12	11',8.5'	17',11'	22',17'	34',22'	44',34'
2"=1'-0"	6	5.5',4.25'	8.5',5.5'	11',8.5'	17',11'	22',17'

SCALE	SCALE FACTOR	ANSI "A" 11"x 8.5"	ANSI "B" 17"x11"	ANSI "C" 22"x17	ANSI "D" 34"x22"	ANSI "E" 44"x34"
METRIC SCALES						
1=1	25.4mm	279,216	432,279	559,432	864,559	1118,864
1=10	10cm	110,85	170,110	220,170	340,220	440,340
1=20	20cm	220,170	340,220	440,340	680,440	880,680
1=50	50cm	550,425	850,550	1100,850	1700,1100	2200,1700
1=100	10cm	1100,850	1700,1100	2200,1700	3400,2200	4400,3400

Using the GRID Command

Use grid to get a relative idea as to the size of objects. Grid is also used to define the size of the display screen originally set by the LIMITS command. The dots that make up the grid will never plot out on paper even if they are visible on the display screen. Grid dots may be turned on or off either by using the GRID command or by pressing the F7 function key. By default, the grid is displayed in one-unit intervals similar to Figure 1–38. Use the following command prompt sequence with the GRID command:

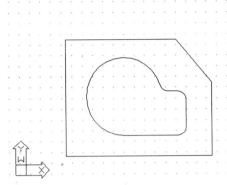

Figure 1–38

Command:**GRID**
Grid spacing(X) or ON/OFF/Snap/Aspect
   <0.0000>: *(Enter a value)*

Illustrated in Figure 1–39 is a grid that has been set to a value of 0.50 or one-half its original size. Use the following command prompt sequence for performing this change.

Command:**GRID**
Grid spacing(X) or ON/OFF/Snap/Aspect
   <0.0000>:**0.50**

While grid is a useful aid for construction purposes, it may reduce the overall performance of the computer system. If the grid is set to a small value and is visible on the display screen, it takes time for the grid to display. If a very small a value is used for grid, a prompt will be displayed warning that the grid value is too small to display on the screen.

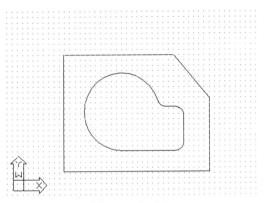

Figure 1–39

Using the SNAP Command

Illustrated in Figure 1–40 is a sample drawing screen with a grid spacing of 1.00 units. The cursor is positioned in between grid dots. It is possible to have the cursor lock onto or snap to a grid dot; this is the purpose of the SNAP command. By default the current snap spacing is 1.00 units. Even though a value is set, the snap must be turned on for the cursor to be positioned on a grid dot. This can be accomplished by using the SNAP command below or by pressing the F9 function key.

Command: **SNAP**
Snap spacing(X) or ON/OFF/Aspect/Rotate/
    Style <1.00>:

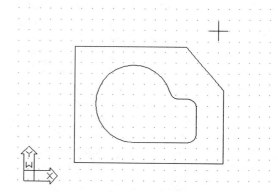

Figure 1–40

The current snap value may even affect the grid. If the current grid value is zero, the Snap value is used for the grid spacing.

Some drawing applications require that the snap be rotated at a specific angular value (see Figure 1–41). Changing the snap in this fashion also affects the cursor. Use the following command prompts for rotating the snap.

Command: **SNAP**
Snap spacing(X) or ON/OFF/Aspect/Rotate/
    Style <1.00>: **ROTATE**
Base point<0.0000,0.0000>: *(Press* [ENTER] *to
    accept this value)*
Rotation angle <0>: **30**

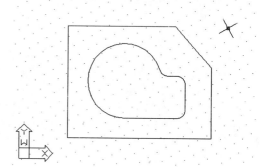

Figure 1–41

One application of rotating snap is for auxiliary views where an inclined surface needs to be projected in a perpendicular direction. This will be explained and demonstrated in a later unit.

Using the Drawing Aids Dialog Box

Selecting "Tools" from the pull-down menu area and then selecting "Drawing Aids…" displays the Drawing Aids dialog box shown in Figure 1–42. This is a helpful dialog box used for making dynamic changes to such commands as GRID and SNAP. In addition to these commands, the following command modes may also be changed: Ortho; Solid Fill; Quick Text; Blips; Highlight; Grouping; and Associative Hatching.

Placing a check in the box provided turns on the specific mode. Checking the box again removes the check turning off the mode. Ortho mode is the ability to force only horizontal or vertical movement. Solid Fill is controlled by the FILL command and affects filled-in objects such as polylines and donuts. Quick Text is controlled by the QTEXT command and converts all text on the display screen to rectangles to speed up drawing response time. For plotting purposes, the rectangles need to be converted back into text. The display of blips is

controlled by the BLIPMODE command. Blips are the small specks or marks made on the screen whenever constructing or editing objects. To clean up blips, use a simple REDRAW command. Turning blips off prevents them from appearing on the screen. Highlight is controlled by the system variable Highlight. This determines if a selected object highlights and informs you that it has been selected.

Another mode controls whether isometric grid is present or not. Also, three isometric modes may be toggled on or off.

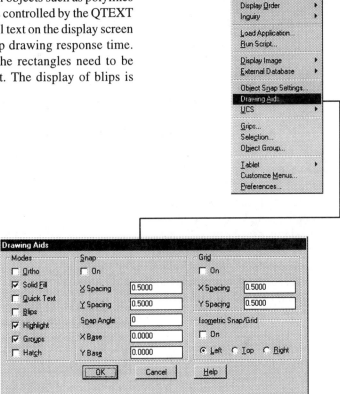

Figure 1–42

Organizing a Drawing Through the Use of Layers

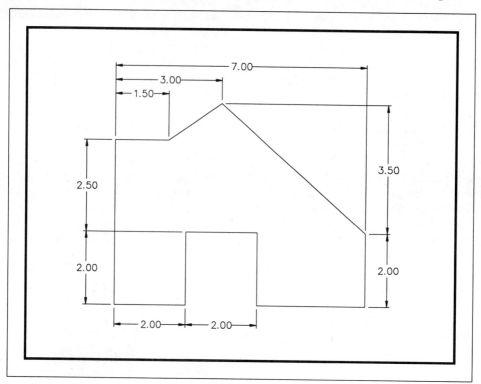

Figure 1–43

As a means of organizing objects, a series of layers should be devised for every drawing. Layers can be thought of as a group of transparent overlays that combine to form the completed drawing. Figure 1–43 shows a drawing consisting of object lines, dimension lines, and border. Creating these three drawing components is illustrated in Figure 1–44. Only the drawing border occupies a layer that could be called "Border." The object lines occupy a layer that could be called "Object" and the dimension lines could be drawn on a layer called "Dim." At times, it may be necessary to turn off the dimension lines for better clarity of the object. Creating all dimensions on a specific layer will allow you to turn off the dimensions while viewing all other objects on layers still turned on.

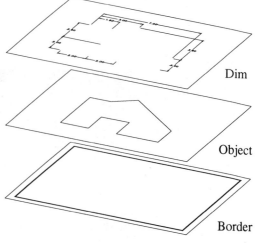

Figure 1–44

Using the -LAYER Command

One method of creating and managing layers is through the use of the -LAYER command which is entered in at the keyboard. The following options display when issuing the -LAYER command:

Command:**-LAYER**
?/Make/Set/New/ON/OFF/Color/Ltype/Freeze/
   Thaw/LOck/Unlock:

A detailed listing of the -LAYER options follow. They are also entered in from the keyboard using the first letter of the option (except for the LOck option which requires the letters "LO" to activate it).

?—Used to give a complete list or partial listing of all layers in the current drawing file.

Make—Used to create a new layer and automatically set the new layer to the current layer.

Set—Allows you to change to a new current layer; all new objects drawn will be drawn on this layer.

New—Used to create a new layer or series of new layers. The Set option is then used to change from one layer to a new current layer.

On—Makes all objects created on a certain layer visible on the display screen.

Off—Turns off or makes invisible all objects created on a certain layer.

Color—Allows you to assign a color to a layer name.

Ltype—Allows you to assign a linetype to a layer name.

Freeze—Similar to the Off option; turns off all objects created on a certain layer. Objects frozen will not be calculated when performing a drawing regeneration. Therefore, Freeze is used as a productivity tool to speed up drawing performance.

Thaw—Similar to On; turns on all objects created on a certain layer that were previously frozen.

LOck—Allows objects on a certain layer to be visible on the display screen while protecting them from accidentally being modified through an EDITING command.

Unlock—Unlocks a previously locked layer.

Using the -LAYER Command to Create New Layers

Illustrated in Figure 1–45 is an object that requires the following layers to be created; Object, Hidden, Center, and Dimension. Use the following prompt sequence to create a layer called "Object" which is assigned the color "Yellow":

Command:**-LAYER**
?/Make/Set/New/ON/OFF/Color/Ltype/Freeze/
  Thaw/LOck/Unlock: **New**
New layer name(s): **Object**
?/Make/Set/New/ON/OFF/Color/Ltype/Freeze/
  Thaw/LOck/Unlock: **Color**
Color: **Yellow**
Layer name for color(2) Yellow <0>: **Object**
?/Make/Set/New/ON/OFF/Color/Ltype/Freeze/
  Thaw/LOck/Unlock: *(Press ENTER to exit this command)*

Now use the following prompt sequence to create a layer called "Hidden" along with the color "Red" and the "Hidden" linetype:

Command: **-LAYER**
?/Make/Set/New/ON/OFF/Color/Ltype/Freeze/
  Thaw/LOck/Unlock: **New**
New layer name(s): **Hidden**
?/Make/Set/New/ON/OFF/Color/Ltype/Freeze/
  Thaw/LOck/Unlock: **Color**
Color: **Red**
Layer name for color(1) Red <0>: **Hidden**
?/Make/Set/New/ON/OFF/Color/Ltype/Freeze/
  Thaw/LOck/Unlock: **Ltype**
Linetype: **Hidden**
Layer name for HIDDEN <0>: **Hidden**
?/Make/Set/New/ON/OFF/Color/Ltype/Freeze/
  Thaw/LOck/Unlock: *(Press ENTER to exit this command)*

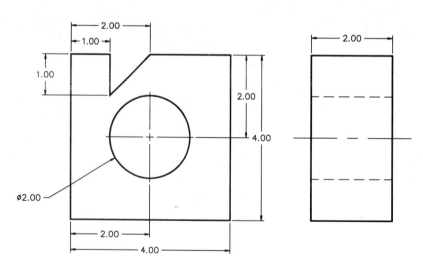

Figure 1–45

Using the Layer & Linetype Properties Dialog Box

A more popular and efficient way to create and manage layers is to use the Layer & Linetype Properties dialog box to perform the following layer operations:

Creating new layers
Making a layer the new current layer
Assigning a color to a layer or group of layers
Assigning a linetype to a layer or group of layers
Turning layers on or off
Freezing or thawing layers
Locking or unlocking layers

Activate this dialog box by first selecting "Format" from the pull-down menu illustrated at the right; then select "Layer..." to display the Layer & Linetype Properties dialog box in Figure 1–46.

The dialog box may also be activated from the keyboard through the LAYER command or by clicking on the Layer Button.

 Command:**LAYER**

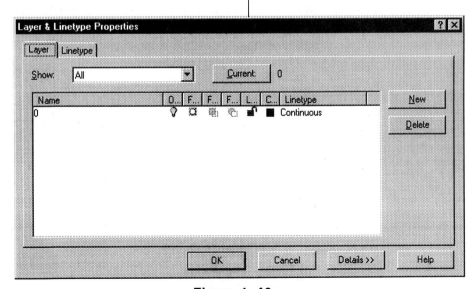

Figure 1–46

Creating New Layers Using the Dialog Box

Clicking on the "New" button of the Layer & Linetype Properties dialog box automatically creates a new layer called "Layer1" which displays itself in the layer list box in Figure 1–47. Since this layer name is completely highlighted, you may elect to change its name to something with more meaning, such as a layer called "Object", or "Hidden" to hold all object or hidden lines in a drawing.

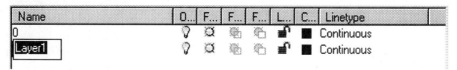

Figure 1–47

Illustrated in Figure 1–48 is the result of changing the name "Layer1" to "Object." As the word "OBJECT" may be entered in with uppercase letters, or "object" in lower case letters, the initial letter of the layer name is capitalized while all other letters appear in lowercase. This is automatically controlled by the dialog box.

Figure 1–48

If more than one layer needs to be created, it is not necessary to continually pick on the "New" button. As this operation will create new layers, a more efficient method would be to perform the following: after creating a new "Layer1", change its name to "Dim" followed by a comma (,); this will automatically create a new "Layer1" (see Figure 1–49). Change its name followed by a comma and a new layer is created and so on.

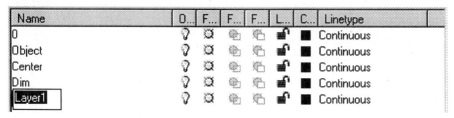

Figure 1–49

Assigning Colors and Linetypes

Once a layer is selected from the list box of the Layer & Linetype Properties dialog box and the color swatch is selected next to the layer name, the Select Color dialog box shown in Figure 1–50 displays. Select the desired color from the "Standard Colors" area, "Gray Shades" area, or from the "Full Color Palette" area. The standard colors consist of the basic colors available on most color display screens. Depending on the video driver, different shades of standard colors may be selected from the "Full Color Palette."

Selecting the name Continuous next to the highlighted layer activates the Select Linetype dialog box shown in Figure 1–51. Use this dialog box to dynamically select preloaded linetypes to be assigned to various layers. If no linetypes are preloaded, click on the Load button of the Select Linetype dialog box to display the Load or Reload Linetypes dialog box shown in Figure 1–52.

Once the Load or Reload Linetypes dialog box is displayed as in Figure 1–52, use the scroll bars to view all linetypes contained in the file ACAD.LIN. Notice in addition to standard linetypes such as Hidden and Phantom, a few linetypes are provided that have text automatically embedded into the linetype. As the linetype is drawn, the text is placed depending how it was originally designed. Notice also three types of Hidden linetypes; one linetype is Hidden, another is Hidden2, and yet another is HiddenX2. Hidden2 represents a linetype where the distances of the dashes and spaces inbetween each dash are half of the original Hidden linetype. HiddenX2 represents a linetype where the distances of the dashes and spaces in between each dash of the original Hidden linetype are doubled.

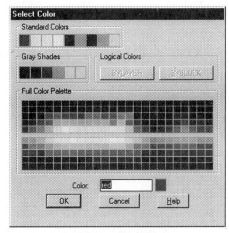

Figure 1–50

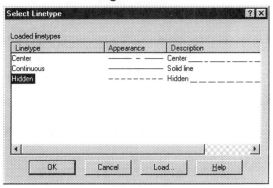

Figure 1–51

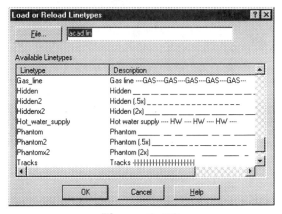

Figure 1–52

Once layers have been created along with color and linetype assignments, the display of the Layer & Linetype properties dialog box will be similar to Figure 1–53. Initially when layers are created, they are placed in the dialog box in the exact order they were created in. Once the dialog box is dismissed using the OK button and then revisited at a later time, all layer names are reordered to be displayed in alphabetical order.

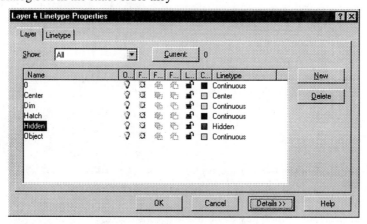

Figure 1–53

Clicking on the Details >> button in Figure 1–53 expands the bottom of the dialog box to include more detailed information about the selected layer (see Figure 1–54). Name, Color, and Linetype information is isolated in individual edit boxes. Also, the properties of the layer (On, Freeze in all viewports, etc) are displayed in a column to the right of the layer name. Clicking on the Details<< button compresses the dialog box back to the illustration in Figure 1–53.

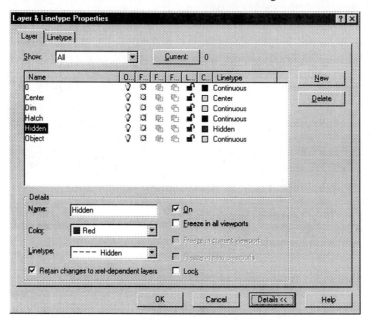

Figure 1–54

Control of Layer Properties

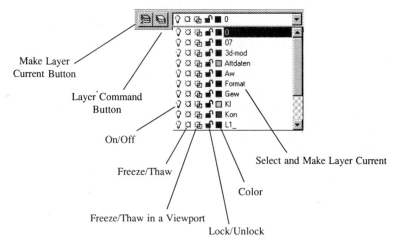

Make Layer
Current Button

Layer Command
Button

On/Off

Freeze/Thaw

Freeze/Thaw in a Viewport

Lock/Unlock

Color

Select and Make Layer Current

Figure 1–55

The Object Properties toolbar displays extra buttons for the control and manipulation of layers in a drawing. The Make Layer Current Button in Figure 1–55 allows you to make a layer the new current layer by just clicking on an object in the drawing. The layer is now made current based on the layer of the selected object. Clicking in the long edit box next to the LAYER command button cascades all layers defined in the drawing in addition to their properties identified by symbols. The presence of the lightbulb signifies that the layer is turned on. Clicking on the lightbulb symbol will turn the layer off. The sun symbol signifies that the layer is thawed. Clicking on the sun turns it into a snowflake symbol signifying that the layer is now frozen. The padlock symbol controls whether a layer is locked or unlocked. By default, all layers are unlocked. Clicking on the padlock changes the symbol to display the image of a locked symbol signifying that the layer is locked. Freezing and thawing of layers per viewport is also possible through this area; this is only made apparent if you have entered the Paper Space environment.

Study Figure 1–56 for a better idea on how the symbols affect the state of certain layers.

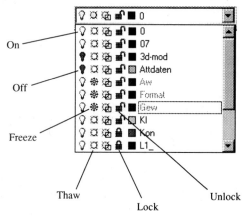

On

Off

Freeze

Thaw

Lock

Unlock

Figure 1–56

Controlling the Linetype Scale

Once linetypes are associated with layers and placed in a drawing, their scale may be controlled by the LTSCALE command. In Figure 1–57, the default Ltscale value of 1.00 is in effect. This scale value acts as a multiplier for all linetype distances. In other words, if the hidden linetype is designed to have dashes 0.125 units long, a Ltscale value of 1.00 displays the dash of the hidden line at a value of 0.125 units. The LTSCALE command displays the following command sequence:

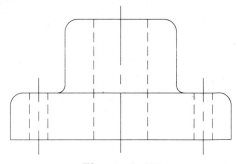

Figure 1–57

Command: **LTSCALE**
New scale factor <1.0000>: *(Press* ENTER *to accept the default or enter another value)*

In Figure 1–58, a Ltscale value of 0.50 units has been applied to all linetypes. As a result, instead of all hidden line dashes measuring 0.125 units, they now measure 0.0625 units as a result of the 0.50 multiplier.

Command: **LTSCALE**
New scale factor <1.0000>: **0.50**

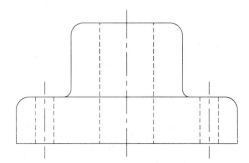

Figure 1–58

In Figure 1–59, a Ltscale value of 2.00 units has been applied to all linetypes. As a result, instead of all hidden line dashes measuring 0.125 units, they now measure 0.25 units as a result of the 2.00 multiplier. Notice how a large multiplier displays the centerlines as what appears to be a continuous linetype.

Command: **LTSCALE**
New scale factor <0.5000>: **2.00**

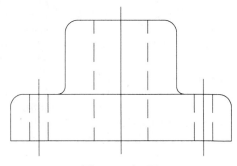

Figure 1–59

Illustrated in Figure 1–60 is a facilities drawing that is designed to plot out at a scale factor of 1/4"=1'0." This creates a multiplier of 48 units. For all linetypes to show as hidden or centerlines, this multiplier must be applied to the drawing through the LTSCALE command.

Since the drawing was constructed in real world units or full size, the linetypes must be converted to these units using the multiplier of 48 through the LTSCALE command.

Command: **LTSCALE**
New scale factor <1.0000>: **48**

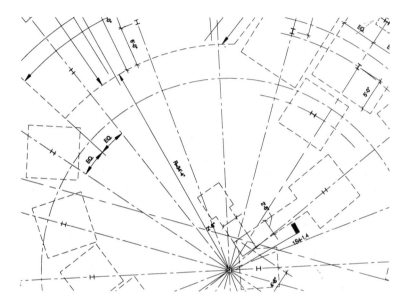

Figure 1–60

The Alphabet of Lines

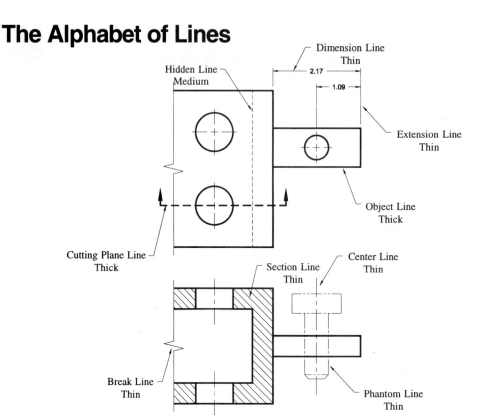

Figure 1–61

Before constructing engineering drawings, the quality of the lines that make up the drawing must first be discussed. Some lines of a drawing should be made thick; others need to be made thin. This is to emphasize certain parts of the drawing and is controlled through a line quality system. Illustrated in Figure 1–61 is a two-view drawing of an object complete with various lines that will be explained further.

The most important line of a drawing outlines the shape of the object and for this reason is referred to as an Object line. Because of their importance, Object lines are made thick and continuous so they can stand out among the other lines in the drawing. It does not mean the other lines are considered unimportant; rather, the Object line takes precedence over all other lines. Cutting Plane line is another thick line; it is used to determine the placement in the drawing where an imaginary saw will cut into

the drawing to expose interior details. It stands out by being drawn as a series of long dashes separated by spaces. Arrowheads determine how the adjacent view will be looked at. This line will be discussed in greater detail in Unit 8, Section Views. The Hidden linetype is a medium weight line used to identify edges that become invisible in a view. The Hidden line consists of a series of dashes separated by spaces. Whether an edge is visible or invisible, it still must be shown with a line. The Dimension line is a thin line used to show the numerical distance between two points. The dimension text is placed in between the dimension line and arrowheads are placed at opposite ends of the dimension line. The Extension line is another thin continuous line and is used as a part of the overall dimension. Extension lines show the distance being dimensioned. In Figure 1–61, the vertical extension lines are used to indicate the horizontal dimension distance. When using the Cutting Plane line to create an area to cut or slice, the surfaces in the adjacent view are sectioned lined using the Section line. This is a thin continuous line. Another important line used to identify the centers of circles is the Center line. It is a thin line consisting of a series of long and short dashes. It is a good practice to dimension to centerlines; this is the reason centerlines, extension lines, and dimensions are made the same line thickness. The Phantom line consists of a thin line made with a series of two dashes and one long dash. It is used to simulate the placement or movement of a part or component without actually detailing the component. The BREAK line is a thin line with a "zigzag" symbol used to establish where an object is broken to simulate a continuation of the object.

Basic Drawing Techniques Using the LINE Command

Use the LINE command to construct a line from one endpoint to the other. As the first point of the line is marked, the rubber band cursor is displayed along with the normal crosshair to help see where the next line segment will be drawn to. The LINE command stays active until either the Close option is used or a null response is issued by pressing the ENTER key at the prompt "To point." Study Figure 1–62 and the following prompt sequence below for using the LINE command.

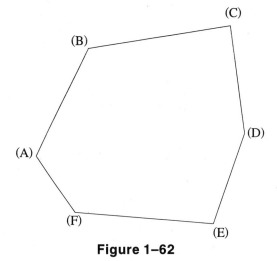

Figure 1–62

 Command: **LINE**
From point: *(Mark a point at "A")*
To point: *(Mark a point at "B")*
To point: *(Mark a point at "C")*
To point: *(Mark a point at "D")*
To point: *(Mark a point at "E")*
To point: *(Mark a point at "F")*
To point: **Close**

From time to time, mistakes are made in the LINE command by drawing an incorrect segment. As illustrated in Figure 1–63, segment DE is drawn incorrectly. Instead of exiting the LINE command and erasing the line, a built-in Undo is used inside of the LINE command. This removes the previously drawn line while still remaining in the LINE command. Refer to Figure 1–63 and the following prompts to use the Undo option of the LINE command.

 Command:**LINE**

From point: *(Mark a point at "A")*
To point: *(Mark a point at "B")*
To point: *(Mark a point at "C")*
To point: *(Mark a point at "D")*
To point: *(Mark a point at "E")*
To point: Undo (*To remove the segment from
    "D" to "E" and still remain in the LINE
    command)*
To point: *(Mark a point at "F")*
To point: **Endp**
of *(Select the endpoint of the line segment at
    "A")*
To point: *(Press* ENTER *to exit this command)*

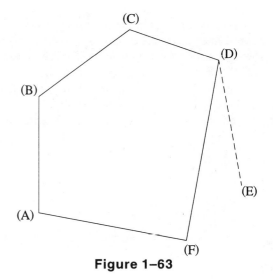

Figure 1–63

Another option of the LINE command is the Continue option. The dashed line segment in Figure 1–64 was the last segment drawn before exiting the LINE command. To pick up at the last point of a previously drawn line segment, enter the LINE command and press the ENTER key. This will activate the Continue option of the LINE command.

 Command:**LINE**

From point: *(Press* ENTER *to activate Continue
    Mode)*
To point: *(Mark a point at "B")*
To point: *(Mark a point at "C")*
To point: **Endp**
of *(Select the endpoint of the vertical line
    segment at "A")*
To point: *(Press* ENTER *to exit this command)*

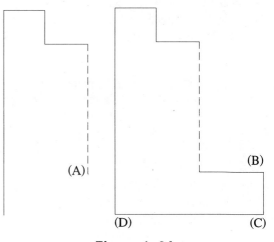

Figure 1–64

Cartesian Coordinates

Before drawing precision geometry such as lines and circles, you must have an understanding of coordinate systems. The Cartesian, or rectangular, coordinate system is used to place geometry at exact distances through a series of coordinates. A coordinate is made up of an ordered pair of numbers usually identified as X and Y. The coordinates are then plotted on a type of graph or chart. The graph, shown in Figure 1–65, is made up of two perpendicular number lines called coordinate axes. The horizontal axis is called the X-axis. The vertical axis is called the Y-axis.

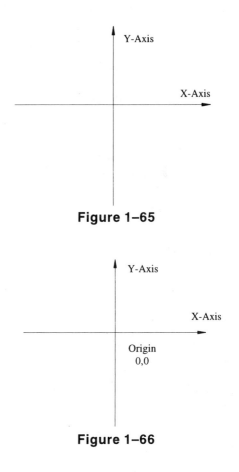

Figure 1–65

As shown in Figure 1–66, the intersection of the two coordinate axes forms a point called the origin. Coordinates used to describe the origin are 0,0. From the origin, all positive directions move up and to the right. All negative directions move down and to the left.

Figure 1–66

The coordinate axes are divided into four quadrants that are labeled I, II, III, and IV as shown in Figure 1–67. In Quadrant I, all X and Y values are positive. Quadrant II has a negative X value and positive Y value. Quadrant III has negative values for X and Y. Quadrant IV has positive X values and negative Y values.

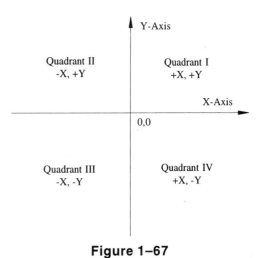

Figure 1–67

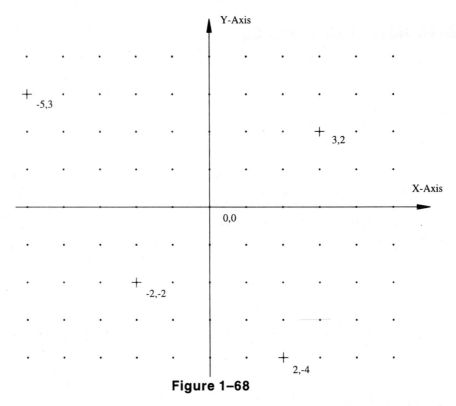

Figure 1–68

For each ordered pair of *(X, Y)* coordinates, X means to move from the origin to the right if positive and to the left if negative. Y means to move from the origin up if positive and down if negative. Figure 1–68 shows a series of coordinates plotted on the number lines. One coordinate is identified in each quadrant to show the positive and negative values. As an example, coordinate 3,2 located in the Quad-rant I means to move 3 units to the right of the origin and up 2 units. The coordinate -5,3 located in Quadrant II means to move 5 units to the left of origin and up 3 units. Coordinate -2,-2 located in Quadrant III means to move 2 units to the left of the origin and down 2. Lastly, coordinate 2,-4 located in Quadrant IV means to move 2 units to the right of the origin and down -4.

When beginning a drawing in AutoCAD, the screen display reflects Quadrant I of the Cartesian coordinate system. As shown in Figure 1–69, the origin 0,0 is located in the lower left corner of the drawing screen. The current screen size is measured by the upper right coordinate of the screen which is, by default, 12,9. This value may be changed using the LIMITS command to accommodate any drawing including architectural and civil engineering.

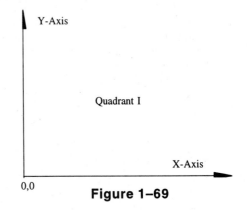

Figure 1–69

Using Absolute Coordinate Mode to Draw Lines

When drawing geometry such as lines, a method of entering precise distances must be used especially when accuracy is important. This is the main purpose of using coordinates. The simplest and most elementary form of coordinate values is Absolute coordinates. Absolute coordinates conform to the following format:

$$X,Y$$

One problem with using absolute coordinates is that all coordinate values refer back to the origin 0,0. This origin on the AutoCAD screen is usually located in the lower left corner of a brand new drawing. The origin will remain in this corner unless it is altered using the LIMITS command. Study the following LINE command prompts as well as Figure 1–70.

 Command: **LINE**
From point: **2,2** *(at "A")*
To point: **2,7** *(at "B")*
To point: **5,7** *(at "C")*
To point: **7,4** *(at "D")*
To point: **10,4** *(at "E")*
To point: **10,2** *(at "F")*
To point: **C** *(To close the figure)*

As you can see, all points on the object make reference to the origin at 0,0. Even though absolute coordinates are useful in starting lines, there are more efficient ways to continue lines and draw objects.

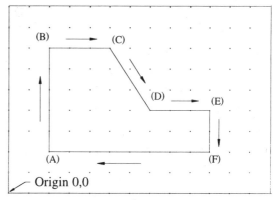

Figure 1–70

Using Relative Coordinate Mode to Draw

In absolute coordinates, the origin at 0,0 must be kept track of at all times in order to enter the correct coordinate. With complicated objects, this is sometimes difficult to accomplish and as a result, the wrong coordinate is entered. It is possible to reset the last coordinate to become a new origin or 0,0 point. The new point would be relative to the previous point and for this reason this point is called

a Relative Coordinate. The format is as follows:

$$@X,Y$$

In this format, we use the save X and Y values with one exception: the "At" symbol or @ resets the previous point to 0,0 and makes entering coordinates less confusing. Study the following LINE command prompts as well as Figure 1–71.

Command: **LINE**

From point: **2,2** *(at "A")*
To point: **@0,4** *(to "B")*
To point: **@4,2** *(to "C")*
To point: **@3,0** *(to "D")*
To point: **@3,-4** *(to "E")*
To point: **@-3,-2** *(to "F")*
To point: **@-7,0** *(back to "A")*
To point: *(Press [ENTER] to exit this command)*

In each command prompt, the @ symbol resets the previous point to 0,0.

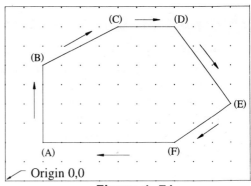

Figure 1–71

The Polar Coordinate Mode to Draw

Another popular method of entering coordinates is by Polar mode. The format is as follows:

@Distance<Direction

As the preceeding format implies, the polar coordinate mode requires a known distance and a direction. The @ symbol resets the previous point to 0,0. The direction is preceded by the < symbol, which reads the next number as a polar direction. In Figure 1–72A is an illustration describing the directions supported by the polar mode. Study the following LINE command prompts as well as Figure 1–72B for the polar coordinate mode.

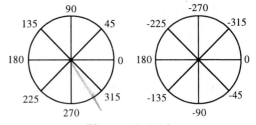

Figure 1–72A

 Command: **LINE**

From point: **3,2** *(at "A")*
To point: **@8<0** *(to "B")*
To point: **@5<90** *(to "C")*
To point: **@5<180** *(to "D")*
To point: **@4<270** *(to "E")*
To point: **@2<180** *(to "F")*
To point: **@2<90** *(to "G")*
To point: **@1<180** *(to "H")*
To point: **@3<270** *(to close back to "A")*
To point: *(Press [ENTER] to exit this command)*

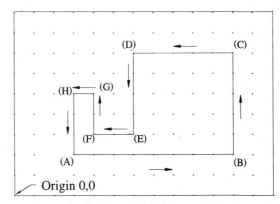

Figure 1–72B

Combining Coordinate Modes to Draw

So far, the preceeding pages concentrated on using each example of coordinate modes (absolute, relative, and polar) to create geometry. At this point, we do not want to give the impression that once you start with a particular coordinate mode you must stay with the mode. Rather, drawings are created using one, two, or three coordinate modes in combination with each other. In Figure 1–73, the drawing starts with an absolute coordinate, changes to a polar coordinate, and changes again to a relative coordinate. It is the responsibility of the CAD operator to choose the most efficient coordinate mode to fit the drawing. Study the following LINE command prompts as well as Figure 1–73.

 Command: **LINE**

From point: **2,2** *(at "A")*
To point: **@3<90** *(to "B")*
To point: **@2,2** *(to "C")*
To point: **@6<0** *(to "D")*
To point: **@5<270** *(to "E")*
To point: **@3<180** *(to "F")*
To point: **@3<90** *(to "G")*
To point: **@2<180** *(to "H")*
To point: **@-3,-3** *(back to "A")*
To point: *(Press* ⏎ *to exit this command)*

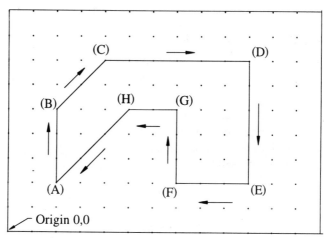

Figure 1–73

Using the Direct Distance Mode

Another method is available for constructing lines and it is called drawing by Direct Distance mode. In this method, the direction a line will be drawn is guided by the location of the cursor. A value is entered and the line is drawn at the specified distance at the angle specified by the cursor. This mode works especially well when drawing horizontal and vertical lines. Illustrated in Figure 1–74 is an example of how the Direct Distance mode is used. To begin, Ortho mode is turned on. Use the following prompt sequence to construct the line segments.

 Command: **LINE**

From point: **2.00,2.00**
To point: *(Move the cursor to the right and enter a value of 7.00 units)*
To point: *(Move the cursor up and enter a value of 5.00 units)*
To point: *(Move the cursor to the left and enter a value of 4.00 units)*
To point: *(Move the cursor down and enter a value of 3.00 units)*
To point: *(Move the cursor to the left and enter a value of 2.00 units)*
To point: **C** *(To close the shape and exit the command)*

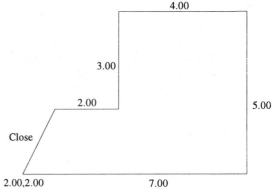

Figure 1–74

Illustrated in Figure 1–75 is another example of an object drawn using Direct Distance mode. Each angle was constructed from the location of the cursor. In this example, Ortho mode is turned off.

 Command: **LINE**

From point: *(Pick a point at "A")*
To point: *(Move the cursor and enter 3.00)*
To point: *(Move the cursor and enter 2.00)*
To point: *(Move the cursor and enter 1.00)*
To point: *(Move the cursor and enter 4.00)*
To point: *(Move the cursor and enter 2.00)*
To point: *(Move the cursor and enter 1.00)*
To point: *(Move the cursor and enter 1.00)*
To point: **C** *(To close the shape)*

This may not be an accurate method of constructing lines at angles. It would be considered good practice to use either Relative or Polar Coordinate mode for better accuracy.

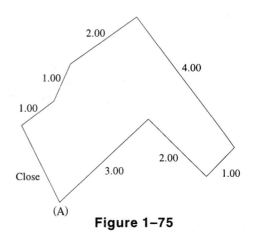

Figure 1–75

Using the OSNAP Command

Snap and Grid are two modes used to aid you in creating and editing objects. However all objects cannot be guaranteed that they will fall exactly on a grid dot. Rather than rely entirely on Grid or Snap, the Object Snap mode is one of the most important modes for locking onto key object points. Figure 1–76 is an example of constructing a vertical line connecting up the endpoint of the fillet with the endpoint of the line at "A." The LINE command is entered and the endpoint mode activated. When the cursor moves over a valid endpoint, an Object Snap symbol appears along with a tool tip telling you which Osnap mode is currently being used. Another example of how Object Snap is used may be in dimensioning applications where exact endpoints and intersections are needed.

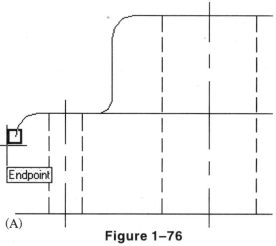

Figure 1–76

    Command: **LINE**

From point: **Endp** *(For endpoint mode)*
of *(Pick the endpoint of the fillet illustrated in Figure 1–76)*
To point: **Endp** *(For endpoint mode)*
of *(Pick the endpoint of the line at "A")*
To point: *(Press* ⌷ENTER⌷ *to exit this command)*

Object Snap modes may be selected from a number of different areas. Illustrated in Figure 1–77A is the standard Object Snap toolbar. In Figure 1–77B is an illustration of the Object Snap modes that were activated by holding down the Shift key and pressing the right mouse button. This popdown menu will display on the screen wherever the cursor is currently positioned. Another method of entering Object Snap modes is from the keyboard. When entering Object Snap modes from the keyboard, only the first three letters are required. The following pages give examples on the applications of all Object Snap modes.

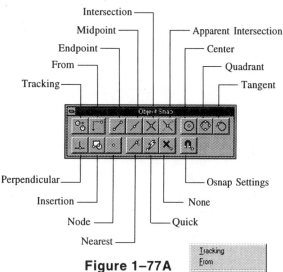

Figure 1–77A

Figure 1–77B

Using Options of the OSNAP Command

 ## Osnap-Center

Use this mode to snap to the center of a circle or arc. To accomplish this, activate the mode by clicking on the Osnap-Center button and moving the cursor along the edge of the circle or arc similar to Figure 1–78. Notice the AutoSnap symbol appearing at the center of the circle or arc.

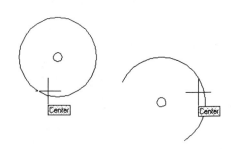

Figure 1–78

 ## Osnap-Endpoint

This is one of the more popular Object Snap modes that is very helpful in snapping to the endpoints of lines or arcs. One application of Osnap-Endpoint is during the dimensioning process where exact distances are needed to produce the desired dimension. Activate this mode by clicking on the Osnap-Endpoint button and move the cursor along the edge of the object to snap to the endpoint. In the case of the line or arc shown in Figure 1–79, the cursor does not actually have to be positioned at the endpoint; favoring one end automatically snaps to the closest endpoint.

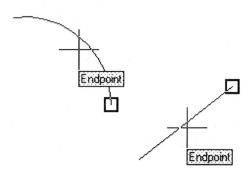

Figure 1–79

 ## Osnap-From

Use this Object Snap mode along with a Secondary Object Snap mode to establish a reference point and construct an offset from that point. In Figure 1–80, the circle needed to be drawn 1.50 units in the X and Y directions from point "A." The CIRCLE command is activated and the Osnap-From mode is used in combination with the Osnap-Intersection mode. The From option requires a base point. Identify the base point at the intersection of corner "A." The next prompt asks for an offset value; enter the relative coordinate value of @1.50,1.50. This completes the use of the From option and identifies the center of the circle at "B."

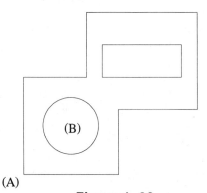

Figure 1–80

Osnap-Insert

This Object Snap option snaps to the insertion point of an object. In the case of the text object in Figure 1–81, activating the Osnap-Insert mode and positioning the cursor any place on the text snaps to its insertion point, in this case at the lower left corner of the text at "A." The other object illustrated in Figure 1–81 is called a block. It appears to be constructed of numerous line objects; however, all objects that make up the block are considered to be a single object. Blocks are then inserted into a drawing. Typical types of blocks are symbols such as doors, windows, bolts, etc., anything that is used numerous times in a drawing. In order for a block to be brought into a drawing, it needs an insertion point, or a point of reference. The Osnap-Insert option, when positioning the cursor on a block, will snap to the insertion point of a block.

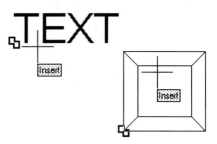

Figure 1–81

 Osnap-Intersection

Another very popular Object Snap mode is Intersection. Use this mode to snap to the intersection of two objects. Position the cursor anywhere near the intersection of two objects and the intersection symbol appears. See Figure 1–82.

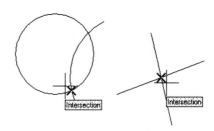

Figure 1–82

 Osnap-Apparent Intersection

Use the Osnap-Apparent Intersection mode to snap to an intersection not considered obvious from the previous example. Figure 1–83 shows two lines that do not intersect. Activate the Osnap-Apparent Intersection mode and click on both lines. Notice the intersection symbol present where the two lines apparently intersect.

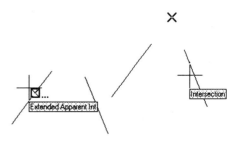

Figure 1–83

Osnap-Midpoint

This Object Snap mode snaps to the midpoint of objects. Line and arc examples are shown in Figure 1–84. When activating the Osnap-Midpoint mode, touch the object anywhere with some portion of the cursor; the midpoint symbol will appear at the exact midpoint of the object.

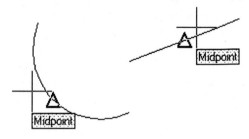

Figure 1–84

 Osnap-Nearest

This Object Snap mode snaps to the nearest point it finds on an object. Use this mode when you need to grab onto an object for the purposes of further editing. The nearest point is calculated based on the closest distance from the intersection of the crosshairs perpendicular to the object; or the shortest distance from the crosshairs to the object. In Figure 1–85, the appearance of the Nearest symbol helps to see where the point identified by this mode is actually located.

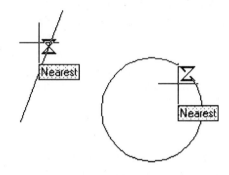

Figure 1–85

 Osnap-NODE

This Object Snap mode snaps to a node or point. Touching the point in Figure 1–86 snaps to its center. The Osnap-Nearest mode shown in Figure 1–85 can also be used to snap to a point.

Figure 1–86

Osnap-Perpendicular

This is a helpful Object Snap mode for snapping to an object normal or perpendicular from a previously identified point. Figure 1–87 shows a line segment drawn perpendicular from the point at "A" to the inclined line "B." A 90-degree angle is formed with the perpendicular line segment and the inclined line "B." With this mode, you are also able to construct lines perpendicular to circles.

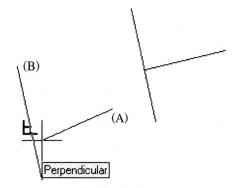

Figure 1–87

 ## Osnap-Quadrant

Circle quadrants are defined as points located at the 0-, 90-, 180-, and 270-degree positions of a circle, as in Figure 1–88. Using the Osnap-Quadrant option will snap to one of these four positions as the edge of a circle or arc is selected. In the example of the circle in Figure 1–88, the edge of the circle is selected by the cursor location. The closest quadrant to the cursor is selected.

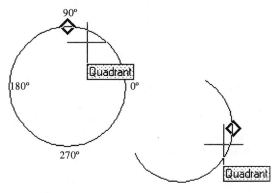

Figure 1–88

 ## Osnap-Tangent

This Object Snap mode is very helpful in constructing lines tangent to other objects such as the two circles in Figure 1–89. In this case, the Osnap-Tangent mode is being used in conjuction with the LINE command. The point at "A" is anchored to the top of the circle using the Osnap-Quadrant mode. Follow this command prompt sequence for constructing a line segment tangent to two circles:

 Command:**LINE**

From point: **Qua**
from *(Select the circle near "A")*
To point: **Tan**
to *(Select the circle near "B")*

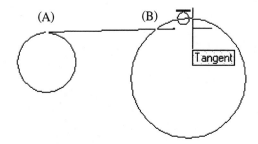

Figure 1–89

Osnap-Tangent

When constructing a line tangent from one point and tangent to another, the Deferred Tangent mode is automatically activated (see Figure 1–90). This means instead of activating the rubber band cursor mode where a line is present, the deferred tangent must have two tangent points identified before the line is constructed and displayed.

 Osnap-Tracking

In Figure 1–91A, a hole represented in the front view as a series of hidden lines needs to display as a circle in the right side view. The hole will be placed in the center of the rectangle identified by "A" and "B." Tracking mode will provide a quick and easy way to locate the center of the rectangle without drawing any construction geometry.

When tracking is started and a point is picked on the screen, AutoCAD forces the next point to conform to an orthogonal path that could extend vertically or horizontally from the first point. The first point may retain an X or Y value and the second point may retain separate X or Y values. Use the following prompt sequence to get an idea of the operation of tracking mode. For best results, be sure Ortho mode is turned off.

 Command: **CIRCLE**
3P/2P/TTR/<Center point>: **TK** *(For tracking)*
First tracking point: **Mid**
of *(Select the midpoint at "A" in Figure 1–91A)*
Next point *(Press [ENTER] to end tracking):* **Mid**
of *(Select the midpoint at "B" in Figure 1–91A)*
Next point *(Press [ENTER] to end tracking):* (Press [ENTER])
Diameter/<Radius>: **D** *(For diameter)*
Diameter: **0.50**

The result is illustrated in Figure 1–91B with the hole drawn in the center of the rectangular shape.

 Command: **LINE**
From point: **Tan**
from *(Select the circle near "A")*
To point: **Tan**
to *(Select the circle near "B")*

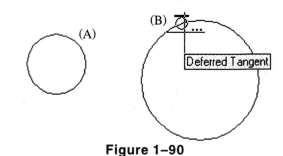

Figure 1–90

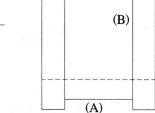

Figure 1–91A

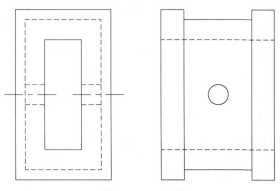

Figure 1–91B

Alternate Methods of Choosing Osnap Options

There is an additional aid to bring up all options of Object Snap. Holding down the "Shift" key located on the keyboard and pressing the ENTER button of the mouse automatically brings up a pull-down menu containing all Object Snap modes (See Figure 1–92A). This menu is displayed wherever the cursor was last positioned. This provides you with an even quicker way of getting to the Object Snap modes. Figure 1–92B shows a typical drawing screen with the Object Snap modes activated with the "Shift-Mouse Enter" operation.

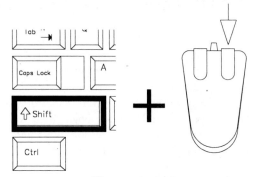

Figure 1–92A

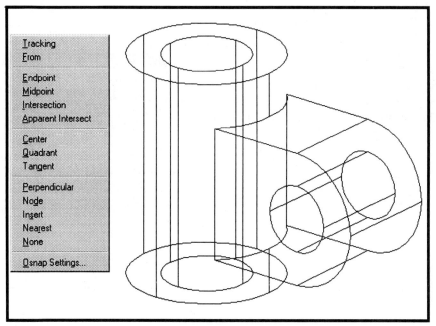

Figure 1–92B

Choosing Running Osnap from the Osnap Settings Dialog Box

So far, all Osnap modes have continuously been selected from the popdown menu or entered in at the keyboard. The problem with these methods is that if a certain mode was used over a series of commands, the Osnap mode has to be selected every time. It is possible to make the Object Snap mode or modes automatically present through Running Osnap. Running Osnap mode may be selected from the "Tools" area of the pull-down menu followed by selecting "Object Snap Settings…." This activates the "Osnap Settings" dialog box illustrated in Figure 1–93A. One or more Object Snap modes may be selected by checking their appropriate boxes in the dialog box. These Osnap modes remain in effect during the drawing until the Osnap button illustrated in Figure 1–93B is clicked on twice; this turns off the current Running Osnap modes and returns to the regular method of using Osnap options. To reinstate the running Osnap modes, click twice again on the Osnap button and the previously set of Osnap modes will be back in effect.

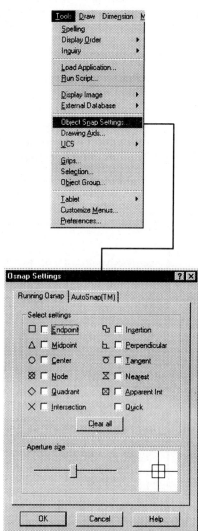

Figure 1–93A

Figure 1–93B

Using the AutoSnap Feature of OSNAP

Clicking on the AutoSnap™ tab displays the dialog box shown in Figure 1–94. This dialog box controls the display of the AutoSnap symbols that appear when an Osnap mode is active. The Select settings area of the dialog box allows you to control if the marker is present during Osnap modes, if the cursor snaps to the symbol, or if a tool tip of the active Osnap mode is present. By default, the marker color is yellow and has been changed to blue for this unit. The yellow marker color is ideal when the screen background is in a dark color such as black or green. You are also able to change the marker size by manipulating the slider bar in Figure 1–94.

Placing a check in the Display aperture box displays the Osnap aperture box (see Figure 1–95). When activating an Osnap mode, a target box appears around the cursor similar to Figure 1–96. In Figure 1–96, the Osnap-Center mode, the target box must touch the edge of the circle before the snap point is identified. In Figure 1–96, the Osnap-Endpoint mode, the edge of the line must be touched by the Osnap target box. The endpoint nearest to the target box is identified.

The size of the aperture box is controlled by the APERTURE command. Use the dialog box illustrated in Figure 1–93A to dynamically change the aperture box using the slider bar under "Aperture size."

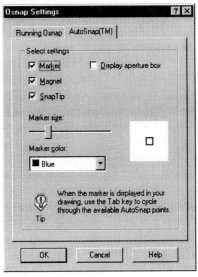

Figure 1–94

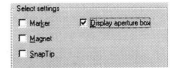

Figure 1–95

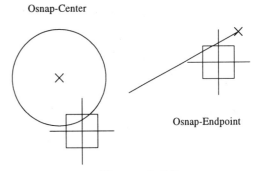

Figure 1–96

Using the ZOOM Command

The ability to magnify details in a drawing or demagnify the drawing to see it in its entirety is a function of the ZOOM command (see Figure 1–97). It does not take much for a drawing to become very busy, complicated, or dense when displayed in the drawing editor. Therefore, use the ZOOM command to work on details or view different parts of the drawing. One area to select this command is from the pull-down menu area shown in Figure 1–97. Clicking on "View" and then "ZOOM" displays the various options of the ZOOM command including zooming in real time, zooming the previous display, using a window to define a boxed area to zoom to, dynamic zooming, zooming to a user-defined scale factor, zooming based on a center point and a scale factor, or performing routine operations such as zooming in or out, zooming all, or zooming the extents of the drawing. All modes will be discussed in the following pages.

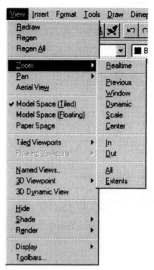

Figure 1–97

A second area to choose ZOOM command options is through the Standard toolbar area at the top of the display screen. Figure 1–98 shows a series of buttons that perform various options of the ZOOM command. Four main buttons initially appear in the Standard toolbar; namely Realtime Pan, Realtime ZOOM, ZOOM-Window, and ZOOM-Previous. Pressing down on the ZOOM-Window option displays a series of zoom option buttons that cascade down. Click on the button to perform the desired zoom operation.

Figure 1–99 a third method of performing ZOOM command options using the dedicated Zoom floating toolbar. This toolbar contains the same identical buttons found in the Standard toolbar; it differs by the fact that the toolbar in Figure 1–99 can be moved to different positions around the display screen.

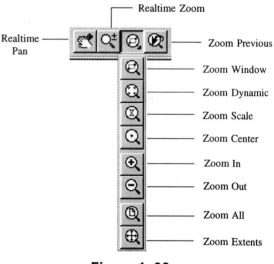

Figure 1–98

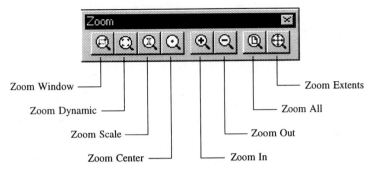

Figure 1–99

The ZOOM command may also be activated from the keyboard by entering either "ZOOM" or the letter "Z" which is the command alias of zoom.

Figure 1–100 is a complex drawing of a part that consists of all required orthographic views. To work on details of this and other drawings, the ZOOM command is used to magnify or demagnify the display screen. The following are options of the ZOOM command:

Command: **ZOOM**
All/Center/Dynamic/Extents/Previous/Scale(X/ XP)/Window/<Realtime>: *(Enter one of the listed options)*

Executing the ZOOM command and picking a blank part of the screen places you in automatic ZOOM-Window mode. Selecting another point zooms into the specified area. Refer to the following command sequence to use this mode of the ZOOM command on the object illustrated in Figure 1–100.

Command: **ZOOM**
All/Center/Dynamic/Extents/Previous/Scale(X/ XP)/Window/<Realtime>: *(Mark a point at "A")*
Other corner: *(Mark a point at "B")*

The ZOOM-Window option is automatically invoked once a blank part of the screen is selected followed by a second point. The results of the magnified screen appear in Figure 1–101.

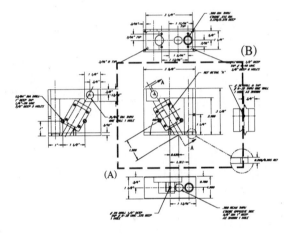

Figure 1–100

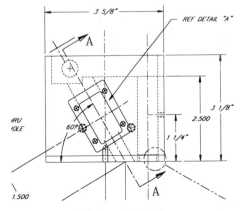

Figure 1–101

Using Realtime ZOOM

A very dynamic option of the ZOOM command is performing screen magnifications or demagnifications in real time. This is the default option of the ZOOM command. Issuing the Realtime option of the ZOOM command displays a magnifying glass icon with a positive sign above the magnifier and a negative sign below the magnifier. Identifying a blank part of the drawing editor, pressing down the select button of the mouse, and moving in an upward direction zooms into the drawing in real time. Identifying a blank part of the drawing editor, pressing down the select button of the mouse, and moving in a downward direction zooms out of the drawing in real time. Use the following command sequence and also see Figure 1–102.

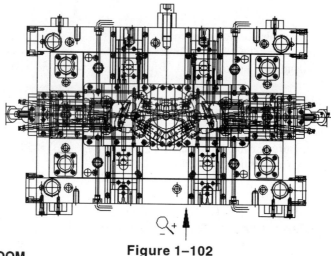

Figure 1–102

 Command:**ZOOM**

All/Center/Dynamic/Extents/Previous/Scale(X/
XP)/Window/<Realtime>: *(Press* ENTER *to
accept Realtime as the default)*

Press ESC or ENTER to exit, or right-click to activate the pop-up menu. (Identify the lower portion of the drawing editor, press the select button of the mouse, and move the realtime cursor up; notice the image zooming in realtime).

Once inside of the Realtime mode of the ZOOM command, pressing on the right mouse button activates the cursor menu shown in Figure 1–103. Use this menu to switch between Realtime ZOOM and Realtime Pan, which gives you the ability to pan across the screen in real time. The ZOOM Window, Previous, and Extents options are also available in the cursor menu in Figure 1–103.

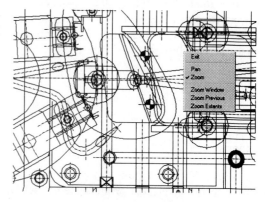

Figure 1–103

Using Aerial View Zooming

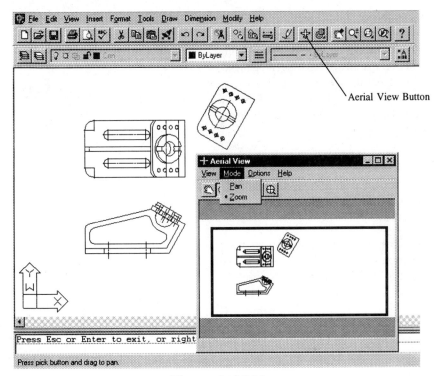

Aerial View Button

Figure 1–104A

Another dynamic way of performing zooms is through the Aerial View dialog box shown in Figures 1–104A and 1–104B. Clicking on the Aerial View button activates a dialog box that displays in the lower right corner of the display screen. As with all dialog boxes, this may be moved to a better location if necessary. Clicking inside of this dialog to make it active and creating a rectangle performs a zoom in the background, yet you can still view the entire drawing in the Aerial View dialog box shown in Figure 1–104B.

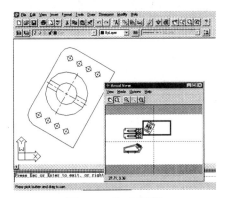

Figure 1–104B

Notice in the illustration above of the Aerial View dialog box, that a Pan option is present. Use it to pan around the Aerial View while viewing the results in the background.

Using ZOOM All

Another option of the ZOOM command is All. Use
this option to zoom to the current limits of the drawing as set by the LIMITS command. In fact, right
after the limits of a drawing have been changed,
issuing a ZOOM-All updates the drawing file to
reflect the latest screen size. To use the ZOOM-All
option, refer to the following command sequence.

 Command:**ZOOM**
All/Center/Dynamic/Extents/Previous/Scale(X/
    XP)/Window/<Realtime>:**All**
Regenerating drawing.

In Figure 1–105A, the top illustration shows a
zoomed in portion of a part. Use the ZOOM-All
option to zoom to the drawing's current limits in
Figure 1–105B.

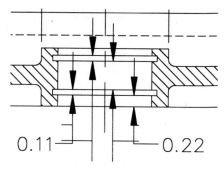

Figure 1–105A

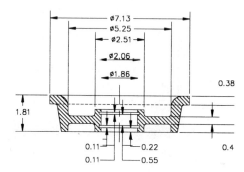

Figure 1–105B

Using ZOOM Center

The ZOOM-Center option allows you to specify a new display based on a selected center point (see Figure 1–106). A window height controls whether the image on the display screen is magnified or demagnified. If a smaller value is specified for the magnification or height, the magnification of the image is increased or the object is zoomed into. If a larger value is specified for the magnification or height, the image gets smaller, or a zoom out is performed.

 Command:**ZOOM**

All/Center/Dynamic/Extents/Previous/Scale(X/
    XP)/Window/<Realtime>:**Center**
Center point: *(Mark a point at the center of
    circle "A" shown in Figure 1–106)*
Magnification or Height <7.776>:**2**

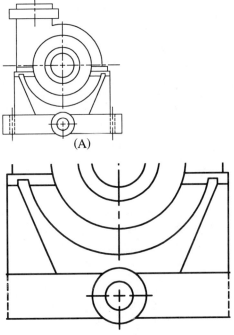

(A)

Figure 1–106

 # Using ZOOM-Extents

The image of the pump in Figure 1–107 reflects a ZOOM-All operation. This displays the entire drawing area based on the drawing limits even if the objects that make up the image appear small. Instead of performing a zoom based on the drawing limits, ZOOM-Extents uses only the extents of the image on the display screen to perform the zoom. Figure 1–107 shows the largest possible image displayed as a result of using the ZOOM command and the Extents option.

 Command:**ZOOM**

All/Center/Dynamic/Extents/Previous/Scale(X/
    XP)/Window/<Realtime>:**Extents**

Zoom-All

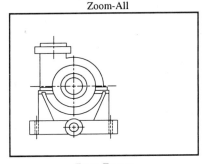

Zoom Extents

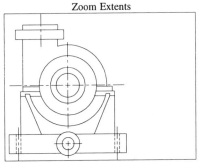

Figure 1–107

Using ZOOM-Window

As shown in Figure 1–108, ZOOM-Window allows you to specify the area to be magnified by marking two points representing a rectangle. The center of the rectangle becomes the center of the new image display; the image inside of the rectangle is either enlarged or reduced. The following prompt sequence demonstrates ZOOM-Window:

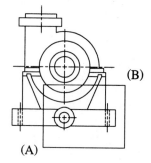

(B)

(A)

 Command:**ZOOM**

All/Center/Dynamic/Extents/Previous/Scale(X/
XP)/Window/<Realtime>:**Window**
First corner: (*Mark a point at "A"*)
Other corner: (*Mark a point at "B"*)

By default, the window option of zoom is considered automatic; in other words, without entering the "Window" option, a point is marked instead. This identifies the first corner of the window box; the prompt "Other corner:" completes ZOOM-Window as indicated in the following prompts:

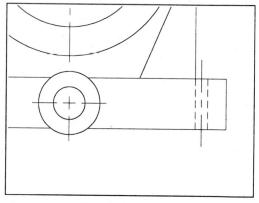

 Command:**ZOOM**

All/Center/Dynamic/Extents/Previous/Scale(X/
XP)/Window/<Realtime>: (*Mark a point at
"A"*)
Other corner: (*Mark a point at "B"*)

Figure 1–108

Using Other Options of the ZOOM Command

 ZOOM-Previous

After magnifying a small area of the display screen, use the Previous option of the ZOOM command to return to the previous display. The system automatically saves up to ten views when zooming. This means you can begin with an overall display, perform two zooms, and use the ZOOM-Previous command twice to return to the original display. Zoom-Previous is also less likely to create a drawing regeneration.

 Command:**ZOOM**

All/Center/Dynamic/Extents/Previous/Scale(X/XP)/Window/<Realtime>:**Previous**

 ZOOM-Scale

Clicking on this button prompts you to enter a zoom scale factor. If a scale factor of 0.5 is used, the zoom is performed into the drawing at a factor of 0.5 based on the original limits of the drawing. If a scale factor of 0.5X is used, the zoom is performed into the drawing again at a factor of 0.5; however, the zoom is based on the current display screen.

 ZOOM-In

Clicking on this button automatically performs a zoom-in operation at a scale factor of 0.5X; the "X" uses the current screen to perform the zoom-in operation.

ZOOM-Out

Clicking on this button automatically performs a zoom-out operation at a scale factor of 2X; the "X" uses the current screen to perform the zoom-in operation.

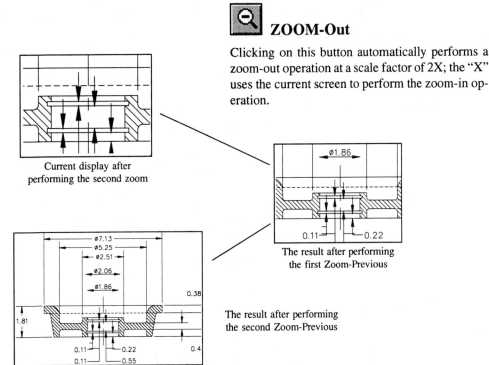

Current display after performing the second zoom

The result after performing the first Zoom-Previous

The result after performing the second Zoom-Previous

Figure 1–109

Using the Pan Command

As you perform numerous ZOOM-Window and ZOOM-Previous operations, it becomes apparent that it would be nice to zoom into a detail of a drawing and simply slide the drawing to a new area without changing the magnification; this is the purpose of the PAN command. In Figure 1–110, the top view is magnified using ZOOM-Window; the result is shown in Figure 1–111. Now, the bottom view needs to be magnified to view certain dimensions. Rather than use ZOOM-Previous and then ZOOM-Window again to magnify the bottom view, use the PAN command.

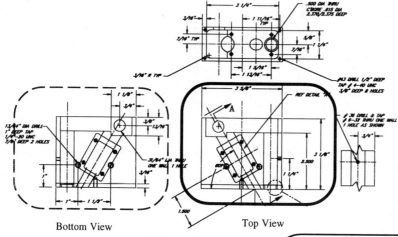

Bottom View   Top View

Figure 1–110

Command:**PAN**

Issuing the PAN command displays the "Hand" symbol. In the illustration in Figure 1–111, pressing the pick button down at "A" and moving the hand symbol to the right at "B" pans the screen and displays a new area of the drawing in the current zoom magnification.

In Figure 1–112, the bottom view is now visible after panning the drawing from the top view to the bottom view while keeping the same display screen magnification. PAN can also be used transparently; that is, while in a current command, the PAN command may be selected, which temporarily interrupts the current command, performs the pan, and restores the current command. PAN may be selected from the pull-down menu area under "View" or may be typed in from the keyboard by entering "P," which is the command alias for the PAN command.

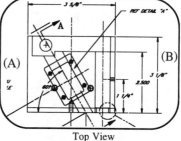

Top View

Figure 1–111

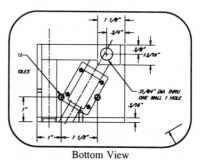

Bottom View

Figure 1–112

Using the VIEW Command

An alternate method of performing numerous zooms is to create a series of views of key parts of a drawing. Then, instead of using the ZOOM command, restore the named view to perform detail work. This named view is saved in the database of the draw- ing for use in future editing sessions. With Figure 1–113 as an example, use the following command sequence to create a series of views:

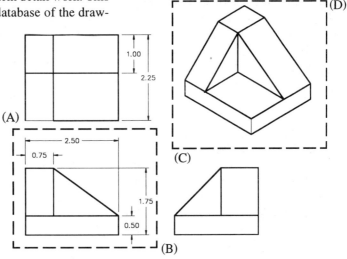

Figure 1–113

Command: **VIEW**
?/Delete/Restore/Save/Window:**Window**
View name: **Front**
First corner: *(Pick a point at "A")*
Other corner: *(Pick a point at "B")*

Command: **VIEW**
?/Delete/Restore/Save/Window:**Window**
View name: **Iso**
First corner: *(Pick a point at "C")*
Other corner: *(Pick a point at "D")*

Once views are created, use the VIEW command along with the Restore option to display the view, as in the following sequence:

Command: **VIEW**
?/Delete/Restore/Save/Window:**Restore**
View name: **Iso** *(Shown in Figure 1–114)*

Command: **VIEW**
?/Delete/Restore/Save/Window:**Restore**
View name: **Front** *(Shown in Figure 1–114)*

View Name = ISO

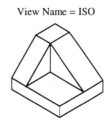

View Name = FRONT

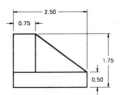

Figure 1–114

Using the DDVIEW Dialog Box

Selecting "Named views…" from the "View" area of the pull-down menu area in Figure 1–115 activates the View Control dialog box shown in Figure 1–116. This same dialog box is activated through the keyboard by entering the following at the command prompt:

 Command:**DDVIEW**

The same options of the VIEW command apply to the operation of this dialog box; that is, you may create a new view, restore an existing view, delete an existing view, or be provided with a description of the current view. Simply pick the button that applies to the appropriate VIEW command option. The View Control dialog box provides a more user-friendly way to manipulate views. Notice the five existing view names in Figure 1–116. To display the contents of one of the views listed, select the view until it highlights, as in the view "FRONT." Pick the "Restore" button to display the view.

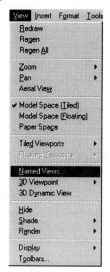

Figure 1–115

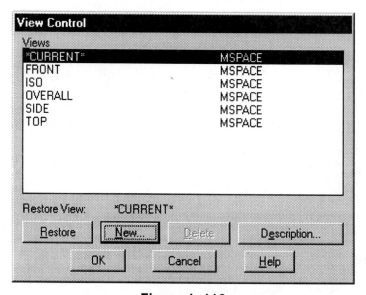

Figure 1–116

Selecting the "New..." button from the View Control dialog box activates the "Define New View" dialog box shown in Figure 1–117. Use this dialog box to guide you in creating a new view. By definition, a view is created from the current display screen. This is the purpose of the "Current Display" radio button. Many views are created using the "Define Window" radio button which will create a view based on the contents of a window that you define. Selecting the "Window <" button prompts you for the first corner and other corner required to create a new view by window. When the window

has been created, the Define New View dialog box redisplays. The absolute coordinate values of the first corner and other corner are displayed in the dialog box. Before selecting the "Save View" button, enter the name of the view you want in the "New Name:" edit box. Finally, select the "Save View" button and the view name, and window coordinates are automatically added to the database of the current drawing. This view may now be restored using the "Restore" button located in the View Control dialog box shown in Figure 1–116.

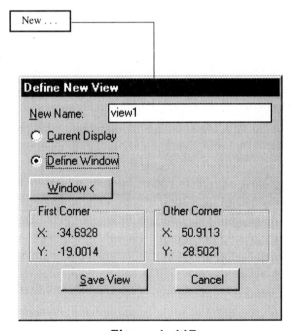

Figure 1–117

Creating Object Selection Sets

Selection sets are used to group a number of objects together for the purpose of editing. Applications of selection sets are covered in the following pages in addition to being illustrated in the next unit. Once a selection set has been created, the group of objects may all be moved, copied, or mirrored. These are just a few of the operations supported through selection sets. An object manipulation command supports the creation of selection sets if it prompts you to "Select objects:." Any command displaying this prompt supports the use of selection sets. Options of selection sets, or how a selection set is made, appear in Figure 1–118. Figures 1–119 through 1–125 illustrate a few applications of how selection sets are used for manipulating groups of objects.

Add
All
CPolygon
Crossing
Fence
Last
Previous
Remove
Window
WPolygon
Undo

Figure 1–118

Object Selection by Individual Picks

When prompted with "Select objects:", a pickbox appears as the cursor is moved across the display screen. Any object enclosed by this small box when picked will be considered selected. To show the difference between a selected and unselected object, the selected object highlights on the display screen. In Figure 1–119, the small box is placed over the arc segment at "A" and picked. To signify that the object is selected, the arc highlights.

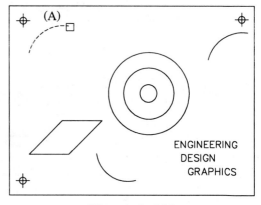

ENGINEERING
DESIGN
GRAPHICS

Figure 1–119

Object Selection by Window

The individual pick method previously outlined works fine for small numbers of objects; however, when numerous objects need to be edited, selecting each individual object could prove time consuming. Instead, all objects desired to become part of a selection set can be selected by the Window selection mode. This mode requires you to create a rectangular box by picking two points. In Figure 1–120, a selection window has been created with point "A" as the first corner and "B" as the other corner. When using this selection mode, only those objects completely enclosed by the window box are selected. The window box selected four line segments, two arcs, and two points (too small to display highlighted). The three circles, even though they are touched by the window, are not completely surrounded by the window and therefore are not selected.

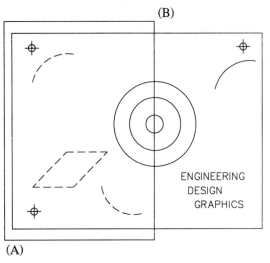

Figure 1–120

Object Selection by Crossing Window

In the previous example of window selection sets, the window only selected those objects completely enclosed by it. Figure 1–121 is an example of selecting objects by a crossing window. The crossing window requires two points to define a rectangle as does the window selection option. In Figure 1–121, a dashed rectangle is used to select objects using "C" and "D" as corners for the rectangle; however this time the crossing window was used. The results are illustrated by the highlighted objects. As objects highlight only if completely enclosed by a window, all objects that are touched by or enclosed by the crossing rectangle are selected. As the crossing rectangle passes through the three circles without enclosing them, they are still selected by this object selection mode.

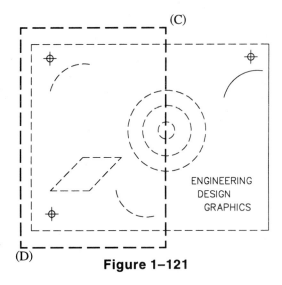

Figure 1–121

Object Selection by Fence

Use this mode to create a selection set by drawing a line called a fence. Any object touched by the fence is selected. The fence does not have to end exactly where it was started. In Figure 1–122, all lines contacted by the fence are selected as represented by the dashed lines.

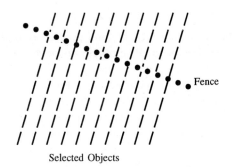

Selected Objects

Figure 1–122

Removing Objects from a Selection Set

All previous examples of creating selection sets have shown you how to create new selection sets. What if you select too many objects? What if you select the wrong object or object set? Rather than select the wrong objects, cancel out of the command, and try selecting the correct objects; the Remove option has been created to remove objects from an existing selection set. In Figure 1–123, a selection set has been created and made up of all of the highlighted objects. However, the large circle was mistakenly selected as part of the selection set. The Remove option allows the operator to remove highlighted objects from a selection set. To activate Remove, press the Shift key and pick on the object; this only works if the "Select objects:" prompt is present. When a highlighted object is removed from the selection set, as shown in circle "A" in Figure 1–124, it deselects and regains its original display intensity.

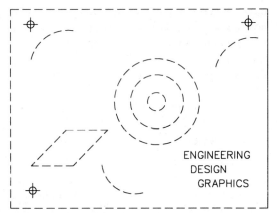

ENGINEERING
DESIGN
GRAPHICS

Figure 1–123

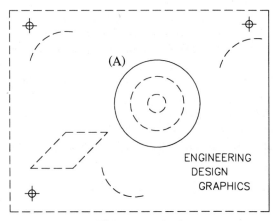

(A)

ENGINEERING
DESIGN
GRAPHICS

Figure 1–124

Object Selection by Crossing Polygon

Whenever using window or crossing window modes of creating selection sets, two points specify a rectangular box to select objects. At times, it is difficult to select objects by the rectangular window or crossing box because in more cases than not, extra objects are selected and have to be removed from the selection set. Figure 1–125 shows a mechanical part with a "C"-shaped slot. Rather than use window or crossing window modes, the crossing polygon mode is used. An operator simply picks points representing a polygon. Any object that touches or is inside the polygon is added to a selection set of objects. In Figure 1–125, the crossing polygon is constructed using points "A" through "F." A similar but different selection set mode is the window polygon. Objects are selected using this mode when they lie completely inside the window polygon similar to the regular window mode.

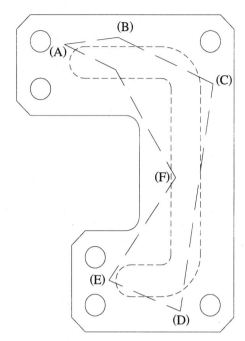

Figure 1–125

Applications of Selecting Objects Using the ERASE Command

Use the ERASE command to delete objects from the database of a drawing. Objects to be erased may be selected individually or through one of the many selection set modes. Figures 1–126 and 1–127 illustrate the use of the Window and Crossing options of deleting the circle and center marker.

Erase—Window

When erasing objects by Window, be sure the objects to be erased are completely enclosed by the window as in Figure 1–126.

 Command:**ERASE**

Select objects: *(Pick a point at "A" and move the cursor to the right; this automatically invokes the window option)*
Other corner: *(Mark a point at "B" and notice the objects that highlight)*
Select objects: *(Press* ⌜ENTER⌝ *to execute the ERASE command)*

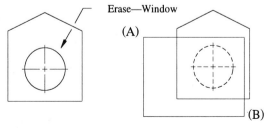

Figure 1–126

Erase—Crossing

Erasing objects by a crossing box is similar to the window box mode; however, any object that touches the crossing box or is completely enclosed by the crossing box is selected. See Figure 1-127.

 Command:**ERASE**

Select objects: *(Pick a point at "A" and move the cursor to the left; this automatically invokes the crossing option)*
Other corner: *(Mark a point at "B" and notice the objects that highlight)*
Select objects: *(Press* ENTER *to execute the ERASE command)*

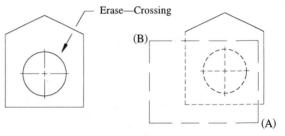

Figure 1–127

Figure 1–128 displays the objects that were selected by the Window option. However, the group of objects on the far right and left have mistakenly been selected and should not be affected by the ERASE command. These objects need to be removed or de-selected from the current selection set of objects by the Remove option of select objects.

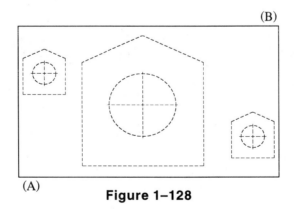

Figure 1–128

Command:**ERASE**

Select objects: *(Pick a point at "A" in Figure*
1–128 and move the cursor to the right; this
automatically invokes the window option)
Other corner: *(Mark a point at "B" in Figure*
1–128 and notice the objects that highlight)

Before performing the ERASE command on the objects in Figure 1–128, issue the Remove option to unhighlight the group of objects shown in Figure 1–129. This is easily accomplished by holding down the Shift key and picking the objects you do not want to erase.

Select objects: *(While pressing the Shift key,*
begin picking all objects in Figures "A" and
"B" in Figure 1–129. Notice the objects
deselecting)
Select objects: *(Press* [ENTER] *to execute the*
ERASE command)

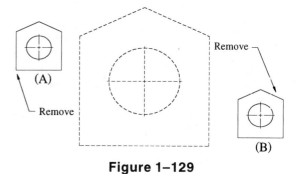

Figure 1–129

By deselecting objects through Shift-Pick, you remain in the command instead of canceling the command for picking the wrong objects and starting over. See Figure 1–130.

Use the Shift key and then select the objects to deselect with any command that displays the "Select objects:" prompt.

The middle group of
objects is erased.

Figure 1–130

Object Selection Cycling

At times, the process of selecting objects can become quite tedious. This can be compounded when you attempt to select an object and another object selects instead. Often, objects lie directly on top of each other. As you select the object to delete, the other object selects instead. To remedy this, press the "Ctrl" key when prompted to "Select objects:." This activates Object Selection Cycling and enables you to scroll through all objects in the vicinity of the pickbox. A message appears in the prompt area alerting you that cycling is on; objects can now be picked until the desired object is highlighted. Pressing the ENTER key not only accepts the highlighted object but toggles cycling off. In Figure 1–131 and with selection cycling on, the first pick selects the line segment; the second pick selects the circle. Keep picking until the desired object highlights.

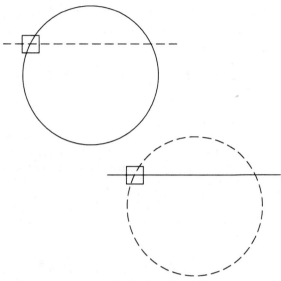

Figure 1–131

Using the Object Selection Settings Dialog Box

Further controls of selecting objects are contained in the Object Selection Settings dialog box in Figure 1–132. Choose this dialog box by first picking Tools from the pull-down menu area and then pick "Selection…." The various selection modes will be discussed.

Noun/Verb Selection—All preceeding examples of selecting objects have required you to first enter the editing command, such as ERASE, and then select the objects to delete. Noun/Verb selection allows you to select the objects while at the Com-

mand prompt. Since the objects have already been selected, entering the edit command, such as ERASE, executes the command immediately.

Use Shift to Add—When selecting numerous objects to be a part of a selection set, the Shift key must be pressed to add objects. If the Shift key is not pressed during object selection, then only the last object or group of objects is selected. Selecting another object replaces all previous selection sets unless you press the Shift key during this process.

Press and Drag—Rather than picking points to define a window or crossing box to select objects, press down the pick button and then surround the objects to select. Lifting off of the pick button creates the selection set of objects using this press and drag technique.

Implied Windowing—This mode allows you to pick a blank part of the drawing screen when confronted with the "Select objects:" prompt; moving the pickbox anywhere to the right automatically switches you to window mode; moving the pickbox anywhere to the left switches to crossing mode.

Object Grouping—This is an advanced feature of toggling group mode on or off. Groups are predefined selection sets formed with the GROUP command.

Associative Hatch—This mode controls which objects will be selected when manipulating an associative hatch pattern. If a check is placed in the box next to Associative Hatch, this mode turns on. This means the boundary objects that define the associative hatch area select along with the associative hatch pattern.

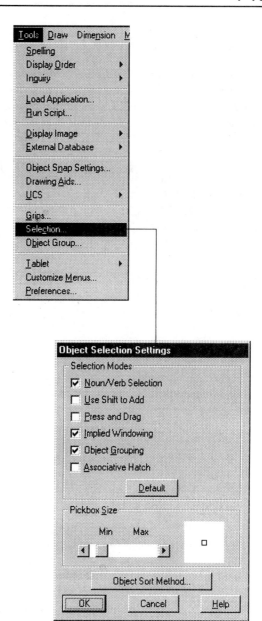

Figure 1–132

Saving a Drawing File

Drawings may be saved using the QSAVE and SAVE commands. Choose two of the three commands from the "File" pull-down menu area in Figure 1–133.

Save As...

Using this command always displays the dialog box shown in Figure 1–133. Simply pick the "Save" button or press the ENTER key to save the drawing under the current name displayed in the edit box. This command is more popular for saving the current drawing under an entirely different name. Simply enter the new name in place of the highlighted name in the edit box.

Save...

This command is short for QSAVE, which stands for Quick Save. If a drawing file has never been saved and this command is selected, the dialog box shown in Figure 1–133 displays. Once a drawing file has been initially saved, selecting this command performs an automatic save and no longer displays this "Save Drawing As" dialog box.

Exit

Use this command to exit AutoCAD and return to the operating system.

The ability to exchange drawings with past releases of AutoCAD is still important to industry. When clicking in the "Save as type:" edit box in Figure 1–133, a popdown menu appears. Use this area to save a drawing file in the following formats: Release 12, Release 13, AutoCAD LT 2.0, or AutoCAD LT 3.0. (See Figure 1–134).

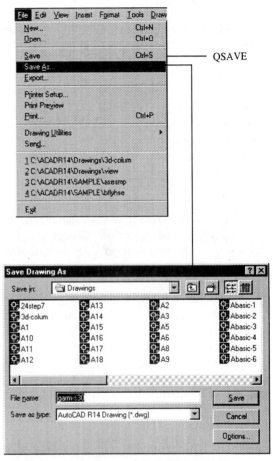

Figure 1–133

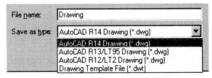

Figure 1–134

? **Using the Help Command**

Selecting "Help" from the "Help" pull-down menu area activates the Help Topics: AutoCAD Help dialog box shown in Figure 1–135. Three tabs can be used to navigate around Help. When first entering Help, the Contents tab displays giving various topics to switch to. Clicking on the Index tab displays the dialog box shown in Figure 1–136. Use this dialog box to enter in any AutoCAD command at the edit box to receive help on the command. Use the Find tab as another means of looking up an AutoCAD command for help.

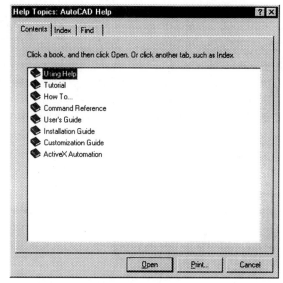

Figure 1–135

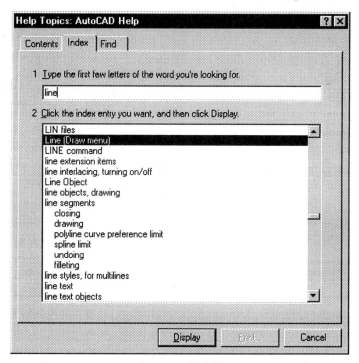

Figure 1–136

Exiting an AutoCAD Drawing Session

It is considered good practice to properly exit any drawing session. One way of exiting is by selecting the "Exit" option located in the "File" pull-down menu area (see Figure 1–137A). Another command that exits AutoCAD is QUIT. This command should be used with extreme caution because QUIT exits the drawing file without saving any changes. This command is useful when just browsing drawing files where no changes were made. If you are a new user of AutoCAD, do not use this command until you have gained sufficient experience.

Whichever command you use to exit AutoCAD, a built-in safe-guard gives you a second chance to save the drawing before exiting especially if changes were made and a Save was not performed. You may be confronted with three options illustrated in the "AutoCAD" alert dialog box shown in Figure 1–137B. By default, the "Yes" button highlights.

If changes were made to the drawing but not saved, you may want to pick the "Yes" button before exiting the drawing. Changes to the drawing will be saved and the software exits back to the operating system.

If changes were made but do not have to be saved, you may want to pick the "No" button. Changes to the drawing will not be saved and the software exits back to the operating system.

If changes were made to the drawing and the Exit option was mistakenly picked, choosing the "Cancel" button cancels the Exit option and returns you to the current drawing.

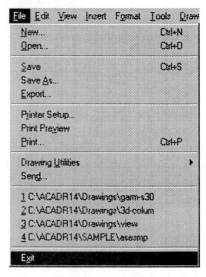

Figure 1–137A

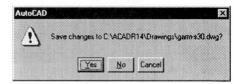

Figure 1–137B

Tutorial Exercise
Creating Layers Using the -LAYER Command

Layer Name	Color	Linetype
Object	White	Continuous
Hidden	Red	Hidden
Center	Yellow	Center
Dim	Yellow	Continuous
Section	Blue	Continuous

Use the command sequences below to create the following layers according to the above specifications:

Command: **-LAYER**
?/Make/Set/New/ON/OFF/Color/Ltype/Freeze/
   Thaw/LOck/Unlock: **New**
New layer name(s):
   Object,Hidden,Center,Dim,Section
?/Make/Set/New/ON/OFF/Color/Ltype/Freeze/
   Thaw/LOck/Unlock: **Color**
Color: **Red**
Layer name(s) for color 1 (RED) <0>: **Hidden**
?/Make/Set/New/ON/OFF/Color/Ltype/Freeze/
   Thaw/LOck/Unlock: **Ltype**
Linetype (or ?) <CONTINUOUS>: **Hidden**
Layer name(s) for linetype HIDDEN <0>:
   Hidden
?/Make/Set/New/ON/OFF/Color/Ltype/Freeze/
   Thaw/LOck/Unlock: **Color**

Color: **Yellow**
Layer name(s) for color 2 (YELLOW) <0>:
   Center,Dim
?/Make/Set/New/ON/OFF/Color/Ltype/Freeze/
   Thaw/LOck/Unlock: **Ltype**
Linetype (or ?) <CONTINUOUS>: **Center**
Layer name(s) for linetype CENTER <0>:
   Center
?/Make/Set/New/ON/OFF/Color/Ltype/Freeze/
   Thaw/LOck/Unlock: **Color**
Color: **Blue**
Layer name(s) for color 5 (BLUE) <0>: **Section**
?/Make/Set/New/ON/OFF/Color/Ltype/Freeze/
   Thaw/LOck/Unlock: *(Press* ENTER *to Exit this
   command)*

Enter the -LAYER command in from the keyboard again followed by a question mark (?) to list the layers just created.

Command: **-LAYER**
?/Make/Set/New/ON/OFF/Color/Ltype/Freeze/
   Thaw/LOck/Unlock: **?**
Layer name(s) to list <*>: *(Press* ENTER *to list all
   layers)*

Layer Name	State	Color	Linetype
0	On	7 (white)	Continuous
Center	On	2 (yellow)	Center
Dim	On	2 (yellow)	Continuous
Hidden	On	1 (red)	Hidden
Object	On	7 (white)	Continuous
Section	On	5 (blue)	Continuous

Tutorial Exercise
Creating Layers Using the Layer & Linetype Properties Dialog Box

Create the same layers outlined in the previous tutorial exercise using the Layer and Linetype Properties dialog box.

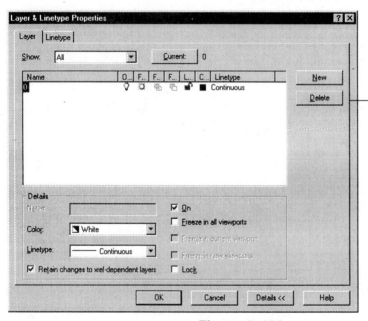

Figure 1–138

Step #1

The previous tutorial exercise dealt with creating layers using a manual method where the -LAYER command is used, prompts are carefully followed, and the result is the desired layer names complete with correct color and linetype assignments. The manual method, however, is open to numerous errors in the assignment of color and linetype. A typical example is to assign the color "Red" to a layer called "Hidden"; unfortunately the prompt states that Layer 0 will be changed to red. This is quite a surprise when the color of Layer 0 changes from "White" to

"Red." A more direct method of creating layers is to use the Layer & Linetype Properties dialog box in showin in Figure 1–138. This graphically displays the creation of layers; also, it is easier to detect mistakes during the layer creation process and even easier to correct these mistakes. Select "Layer…" from the "Format" area of the pull-down menu (top portion of Figure 1–138) to display the Layer & Linetype Properties dialog (shown in the lower portion of Figure 1–138).

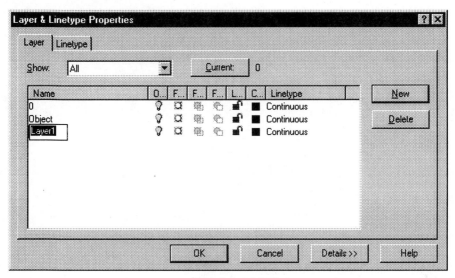

Figure 1–139

Step #2

Figure 1–139 of the Layer & Linetype Properties dialog box shows an abbreviated dialog slightly different from that in Figure 1-138. Clicking on the Details >> button displays the shortened version of the dialog box in shown in Figure 1–139. Once the Layer & Linetype Properties dialog box displays as in Figure 1–139, notice only one layer is currently listed in the large edit box, namely Layer 0. This happens to be the default layer; the layer that is automatically assigned to all new drawings. Since it is considered poor practice to construct any objects on Layer 0, new layers will be created not only for object lines but for hidden lines, centerlines, dimension objects, and section lines as well. To create these layers, click on the "New" button and notice a layer is automatically added to the list of layers. This layer is called "Layer 1." While this layer is highlighted, enter in the first layer called "Object."

Entering a comma after the name of the layer allows more layers to be added to the listing of layers. Once the comma was entered after the layer "Object" and the new layer appears, enter the new name of the layer as "Hidden." Repeat this procedure of using the comma to create multiple layers for "Center," "Dim," and "Section." Press the ENTER key after typing in "Section." The complete layer listing is shown in Figure 1–140.

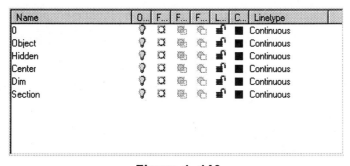

Figure 1–140

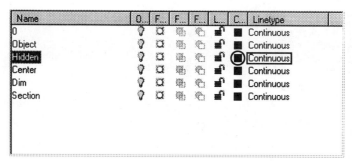

Figure 1–141

Step #3

As all layers are displayed, the names may be different, but they all have the same color and linetype assignments (see Figure 1–141). At this point, the dialog box comes in real handy to assign color and linetypes to layers in a quick and easy manner. First, highlight the desired layer to add color or linetypes by picking the layer. A horizontal bar displays signifying that this is the selected layer. Click on the color swatch identified by the circle in Figure 1–141 to and follow the next step to assign the color "Red" to the "Hidden" layer name.

Step #4

Clicking on the color swatch in the previous step displays the "Select Color" dialog box shown in Figure 1–142. Select the desired color from one of the following areas: Standard Colors; Gray Shades; Full Color Palette. The standard colors represent colors 1 through 9. On display terminals with high resolution cards, the full color palette displays different shades of the standard colors, which gives you a greater variety of colors to choose from. For the purposes of this tutorial, the color "Red" will be assigned to the "Hidden" layer. Select the box displaying the color red; a box outlines the color and echoes the color in the bottom portion of the dialog box. Click on the "OK" button to complete the color assignment; if "Cancel" is selected, the color assignment will be removed requiring this operation to be used again. Continue with this step by assigning the color Yellow to the "Center" and "Dim" layers; assign the color Blue to the "Section" layer.

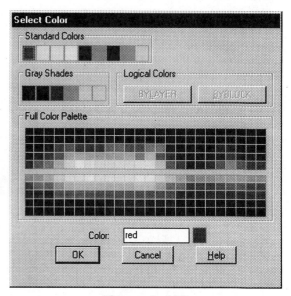

Figure 1–142

Step #5

Once the color has been assigned to a layer, the next step is to assign a linetype, if any, to the layer. The "Hidden" layer requires a linetype called "Hidden." Click on "Continuous" identified by the rectangle in Figure 1–141 to display the "Select Linetype" dialog box illustrated in Figure 1–143. By default, Continuous is the only linetype loaded. Clicking on the Load button displays the next dialog box illustrated in Figure 1–144.

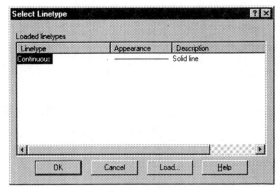

Figure 1–143

Pick the desired linetype to load from the Load or Reload the Linetype dialog box shown in Figure 1–144. Scroll through the linetypes until the "Hidden" linetype is found. Click on the OK button to return to the Select Linetype dialog box.

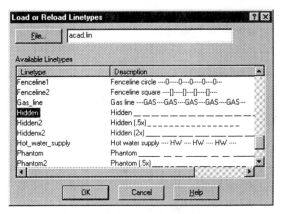

Figure 1–144

Once back in the Select Linetype dialog box shown in Figure 1–145, notice the "Hidden" linetype listed along with "Continuous." As this linetype has just been loaded, it still has not been assigned to the "Hidden" layer. Click on the Hidden linetype listed in the Select Linetype dialog box and click on the OK button. Once the Layer & Linetype Properties dialog box reappears, notice the "Hidden" linetype has been assigned to the "Hidden" layer. Repeat this procedure to assign the "Center" linetype to the "Center" layer.

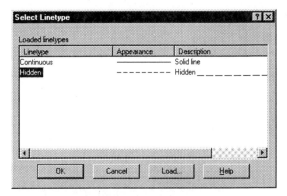

Figure 1–145

Step #6

Once completing all color and linetype assignments, the Layer & Linetype Properties dialog box should appear similar to Figure 1–146. Click on the OK

button to save all layer assignments and return to the drawing editor.

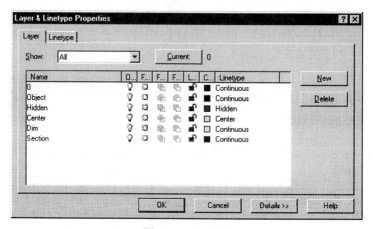

Figure 1–146

Step #7

There is one additional point to make with the creation of layers. When layers are first created, they are listed in the order they were entered. When the

layers are saved and the Layer & Linetype dialog box is displayed again, all layers are reordered alphabetically. See Figure 1–147.

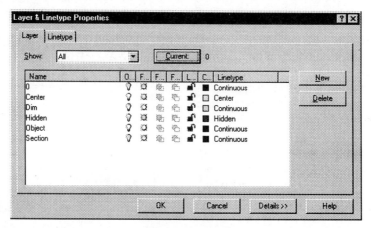

Figure 1–147

Tutorial Exercise
Template.Dwg

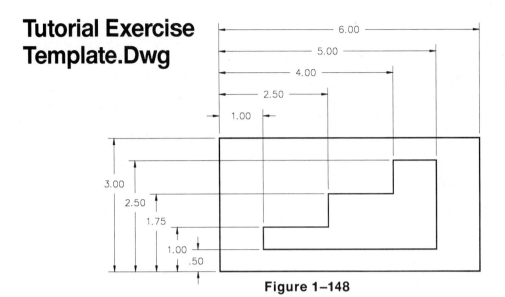

Figure 1–148

Purpose

This tutorial is designed to allow you to construct a one-view drawing of the Template using the Absolute, Relative, and Polar Coordinate modes. See Figure 1–148. The Direct Distance mode can also be used to perform this exercise.

System Settings

Use the current default settings for the limits of this drawing, (0,0) for the lower left corner and (12,9) for the upper right corner.

Use the GRID command and change the grid spacing from 1.0000 to 0.25 units. The grid will be used only as a guide for constructing this object. Do not turn the SNAP or Ortho commands on.

Layers

Create the following layers with the format :

Name	Color	Linetype
Object	White	Continuous

Suggested Commands

The LINE command will be used entirely for this tutorial in addition to a combination of coordinate systems. The ERASE command could be used although a more elaborate method of correcting a mistake would be to use the LINE-UNDO command to erase a previously drawn line and still stay in the LINE command. The OSNAP-From mode will also be used to construct lines from a point of reference. The coordinate mode of entry and the direct distance mode will be used throughout this tutorial exercise.

Whenever possible, substitute the appropriate command alias in place of the full AutoCAD command in each tutorial step. For example, use "Co" for the COPY command, "L" for the LINE command, and so on. The complete listing of all command aliases is located on pages 1–9 and 1–10.

Step #1

Begin this tutorial exercise with the LINE command and draw the outer perimeter of the box. One method of constructing the box is to use an absolute coordinate point followed by polar coordinates using the following prompt sequence and the illustration in Figure 1–149 as guides.

 Command: **LINE**

From point: **2,2**
To point: **@6<0**
To point: **@3<90**
To point: **@6<180**
To point: **@3<270**
To point: *(Press* ⌷ENTER⌷ *to exit this command)*

An alternate mode is to use the Direct Distance mode. Since the box consists of horizontal and vertical lines, Ortho mode is first turned on; this will force all movements to be in the horizontal or vertical directions. To construct a line segment, move the cursor in the direction the line is to be drawn in and enter the exact value of the line. The line is drawn at the designated distance in the current direction of the cursor. Repeat this procedure for the other lines that make up the box, as shown in Figure 1–150.

Command: **ORTHO**
ON/OFF <OFF>: **ON**

 Command: **LINE**

From point: **2,2**
To point: *(Move the cursor to the right and enter a value of 6.00 units)*
To point: *(Move the cursor up and enter a value of 3.00 units)*
To point: *(Move the cursor to the left and enter a value of 6.00 units)*
To point: **C** *(To close the shape)*

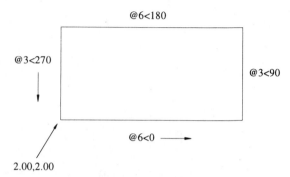

@6<180
@3<270
@3<90
@6<0
2.00,2.00

Figure 1–149

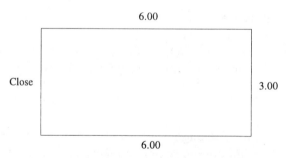

6.00
Close
3.00
6.00

Figure 1–150

Step #2

The next step will be to draw the stair step outline of the template using the LINE command again. However we first need to identify the starting point of the template. Absolute coordinates could be calculated but in more complex objects this would be difficult. A more efficient method would be to use the Osnap-From mode along with the Osnap-Intersection mode to start the line relative to another point. Use the following prompt sequence and Figure 1–151 as guides for performing this operation.

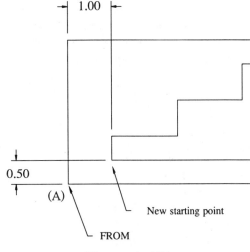

Figure 1–151

 Command: **LINE**
From point: **From**
Base point: **Int**
of *(Pick the intersection at "A")*
<Offset>:**@1.00,0.50**

The relative coordinate offset value begins a new line a distance of 1.00 units in the "X" direction and 0.50 units in the "Y" direction. Continue with the LINE command to construct the stair step outline as shown in Figure 1–152.

To point: **@4.00<0**
To point: **@2.00<90**
To point: **@1.00<180**
To point: **@0.75<270**
To point: **@1.50<180**
To point: **@0.75<270**
To point: **@1.50<180**
To point: **@0.50<270**
To point: *(Press* ENTER *to exit this command)*

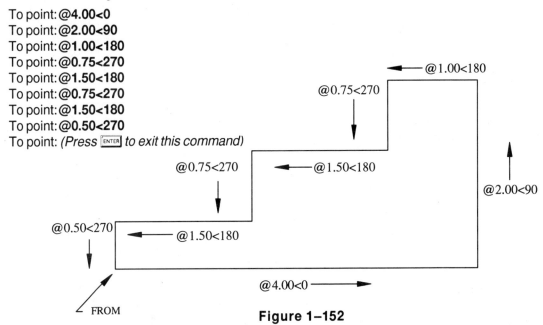

Figure 1–152

An alternate step would be to use the Osnap-From and Osnap-Intersect modes in combination with the Direct Distance mode to construct the inner stair step outline (see Figure 1–153). Again, Direct Distance mode is a good choice to use on this object especially since all lines are either horizontal or vertical. Use the following command sequence to construct the object with this alternate method.

Command: **LINE**

From point: **From**
Base point: **Int**
of *(Pick the intersection at "A" in Figure 1–151)*
<Offset>:**@1.00,0.50**
To point: *(Move the cursor to the right and enter a value of 4.00 units)*
To point: *(Move the cursor up and enter a value of 2.00 units)*
To point: *(Move the cursor to the left and enter a value of 1.00 units)*
To point: *(Move the cursor down and enter a value of 0.75 units)*
To point: *(Move the cursor to the left and enter a value of 1.50 units)*
To point: *(Move the cursor down and enter a value of 0.75 units)*
To point: *(Move the cursor to the left and enter a value of 1.50 units)*
To point: **C** *(To close the shape)*

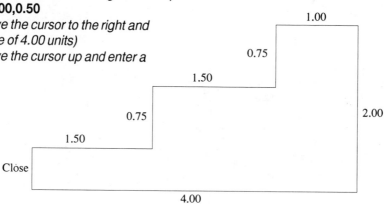

Figure 1–153

Step #3

The completed problem is shown in Figure 1–154. Dimensions may be added at a later time upon the request of your instructor.

Figure 1–154

Problems for Unit 1

Directions for Problems 1–1 through 1–6:

Supply the appropriate absolute, relative, and/or polar coordinates for these figures in the matrix that follows each object.

Problem 1–1

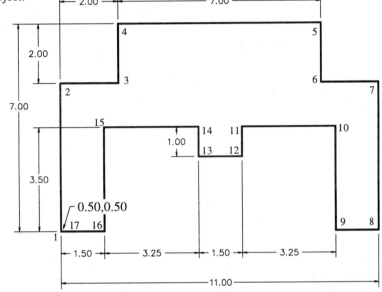

	Absolute	Relative	Polar
From Pt (1)	0.50,0.50	0.50,0.50	0.50,0.50
To Pt (2)			
To Pt (3)			
To Pt (4)			
To Pt (5)			
To Pt (6)			
To Pt (7)			
To Pt (8)			
To Pt (9)			
To Pt (10)			
To Pt (11)			
To Pt (12)			
To Pt (13)			
To Pt (14)			
To Pt (15)			
To Pt (16)			
To Pt (17)			
To Pt	Enter	Enter	Enter

Problem 1–2

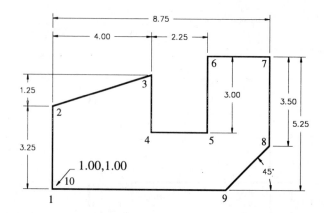

	Absolute	Relative
From Pt (1)	1.00,1.00	1.00,1.00
To Pt (2)		
To Pt (3)		
To Pt (4)		
To Pt (5)		
To Pt (6)		
To Pt (7)		
To Pt (8)		
To Pt (9)		
To Pt (10)		
To Pt	Enter	Enter

Problem 1–3

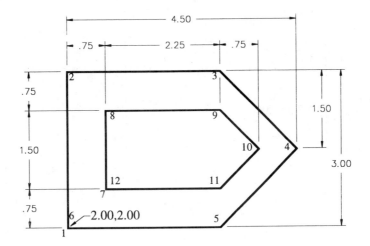

	Absolute	Relative
From Pt (1)	2.00,2.00	2.00,2.00
To Pt (2)		
To Pt (3)		
To Pt (4)		
To Pt (5)		
To Pt (6)		
To Pt	Enter	Enter
From Pt (7)		
To Pt (8)		
To Pt (9)		
To Pt (10)		
To Pt (11)		
To Pt (12)		
To Pt	Enter	Enter

Problem 1–4

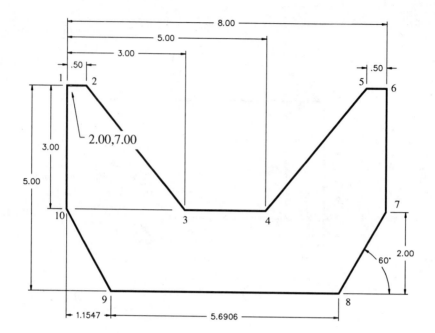

	Absolute
From Pt (1)	2.00,7.00
To Pt (2)	
To Pt (3)	
To Pt (4)	
To Pt (5)	
To Pt (6)	
To Pt (8)	
To Pt (7)	
To Pt (9)	
To Pt (10)	
To Pt	

Problem 1–5

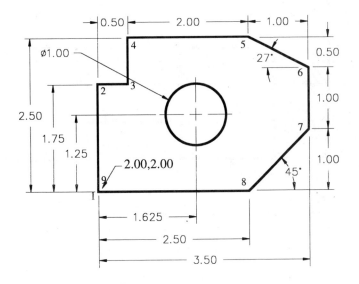

	Absolute	Relative
From Pt (1)	2.00,2.00	2.00,2.00
To Pt (2)		
To Pt (3)		
To Pt (4)		
To Pt (5)		
To Pt (6)		
To Pt (7)		
To Pt (8)		
To Pt (9)		
To Pt	Enter	Enter
Center Pt (10)		

Problem 1–6

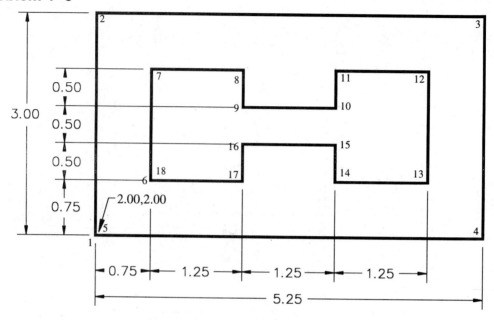

	Absolute	Relative	Polar
From Pt (1)	2.00,2.00	2.00,2.00	2.00,2.00
To Pt (2)			
To Pt (3)			
To Pt (4)			
To Pt (5)			
To Pt	Enter	Enter	Enter
From Pt (6)			
To Pt (7)			
To Pt (8)			
To Pt (9)			
To Pt (10)			
To Pt (11)			
To Pt (12)			
To Pt (13)			
To Pt (14)			
To Pt (15)			
To Pt (16)			
To Pt (17)			
To Pt (18)			
To Pt	Enter	Enter	Enter

Directions for Problems 1–7 through 1–20
   Construct one-view drawings of the following figures using the LINE command along with coordinate
   or direct distance modes.

Problem 1–7

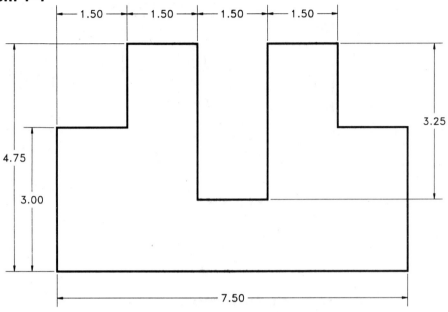

Problem 1–8

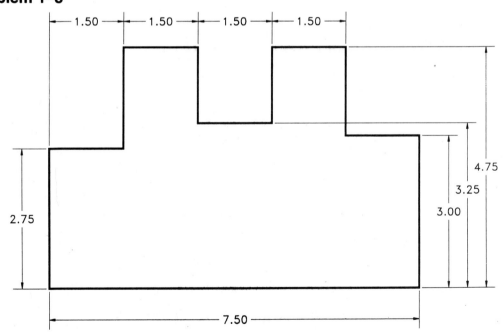

Problem 1–9

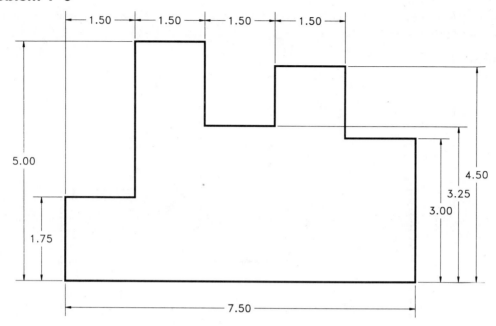

Problem 1–10

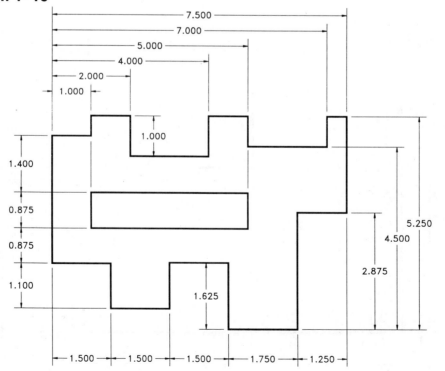

Problem 1–11

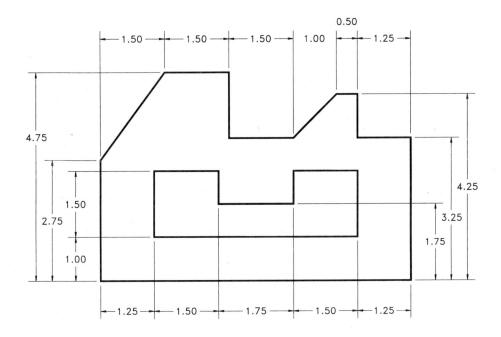

Problem 1–12

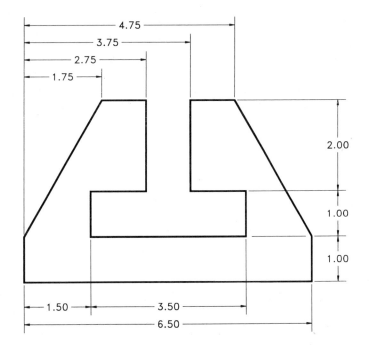

Problem 1–13

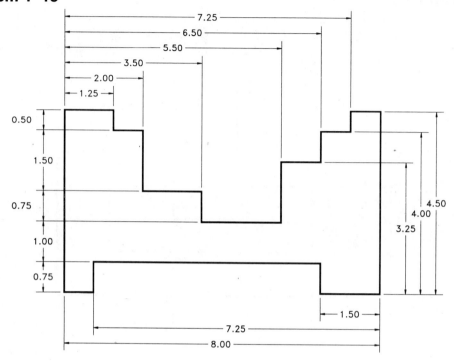

Problem 1–14

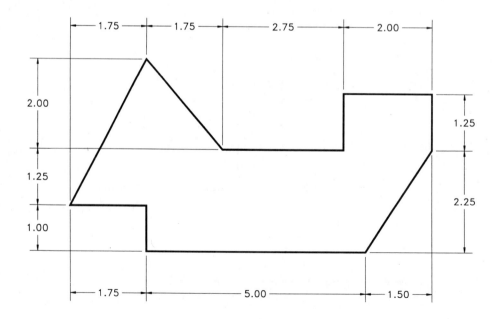

Problem 1–15

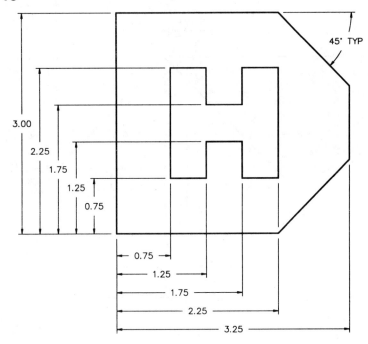

Problem 1–16

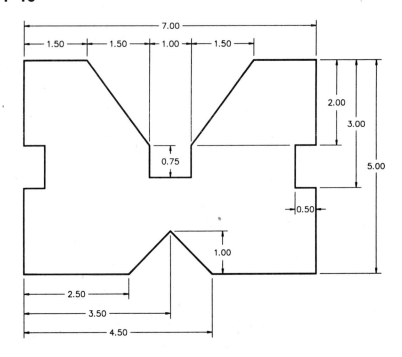

Problem 1–17

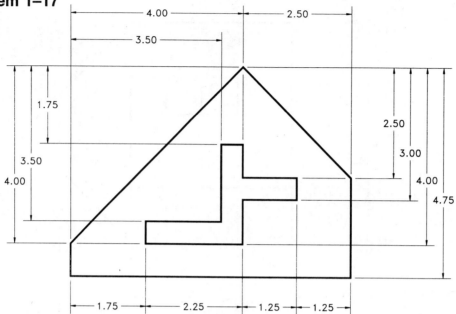

Problem 1–18

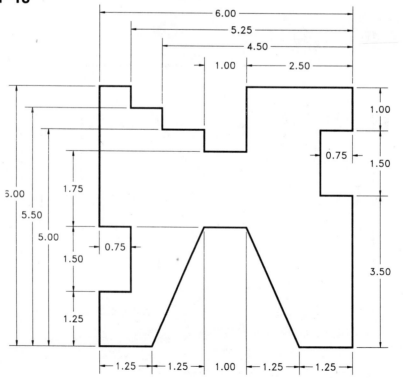

Problem 1–19

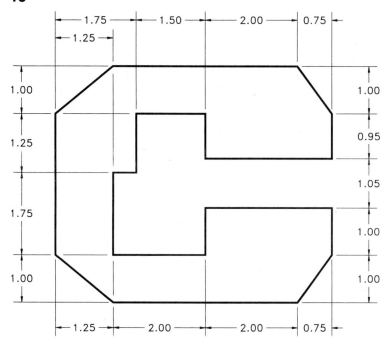

Problem 1–20

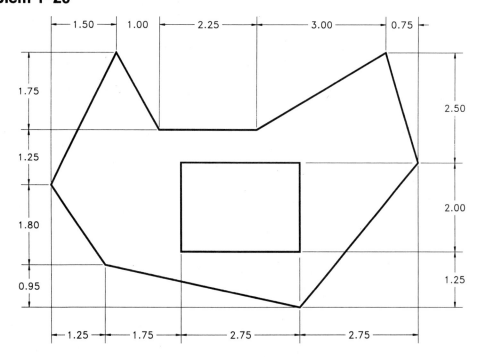

12X ∅0.50

4X ∅1.00

45°

R1.00

40°

R1.75   R3.00

0.125

Object Construction
and Manipulation

This chapter begins the study of the drawing and editing commands used in the creation of a drawing. Blocks and their role in object creation is also discussed.

The heart of any CAD system is its ability to edit and manipulate existing geometry and AutoCAD is no exception. Many editing commands relieve the designer of the drudgery and mundane tasks and this allows more productive time for conceptualizing the design.

Following the explanation of these commands, a series of tutorial exercises are provided to give directions on system preparation and suggested command usage before the main body of each tutorial is presented. A series of one-view drawings are presented at the end of the unit; these drawings are usually less complicated than multiple-view drawings, section views, or drawings involving auxiliary views. Typical examples of these drawings include such items as automotive gaskets and thin sheet metal parts where the thickness of these items is too thin to represent in multiview drawing form.

Methods of Selecting Draw Commands

This unit introduces the commands used for object creation. The following commands will be explained in this section:

> ARC
> CIRCLE
> DONUT
> DTEXT
> ELLIPSE
> MLINE
> MTEXT
> POINT
> POLYGON
> RECTANG
> SPLINE
> XLINE

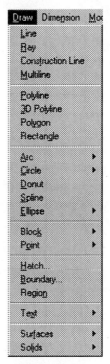

Figure 2–1

The LINE command was already discussed in Unit 1. The bulk of most drawing commands can be found in the Draw area of the pull-down menu shown in Figure 2–1. Arrowheads displayed to the right of the command support a cascading menu that holds additional options of the main command.

Figure 2–2 is the Draw toolbox that holds most drawing commands.

DRAW commands may also be entered directly from the keyboard by using their entire name such as "POINT" for the POINT command. The following commands may be entered by using only the first letter of the command as part of AutoCAD's command aliasing feature:

> Enter "A" for the ARC command
> Enter "C" for the CIRCLE command
> Enter "L" for the LINE command

See pages 1–9 and 1–10 for the complete listing of all command aliases supported in AutoCAD Release 14.

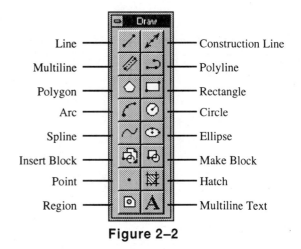

Figure 2–2

Using the ARC Command

Clicking on "Draw" in the pull-down menu area and then selecting "Arc" displays the cascading dialog box shown in Figure 2–3. All supported methods of constructing arcs are displayed in the list. By default, the 3 POINT arc mode supports arc constructions in the clockwise as well as counterclockwise directions. All other arc modes support the ability to construct arcs only in the counterclockwise direction. The following pages detail most of the arc modes labeled in Figure 2–3.

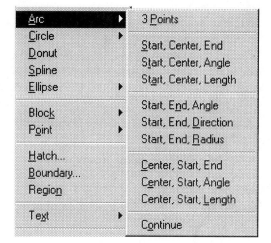

Figure 2–3

3 Point Arc Mode

By default, arcs are drawn using the 3 point method. The first and third points identify the endpoints of the arc. This arc may be drawn either in the clockwise or counterclockwise directions. Use the following prompt sequence along with Figure 2–4 to construct a 3 point arc.

 Command: **ARC**

Center/<Start point>: *(Mark a point at "A")*
Center/End/<Second point>: *(Mark a point at "B")*
End point: *(Mark a point at "C")*

Start, Center, End Mode

Use this arc mode to construct an arc by defining its start point, center point, and end point. This arc will always be constructed in a counterclockwise direction. Use the following prompt sequence along with Figure 2–5 for constructing an arc by start, center, and end points.

 Command: **ARC**

Center/<Start point>: *(Mark a point at "A")*
Center/End/<Second point>: **C** *(for Center)*
Center: *(Mark a point at "B")*
Angle/Length of chord/<End point>: *(Mark a point at "C")*

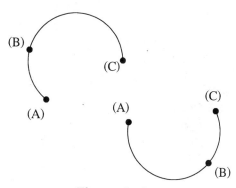

Figure 2–4

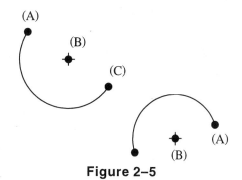

Figure 2–5

Start, End, Angle Mode

Use this arc mode to construct an arc by defining its starting point, end point, and included angle. This arc is draw in a counterclockwise direction when a positive angle is entered; it the angle is negative, the arc is drawn clockwise. Use the following prompt sequence along with Figure 2–6 for constructing an arc by start point, end point, and included angle.

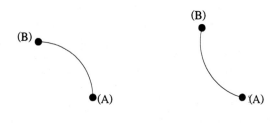

Angle = 90 degrees   Angle = -90 degrees

Figure 2–6

 Command: **ARC**

Center/<Start point>: *(Mark a point at "A")*
Center/End/<Second point>: **E** *(for End)*
End point: *(Mark a point at "B")*
Angle/Length of chord/<End point>: **Angle**
Included angle: **90**

Start, End, Direction Mode

Use this method to create an arc in a specified direction. This method is especially helpful for drawing arcs tangent to other objects. Use the following prompt sequence along with Figure 2–7 for constructing an arc by direction.

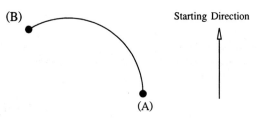

Figure 2–7

 Command: **ARC**

Center/<Start point>: *(Mark a point at "A")*
Center/End/<Second point>: **E** *(for End)*
End point: *(Mark a point at "B")*
Angle/Direction/Radius/<Center point>: **D** *(for Direction)*
Direction from start point: **@1<90**

Start, Center, Angle Mode

Use this mode to construct an arc by start point, center point, and included angle. If the angle is positive, the arc is drawn in the counterclockwise direction; a negative angle constructs the arc in a clockwise direction. See Figure 2–8.

 Command: **ARC**

Center/<Start point>: *(Mark a point at "A")*
Center/End/<Second point>: **C** *(for Center)*
Center: *(Mark a point at "B")*
Angle/Length of chord/<End point>: **Angle**
Included angle: **135**

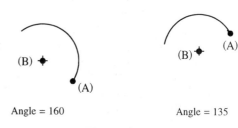

Angle = 160   Angle = 135

Figure 2–8

Start, Center, Length Mode

Use this mode to construct an arc by start point, centerpoint, and length of chord. See Figure 2–9 to show what defines a chord.

 Command: **ARC**

Center/<Start point>: *(Mark a point at "A")*
Center/End/<Second point>: **C** *(for Center)*
Center: *(Mark a point at "B")*
Angle/Length of chord/<End point>: **Length**
Length of chord: **2.50**

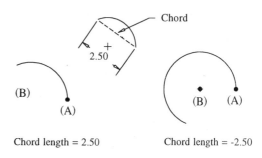

Chord length = 2.50   Chord length = -2.50

Figure 2–9

Start, End, Radius Mode

Use this mode to construct an arc by start point, endpoint, and radius. A positive radius draws a minor arc; a negative radius draws a major arc. Refer to Figure 2–10.

 Command: **ARC**

Center/<Start point>: *(Mark a point at "A")*
Center/End/<Second point>: **E** *(for End)*
End point: *(Mark a point at "B")*
Angle/Direction/Radius/<Center point>:
  Radius
Radius: **1.00**

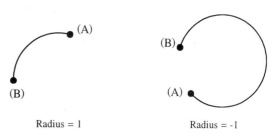

Radius = 1   Radius = -1

Figure 2–10

Using the CIRCLE Command

Clicking on "Draw" in the pull-down menu area and then selecting "Circle" displays the cascading dialog box shown in Figure 2–11. All supported methods of constructing circles are displayed in the list. Three groupings exist in the cascading menu; constructing circles by radius or diameter, defining two or three points to construct a circle, and constructing a circle tangent to other objects in the drawing. Most of these options of the CIRCLE command are explained in the following pages.

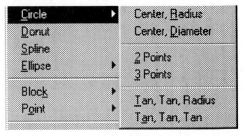

Figure 2–11

Circle by Radius Mode

Use the CIRCLE command and the Radius mode to construct a circle by a radius value that you specify. After selecting a center point for the circle, you are prompted to enter a radius for the desired circle. Study the following prompts and Figure 2–12 for constructing a circle using the Radius mode.

 Command:**CIRCLE**

3P/2P/TTR/<Center point>: *(Mark the center at "A")*
Diameter/<Radius>:**1.50**

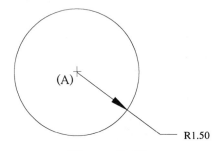

Figure 2–12

Circle by Diameter Mode

Use the CIRCLE command and the Diameter mode to construct a circle by a diameter value that you specify. After selecting a center point for the circle, you are prompted to enter a diameter for the desired circle. Study the following prompts and Figure 2–13 to construct a circle using the Diameter mode.

 Command:**CIRCLE**

3P/2P/TTR/<Center point>: *(Mark the center at a convenient location)*
Diameter/<Radius>: **D** *(For Diameter)*
Diameter:**3.00**

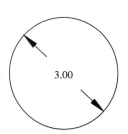

Figure 2–13

3 Point Circle Mode

Use the CIRCLE command and the 3 Point mode to construct a circle by 3 points that you identify. No center point is required when entering the 3 Point mode. Simply select three points and the circle is drawn. Study the following prompts and Figure 2–14 for constructing a circle using the 3 Point mode.

 Command: **CIRCLE**

3P/2P/TTR/<Center point>: **3P**
First point: *(Select the point at "A")*
Second point: *(Select the point at "B")*
Third point: *(Select the point at "C")*

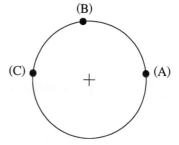

Figure 2–14

2 Point Circle Mode

Use the CIRCLE command and the 2 Point mode to construct a circle by selecting 2 points. These points will form the diameter of the circle. No center point is required after entering the 2 Point mode. Study the following prompts and Figure 2–15 for constructing a circle using the 2 Point mode.

 Command: **CIRCLE**

3P/2P/TTR/<Center point>: **2P**
First point: *(Select the point at "A")*
Second point: *(Select the point at "B")*

Tangent-Tangent-Radius Mode— Method #1

This mode is very powerful for constructing a circle tangent to two objects. In Figure 2–16 shows an application of using the Circle TTR mode to construct a circle tangent to two line segments. Study the following prompts to create this type of circle.

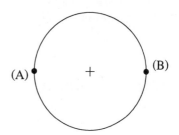

Figure 2–15

 Command: **CIRCLE**

3P/2P/TTR/<Center point>: **TTR**
Enter Tangent spec: *(Select the line at "A")*
Enter second Tangent spec: *(Select the line at "B")*
Radius: **0.75**

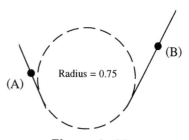

Figure 2–16

Tangent-Tangent-Radius Mode— Method #2

Figure 2–17 shows an application of using the Circle TTR mode to construct a circle tangent to a line segment and another circle. Study the following prompts to create this type of circle.

 Command: **CIRCLE**

3P/2P/TTR/<Center point>: **TTR**
Enter Tangent spec: *(Select the line at "A")*
Enter second Tangent spec: *(Select the circle at "B")*
Radius: **1.00**

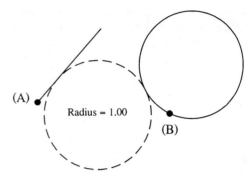

Figure 2–17

Tangent-Tangent-Radius Mode— Method #3

Figure 2–18 shows an application of using the Circle TTR mode to construct a circle tangent to two circles. Study the following prompts to create this type of circle.

 Command: **CIRCLE**

3P/2P/TTR/<Center point>: **TTR**
Enter Tangent spec: *(Select the circle at "A")*
Enter second Tangent spec: *(Select the circle at "B")*
Radius: **1.00**

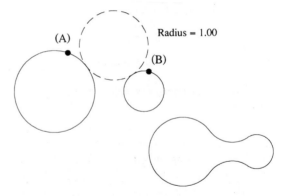

Figure 2–18

Using the DONUT Command

Use the DONUT command to construct a filled-in circle. This object belongs to the polyline family. Figure 2–19A is an example of a donut with an inside diameter of 0.50 units and an outside diameter of 1.00 units. When placing donuts in a drawing, the multiple option is automatically invoked. This means you can place as many donuts as you like until another command is selected from one of the menu areas or a "Cancel" is issued by pressing the ESC key.

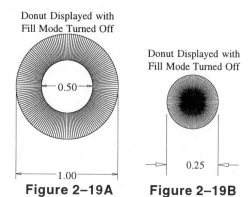

Donut Displayed with Fill Mode Turned Off

Donut Displayed with Fill Mode Turned Off

Figure 2–19A   **Figure 2–19B**

Command: **DONUT**
Inside Diameter<0.50>: *(Press ⌜ENTER⌟ to accept the default)*
Outside Diameter<1.00>: *(Press ⌜ENTER⌟ to accept the default)*
Center of donut: *(Select a point to place the donut)*
Center of donut: *(Select a point to place another donut or press ⌜ENTER⌟ to exit this command)*

Setting the inside diameter of a donut to a value of zero (0) and an outside diameter to any other value constructs a donut representing a dot. See Figure 2–19B.

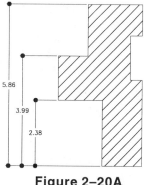

Figure 2–20A

Command: **DONUT**
Inside Diameter<0.50>: **0**
Outside Diameter<1.00>: **0.25**
Center of donut: *(Select a point to place the donut)*
Center of donut: *(Select a point to place another donut or press ⌜ENTER⌟ to exit this command)*

Figures 2–20A and 2–20B show two applications of where donuts could be useful; donuts are sometimes used in place of arrows to act as terminators for dimension lines, as shown in Figure 2–20A. Figure 2–20B illustrates how donuts might be used to act as connection points in an electrical schematic.

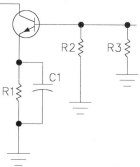

Figure 2–20B

Using the DTEXT Command

The DTEXT command stands for Dynamic Text mode and allows you to place text in a drawing and view the text as you typed it in. This command can be selected by picking "Draw" from the pull-down menu area followed by "Text" and finally "Single Line Text. Multiline Text is a more powerful method of placing text and will be discussed in detail later in this unit. See Figure 2–21.

Command: **DTEXT**
Justify/Style/<Start point>: *(Pick the point at "A")*
Height <0.20>: **0.50**
Rotation angle <0>: *(Press* ENTER *to accept this default)*
Text: **MECHANICAL**
Text: *(Press* ENTER *to exit Dtext and return to the command prompt)*

Figure 2–23 and the following prompt sequences are an example of justifying text by a center point.

Command: **DTEXT**
Justify/Style/<Start point>: **Center**
Center point: *(Pick a point at "A")*
Height <0.20>: **0.50**
Rotation angle <0>: *(Press* ENTER *to accept this default)*
Text: **CIVIL ENGINEERING**
Text: *(Press* ENTER *to exit Dtext and return to the command prompt)*

Figure 2–24 and the following prompt sequences are examples of justifying text by a middle point.

Command: **DTEXT**
Justify/Style/<Start point>: **Middle**
Middle point: *(Pick a point at "A")*
Height <0.20>: **0.50**
Rotation angle <0>: *(Press* ENTER *to accept this default)*
Text: **CIVIL ENGINEERING**
Text: *(Press* ENTER *to exit Dtext and return to the command prompt)*

By default the justification mode used by the DTEXT command is left justified. Study Figure 2–22 and the following prompt sequence to place the text string "MECHANICAL."

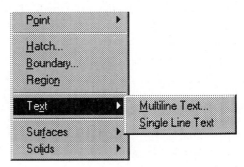

Figure 2–21

MECHANICAL
(A)   **Figure 2–22**

CIVIL ENGINEERING
(A)

Figure 2–23

CIVIL ENGINEERING
(A)

Figure 2–24

Figure 2–25 and the following prompt sequence are examples of justifying text by aligning the text between two points. The text height is automatically scaled depending on the length of the points and the number of letters that make up the text.

Command: **DTEXT**
Justify/Style/<Start point>: **Align**
First text line point: *(Pick the point at "A")*
Second text line point: *(Pick the point at "B")*
Rotation angle <0>: *(Press* ENTER *to accept this default)*
Text: **MECHANICAL**
Text: *(Press* ENTER *to exit Dtext and return to the command prompt)*

Figure 2–26 and the following prompt sequences are examples of justifying text by fitting the text in between two points and specifying the text height. Notice how the text appears compressed due to the large text height and short distance of the text line.

Command: **DTEXT**
Justify/Style/<Start point>: **Fit**
First text line point: *(Pick the point at "A")*
Second text line point: *(Pick the point at "B")*
Height <0.20>: **0.50**
Rotation angle <0>: *(Press* ENTER *to accept this default)*
Text: **MECHANICAL**
Text: *(Press* ENTER *to exit Dtext and return to the command prompt)*

Figure 2–27 and the following prompt sequences are examples of justifying text by a point at the right.

Command: **DTEXT**
Justify/Style/<Start point>: **Right**
End point: *(Pick the point at "A")*
Height <0.20>: **0.50**
Rotation angle <0>: *(Press* ENTER *to accept this default)*
Text: **MECHANICAL**
Text: *(Press* ENTER *to exit Dtext and return to the command prompt)*

MECHANICAL
(A) (B)
Figure 2–25

MECHANICAL
(A) (B)
Figure 2–26

MECHANICAL
(A)
Figure 2–27

Text Justification Modes

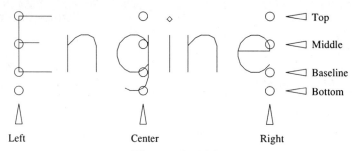

Figure 2–28A

Study Figures 2–28A and 2–28B for the numerous ways of justifying text. By default, text used in the DTEXT command is left justified. Enter one of the following initials in Figure 2–28B to justify text in a different location.

Special Text Characters

Special text characters called control codes enable you to apply certain symbols to text objects. All control codes begin with the double percent sign (%%) followed by the special character that invokes the symbol. These special text characters are given.

Underscore (%%U)

Use the double percent signs followed by the letter "U" to underscore a particular text item. Figure 2–29 shows the word "MECHANICAL," which is underscored. When prompted for entering the text, type the following:

Text: **%%UMECHANICAL**

Diameter Symbol (%%C)

The double percent signs followed by the letter "C" create the diameter symbol shown in Figure 2–30.

LEFT	Align Left (Default)
C	Center
M	Middle
R	Right
TL	Top/Left
TC	Top/Center
TR	Top/Right
ML	Middle/Left
MC	Middle/Center
MR	Middle/Right
BL	Bottom/Left
BC	Bottom/Center
BR	Bottom/Right
A	Align
F	Fit

Figure 2–28B

Figure 2–29

When prompted for entering text, the diameter symbol is displayed by typing the following:

Text: **%%C0.375**

Ø0.375

Figure 2–30

Plus/Minus Symbol (%%P)

The double percent signs followed by the letter "P" create the plus/minus symbol shown in Figure 2–31. When prompted for entering text, the plus/minus symbol is displayed by typing the following:

Text: **%%P0.005**

±0.005

Figure 2–31

Degree Symbol (%%D)

The double percent signs followed by the letter "D" create the degree symbol shown in Figure 2–32. When prompted for entering text, the degree symbol is displayed by typing the following:

Text: **37%%D**

37°

Figure 2–32

Overscore (%%O)

Similar to the underscore, the double percent signs followed by the letter "O" overscore a text object as shown in Figure 2–33. When prompted for entering text, the overscore is displayed by typing the following:

Text: **%%OMECHANICAL**

$\overline{\text{MECHANICAL}}$

Figure 2–33

Combining Special Text Character Modes

Figure 2–34 shows how the control codes are used to toggle on or off the special text characters. Enter the following at the text prompt:

Text: **%%UTEMPERATURE%%U 29%%D F**

$\underline{\text{TEMPERATURE}}$ 29° F

Figure 2–34

where the first "%%U" toggles underscore mode on and the second "%%U" turns underscore off.

Creating Different Font Styles

Use the Format pull-down menu area shown in Figure 2–35 to create different text styles. Select "Text Style...," to bring up the Text Style dialog box shown in Figure 2–35.

Use the Text Style dialog box to create new text styles or make existing styles current in the drawing. By default, the current Style Name is STANDARD. When creating new text styles, the name can be anything up to and including thirty-one characters. Once a new style is created, a font name is matched with the style. Clicking in the edit box for Font Name displays a list of all text fonts supported by the operating system. These fonts have different extensions such as SHX and TTF. TTF or True Type Fonts are especially helpful in AutoCAD since it displays the font in the drawing in its true form. If the font is bold and filled in, the font in the drawing file displays as bold and filled in. When a Font Name is selected, it displays in the Preview area located in the lower right corner of the dialog box. The Effects area allows you to display the text upside down, backwards, or vertical. Other effects include a width factor, explained later, and the oblique angle for text displayed at a slant.

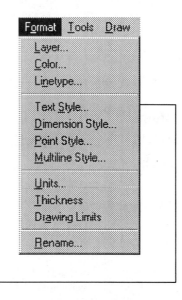

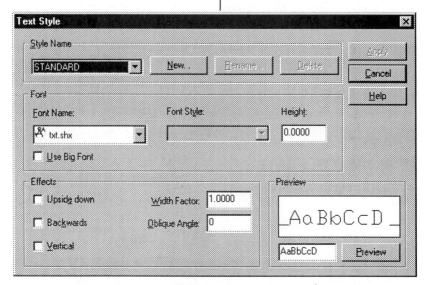

Figure 2–35

Clicking on the "New..." button of the Text Style dialog box displays the New Text Style dialog box shown in Figure 2–36. A new style is created called BLOCK. Clicking on the OK button returns to the Text Style dialog box illustrated shown in Figure 2–37. Clicking in the Font Name edit box displays all supported fonts. Clicking on romand.shx associates the font with the style name BLOCK. Clicking the Apply button saves the font to the database of the current drawing file.

In the Text Style dialog box in Figure 2–37, an edit box is present and deals with the text height. By default, the value is set to 0.0000 units. This allows the DTEXT or MTEXT commands to control the text height. Entering a value for "Height" in the dialog box places all text under this style at that height. When using the DTEXT or MTEXT commands, the text height is already set to a value defined in the Text Style dialog box.

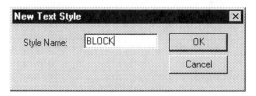

Figure 2–36

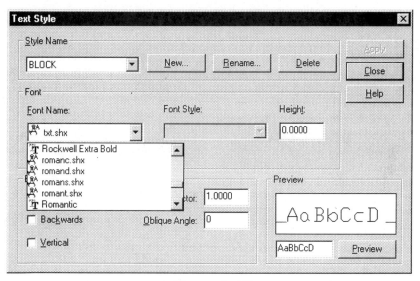

Figure 2–37

The width factor acts as a separator between letters. By default, the width factor is set to a value of 1.00 units. Entering a value smaller than 1.00 such as 0.50 units shown in Figure 2–38 tightens up or compresses the text; entering a value larger than 1.00 such as 2.00 units in Figure 2–38 spreads out or expands the line of text.

Width Factor = 0.50

SIMPLEX

Width Factor = 2.00

SIMPLEX

Figure 2–38

Using the ELLIPSE Command

Use the ELLIPSE command to construct a true elliptical shape. Before studying the three examples for ellipse construction, see Figure 2–39 to view two important parts of any ellipse, namely its major and minor diameters.

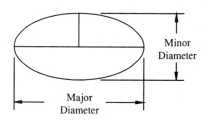

Figure 2–39

You can construct an ellipse by marking two points, which specify one of its axes (see Figure 2–40). These first two points also identify the angle with which the ellipse will be drawn. Responding to the prompt "Other axis distance" with another point identifies half of the other axis. The rubber banded line is added to assist you in this ellipse construction method.

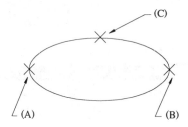

Figure 2–40

 Command:**ELLIPSE**

<Axis endpoint 1>/Center: *(Mark a point at "A")*
Axis endpoint 2: *(Mark a point at "B")*
<Other axis distance>/Rotation: *(Mark a point at "C")*

An ellipse may also be constructed by first identifying its center. Points may be picked to identify its axes or polar coordinates may be used to accurately define the major and minor diameters of the ellipse. See Figure 2–41 and the following prompt sequence to construct this type of ellipse. Use the polar coordinate or direct distance modes for locating the two axis endpoints of the ellipse.

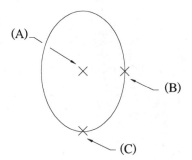

Figure 2–41

 Command:**ELLIPSE**

<Axis endpoint 1>/Center: **Center**
Center of ellipse: *(Mark a point at "A")*
Axis endpoint: @**1.50<0** *(To point "B")*
<Other axis distance>/Rotation: @**2.50<270**
  (To point "C")

This last method in Figure 2–42 illustrates constructing an ellipse by way of rotation. Identify the first two points for the first axis. Reply to the prompt "Other axis distance/Rotation" with Rotation. The first axis defined is now used as an axis of rotation that rotates the ellipse into a third dimension.

 Command:**ELLIPSE**

\<Axis endpoint 1>/Center: *(Mark a point at "A")*
Axis endpoint 2: *(Mark a point at "B")*
\<Other axis distance>/Rotation:**Rotation**
Rotation around major axis: **80**

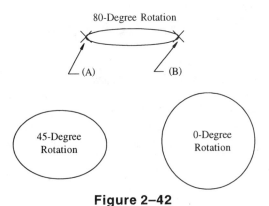

Figure 2–42

Multilines

Multilines consist of a series of parallel line segments. They may be compared with past AutoLisp routines such as Dline. However multilines differ greatly from the Dline routine in that groups of as many of 16 parallel line segments may be grouped together to form a single multiline object. Multilines may take on color, be spaced apart at different increments, and may take on different linetypes. In Figures 2–43A through 2–43C, the floor plan layout was made using the default multiline style of two parallel line segments. The spacing was changed to an offset distance of 4 units. Once the multilines are laid out as in Figure 2–43A, the Multiline Edit (MLEDIT) command cleans up all corners and intersections as displayed in Figure 2–43B. Since multilines cannot be partially deleted using the BREAK command, the MLEDIT command allows you to cut between two points that you identify. Also you have the option of breaking all or only one multiline segment depending on your desired results. Once a multiline segment has been cut, it can be welded back together again using the MLEDIT command. The following pages outline the creation of a multiline style using the Multiline Style (MLSTYLE) command. Two additional dialog boxes are contained inside of the Multiline Style dialog box; the Element Properties dialog box sets the spacing, color, and linetype of the individual multiline segments; the Multiline Properties dialog box provides other information on multilines including capping modes and the ability to fill the entire multiline in a selected color.

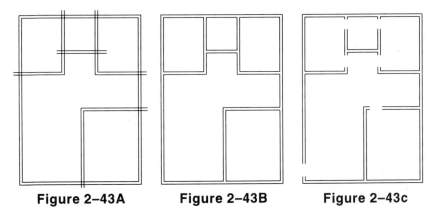

Figure 2–43A   **Figure 2–43B**   **Figure 2–43c**

Using the MLINE Command

Once the element and multiline properties have been set and saved to a multiline style, loading a new current style and picking the "OK" button of the Mutliline Styles dialog box draws the multiline to the current multiline style configuration. Figure 2–44 is the result of setting the offset distances on the previous page. The above configuration may be used for an architectural application where a block wall of 8 units thickness and a footing of 16 units thick is needed.

Multilines are drawn using the MLINE command and the following command sequence.

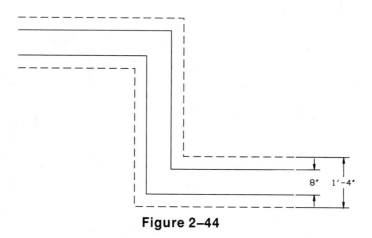

Figure 2–44

Also see Figure 2–45 to change the justification of multilines. Justification modes reference the top of the multiline, the bottom of the multiline, or the zero location of the multiline, which is the multiline's center.

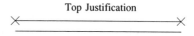

Top Justification

 Command:**MLINE**

Justification = Top, Scale = 1.00, Style = STANDARD
Justification/Scale/STyle/<From point>: **J** (For Justification)
Top/Zero/Bottom <top>:**Zero**
Justification = Zero, Scale = 1.00, Style = STANDARD
Justification/Scale/STyle/<From point>: (Pick a point)
<To point>: (Pick a point)
Undo/<To point>: (Either pick another point or press ⌷ENTER⌷ to exit the MLINE command)

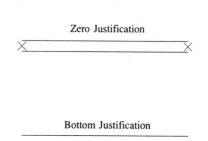

Zero Justification

Bottom Justification

Figure 2–45

Creating Multiline Styles
The MLSTYLE Command

The Multiline Styles dialog box shown in Figure 2–46 is used to create a new multiline style or to make an existing style current. This command is located under the Format pull-down menu area. The concept of creating multiline styles through this dialog box is very similar to that of creating dimension styles and making an existing dimension style current. The dialog box in Figure 2–46 lists the current multiline style which by default is named "STANDARD." For creating a new multiline style, click on the Element Properties... button or Multiline Properties... button, make the desired changes, and when completed, return to the main Multiline Styles dialog box. Change the name STANDARD to a name depicting the purpose of the new multiline style and click the ADD button to add the new multiline style to the database of the current drawing.

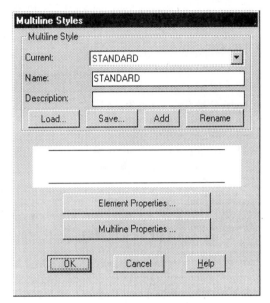

Figure 2–46

Mlstyle—Element Properties

You can use the Element Properties dialog box in Figure 2–47 to make offset, color, and linetype assignments to the various multiline segments. Picking the Add button adds a new multiline listing to the Elements box. Offset distances can now be changed along with Color and Linetype assignments to the selected multiline object. If a linetype is not present, a special Load button is available at the bottom of the Linetype dialog box. This in turn displays all supported linetypes. Picking the "OK" button in the Element Properties dialog box returns you to the main Multiline Styles dialog box where a new multiline style name may be entered and the changes added to the new multiline style.

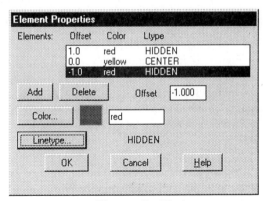

Figure 2–47

MLSTYLE—Multiline Properties

Additional control of multilines can be achieved through the Multiline Properties dialog box in Figure 2–48. This dialog box controls whether line segments are drawn at each multiline joint, whether the beginning and/or end of the multiline is capped with a line segment, whether the beginning and/or end of the outer parts of a multiline are capped by arcs, whether the beginning and/or end of the inner parts of a multiline are capped by arcs, or whether a user-defined angle is used to cap the start and end of the multiline. With Multiline Fill mode turned on, the entire multiline will be filled in with the selected color.

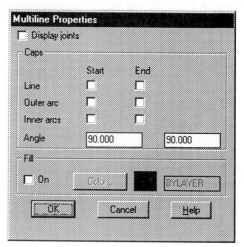

Figure 2–48

The MLEDIT Command

The MLEDIT command is a special feature designed for editing or cleaning up intersections of multiline objects (see Figure 2–49 and Figure 2–50A through 2–50E). This command is found under the Modify pull-down menu area by picking "Object" and then "Multiline." Various cleanup modes are available such as creating a closed cross, creating an open cross, or creating a merged cross. Options are available for creating "Tee" sections in multilines. Corner joints may be created along with adding or deleting a vertex of a multiline. In the vertex example in Figure 2–50B, deleting a vertex forces the multiline segment to straighten out. When adding a vertex as in Figure 2–50E, the results may not be as evident. After the vertex was added to the straight multiline segment, the STRETCH command was used to create the shape in Figure 2–50E. Multiline segments cannot be broken. They must be cut using the option illustrated at the far right of the Multiline Edit Tools dialog box, shown in Figure 2–49. To mend the cut, the multiline Weld option is used. For Figures 2–50A, 2–50C and 2–50D depicting the creation of open intersections and tees or the corner operation, the identifying letters "A" and "B" show where to pick the multiline when using the Multiline Edit Tools dialog box.

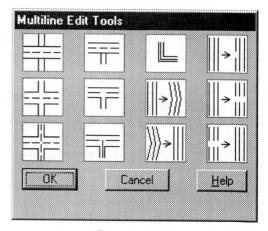

Figure 2–49

Open Cross

Delete Vertex

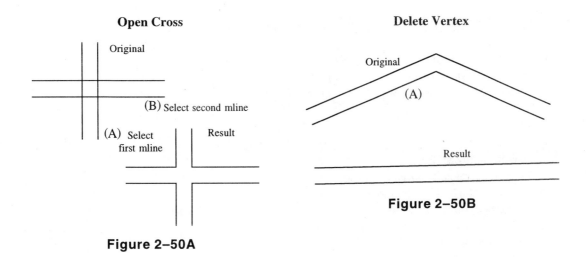

Figure 2–50A

Figure 2–50B

Open Tee

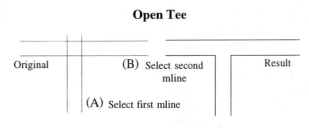

Figure 2–50C

Corner Joint

Add Vertex

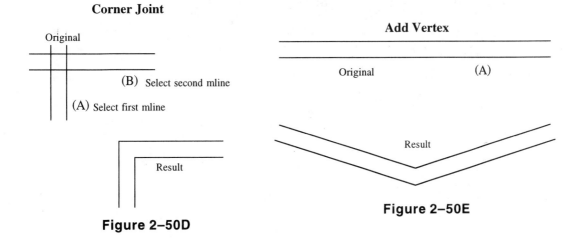

Figure 2–50D

Figure 2–50E

Using the MTEXT Command

The MTEXT command allows for the placement of text in multiple lines. Entering the MTEXT command from the command line displays the following prompts:

 Command:**MTEXT**

Attach/Rotation/Style/Height/Direction/
 <Insertion point>: *(Mark a point to identify
 one corner of the MTEXT box)*
Attach/Rotation/Style/Height/Direction/Width/
 2Points/<Other corner>: *(Mark another
 corner forming a box)*

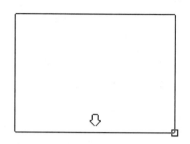

Figure 2–51A

Clicking on the first insertion point displays a user defined box with an arrow at the bottom similar to that shown in Figure 2–51A. This box will define the area that the multiline text will be placed in. If the text cannot fit in one line, it will wrap to the next line automatically. After clicking on a second point marking the other corner of the insertion box, the Multiline Text Editor dialog box appears similar to Figure 2–51B. Begin typing in text and it appears in this box.

In the dialog box shown in Figure 2–51B, as the multiline text is entered, it automatically wraps to the next line. The current text font is displayed in addition to the text height. You have the ability to make the text bold, italicized, or underlined through the "B," "I," and "U" buttons present at the top of the dialog box. When the text has been entered, click the OK button to dismiss the dialog box and place the text in the drawing, as shown in Figure 2–51C.

Figure 2–51B

THIS IS AN EXAMPLE OF PLACING TEXT
IN AN AUTOCAD DRAWING USING THE
MTEXT COMMAND

Figure 2–51C

Using DDEDIT on an Mtext Object

Figure 2–52A

When issuing the DDEDIT command and selecting an mtext object, the Multiline Text Editor dialog box appears as in Figure 2–52A. Use this to edit the text font, text height, or to make the text bold, italic, or underlined. Clicking on the Properties tab allows you to change such items as the text justification. Clicking on the Find/Replace tab allows for certain text items or phrases to be located and replaced with another text item. When editing mtext through the dialog box, highlight the desired text to edit before making changes in text font and text height. The text in Figure 2–52A was first highlighted; then the text font RomanD was located in the text font edit box. Clicking on this new font changes all highlighted text. The result is in Figure 2–52B.

MTEXT does not have to be edited as a whole group. Seleted text items may be highlighted and edited as in Figure 2–52C of the Multiline Text Editor

THIS IS AN EXAMPLE OF PLACING TEXT IN
AN AUTOCAD DRAWING USING THE MTEXT
COMMAND

Figure 2–52B

dialog box. The DDEDIT command was issued and the mtext item selected to display the dialog box above. The text "PLACING" in Figure 2–52C was already underlined. Also, the word "MTEXT" was increased to a new text height of 0.30 units. To make changes to individual words or letters in an mtext object, first highlight the text to edit, as in the word "AUTOCAD." Next make the change; in Figure 2–52C, the font was changed to reflect the true type Swiss font. Clicking on the OK button displays the results in Figure 2–52D.

THIS IS AN EXAMPLE OF PLACING TEXT
IN AN **AUTOCAD** DRAWING USING THE
MTEXT COMMAND

Figure 2–52D

Figure 2–52C

Using the SPELL Command

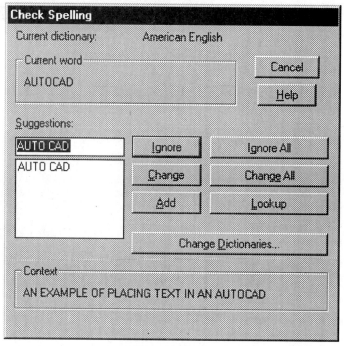

Figure 2–53

Issuing the SPELL command and selecting the multitext object displays the Check Spelling dialog box shown in Figure 2–53, and along with the following components: the Current dictionary; the Current word identified by the spell checker as being misspelled. The presence of the word does not necessarily mean that the word is misspelled, such as AUTOCAD. The word is not part of the current dictionary. The Suggestions area displays all possible alternatives to the word identified as being misspelled. Ignore allows you to skip the current word; this would be applicable especially in the case of acronyms such as CAD and GDT. Ignore All skips all remaining words that match the current word. If the word "AUTOCAD" keeps coming up as a misspelled word, instead of constantly picking the Ignore button, Ignore All will skip the word "AUTOCAD" in any future instances. The Change option replaces the word in the Current word box

with a word in the Suggestions box. The Add option adds the current word to the current dictionary. The Lookup button checks the spelling of a selected word found in the Suggestions box. The Change Dictionaries button allows you to change to a different dictionary containing other types of words. The Context area displays a phrase on how the current word is being used in a sentence. Use this area to check proper sentence structure. In Figure 2–53, the word "AUTOCAD" was identified as being misspelled. Clicking the Ignore button continues with the spell checking operation until completed. The MTEXT object after undergoing spell checking is displayed in Figure 2–54.

THIS IS AN EXAMPLE OF PLACING TEXT IN AN AUTOCAD DRAWING USING THE MTEXT COMMAND

Figure 2–54

Using the PLINE Command

Polylines are similar to individual line segments except that a polyline may consist of numerous segments and still be considered a single object. Width may also be assigned to a polyline compared to regular line segments; this makes polylines perfect for drawing borders and title blocks. Study Figures 2–55 and 2–56 and both command sequences that follow to use the PLINE command.

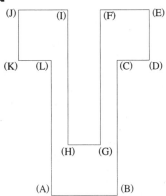

Figure 2–56

 Command:**PLINE**

From point: *(Select a point at "A" in Figure 2–55)*
Current line-width is 0.0000
Arc/Close/Halfwidth/Length/Undo/Width/
  <Endpoint of line>: *(Mark a point at "A")*
Arc/Close/Halfwidth/Length/Undo/Width/
  <Endpoint of line>:**Width**
Starting width <0.0000>:**0.10**
Ending width <0.1000>: *(Press* ENTER *to accept default)*
Arc/Close/Halfwidth/Length/Undo/Width/
  <Endpoint of line>: *(Mark a point at "B")*
Arc/Close/Halfwidth/Length/Undo/Width/
  <Endpoint of line>: *(Mark a point at "C")*
Arc/Close/Halfwidth/Length/Undo/Width/
  <Endpoint of line>: *(Mark a point at "D")*
Arc/Close/Halfwidth/Length/Undo/Width/
  <Endpoint of line>: *(Mark a point at "E")*
Arc/Close/Halfwidth/Length/Undo/Width/
  <Endpoint of line>: *(Press* ENTER *to exit this command)*

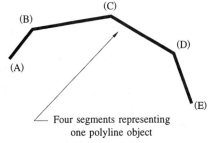

Four segments representing one polyline object

Figure 2–55

 Command:**PLINE**

From point: *(Select a point at "A" in Figure 2–56)*
Current line-width is 0.0000
Arc/Close/Halfwidth/Length/Undo/Width/
  <Endpoint of line>: **@1.00<0** (To "B")
Arc/Close/Halfwidth/Length/Undo/Width/
  <Endpoint of line>: **@2.00<90** (To "C")
Arc/Close/Halfwidth/Length/Undo/Width/
  <Endpoint of line>: **@0.50<0** (To "D")
Arc/Close/Halfwidth/Length/Undo/Width/
  <Endpoint of line>: **@0.75<90** (To "E")
Arc/Close/Halfwidth/Length/Undo/Width/
  <Endpoint of line>: **@0.75<180** (To "F")
Arc/Close/Halfwidth/Length/Undo/Width/
  <Endpoint of line>: **@2.00<270** (To "G")
Arc/Close/Halfwidth/Length/Undo/Width/
  <Endpoint of line>: **@0.50<180** (To "H")
Arc/Close/Halfwidth/Length/Undo/Width/
  <Endpoint of line>: **@2.00<90** (To "I")
Arc/Close/Halfwidth/Length/Undo/Width/
  <Endpoint of line>: **@0.75<180** (To "J")
Arc/Close/Halfwidth/Length/Undo/Width/
  <Endpoint of line>: **@0.75<270** (To "K")
Arc/Close/Halfwidth/Length/Undo/Width/
  <Endpoint of line>: **@0.50<0** (To "L")
Arc/Close/Halfwidth/Length/Undo/Width/
  <Endpoint of line>:**Close**

Using the POINT Command

Use the POINT command to identify the location of a point on a drawing. This point may be used for reference purposes. The OSNAP-Node or Nearest options are used to snap to points. By default, a point is displayed as a dot on the screen. This dot may be confused with the existing grid dots already on the screen. To distinguish point objects from grid dots, use Figure 2–57 to assign a new point type; this is accomplished through the PDMODE system vari-able. Entering a value of 3 for PDMODE displays the point as an "X." The PDSIZE system variable controls the size of the point. Use the following prompts for changing the point mode to a value of 3.

Command: **PDMODE**
New value for variable PDMODE <0>: **3**

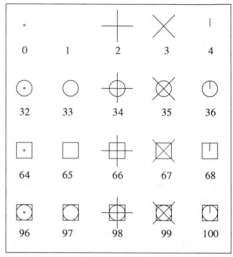

Command: **POINT**

Point: *(Mark the new position of a point using the cursor or one of the many coordinate systems)*
Point: *(Either mark another point or press [ENTER] to exit to the command prompt)*

Figure 2–57

Using the Point Style Dialog Box

Selecting "Point Style" from the Format pull-down menu area displays the Point Style dialog box (see Figure 2–58). Use this icon menu to set a different point mode and point size. This is similar to using the PDMODE and PDSIZE system variable commands.

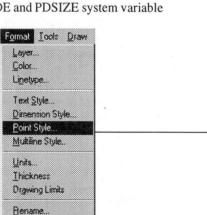

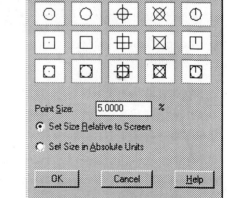

Figure 2–58

Using the POLYGON Command

The POLYGON command is used to construct a regular polygon. Polygons are defined by the radius of circle that classifies the polygon as either being inscribed or circumscribed. Polygons consist of a closed polyline object with width set to zero. The following prompt sequence is used to construct an inscribed polygon using Figure 2–59 as a guide.

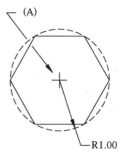

An Inscribed Polygon

(A)

R1.00

Figure 2–59

 Command: **POLYGON**

Number of sides: **6**
Edge/<Center of polygon>: *(Select a point at "A")*
Inscribed in circle/Circumscribed about circle
   (I/C): **I** *(For Inscribed)*
Radius of circle: **1.00**

The following prompt sequence is used to construct a circumscribed polygon with the in Figure 2–60 as a guide.

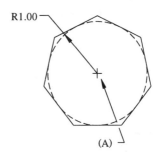

A Circumscribed Polygon

R1.00

(A)

Figure 2–60

 Command: **POLYGON**

Number of sides: **7**
Edge/<Center of polygon>: *(Select a point at "A")*
Inscribed in circle/Circumscribed about circle
   (I/C): **C** *(or Circumscribed)*
Radius of circle: **1.00**

Polygons may be specified by locating the endpoints of one of its edges. The polygon is then drawn in a counterclockwise direction. Study Figure 2–61 and the following prompt sequence to construct a polygon by one of its edges.

Command: **POLYGON**
Number of sides<4>: **5**
Edge/<Center of polygon>: **E** *(For Edge)*
First endpoint of edge: *(Select a point at "A")*
Second endpoint of edge: *(Select a point at "B")*

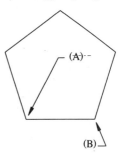

A Polygon by Edge

(A)

(B)

Figure 2–61

Using the RAY Command

A Ray is a type of construction line object that begins at a user-defined point and extends to infinity only in one direction. In Figure 2–62, the quadrants of the circles identify all points where the Ray objects begin and are drawn to infinity to the right. You should organize Ray objects on specific layers. You should also exercise care in the editing of Rays, and take special care not to leave segments of objects in the drawing database as a result of breaking Ray objects. Breaking the Ray object at the right at "A" in Figure 2–62 converts one object into an individual line segment, the other object remains as a Ray.

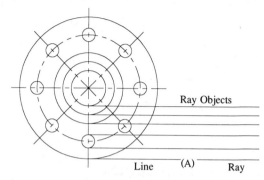

Figure 2–62

Using the RECTANG Command

Use the RECTANG command to construct a rectangle by defining two points. In Figure 2–63, two diagonal points are picked to define the rectangle. The rectangle drawn is as a single polyline object.

 Command: **RECTANG**
Chamfer/Elevation/Fillet/Thickness/Width/
  <First corner>: *(Mark a point at "A")*
Other corner: *(Mark a point at "B")*

Other options of the RECTANG command enable you to construct a chamfer or fillet at all corners of the rectangle, the ability to assign a width to the rectangle, and to have the rectangle drawn at a specific elevation and at a thickness for 3D purposes. In Figure 2-64 and the following prompt sequence, a rectangle is constructed with a chamfer distance of 0.20 units; the width of the rectangle is also set at 0.05 units. A relative coordinate value of 1.00,2.00 will used to construct the rectangle 1 unit in the "X" direction and 2 units in the "Y" direction. The "@" symbol resets the previous point at "A" to zero.

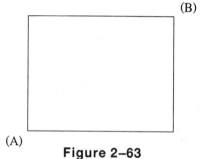

Figure 2–63

Figure 2–64

Command:**RECTANG**

Chamfer/Elevation/Fillet/Thickness/Width/
   <First corner>: **C** *(For Chamfer)*
First chamfer distance for rectangles
   <0.0000>: **0.20**
Second chamfer distance for rectangles
   <0.2000>: *(Press* [ENTER] *to accept this default)*
Chamfer/Elevation/Fillet/Thickness/Width/

<First corner>: **W** *(For Width)*
Width for rectangles <0.0000>: **0.05**
Chamfer/Elevation/Fillet/Thickness/Width/
   <First corner>: *(Mark a point at "A")*
Other corner: **@1.00,2.00** *(Identifying the other corner at "B")*

Using the SPLINE Command

Use the SPLINE command to construct a smooth curve given a sequence of points. You have the option of changing the accuracy of the curve given a tolerance range. The basic command sequence follows, which constructs the Spline segment shown in Figure 2–65.

 Command:**SPLINE**

Object/<Enter first point>: *(Pick a first point)*
Enter point: *(Pick another point)*
Close/Fit Tolerance/<Enter point>: *(Pick another point)*
Close/Fit Tolerance/<Enter point>: *(Pick another point)*
Close/Fit Tolerance/<Enter point>: *(Press* [ENTER] *to continue)*
Enter start tangent: *(Press* [ENTER] *to accept)*
Enter end tangent: *(Press* [ENTER] *to accept the end tangent position which exits the command and places the spline)*

Once you construct the spline curve, you can use the LIST command to display all control and fit points that make up the curve. Other information provided includes the length of the spline.

The Spline may be closed to display a continuous segment shown in Figure 2–66. Entering a different tangent point at the end of the command changes

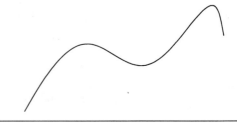

SPLINE Layer: 0
      Space: Model space
      Handle = 60
      Length: 13.0251
      Order: 4
      Properties: Planar, Non-Rational, Non-Periodic
      Parametric Range:Start 0.0000
                 End 12.4959
Control Points:
      X = 21.5709, Y = 14.4679, Z = 0.0000
      X = 22.5570, Y = 16.1296,Z = 0.0000
User Data: Fit Points
      X = 21.5709, Y = 14.4679, Z = 0.0000
      X = 24.8050, Y = 17.4561, Z = 0.0000

Figure 2–65

the shape of the curve connecting the beginning and end of the spline.

 Command:**SPLINE**

Object/<Enter first point>: *(Pick a first point)*
Enter point: *(Pick another point)*
Close/Fit Tolerance/<Enter point>: *(Pick another point)*
Close/Fit Tolerance/<Enter point>: *(Pick another point)*
Close/Fit Tolerance/<Enter point>: *(Pick another point)*
Close/Fit Tolerance/<Enter point>: **C** *(To Close)*
Enter tangent: *(Press* ⟦ENTER⟧ *to exit the command and place the spline)*

Grips provide an excellent means of editing a Spline curve. With the Spline selected, grips appear at all control points. As a grip is selected, you can dynamically see how much to pull the selected control point to achieve the desired results as shown in the examples in Figure 2–67. Grips will be discussed in greater detail at the end of this unit.

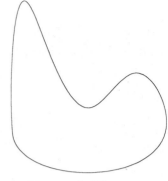

Figure 2–66

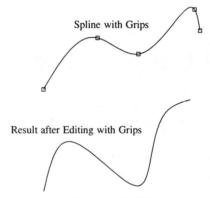

Spline with Grips

Result after Editing with Grips

Figure 2–67

 # Using the XLINE Command

Xlines are construction lines drawn from a user-defined point. You are not prompted for any length information as the Xline is drawn in an unlimited length beginning at the user defined point and going off to infinity in opposite directions from the point. Xlines can be drawn horizontal, vertical, and angular. You can bisect an angle using an Xline or offset the Xline at a specific distance. In Figure 2–68A, the circular view represents the front view of a flange. To begin the creation of the side views, lines are usually projected from key features on the adjacent view. In the case of the front view in Figure 2–68A, the key features are the top of the plate

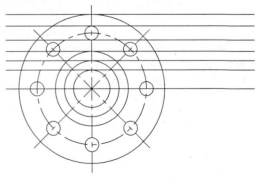

Figure 2–68A

in addition to the other circular features. In this case, the Xlines were drawn using the Horizontal mode from the Quadrant of all circles. The following prompts outline the XLINE command sequence:

 Command:**XLINE**

Hor/Ver/Ang/Bisect/Offset/<From point>:
Through point: *(Pick a point on the display screen to place the first Xline)*
Through point: *(Pick a point on the display screen to place the second Xline)*

Since the Xlines continue to be drawn in both directions, care must be taken to manage these objects. Construction management techniques of Xlines could take the form of placing all Xlines on a specific layer to be turned off or frozen when not needed. When editing Xlines (especially with the BREAK command), you need to take special care to remove all excess objects that will still remain on the drawing screen. In Figure 2–68B, breaking the Xline converts the object into a Ray object. Use the ERASE command to remove any access Xlines.

Figure 2–69 is an example of how Xlines are used to bisect or divide an angle in half. After entering the Bisect option, three endpoints are identified on the angle; namely the vertex of the angle, angle starting point, and angle ending point. Pressing the ENTER key at the second "Angle end point" prompt bisects the angle with an Xline, which has an unspecified length.

 Command:**XLINE**

Hor/Ver/Ang/Bisect/Offset/<From point>:
  Bisect
Angle vertex point: *(Pick the endpoint at "A")*
Angle start point: *(Pick the endpoint at "B")*
Angle end point: *(Pick the endpoint at "C")*
Angle end point: *(Press* ⌷ENTER *to exit this command)*

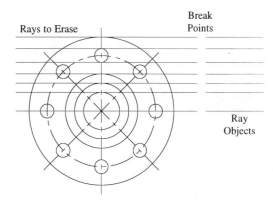

Figure 2–68B

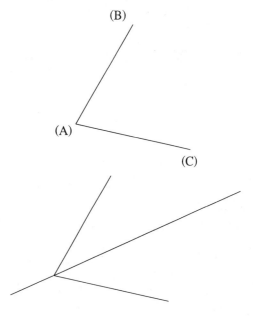

Figure 2–69

Methods of Selecting Block Commands

Another way of creating objects is through the
BLOCK command. Typical blocks take the form
of symbol libraries consisting of doors, windows,
and appliances for architectural applications and
nuts, bolts, and screws for mechanical applications.
When creating a block, numerous objects that make
up the block are all considered a single object. Once
you save the block under a name and insertion point,
you can merge the block into the drawing using the
INSERT command. When defining blocks using the
BLOCK command, you can use only the newly created object in the current drawing file. This means
a block consisting of a bolt is only able to be used
in the current drawing file; it cannot be shared with
other drawings. To create global blocks to be merged
in any drawing file, the WBLOCK command is used.
The same INSERT command is used to merge global blocks into drawings. The BLOCK, INSERT,
and WBLOCK commands will be explained in the
following pages.

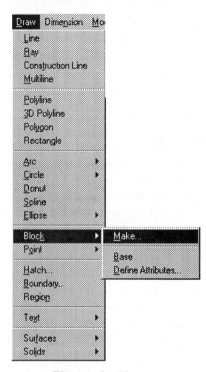

Figure 2–70

The BLOCK command is chosen from the Draw
pull-down menu area in Figure 2–70. The INSERT
command is selected from the Insert pull-down menu
area in Figure 2–71. Clicking on "Block..." activates the Insert dialog box. Activating the WBLOCK
command will be discussed in the following pages.

BLOCK commands may also be entered in from
the keyboard. For a complete listing of other two-
letter commands, refer to the list located on pages
1–9 and 1–10.

Figure 2–71

Using the BLOCK Command

Figure 2–72 is a drawing of a hex head bolt. This drawing consists of one polygon representing the hexagon, a circle indicating that the hexagon is circumscribed about the circle, and two centerlines. The centerlines were constructed by setting the dimension variable DIMCEN to a -0.09 value and using the DIMCENTER command. Rather than copy these individual objects numerous times throughout the drawing, you can create a symbol or block using the BLOCK command and the following command prompt sequence:

 Command: **BLOCK**

Block name (or ?): **Hex-hd**
Insertion base point: **Int**
of *(Select the intersection of the two lines at "A")*
Select objects: *(Mark a point at "B")*
Other corner: *(Mark a point at "C")*
Select objects: *(Press* ⌨ENTER *to create the block)*

As the block is written to the database of the current drawing, the individual objects just used to create the block disappear from the screen. These may be retrieved back to the screen using the OOPS command.

You may use up to 31 alphanumeric characters when naming a block. The insertion point is considered a point of reference. When the block is inserted into a drawing, it will be brought in relation to the insertion point.

Picking the Draw area of the pull-down menu area in Figure 2–73 activates the Block > Make... menu. Clicking on "Make..." displays the Block Definition dialog box in Figure 2–73 which enables you to define blocks through the use of a dialog box. The Block Definition dialog box also appears when clicking on the block button.

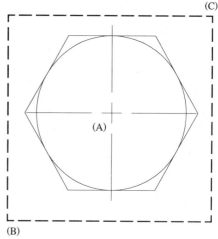

Figure 2–72

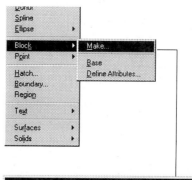

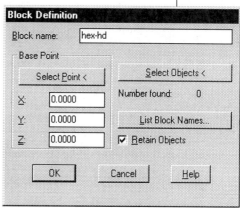

Figure 2–73

Using the WBLOCK Command

In the previous example of using the BLOCK command, a group of objects were grouped into a single
object for insertion into a drawing. As this type of
symbol is created, it is only able to be inserted into
the original drawing it was created in. Using the
WBLOCK command creates a symbol similar to
the creation process used in the BLOCK command.
Creating the symbol using the WBLOCK command
writes the objects to disk, which allows the symbol
to be inserted into any type of drawing. Use the
following prompt sequence and Figure 2–74 for creating a WBLOCK from the command prompt:

Command: **WBLOCK**

The Create Drawing File dialog box appears in Figure 2–75. enter a name in the "File Name:" box
and click on the Save button. Continue with the
following prompt sequence.

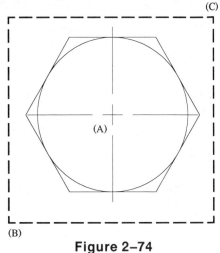

Figure 2–74

Figure 2–75

Block name: *(Press* ⟦ENTER⟧ *to make the block
   name the same as the file name)*
Insertion base point: **Int**
of *(Select the intersection of the two lines at "A"
   in Figure 2–74)*
Select objects: *(Mark a point at "B" in
   Figure 2–74)*
Other corner: *(Mark a point at "C" in
   Figure 2–74)*
Select objects: *(Press* ⟦ENTER⟧ *to create the object)*

Another method of creating a global block is to select the EXPORT... command located in the File pull-down menu area in Figure 2–76 to activate the Export Data dialog box illustrated in Figure 2–76. A number of file types are supported for exporting data. Click in the "Save As type:" edit box to expose the Block (*.dwg) format. This is the same as using the WBLOCK command. After entering a name in the "File name:" area of the dialog box, the EXPORT command returns to the command prompt area where the global symbol definition can be completed.

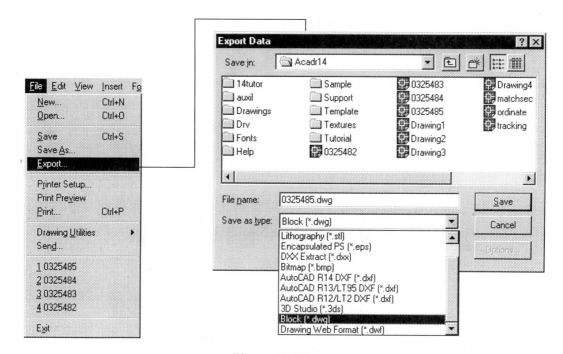

Figure 2–76

Using the Insert Dialog Box

Both BLOCK and WBLOCK commands are used to create symbols out of individual objects. To merge these symbols into drawings, use the Insert dialog box illustrated in Figure 2–79. This dialog box is used to dynamically insert blocks made with the BLOCK command or global symbols made with the WBLOCK command. Activate this dialog box by first selecting the "Insert" area of the pull-down menu and then picking "Block..." in Figure 2–77. By default, the "Specify Parameters on Screen" box is checked. All prompts for the insertion point, scale, and rotation of symbols will occur at the command prompt area of the drawing editor. The "Explode" box determines if the symbol is inserted as one object or if the symbol is inserted and then exploded back to its individual objects. Once the symbol is listed, click on the OK button; the following prompts complete the block insertion operation:

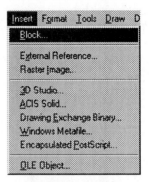

Figure 2–77

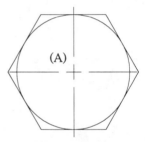

Figure 2–78

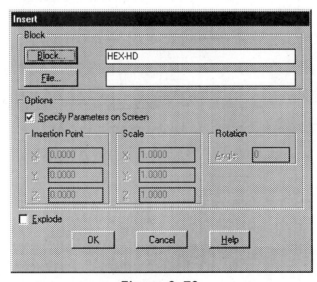

Figure 2–79

Insertion point: *(Mark a point at "A" in the in Figure 2–78 to insert the block)*
X scale factor <1>/Corner/XYZ: *(Press* ⏎ *to accept default X scale factor)*
Y scale factor (default=X): *(Press* ⏎ *to accept default)*
Rotation angle <0>: *(Press* ⏎ *to accept the default rotation angle and insert the block in Figure 2–78)*

To assist you in selecting the blocks and wblocks, two additional dialog boxes are present in the main Insert dialog box in Figure 2–79. Illustrated in Figure 2–80 is the result of picking "Block...." If blocks are defined as part of the database of the current drawing, they may be selected from the dialog box in Figure 2–80.

Picking "File…" from the Insert dialog box in Figure 2–79 displays the Select Drawing File dialog box illustrated in Figure 2–81 for inserting global symbols into a drawing. This is the same dialog box associated with opening up drawing files. Since the WBLOCK command writes the group of objects to disk, this makes it possible to insert any valid AutoCAD drawing file into the current drawing. Select the desired subdirectory where the global symbol or drawing file is located followed by the name of the drawing. This will return you to the main Insert dialog box found in Figure 2–79 on the previous page.

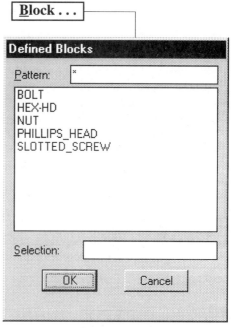

Figure 2–80

Figure 2–81

Methods of Selecting Editing Commands

The heart of any CAD system is its ability to modify or edit existing geometry. This is the function of the numerous editing commands available in AutoCAD.

As with all commands, the main body of editing commands may be selected from the pull-down menu area shown in Figure 2–82. Another convenient way of selecting edit commands is through the Modify toolbox shown in Figure 2–83. All editing commands may also be entered in directly from the keyboard either using their entire name or through command aliasing as in the following examples:

Enter "E" for the ERASE command
Enter "M" for the MOVE command

For a complete listing of other two-letter commands, refer to the list located on pages 1–9 and 1–10.

These edit commands will be discussed in greater detail in the following pages:

ARRAY	FILLET
BREAK	MATCHPROP
CHAMFER	MEASURE
CHPROP	MIRROR
COPY	MOVE
DDCHPROP	OFFSET
DDEDIT	PEDIT
DIVIDE	ROTATE
EXPLODE	SCALE
EXTEND	STRETCH
TRIM	

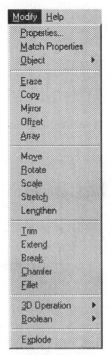

Figure 2–82

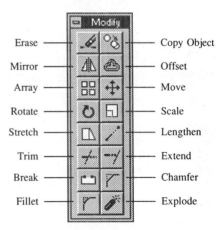

Figure 2–83

Creating Rectangular Arrays

The ARRAY command allows you to arrange multiple copies of an object or group of objects similar to Figure 2–85 in a rectangular pattern. Suppose a rectangular pattern of the object in Figure 2–84 needs to be made. The final result is to create three rows and three columns of the object. Finally, the spacing between rows is to be 0.50 units and the spacing between columns is 1.25 units. Study the following prompts and Figure 2–85.

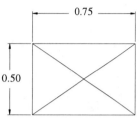

Figure 2–84

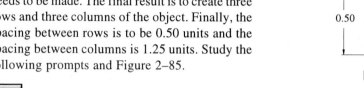 Command: **ARRAY**

Select objects: **All**
(This will select all objects in the original figure)
Select objects: *(Press* ENTER *to continue)*
Rectangular or Polar array (<R>/P): **R** *(For
    Rectangular)*
Number of rows (---): **3**
Number of columns (lll): **3**
Unit cell distance between rows (---): **1.00**
Distance between columns (lll): **2.00**

For rectangular objects to be arrayed, a reference point in the lower left corner of the figure becomes a point where you may calculate the spacing between rows and columns. Not only must the spacing distance be used; the overall size of the object plays a role in coming up with the spacing distances. With the total height of the original object at 0.50 and a required spacing between rows of 0.50, both object height and spacing results in a distance of 1.00 between rows. In the same manner, with the original length of the object at 0.75 and a spacing of 1.25 units between columns, the total spacing results in a distance of 2.00.

Figure 2–85 of the rectangular array illustrates an array that runs to the right and above of the original figure. At times these directions change to the left and below the original object. The only change occurs in the distances between rows and columns where negative values dictate the direction of the rectangular array. See Figure 2–86.

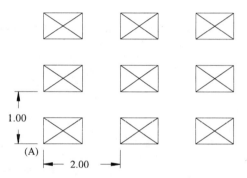

Figure 2–85

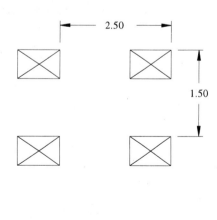

Figure 2–86

Command: **ARRAY**

Select objects: **All**
(This will select all objects in the original figure)
Select objects: *(Press [ENTER] to continue)*
Rectangular or Polar array (<R>/P): **R** *(For Rectangular)*
Number of rows (---): **3**
Number of columns (|||): **2**
Unit cell distance between rows (---): **-1.50**
Distance between columns (|||): **-2.50**

The results for the above prompt sequence are illustrated in Figure 2–86.

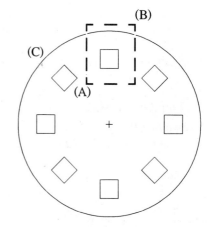

Figure 2–87

Creating Polar Arrays

Polar arrays allow you to create multiple copies of objects in a circular or polar pattern. The following prompts and Figure 2–87 for performing polar arrays:

 Command: **ARRAY**

Select objects: *(Select a point at "A")*
Other corner: *(Select a point at "B")*
Select objects: *(Press [ENTER] to continue)*
Rectangular or Polar array (<R>/P): **P** *(For Polar)*
Base/<Specify center point of array>: **Cen**
of *(Select the edge of the large circle at "C")*
Number of items: **8**
Angle to fill (+=ccw, -=cw)<360>: *(Press [ENTER] to accept)*
Rotate objects as they are copied? <Y>: **Yes**

To array rectangular or square objects in a polar pattern without rotating the objects, the square or rectangle is first converted into a block with an insertion point located in the center of the square. Now all squares lie an equal distance from their common center. See Figure 2–88.

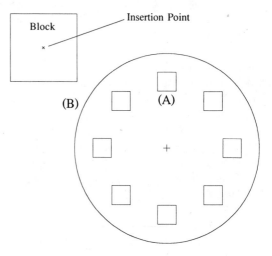

Figure 2–88

Command: **ARRAY**

Select objects: *(Select the square block at "A"
    in Figure 2–88)*
Select objects: *(Press [ENTER] to continue)*
Rectangular or Polar array (<R>/P): **P** *(For
    Polar)*
Base/<Specify center point of array>: **Cen**
of *(Select the edge of the large circle at "B" in
    Figure 2–88)*
Number of items: **8**
Angle to fill (+=ccw, -=cw)<360>: *(Press [ENTER] to
    accept)*
Rotate objects as they are copied? <Y>: **No**

Arraying Bolt Holes—Method #1

Multiple copies of objects such as bolt holes are
easily duplicated in circular patterns using the AR-
RAY command along with the following prompts:

 Command: **ARRAY**

Select objects: *(Select both small circles in
    Figure 2–89)*
Rectangular or Polar array (<R>/P): **P** *(For
    Polar)*
Base/<Specify center point of array>: **Int**
of *(Select the intersection at "A")*
Number of items: **8**
Angle to fill (+=CCW, -=CW)<360>: *(Press [ENTER]
    for default)*
Rotate objects as they are copied? <Y>:
    (Press [ENTER] for default)

As the circles are copied in the circular pattern,
centerlines need to be updated to the new bolt hole
positions (see Figure 2–90). Again the ARRAY command is used to copy and duplicate one centerline
over a 45-degree angle to fill in the counterclockwise direction.

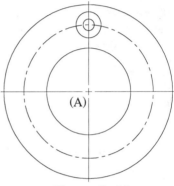

Figure 2–89

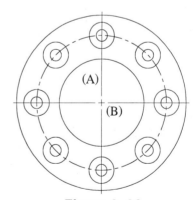

Figure 2–90

Command: **ARRAY**

Select objects: *(Select the vertical centerline at
    "A" in Figure 2–91)*
Rectangular or Polar array (<R>/P): **P** *(For
    Polar)*
Base/<Specify center point of array>: **Int**
of *(Select the intersection at "B" in Figure 2–91)*
Number of items: **2**
Angle to fill (+=CCW, -=CW)<360>: **45**
Rotate objects as they are copied? <Y>:
    (Press ⌷ENTER⌷ for default)

Now use the ARRAY command to copy and duplicate the last centerline and mark the remaining bolt
hole circles as shown in Figure 2–91.

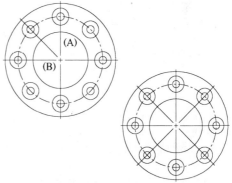

Figure 2–91

 Command: **ARRAY**

Select objects: **L** *(This should select the last
    line)*
Rectangular or Polar array (<R>/P): **P** *(For
    Polar)*
Base/<Specify center point of array>: **Int**
of *(Select the intersection at "A")*
Number of items: **4**
Angle to fill (+=CCW, -=CW)<360>: *(Press ⌷ENTER⌷
    for default)*
Rotate objects as they are copied? <Y>:
    (Press ⌷ENTER⌷ for default)

Forming Bolt Holes—Method #2

From the previous example, we have seen how easily
the ARRAY command can be used for making multiple copies of objects equally spaced around an
entire circle. What if the objects are copied only
partially around a circle, such as the bolt holes in
Figure 2–92. The ARRAY command is used here
to copy the bolt holes in 40-degree increments. Follow the next command sequence for performing the
operation shown in Figure 2–92.

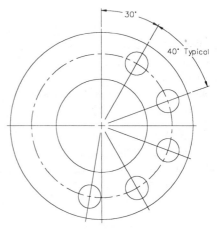

Figure 2–92

First use the ARRAY command to rotate and copy the vertical centerline at a -30-degree angle to fill. The -30 degrees will copy and rotate the centerline in the clockwise direction. Next place the circle at the intersection of the centerlines using the CIRCLE command. Use the ARRAY command, select the circle and centerline, and copy the selected objects at 40-degree increments using the following prompt sequence.

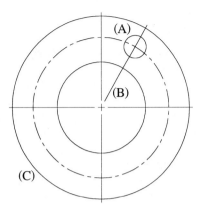

Figure 2–93

 Command: **ARRAY**

Select objects: *(Select the small circle at "A" and line at "B" in Figure 2–93)*
Rectangular or Polar array (<R>/P): **P** *(For Polar)*
Base/<Specify center point of array>: **Cen**
of *(Select the large circle anywhere near "C")*
Number of items: *(Press* ENTER *to continue with this command)*
Angle to fill (+=CCW, -=CW): **-160**
Angle between items: **-40** *(To copy 40 degrees clockwise)*
Rotate objects as they are copied? <Y>: *(Press* ENTER *for default)*

The result is illustrated in Figure 2–94. This method of identifying bolt holes shows how the number can be controlled by specifying the total angle to fill and the angle between items. In both angle specifications, a negative value is entered to force the array to be performed in the clockwise direction.

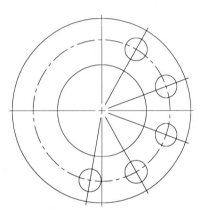

Figure 2–94

Forming Bolt Holes—Method #3

The object in Figure 2–95 is similar to the object shown in Forming Bolt Holes—Method #2 previous example. A new series of holes are to be placed 20 degrees away from each other using the ARRAY command.

Begin placing the holes by laying out one centerline using the ARRAY command with an angle of 15 degrees to fill. Add one circle using the CIRCLE command at the intersection of the centerlines. Use the following prompts to add the remaining holes:

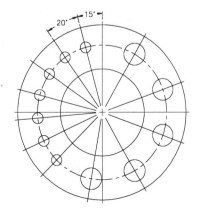

Figure 2–95

Command: **ARRAY**

Select objects: *(Select the small circle at "A" and line at "B" in Figure 2–96)*

Rectangular or Polar array (<R>/P): **P** *(For Polar)*

Base/<Specify center point of array>: **Cen**
of *(Select the large circle anywhere near "C")*

Number of items: *(Press* ENTER *to continue with this command)*

Angle to fill (+=CCW, -=CW)<360>: **120**

Angle between items: **20**

Rotate objects as they are copied? <Y>:
(Press ENTER *for default)*

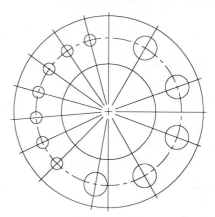

Figure 2–96

Since the direction of rotation for the array is in the counterclockwise direction, all angles are specified in positive values. See Figure 2–97.

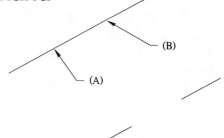

Figure 2–97

Using the BREAK Command

Use the BREAK command to partially delete a segment of an object. The following command sequence and Figure 2–98 refer to using the BREAK command.

Figure 2–98

Command: **BREAK**

Select objects: *(Select the line at "A" in Figure 2–98)*
Enter second point (or F for first point): *(Select the line at "B" in Figure 2–98)*

To select key objects to break using Osnap options, utilize the "First" option of the BREAK command. This option resets the command and allows you to select an object to break followed by two different points that identify the break. See the following command sequence and Figure 2–99 to use the "First" option of the BREAK command:

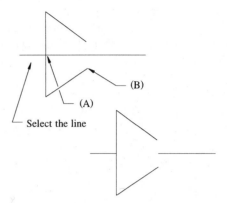

Figure 2–99

 Command: **BREAK**

Select objects: *(Select the line shown in Figure 2–99)*
Enter second point (or F for first point): **First**
Enter first point: **Int**
of *(Select the intersection of the two lines at "A")*
Enter second point: **Endp**
of *(Select the endpoint of the line at "B")*

Breaking circles is always accomplished in the counterclockwise direction. Study the following command sequence and Figure 2–100 for breaking circles.

 Command: **BREAK**

Select objects: *(Select the circle in Figure 2–100)*
Enter second point (or F for first point): **First**
Enter first point: *(Select the point in either in Figure 2–100 at "A")*
Enter second point: *(Select the point in either in Figure 2–100 at "B")*

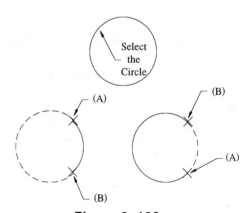

Figure 2–100

Using the CHAMFER Command

The CHAMFER command produces an inclined surface at an edge of two intersecting line segments. Distances determine how far from the corner the chamfer is made. One method of producing chamfers is shown in Figure 2–101.

 Command: **CHAMFER**

(TRIM mode) Current chamfer Dist1=0.5000, Dist2=0.5000
Polyline/Distance/Angle/Trim/Method/<Select first line>: *(Select the line at "A")*
Select second line: *(Select the line at "B")*

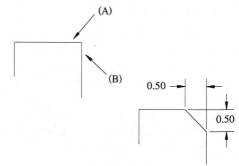

Figure 2–101

If the distances from the corner of the object are different, a beveled surface is formed in Figure 2–102.

 Command: **CHAMFER**

(TRIM mode) Current chamfer Dist1=0.5000, Dist2=0.5000
Polyline/Distance/Angle/Trim/Method/<Select first line>: **D**
Enter first chamfer distance <0.5000>: *(Press* ENTER *to accept the default)*
Enter second chamfer distance <0.5000>: **0.25**

 Command: **CHAMFER**

(TRIM mode) Current chamfer Dist1=0.5000, Dist2=0.2500
Polyline/Distance/Angle/Trim/Method/<Select first line>: *(Select the line at "A")*
Select second line: *(Select the line at "B")*

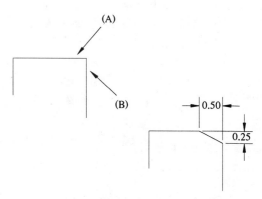

Figure 2–102

With nonintersecting corners, you could use the CHAMFER command to connect both lines. The CHAMFER command distances are both set to a value of 0 to accomplish this task. See Figure 2–103.

 Command: **CHAMFER**

(TRIM mode) Current chamfer Dist1=0.5000, Dist2=0.2500
Polyline/Distance/Angle/Trim/Method/<Select first line>: **D** *(For Distance)*
Enter first chamfer distance <0.5000>: **0**
Enter second chamfer distance <0.0000>: *(Press* ENTER *to accept the default)*

 Command: **CHAMFER**

(TRIM mode) Current chamfer Dist1=0.0000, Dist2=0.0000
Polyline/Distance/Angle/Trim/Method/<Select first line>: *(Select the line at "A")*
Select second line: *(Select the line at "B")*

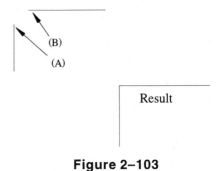

Figure 2–103

Since a polyline consists of numerous segments representing a single object, using the CHAMFER command with the Polyline option produces corners throughout the entire polyline. See Figure 2–104.

 Command: **CHAMFER**

(**TRIM** mode) Current **CHAMFER** Dist1 = 0.00, Dist2 = 0.00
Polyline/Distance/Angle/Trim/Method/<Select first line>: **D** *(For distance)*
Enter first chamfer distance <0.00>: **0.50**
Enter second chamfer distance <0.50>: *(Press* ENTER *to accept the default)*

 Command: **CHAMFER**

Polyline/Distances/<Select first line>: **P** *(For polyline)*
Select 2D Polyline: *(Select the polyline)*

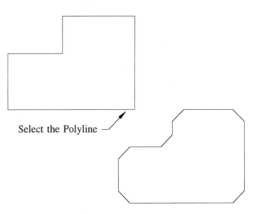

Figure 2–104

The CHAMFER command supports a "Trim/No trim" option enabling a chamfer to be placed with lines trimmed or not trimmed as in the in Figure 2–105.

 Command: **CHAMFER**

(TRIM mode) Current chamfer Dist1 = 0.00, Dist2 = 0.00
Polyline/Distance/Angle/Trim/Method/<Select first line>: **T**
Trim/No trim <Trim>: **N**
Polyline/Distance/Angle/Trim/Method/<Select first line>: **D**
Enter first chamfer distance <0.0000>: **1.00**
Enter second chamfer distance <1.0000>:
    (Press ⌜ENTER⌝ *to accept the default)*

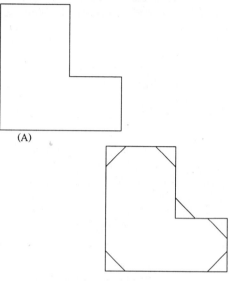

(A)

 Command: **CHAMFER**

(NOTRIM mode) Current chamfer Dist1 = 1.00, Dist2 = 1.00
Polyline/Distance/Angle/Trim/Method/<Select first line>: **P** *(For polyline)*
Select 2D Polyline: *(Select the polyline at "A")*

All edges are chamfered without being trimmed.

Figure 2–105

Using the CHPROP Command

The CHPROP command specifically changes the properties of an object. You can change the color, layer, linetype, and thickness of an object. Use the following command sequence for the CHPROP command:

Command: **CHPROP**
Select objects: *(Mark a point at "A")*
Other corner: *(Mark a point at "B")*
Select objects: *(Press* ENTER *to continue)*
Change what property (Color/LAyer/LType/
   ltScale/Thickness): *(Select Color, LAyer,
   LType, Linetype Scale or Thickness to
   change)*

In the previous command sequence, notice that selecting a blank part of the screen at "A" for the prompt "Select objects" places you in automatic window selection mode when selecting the other corner at "B." If "B" were the first point and "A" the other corner, automatic crossing selection mode is invoked.

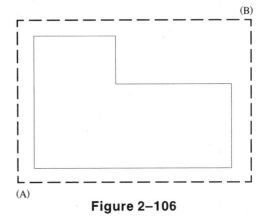

Figure 2–106

▦ Using the Change Properties Dialog Box

Using the DDCHPROP command on an object brings up the dialog box shown in Figure 2–107. Use this dialog box the same way that you would use the normal CHPROP command; namely, to change an object's color, layer, linetype, linetype scale, or thickness. Selecting "Color" from the Change Properties dialog box brings up another dialog box displaying the current color supported by the monitor. In the same way, an additional dialog box is displayed when selecting "Layer." A third dialog box is displayed when selecting the "Linetype" box. This dialog box will show the current linetypes loaded into the drawing file.

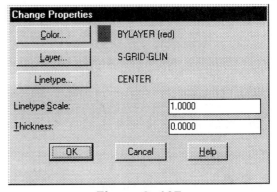

Figure 2–107

Using the COPY Command

The COPY command is used to duplicate an object
or group of objects. In Figure 2–108, the Window
mode is used to select all objects to copy. Point "C"
is used as the base point or displacement, or where
you want to copy the objects from. Point "D" is
used as the second point of displacement, or where
you want to the copy the objects to.

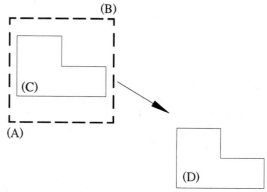

Figure 2–108

 Command: **COPY**

Select objects: **W** *(To invoke the Window
    mode)*
First corner: *(Select near point "A")*
Other corner: *(Select near point "B")*
Select objects: *(Press* ENTER *to continue)*
<Base point or displacement>/Multiple: **Endp**
of *(Select the endpoint of the line at "C")*
Second point of displacement: *(Select a point
    near "D")*

The COPY command is also used to duplicate numerous objects while staying inside the command.
Figure 2–109 are a group of objects copied using
the Multiple option of the COPY command. See
the following command sequence to use the Multiple option of the COPY command:

Second point of displacement: *(Select a point
    near "E")*
Second point of displacement: *(Select a point
    near "F")*
Second point of displacement: *(Press* ENTER *to
    exit this command)*

 Command: **COPY**

Select objects: *(Select all objects that make up
    figure "A")*
Select objects: *(Press* ENTER *to continue)*
<Base point or displacement>/Multiple: **M** *(For
    Multiple)*
Base point: **Endp**
of *(Select the endpoint of the line at "A")*
Second point of displacement: *(Select a point
    near "B")*
Second point of displacement: *(Select a point
    near "C")*
Second point of displacement: *(Select a point
    near "D")*

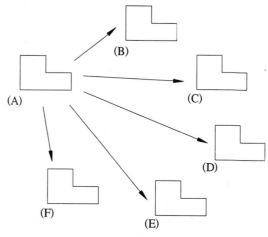

Figure 2–109

Using the DDEDIT Command

Mtext objects are easily modified using the DDEDIT command. Selecting the text object in Figure 2–110A displays the Multiline Text Editor in Figure 2–110B. The text height, font, color, and justification can all be changed inside of the dialog box.

Command: **DDEDIT**
<Select a TEXT or ATTDEF object>/Undo:
  (Select the text and the Multiline Text Editor dialog box will appear)

THE DDEDIT COMMAND CAN BE USED WHEN YOU WANT TO CHANGE EITHER THE TEXT CONTENT OR FORMATTING OF A MULTILINE TEXT. THE CHANGES MADE AFFECT ONLY THE SELECTED TEXT, NOT THE STYLE THE MTEXT WAS DRAWN IN.

Figure 2–110A

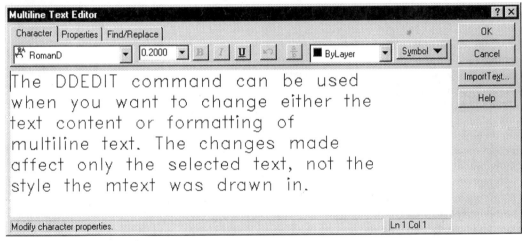

Figure 2–110B

Using the DDEDIT command on a text object created with the DTEXT command as shown in Figure 2–111A displays the Edit Text dialog box shown in Figure 2–111B. Use this to change text inside of the edit box provided. Font, justification, and text height are not supported in this dialog box.

MECHANICAL ENGINEERING

Figure 2–111A

Figure 2–111B

Using the DIVIDE Command

The DIVIDE command will take an object such as a line or arc and divide it equally depending on the number of segments desired. The DIVIDE command does this by placing a point object where a division occurs. It is important to note that as a point is placed along an object during the division process, the object is not automatically broken at the location of the point. Rather the point is commonly used along with the OSNAP-Node option to construct from. Select the DIVIDE command by picking Draw from the pull-down menu area; then pick POINT to expose the DIVIDE command (see Figure 2–112). Refer to the following prompts to use the DIVIDE command.

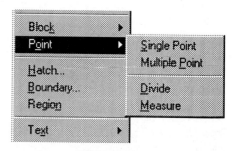

Figure 2–112

Command:**DIVIDE**
Select object to divide: *(Select line "A" shown
   in Figure 2–113)*
<Number of segments>/Block:**8**

If the results of the DIVIDE command are not obvious, it is because of the current point mode which is set by the system variable Pdmode to a value of 0, which places a point similar in appearance to that of a grid dot. When the dot is placed on an object, it is unable to be seen easily. For this reason, use the PDMODE variable to change the point value from 0 to 3. This will produce a point similar to an "X" and make the point visible. Since only one point style may be displayed at any one time, use the REGEN command to regenerate the screen. All points will now take on the current PDMODE value.

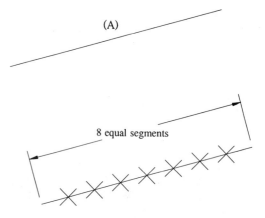

Figure 2–113

Command:**PDMODE**
New value for PDMODE <0>:**3**

Command:**REGEN**

If the points used in the DIVIDE command are still hard to detect because of their size, use the PDSIZE system variable to change the size of the points (see Figure 2–114). Issue another regeneration to affect the size of all points. This can also be accomplished through the Point Style dialog box found in the Format area of the pull-down menu.

Command:**PDSIZE**
New value for PDSIZE <0>: **0.50**

Command:**REGEN**

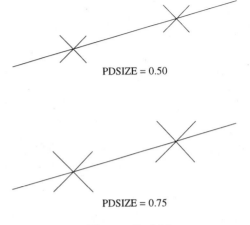

PDSIZE = 0.50

PDSIZE = 0.75

Figure 2–114

 Using the EXPLODE Command

Using the EXPLODE command on a polyline, associative dimension, or block separates the single into its individual parts. Figure 2–115 is a polyline that is considered a single object. Using the EXPLODE command and selecting the polyline breaks the polyline into four individual objects.

 Command:**EXPLODE**

Select objects: *(Select the polyline in Figure 2–115)*
The UNDO command will restore it.

An associative dimension consists of extension lines, dimension lines, dimension text, and arrowheads, all considered a single object. This type of dimension is covered further in Unit 5. Using the EXPLODE command on an associative dimension breaks the dimension down into individual extension lines, dimension lines, arrowheads, and dimension text. This is not advisable since the dimension value will not update if the dimension is stretched along with the object being dimensioned. Also the ability to manipulate dimensions with grips is lost. Grips will be discussed later in this unit.

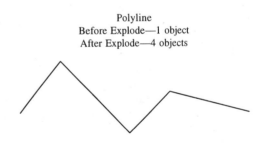

Polyline
Before Explode—1 object
After Explode—4 objects

Figure 2–115

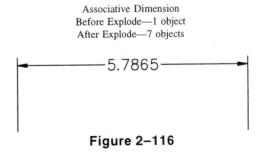

Associative Dimension
Before Explode—1 object
After Explode—7 objects

5.7865

Figure 2–116

Command:**EXPLODE**

Select objects: *(Select the associative dimension shown in Figure 2–116)*
The UNDO command will restore it.

Using the EXPLODE command on a block converts the block into individual objects, lines, and circles. If the block had yet another block nested in it, you would need to be perform an additional Explode operation to break this object into individual objects.

 Command:**EXPLODE**

Select objects: *(Select the block shown in Figure 2–117)*
The UNDO command will restore it.

You can also be use the EXPLODE command to explode non-uniformly scaled block objects as shown in Figure 2–118. In some past versions of AutoCAD, a block of only equal scale factors could be exploded. As shown in Figure 2–118, all blocks will now be affected by the EXPLODE command regardless of their scale factors and the blocks will be broken into individual objects without redefining the block.

Block
Before Explode—1 object
After Explode—15 objects

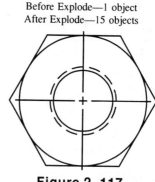

Figure 2–117

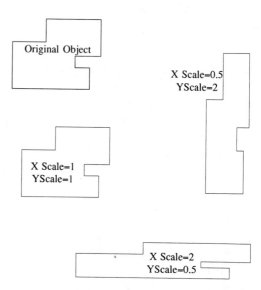

Figure 2–118

Using the EXTEND Command

Use the EXTEND command to extend objects to a specified boundary edge. In Figure 2–119, select the large circle "A" as the boundary edge. After pressing the ENTER key to continue with the command, select the arc at "B," line at "C," and arc at "D" to extend these objects to the circle. If you select the wrong end of an object, use the Undo feature, which is an option of the command to undo the change and repeat the procedure at the correct end of the object.

 Command:**EXTEND**

Select boundary edges: (Projmode = UCS,
   Edgemode = No extend)
Select objects: *(Select the large circle at "A")*
Select objects: *(Press ⌅ to continue)*
<Select object to extend>/Project/Edge/Undo:
   (Select the arc at "B")
<Select object to extend>/Project/Edge/Undo:
   (Select the line at "C")
<Select object to extend>/Project/Edge/Undo:
   (Select the arc at "D")
<Select object to extend>/Project/Edge/Undo:
   (Press ⌅ to exit this command)

An alternate method of selecting boundary edges is to press the ENTER key in response to the prompt "Select objects:." This automatically creates boundary edges out of all objects in the drawing. When using this method, the boundary edges do not hightlight.

To extend multiple objects such as the five line segments shown in Figure 2–120, select the line at "A" as the boundary edge and use the Fence mode to select the ends of the line segments at "B" and "C" to extend. The Fence option allows you to define a crossing line or series of lines that you use to select multiple objects to extend.

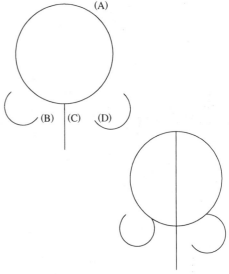

Figure 2–119

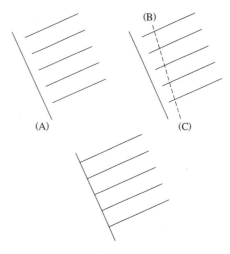

Figure 2–120

Command: **EXTEND**

Select boundary edges: (Projmode = UCS,
   Edgemode = No extend)
Select objects: *(Select the line at "A" in Figure
   2–120)*
Select objects: *(Press* ENTER *to continue)*
<Select object to extend>/Project/Edge/Undo:
   F *(To invoke the fence option)*
Undo/<Endpoint of line>: *(Mark a point at "B")*
Undo/<Endpoint of line>: *(Mark a point at "C")*
Undo/<Endpoint of line>: *(Press* ENTER *to end the
   Fence and execute the EXTEND command)*
<Select object to extend>/Project/Edge/Undo:
   (Press ENTER *to exit this command)*

Certain conditions require the boundary edge to be
extended where an imaginary edge is projected enabling objects not in direct sight of the boundary
edge to still be extended. Study Figure 2–121 and
the following prompts on this special case involving the EXTEND command.

 Command: **EXTEND**

Select boundary edges: (Projmode = UCS,
   Edgemode = No extend)
Select objects: *(Pick line "A")*
Select objects: *(Press* ENTER *to continue)*
<Select object to extend>/Project/Edge/Undo:
   E *(For edge)*
Extend/No extend <No extend>: **E** *(For extend
   the edge)*
<Select object to extend>/Project/Edge/Undo:
   (Pick line "B")

Continue picking the remaining lines to be extended
to the extended boundary edge. You could also use
the Fence mode select all line segments to extend.

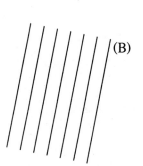

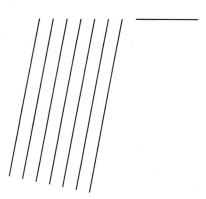

Figure 2–121

Using the FILLET Command

Fillets are corners that are rounded off at a specified radius value and automatically trimmed. If the radius value is too large for the specified objects, an error message states this. By default, the fillet radius is set to 0.50 units. Figure 2–122 shows one method of producing a fillet at a designated corner.

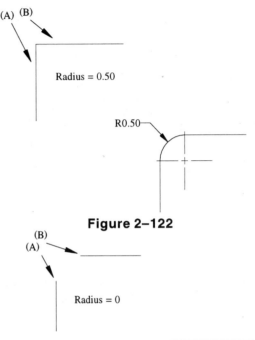

Figure 2–122

 Command:**FILLET**

(TRIM mode) Current fillet radius = 0.5000
Polyline/Radius/Trim/<Select first object>:
   (Select line "A")
Select second object: *(Select line "B")*

A Fillet radius of 0 produces a corner out of two non-intersecting objects. See Figure 2–123.

 Command:**FILLET**

(TRIM mode) Current fillet radius = 0.5000
Polyline/Radius/Trim/<Select first object>:**R**
   (For radius)
Enter fillet radius <0.5000>:**0**

 Command:**FILLET**

(TRIM mode) Current fillet radius = 0.0000
Polyline/Radius/Trim/<Select first object>:
   (Select line "A")
Select second object: *(Select line "B")*

Figure 2–123

Use the FILLET command on a polyline object to produce rounded edges at all corners of the polyline in a single operation (see Figure 2–124).

 Command:**FILLET**

(TRIM mode) Current fillet radius = 0.0000
Polyline/Radius/Trim/<Select first object>:**R**
   (For Radius)
Enter fillet radius <0.0000>:**0.25**

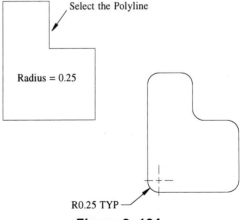

Figure 2–124

Command:**FILLET**

(TRIM mode) Current fillet radius = 0.2500
Polyline/Radius/Trim/<Select first object>:**P**
    (For polyline)
Select 2D polyline: *(Select the polyline in
    Figure 2–124)*.

The FILLET command can also be used to control
whether or not to trim the excess corners after a
fillet is placed. Figure 2–125A is a typical fillet
operation where the polyline at "A" is selected. However, instead of automatically trimming the corners
of the polyline, a new "Trim/No trim" option allows for the lines to remain. Use the following
prompts that illustrate this operation.

 Command:**FILLET**

(TRIM mode) Current fillet radius = 0.0000
Polyline/Radius/Trim/<Select first object>:
    Trim
Trim/No trim <Trim>:**N** *(For no trim)*
Polyline/Radius/Trim/<Select first object>:**R**
    (For radius)
Enter fillet radius <0.0000>:**1.00**

 Command:**FILLET**

(NOTRIM mode) Current fillet radius = 1.00
Polyline/Radius/Trim/<Select first object>:**P**
    (For polyline)
Select 2D polyline: *(Select polyline "A")*

Filleting two parallel lines in Figure 2–125B automatically constructs a semicircular arc object connecting both lines at their endpoints.

 Command:**FILLET**

(NOTRIM mode) Current fillet radius = 0.0000
Polyline/Radius/Trim/<Select first object>:
    (Select line "B")
Select second object: *(Select line "C")*

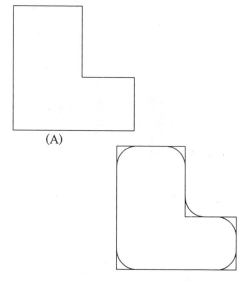

(A)

Figure 2–125A

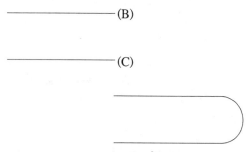

(B)

(C)

Figure 2–125B

Using the LENGTHEN Command

Use the LENGTHEN command to change the length of a selected object without disturbing other object qualities such as angles of lines or radii of arcs. See Figure 2–126.

 Command:**LENGTHEN**

DElta/Percent/Total/DYnamic/<Select object>: (Select line "A")
Current length: 12.3649
DElta/Percent/Total/DYnamic/<Select object>: **Total**
Angle/<Enter total length (1.0000)>:**20**
<Select object to change>/Undo: (Select the line at "A")
<Select object to change>/Undo: (Press ENTER to exit)

When using the LENGTHEN command on an arc segment, both the length and included angle information are displayed before making any changes as shown in Figure 2–127. After supplying the new total length of any object, be sure to select the desired end to lengthen when prompted for "Select object to change."

 Command:**LENGTHEN**

DElta/Percent/Total/DYnamic/<Select object>: (Select arc "B")
Current length: 1.4459, included angle: 54.9597
DElta/Percent/Total/DYnamic/<Select object>: **Total**
Angle/<Enter total length (1.0000)>:**5**
<Select object to change>/Undo: (Pick the arc at "B")
<Select object to change>/Undo: (Press ENTER to exit)

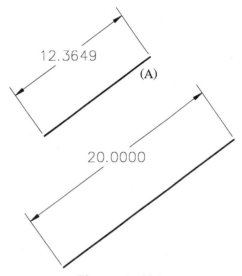

12.3649
(A)
20.0000

Figure 2–126

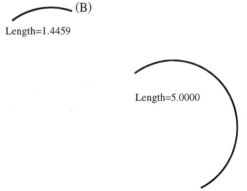

(B)
Length=1.4459

Length=5.0000

Figure 2–127

Using the MATCHPROP Command

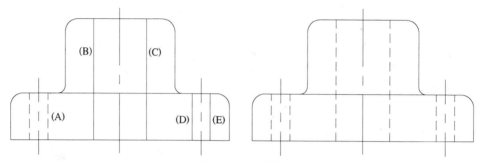

Figure 2–128

At times objects are drawn in the wrong layers or the wrong color scheme is applied to a group of objects. Text objects are sometimes drawn using an incorrect text style. As the DDCHPROP and DDMODIFY commands provide quick ways to fix the above problems, a more efficient command is available to change the properties of objects—the MATCHPROP command. This command stands for "Match Properties." When starting the command sequence, a source object is required. This source object transfers all of its current properties to other objects designated as "Destination Objects." In Figure 2–128, the flange on the left requires the object lines located at "B," "C," "D," and "E" be converted to hidden lines. Using the MATCHPROP command, select the existing hidden line "A" as the source object; select lines "B" through "E" as the destination objects. The results appear in the flange on the right where the continuous object lines were converted into hidden lines. Not only did the linetype change, but also the color and layer information.

 Command:**MATCHPROP**

Select Source Object: *(Select the hidden line at "A")*
Current active settings = color layer ltype ltscale thickness text dim hatch
Settings/<Select Destination Object(s)>: *(Select line "B")*

Settings/<Select Destination Object(s)>: *(Select line "C")*
Settings/<Select Destination Object(s)>: *(Select line "D")*
Settings/<Select Destination Object(s)>: *(Select line "E")*
Settings/<Select Destination Object(s)>: *(Press ENTER to exit this command)*

To get a better idea on what object properties are affected when using the MATCHPROP command, reenter the command, pick a source object, and instead of picking a destination object immediately, enter the letter "S" for settings. This will display the dialog box shown in Figure 2–129.

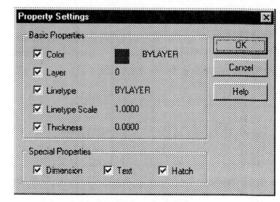

Figure 2–129

Command:**MATCHPROP**

Select Source Object: *(Select the hidden line at "A")*

Current active settings = color layer ltype ltscale thickness text dim hatch

Settings/<Select Destination Object(s)>:**S** *(For settings). This displays the dialog box in Figure 2–129.*

The MATCHPROP command can control special properties of dimensions, text, and hatch patterns (see Figure 2–130). Figure 2–131A shows two blocks: the block assigned a dimension value of 46.6084 was dimensioned using a metric dimension style with the ROMAND font applied. The block assigned a dimension value of 2.3872 was dimensioned using the STANDARD dimension style with the TXT font applied. Both blocks need to be dimensioned using the METRIC dimension style. Issue the MATCHPROP command and select the 46.6084 dimension as the source object; select the 2.3872 dimension as the destination object.

 Command:**MATCHPROP**

Select Source Object: *(Select the dimension at "A")*

Current active settings = color layer ltype ltscale thickness text dim hatch

Settings/<Select Destination Object(s)>: *(Select the dimension at "B")*

Settings/<Select Destination Object(s)>: *(Press [ENTER] to exit this command)*

The results are in Figure 2–131B with the METRIC dimension style being applied to the STANDARD dimension style through the use of the MATCHPROP command. Since the text font was associated with the dimension style, it also changed in the destination object.

Any box with a check displayed in it will transfer that property from the source object to all destination objects. If you need to transfer only the layer information and not the color and linetype properties of the source object, remove the checks from the Color and Linetype properties before you select the destination objects. These properties will not transfer to any destination objects.

Figure 2–130

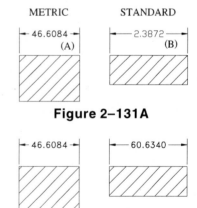

Figure 2–131A

Figure 2–131B

The following is an example of how the MATCHPROP command affects a text object using special properties in Figure 2–132.

Figure 2–133A shows two text items displayed in different fonts. The text "Coarse Knurl" at "A" was constructed using a text style called ROMAND. The text "Medium Knurl" at "B" was constructed using the default text style called STANDARD. Use the command prompt sequence to match the STANDARD text style with the ROMAND text style using the MATCHPROP command.

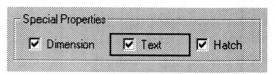

Figure 2–132

 Command:**MATCHPROP**

Select Source Object: *(Select the text at "A")*
Current active settings = color layer ltype
   ltscale thickness text dim hatch
Settings/<Select Destination Object(s)>:
   (Select the text at "B")
Settings/<Select Destination Object(s)>:
   (Press ENTER *to exit this command)*

The result is shown in Figure 2–133B. Both text items now share the same text style. Notice the text string stays intact when matching text properties. Only the text style of the source object is applied to the destination object.

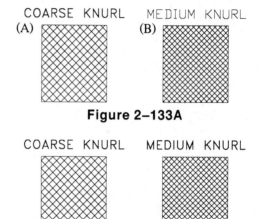

Figure 2–133A

Figure 2–133B

A source hatch object can also be matched to a destination pattern using the MATCHPROP command (see Figure 2–134). In Figure 2–135A, the cross-hatch patterns at "B" and "C" are at the wrong angle and scale. They should reflect the pattern at "A" since it is the same part. Use the MATCHPROP command, select the hatch pattern at "A" as the source object, and select the patterns at "B" and "C" as the destination objects.

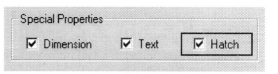

Figure 2–134

 Command:**MATCHPROP**

Select Source Object: *(Select the hatch pattern at "A")*
Current active settings = color layer ltype
  ltscale thickness text dim hatch
Settings/<Select Destination Object(s)>:
  (Select the hatch pattern at "B")
Settings/<Select Destination Object(s)>:
  (Select the hatch pattern at "C")
Settings/<Select Destination Object(s)>:
  (Press ⟨ENTER⟩ to exit this command)

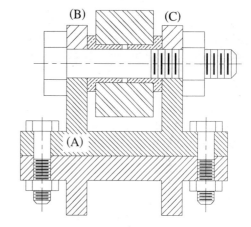

Figure 2–135A

The results appear in Figure 2–135B where the source hatch pattern property was applied to all destination hatch patterns.

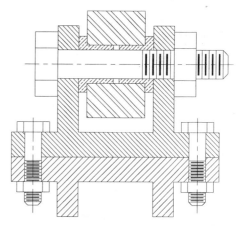

Figure 2–135B

Using the MEASURE Command

The MEASURE command will take an object such as a line or arc and measure along it depending on the length of the segment. The MEASURE command, similar to the DIVIDE command, places a point object at a specified distance given in the MEASURE command (see Figure 2–136). It is important to note that as a point is placed along an object during the measuring process, the object is not automatically broken at the location of the point. Rather, the point is commonly used to construct from along with the OSNAP-Node option. Select the MEASURE command by picking Draw from the pull-down menu area; then pick Point to expose the MEASURE command. Use the following prompts for the MEASURE command.

Command:**MEASURE**
Select object to measure: *(Select the designated end shown in Figure 2–137)*
<Segment length>/Block:**0.50**

If the results of the MEASURE command are not obvious, it is because of the current point mode which is set by the system variable PDMODE. By default, this variable is set to a value of 0 which places a point similar in appearance to that of a grid dot. When the dot is placed on an object, it is unable to be seen easily. For this reason, use the PDMODE variable to change the point value from 0 to 3. This will produce a point similar to an "X" and thus make the point visible. Since only one point may be displayed at any one time, use the REGEN command to regenerate the screen. All points will now take on the current PDMODE value. Changing to a different type of point can also be accomplished through the Point Style dialog box found in the Format area of the pull-down menu.

Command:**PDMODE**
New value for PDMODE <0>: **3**

Command:**REGEN**

It is important to note that the measuring starts at the endpoint closest to the point you used to select the object.

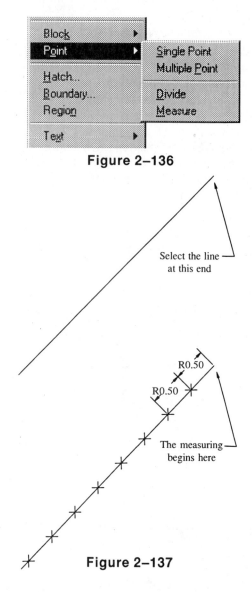

Figure 2–136

Figure 2–137

Using the MIRROR Command—Method #1

Use the MIRROR command to create a mirrored copy of an object or group of objects. When performing a mirror, the operator has the option of deleting the original object, which would be the same as flipping the object, or keeping the original object along with the mirror image, which would be the same as flipping and copying. Refer to the following prompts along with Figure 2–138 for using the MIRROR command:

 Command:**MIRROR**

Select objects: *(Select a point near "X")*
Other corner: *(Select a point near "Y")*
Select objects: *(Press the [ENTER] key to continue)*
First point of MIRROR line: **Endp**
of *(Select the endpoint of the centerline at "A")*
Second point: **Endp**
of *(Select the endpoint of the centerline at "B")*
Delete old objects? <N> **No**

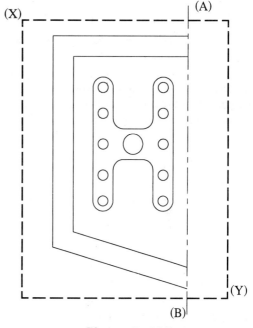

Figure 2–138

Since the original object was required to be retained by the mirror operation, the image result is shown in Figure 2–139. The MIRROR command works well when symmetry is required.

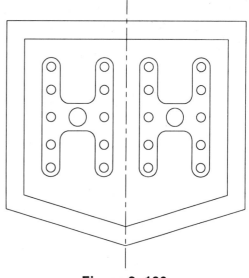

Figure 2–139

Using the MIRROR Command—Method #2

The illustration in Figure 2–140A is a different application of the MIRROR command. It is required to have all items that make up the bathroom plan flip but not copy to the other side. This is a typical process involving "What if" situations. Use the following command prompts to perform this type of mirror operation. The results are displayed in Figure 2–140B.

(A)

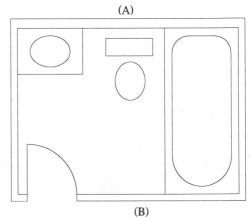

(B)

Figure 2–140A

 Command: **MIRROR**

Select objects: **All** *(This will select all objects shown in Figure 2–140A)*
Select objects: *(Press the* [ENTER] *key to continue)*
First point of mirror line: **Mid**
of *(Select the midpoint of the line at "A")*
Second point: **Per**
to *(Select line "B," which is perpendicular to point "A")*
Delete old objects? <N>**Yes**

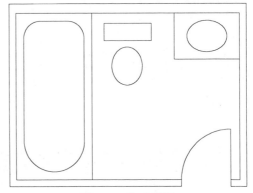

Figure 2–140B

Using the MIRROR Command—Method #3

Situations sometimes involve mirroring text as in Figure 2–141A. A system variable called MIRRTEXT controls this occurrence and by default is set to a value of "1" or "on." Use the following prompts to see the results of the MIRROR command on text.

 Command: **MIRROR**

Select objects: *(Select a point near "X")*
Other corner: *(Select a point near "Y")*
Select objects: *(Press [ENTER] to continue)*
First point of mirror line: **Endp**
of *(Select the endpoint of the centerline at "A")*
Second point: **Endp**
of *(Select the endpoint of the centerline at "B")*
Delete old objects? <N> **No**

Notice that with MIRRTEXT turned "on," text is mirrored, which lends itself unreadable (see Figure 2–141B). To mirror text and have it right reading, set the MIRRTEXT system variable to a value of "0" or "off," as in the following prompt sequence. Also see Figure 2–141C.

 Command: **MIRRTEXT**

New value for MIRRTEXT <1>: **0**

Using the MIRROR command with the MIRRTEXT system variable set to a value of "0" results in text being able to be read similar to Figure 2–141C. Follow this prompt sequence to accomplish this.

 Command: **MIRROR**

Select objects: **P** *(For Previous)*
Select objects: *(Press the [ENTER] key to continue)*
First point of mirror line: **Endp**
of *(Select the endpoint of the centerline at "A")*
Second point: **Endp**
of *(Select the endpoint of the centerline at "B")*
Delete old objects? <N> **No**

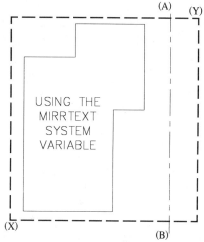

Figure 2–141A

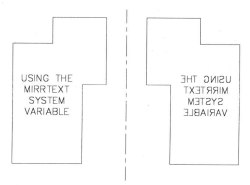

Figure 2–141B

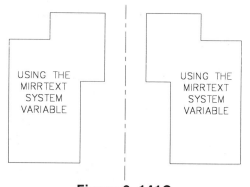

Figure 2–141C

Using the MOVE Command

The MOVE command repositions an object or group of objects to a new location. Once the objects to move are selected, a base point or displacement (where the object is to move from) is found. Next, a second point of displacement (where the object is to be moved to) is needed.

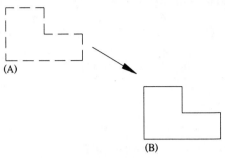

Figure 2–142

 Command: **MOVE**

Select objects: *(Select all dashed objects in Figure 2–142)*
Select objects: *(Press* ⌣ENTER *to continue)*
Base point or displacement: **Endp**
of *(Select the endpoint of the line at "A")*
Second point of displacement: *(Mark a point at "B")*

The slot shown in Figure 2–143 is incorrectly positioned; it needs to be placed 1.00 unit away from the left edge of the object. The MOVE command in combination with a polar coordinate or direct distance mode can be used to perform this operation.

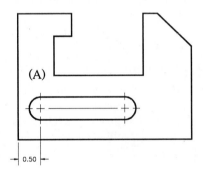

Figure 2–143

 Command: **MOVE**

Select objects: *(Select the slot and all centerlines in Figure 2–143)*
Select objects: *(Press* ⌣ENTER *to continue)*
Base point or displacement: **Cen**
of (Select the edge of arc "A")
Second point of displacement: **@0.50<0**

As the slot is moved into a new position using the MOVE command, a new horizontal dimension must be placed to reflect the correct distance from the edge of the object to the centerline of the arc. See Figure 2–144. Another command will be explained in the following pages to affect a group of objects along with the dimension.

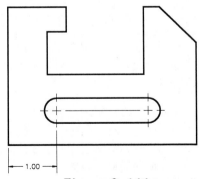

Figure 2–144

Using the OFFSET Command

The OFFSET command is commonly used for creating one object parallel to another. One method of offsetting is to identify a point to offset through, called a through point. Once an object is selected to offset, a through point is identified. The selected object offsets to the point shown in Figure 2–145. Refer to the following prompt sequence to use this method of the OFFSET command.

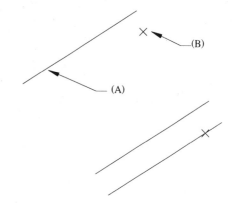

 Command: **OFFSET**

Offset distance or Through <>: **Through**
Select object to offset: *(Select the line at "A")*
Through point: **Nod**
of *(Select the point at "B")*
Select object to offset: *(Press* ENTER *to exit this command)*

Figure 2–145

Another method of offsetting is by a specified offset distance. In Figure 2–146, an offset distance of 0.50 is set. The line segment "A" is identified as the object to offset. To complete the command, you must identify a side to offset to give the offset a direction in which to operate. Follow the command sequence below for using this method of offsetting objects.

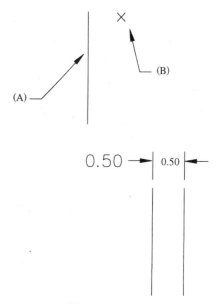

 Command: **OFFSET**

Offset distance or Through <>: **0.50**
Select object to offset: *(Select the line at "A")*
Side to offset: *(Mark a point at "B")*
Select object to offset: *(Press* ENTER *to exit this command)*

Figure 2–146

Another method of offsetting is shown in Figure 2–147 where the objects need to be duplicated at a set distance away from existing geometry. The COPY command could be used for this operation; a better command would be OFFSET. This allows you to specify a distance and a side for the offset to occur. The result is an object parallel to the original object at a specified distance. All objects in Figure 2–147 need to be offset 0.50 toward the inside of the original object. See the prompt sequences below.

Command:**OFFSET**

Offset distance or Through <0.00>: **0.50**
Select object to offset: *(Select the horizontal line at "A")*
Side to offset: *(Mark a point anywhere on the inside near "B")*
Repeat the above procedure for lines C through J.

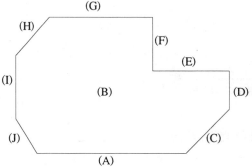

Figure 2–147

Notice that when all lines were offset, the entire original lengths of all line segments were maintained. Since all offsetting occurred inside, this resulted in segments that overlap at their intersection points (see Figure 2–148). In one case, at A and B, the lines did not even meet at all. The CHAMFER command may be used to edit all lines to form a sharp corner. This is accomplished by assigning chamfer distances of 0.00 units.

Command:**CHAMFER**

(TRIM mode) Current chamfer Dist1 = 0.5000, Dist2 = 0.5000
Polyline/Distance/Angle/Trim/Method/<Select first line>: **D** *(For Distance)*
Enter first chamfer distance <0.5000>: **0**
Enter second chamfer distance <0.0000>: *(Press ENTER)*

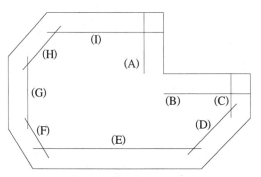

Figure 2–148

Command: **CHAMFER**

(TRIM mode) Current chamfer Dist1 = 0.0000,
 Dist2 = 0.0000
Polyline/Distance/Angle/Trim/Method/<Select
 first line>: *(Select the line at "A" in
 Figure 2–148)*
Select second line: *(Select the line at "B")*
Repeat the above procedure for lines "B"
 through "I" in Figure 2–148.

Using the OFFSET command along with the
CHAMFER command produces the result shown in
Figure 2–149. The Chamfer distances must be set to
a value of 0 for this special effect. The FILLET command performs the same result when set to a radius
value of 0.

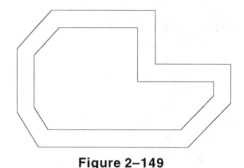

Figure 2–149

 # Using the PEDIT Command

Editing of polylines can lead to interesting results.
A few of these options will be explained in the following pages. Figure 2–150 is a polyline of width
0.00. The PEDIT command was used to change the
width of the polyline to 0.10 units. Refer to the following command sequence to use the PEDIT command along with the Width option.

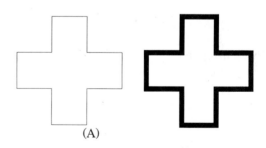

(A)

Figure 2–150

 Command: **PEDIT**

Select polyline: *(Select the polyline at "A")*
Close/Join/Width/Edit vertex/Fit/Spline/
 Decurve/Ltype gen/Undo/eXit <X>: **W** *(To
 edit the width of the polyline)*
Enter new width for all segments: **0.10**
Close/Join/Width/Edit vertex/Fit/Spline/
 Decurve/Ltype gen/Undo/eXit <X>: *(Press
 ENTER to exit this command)*

It is possible to convert regular objects into polylines. In Figure 2–151, the arc segment and individual line segments may be converted into a polyline. The circle is unable to be converted unless part of the

 Command: **PEDIT**

Select polyline: *(Select the line at "A")*
Object selected is not a polyline.
Do you want it to turn into one? **Yes**
Close/Join/Width/Edit vertex/Fit/Spline/
   Decurve/Ltype gen/Undo/eXit <X>: **J**
Select objects: *(Select lines "B" through "D")*
3 lines added to polyline.
Close/Join/Width/Edit vertex/Fit/Spline/
   Decurve/Ltype gen/Undo/eXit <X>: *(Press*
   ENTER *to exit this command)*

circle is broken resulting in an arc segment. Use the following prompts for converting the line segments into a polyline.

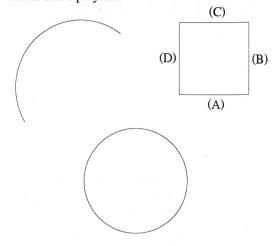

Figure 2–151

In the previous example, regular objects were selected individually before being converted into a polyline. For more complex objects, use the Window option to select numerous objects and perform

 Command: **PEDIT**

Select polyline: *(Select the line at "A")*
Object selected is not a polyline.
Do you want it to turn into one? **Yes**
Close/Join/Width/Edit vertex/Fit/Spline/
   Decurve/Ltype gen/Undo/eXit <X>: **J**
Select objects: *(Pick a point at "B")*
Other corner: *(Pick a point at "C")*
56 lines added to polyline.
Close/Join/Width/Edit vertex/Fit/Spline/
   Decurve/Ltype gen/Undo/eXit <X>: *(Press*
   ENTER *to exit this command)*

the PEDIT-Join operation which is faster. Refer to the following prompt sequence and Figure 2–152 to use this command.

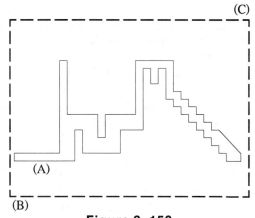

Figure 2–152

The polyline in Figure 2–153 will be used as an example of using the PEDIT command along with various curve fitting utilities. In Figures 2–154A and 2–154B, the Spline option and Fit Curve option are shown. The Spline option produces a smooth fitting curve based on control points in the form of

the vertices of the polyline. The Fit Curve option passes entirely through the control points producing a less desirable curve. Study Figures 2–154A and 2–154B that illustrate both curve options of the PEDIT command.

Spline Curve Generation

 Command:**PEDIT**

Select polyline: *(Select the polyline at "A")*
Close/Join/Width/Edit vertex/Fit/Spline/
   Decurve/Ltype gen/Undo/eXit <X>:**Spline**
Close/Join/Width/Edit vertex/Fit/Spline/
   Decurve/Ltype gen/Undo/eXit <X>: *(Press
   [ENTER] to exit this command)*

The original polyline frame is usually not visible when creating a spline and is shown only for illustrative purposes.

Figure 2–153

(A)

Figure 2–154A

Fit Curve Generation

 Command:**PEDIT**

Select polyline: *(Select the polyline at "B")*
Close/Join/Width/Edit vertex/Fit/Spline/
   Decurve/Ltype gen/Undo/eXit <X>: **Fit**
Close/Join/Width/Edit vertex/Fit/Spline/
   Decurve/Ltype gen/Undo/eXit <X>: *(Press
   [ENTER] to exit this command)*

(B)

Figure 2–154B

The Linetype Generation option of the PEDIT command controls the pattern of the linetype from polyline vertice to vertice. In the polyline at "C" in Figure 2–154C, the hidden linetype is generated from the first vertice to the second vertice. An en-

tirely different pattern is formed from the second vertice to the third vertice and so on. The polyline at "C" has the linetype generated throuthout the entire polyline. In this way, the hidden linetype is smoothed throughout the polyline.

 Command:**PEDIT**

Select polyline: *(Select the polyline at "D" in
   Figure 2–154C)*
Close/Join/Width/Edit vertex/Fit/Spline/
   Decurve/Ltype gen/Undo/eXit <X>: **Lt**
Full PLINE linetype ON/OFF <Off>: **On**
Close/Join/Width/Edit vertex/Fit/Spline/
   Decurve/Ltype gen/Undo/eXit <X>: *(Press
   [ENTER] to exit this command)*

(C)

(D)

Figure 2–154C

The object shown in Figure 2–155 is identical to the Figure 2–147. Also in Figure 2–147, each individual line had to be offset to copy the lines parallel at a specified distance. Then the CHAMFER command was used to clean up the corners. There is an easier way to perform this operation. First convert all individual line segments into one polyline using the PEDIT command.

 Command:**PEDIT**

Select polyline: *(Select the line at "A")*
Object selected is not a polyline.
Do you want it to turn into one? **Yes**
Close/Join/Width/Edit vertex/Fit/Spline/
    Decurve/Ltype gen/Undo/eXit <X>: **Join**
Select objects: *(Mark a point at "X")*
Other corner: *(Mark a point at "Y")*
Select objects: *(Press* ENTER *to continue)*
8 lines added to polyline.
Close/Join/Width/Edit vertex/Fit/Spline/
    Decurve/Ltype gen/Undo/eXit <X>: *(Press*
    ENTER *to exit this command)*

The OFFSET command is used to copy the shape 0.50 units on the inside. Since the object was converted into a polyline, all objects are offset at the same time. This procedure bypasses the need to use the CHAMFER or FILLET commands to corner all intersections. See Figure 2–156.

 Command:**OFFSET**

Offset distance or Through <0.00>: **0.50**
Select object to offset: *(Select the polyline at "A")*
Side to offset: *(Select a point anywhere near "B")*
Select object to offset: *(Press* ENTER *to exit this
    command)*

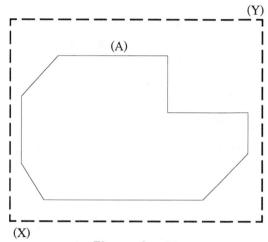

(Y)

(A)

(X)

Figure 2–155

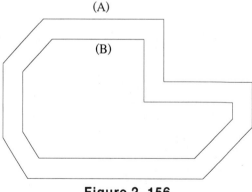

(A)

(B)

Figure 2–156

Using the ROTATE Command

The ROTATE command changes the orientation of an object or group of objects by identifying a base point and a rotation angle which completes the new orientation. Figure 2–157 shows an object complete with crosshatch pattern, which needs to be rotated to a 30-degree angle using point "A" as the base point. Use the following prompts and Figure 2–157 to perform the rotation.

ROTATE—Reference

At times it is necessary to rotate an object to a desired angular position. However, this must be accomplished even if the current angle of the object is unknown. To maintain the accuracy of the rotation operation, the Reference option of the ROTATE command is used. Figure 2–158 shows an object that needs to be rotated to the 30-degree angle position. Unfortunately, we do not know the angle the object currently lies in. Entering the Reference angle option and identifying two points creates a known angle of reference. Entering a new angle of 30 degrees rotates the object to the 30 degree position from the reference angle. Use the following prompts and Figure 2–158 to accomplish this.

 Command:**ROTATE**

Select objects: *(Select the object in Figure 2–158)*
Select objects: *(Press* ENTER *to continue)*
Base point: **Cen**
of *(Pick either the circle or two arc segments)*
<Rotation angle>/Reference:**R** *(For Reference)*
Reference angle <0>: **Cen**
of *(Again pick either the circle or the two arc segments)*
Second point: **Mid**
of *(Pick the line "A" to establish the reference angle)*
New angle: **30**

 Command:**ROTATE**

Select objects: **All** *(This will select all objects in Figure 2–157)*
Select objects: *(Press* ENTER *to continue)*
Base point: **Endp**
of *(Select the endpoint of the line at "A")*
<Rotation angle>/Reference:**30**

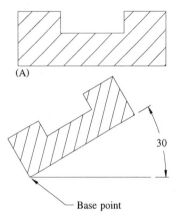

(A)

30

Base point

Figure 2–157

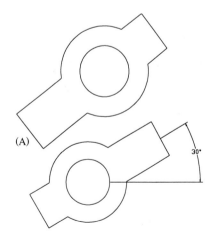

(A)

30°

Figure 2–158

Using the SCALE Command

Use the SCALE command to change the overall size of an object. The size may be larger or smaller in relation to the original object or group of objects. The SCALE command requires a base point and scale factor to complete the command. Study Figures 2–159A through 2–159C to perform the SCALE command.

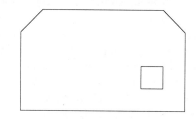

Figure 2–159A

Using a base point at "A" and a scale factor of 0.50, the results of using the SCALE command on a group of objects are shown in Figures 2–159B and 2–159C.

 Command:**SCALE**

Select objects: **All** *(This will select all dashed objects in Figure 2–159B)*
Select objects: *(Press* ENTER *to continue)*
Base point: **Endp**
of *(Select the endpoint of the line at "A")*
<Rotation angle>/Reference:**0.50**

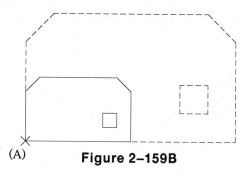

Figure 2–159B

The example in Figure 2–159C shows the effects of identifying a new base point in the center of the object.

 Command:**SCALE**

Select objects: **All** *(This will select all dashed objects shown in Figure 2–159C)*
Select objects: *(Press* ENTER *to continue)*
Base point: *(Mark a point near "A")*
<Rotation angle>/Reference:**0.40**

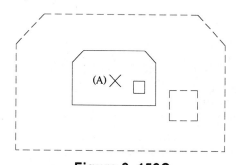

Figure 2–159C

Scale—Reference

Suppose the length of a line needs to be increased to a known length; however, you do not know the exact length of the line. The Reference option of the SCALE command can be used to identify endpoints of a line segment that act as a reference length. Entering a new length value could increase or decrease the line depending on the desired effect. Study Figure 2–160 and the following prompts.

 Command:**SCALE**
Select objects: *(Select line segment "A")*
Select objects: *(Press [ENTER] to continue)*
Base point: **Endp**
of *(Pick the endpoint at "A")*
<Scale factor>/Reference: **R** *(For Reference)*
Reference length <1>: **Endp**
of *(Pick the endpoint at "A")*
Second point: **Endp**
of *(Pick the endpoint at "B")*
New length: **10.0000**

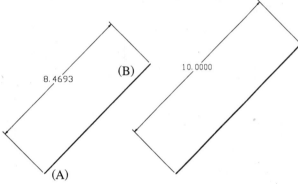

Figure 2–160

Using the STRETCH Command

Use the STRETCH command to move a portion of a drawing while still preserving the connections to parts of the drawing remaining in place. To perform this type of operation, the Crossing option of "Select objects" must be used. In Figure 2–161, a group of objects is selected using the crossing box. Next a base point is identified by the endpoint at "C." Finally, a second point of displacement is identified using a polar coordinate; the direct distance mode could also be used. Once the objects selected in the crossing box are stretched, the objects not only move to the new location but also mend themselves.

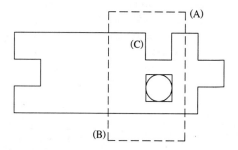

 Command:**STRETCH**

Select objects to stretch by crossing-window or
    crossing-polygon...
Select objects: *(Mark a point at "A")*
Other corner: *(Mark a point at "B")*
Base point of displacement: **Endp**
of *(Select the endpoint of the line at "C")*
Second point of displacement:**@1.75<180**

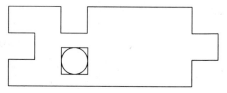

Figure 2–161

Figure 2–162 is another example of using the STRETCH command. The crossing window is employed along with a base point at "C" and a polar coordinate.

 Command:**STRETCH**

Select objects to stretch by crossing-window or
    crossing-polygon...
Select objects: *(Mark a point at "A")*
Other corner: *(Mark a point at "B")*
Base point of displacement: **Endp**
of *(Select the endpoint of the line at "C")*
Second point of displacement:**@2.75<0**

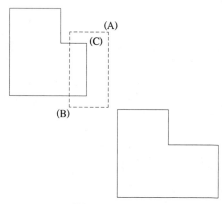

Figure 2–162

Applications of the STRETCH command include Figure 2–163 where a window needs to be positioned at a new location. Use the following command sequence to stretch the window at a set distance using a polar coordinate. Also see Figure 2–163.

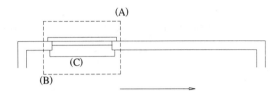

 Command:**STRETCH**

Select objects to stretch by crossing-window or
    crossing-polygon...
Select objects: *(Mark a point at "A")*
Other corner: *(Mark a point at "B")*
Base point of displacement: **Mid**
of *(Select the endpoint of the line at "C")*
Second point of displacement: **@10'6<0**

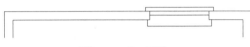

Figure 2–163

 Using the TRIM Command

Use the TRIM command to partially delete an object or group of objects based on a cutting edge. In Figure 2–164, the four dashed lines are selected as cutting edges. Next, segments of the circles are selected that trim out between the line segment cutting edges.

 Command:**TRIM**

Select cutting edges:(Projmode = UCS,
    Edgemode = No extend)
Select objects: *(Select the 4 dashed lines
    shown in Figure 2–164)*
Select objects: *(Press* ENTER *to continue)*
<Select object to trim>/Project/Edge/Undo:
    (Select the circle at "A")
<Select object to trim>/Project/Edge/Undo:
    (Select the circle at "B")
<Select object to trim>/Project/Edge/Undo:
    (Select the circle at "C")
<Select object to trim>/Project/Edge/Undo:
    (Select the circle at "D")
<Select object to trim>/Project/Edge/Undo:
    (Press ENTER *to exit this command)*

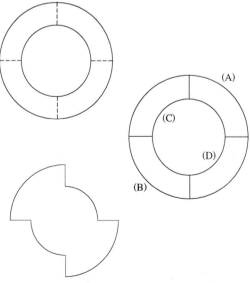

Figure 2–164

An alternate method of selecting cutting edges is to press the ENTER key in response to the prompt "Select objects:." This automatically creates cutting edges out of all objects in the drawing. When using this method, the cutting edges do not highlight.

In Figure 2–165, even though the four dashed lines are selected as cutting edges, the middle of the cutting edge may be trimmed out at all locations identified by "A" through "D."

 Command:**TRIM**

Select cutting edges:(Projmode = UCS,
  Edgemode = No extend)
Select objects: *(Press ENTER to select all four
  dashed lines as cutting edges)*
<Select object to trim>/Project/Edge/Undo:
  (Select the segment at "A")
<Select object to trim>/Project/Edge/Undo:
  (Select the segment at "B")
<Select object to trim>/Project/Edge/Undo:
  (Select the segment at "C")
<Select object to trim>/Project/Edge/Undo:
  (Select the segment at "D")
<Select object to trim>/Project/Edge/Undo:
  (Select ENTER to exit this command)

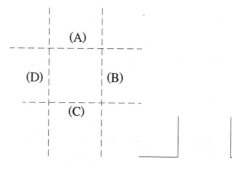

Figure 2–165

Yet another application of the TRIM command uses the Fence option of "Select objects." First, invoke the TRIM command and select the small circle as the cutting edge. Begin the prompt of "Select object to trim" with "Fence." See Figure 2–166.

 Command:**TRIM**

Select cutting edges:(Projmode = UCS,
  Edgemode = No extend)
Select objects: *(Select the small circle)*
Select objects: *(Press ENTER to continue)*
<Select object to trim>/Project/Edge/Undo:**F**
  (For Fence)

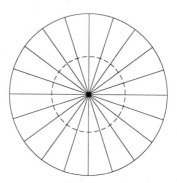

Figure 2–166

Continue with the TRIM command by identifying a Fence. This consists of a series of line segments that take on a dashed appearance. This means the fence will select any object it crosses. When completed with the construction of the desired fence shown in Figure 2–167, Press the ENTER key.

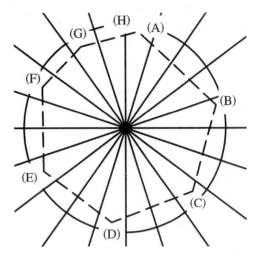

Figure 2–167

First fence point: *(Mark a point at "A" beginning the Fence)*
Undo/<Endpoint of line>: *(Mark a point at "B")*
Undo/<Endpoint of line>: *(Mark a point at "C")*
Undo/<Endpoint of line>: *(Mark a point at "D")*
Undo/<Endpoint of line>: *(Mark a point at "E")*
Undo/<Endpoint of line>: *(Mark a point at "F")*
Undo/<Endpoint of line>: *(Mark a point at "G")*
Undo/<Endpoint of line>: *(Mark a point at "H")*
Undo/<Endpoint of line>: *(Press* ENTER *to end the Fence and execute the TRIM command)*
<Select object to trim>/Project/Edge/Undo: *(Press* ENTER *to exit this command)*

The power of the Fence option of "Select objects" is shown in Figure 2–168. Rather than select each individual line segment inside the small circle to trim, the Fence trims all objects it touches in relation to the cutting edge.

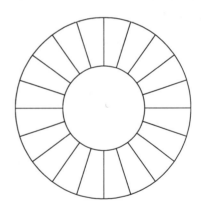

Figure 2–168

The TRIM command also allows you to trim to an extended cutting edge. In extended cutting edge mode, an imaginary cutting edge is formed; all objects sliced along this cutting edge will be trimmed if selected individually or by the Fence mode. Study Figure 2–169 and the following prompts on this feature of the TRIM command.

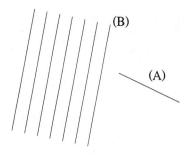

 Command: **TRIM**

Select cutting edges:(Projmode = UCS,
   Edgemode = No extend)
Select objects: *(Pick line segment "A")*
Select objects: *(Press* ⏎ *to continue)*
<Select object to trim>/Project/Edge/Undo: **E**
   (For Edge)
Extend/No extend <No extend>: **E** *(For*
   Extend)
<Select object to trim>/Project/Edge/Undo:
   (Pick line "B" along with the other segments)
<Select object to trim>/Project/Edge/Undo:
   (Press ⏎ *to exit this command)*

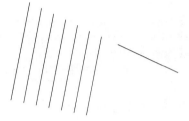

Figure 2–169

The Fence mode can also be used to select all line segments at once to trim.

Using Object Grips

An alternate method of editing may be performed using object grips. The grip is a small rectangle appearing at key object locations such as the endpoints and midpoints of lines and arcs or the center and quadrants of circles. When grips are enabled, a grip pickbox is displayed at the intersection of the crosshairs shown in Figure 2–170. Once grips are selected, the object may be stretched, moved, rotated, scaled, or mirrored. Grips are at times referred to as visual Osnaps since the cursor automatically snaps to all grips displayed along an object. Displayed in Figure 2–171 are other examples of grip locations on various objects.

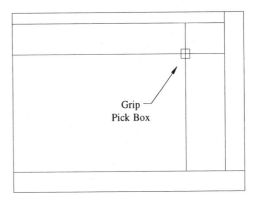

Figure 2–170

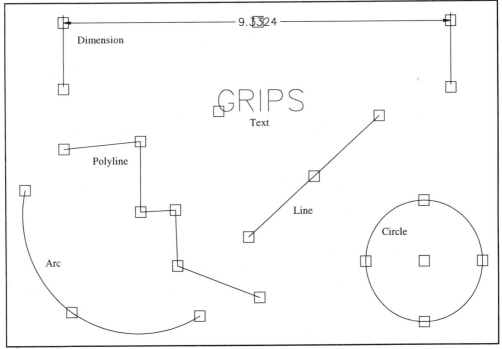

Figure 2–171

The Grips dialog box is available to change settings, color, and grip size. Choose this dialog box from the "Tools" area of the pull-down menu (see Figure 2–172). By default, grips are enabled; the presence of the check in the box "Enable Grips" will display grips when selecting an object. Also by default, a grip is placed at the insertion point when a block is selected. Check "Enable Grips Within Blocks" if you want grips to be displayed along all objects within the block. Color is applied to selected and unselected grips. Selecting "Unselected..." or "Selected..." displays a color dialog box used to change the color of selected or unselected grips. Use the "Grip Size" area to move a slider bar left or right and visually see the size of the grip in the box at the far right. Moving the slider to the left makes the grip smaller; moving to the right makes the grip larger.

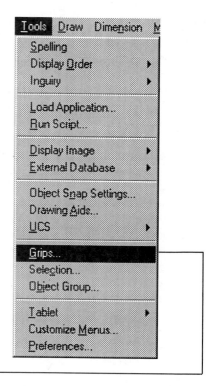

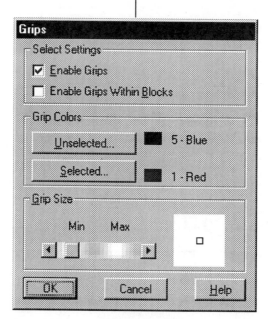

Figure 2–172

Object Grip Modes

Figure 2–173 shows the three types of grips. When an object is first selected using the grip pickbox located at the intersection of the crosshairs, the object highlights and the square grips are displayed; this type of grip is called a warm grip. The entire object is subject to the many grip edit commands. When one of the grips is selected, it fills in with the current selected grip color; this type of grip is called a hot grip. Once a hot grip is selected, the following prompts appear at the command prompt:

STRETCH
<STRETCH to point>/Base point/Copy/Undo/
  eXit: *(Press the spacebar)*
MOVE
<Move to point>/Base point/Copy/Undo/eXit:
  (Press the spacebar)
ROTATE
<Rotation angle>/Base point/Copy/Undo/
  Reference/eXit: *(Press the spacebar)*
SCALE
<Scale factor>/Base point/Copy/Undo/
  Reference/eXit: *(Press the spacebar)*
MIRROR
<Second point>/Base point/Copy/Undo/eXit:

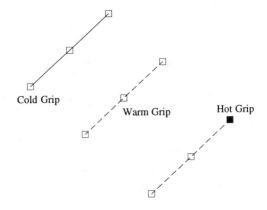

Figure 2–173

To move from one edit command mode to another, press the spacebar. Once an editing command is completed, pressing the ESC key once removes the highlight from the object but leaves the grip; this type of grip is considered cold. Pressing the ESC key again removes the grips from the object. Figure 2–174 shows various examples of each editing mode.

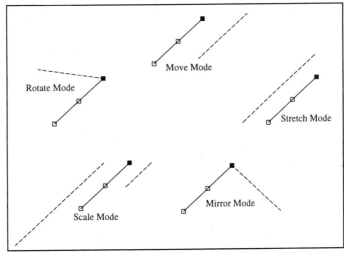

Figure 2–174

Activating the Grip Cursor Menu

In Figure 2–175, a horizontal line has been selected and grips appear. The right most endpoint grip has been selected. Rather than use the Spacebar to scroll through the various grip modes, click the right mouse button. Notice a cursor menu on grips appears. This provides an easier way of navigating from one Grip mode to another.

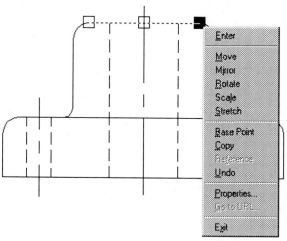

Figure 2–175

Using the Grip—STRETCH Mode

The STRETCH mode of grips operates similar to the normal STRETCH command. Use STRETCH mode to move an object or group of objects and have the results mend themselves similar to Figure 2–176. The line segments "A" and "B" are both too long by one unit. To decrease these line segments by one unit, use the STRETCH mode by selecting lines "A," "B," and "C" with the grip pickbox at the command prompt. Next, while holding down the Shift key, select the warm grips "D," "E," and "F." This will make the grips hot and ready for the Stretch operation. Lift off the SHIFT key and pick the grip at "E" again. The STRETCH mode appears in the command prompt area. Since these hot grips are considered base points, entering a polar coordinate value of @1.00<180 will stretch the three highlighted grip objects to the left at a distance of one unit. The direct distance mode could also be used with grips. To remove the object highlight and grips, press the ESC key twice at the command prompt.

Command: (Select the three dashed lines
　　shown in Figure 2–176. Then, while holding
　　down the Shift key, select the warm grips at
　　"D," "E," and "F" to make them hot. Lift off the
　　SHIFT key and pick the grip at "E" again)
STRETCH
<STRETCH to point>/Base point/Copy/Undo/
　　eXit:@1<180
Command: (Press ESC to remove object
　　highlight)
Command: (Press ESC to remove the grips)

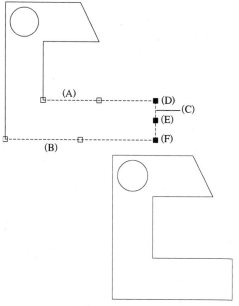

Figure 2–176

Using the Grip—SCALE Mode

Using the SCALE mode of object grips allows an object to be uniformly scaled in both the X and Y directions. This means that a circle, such as the one shown in Figure 2–177, cannot be converted into an ellipse by using different X and Y values. As the warm grip is converted into a hot grip, any cursor movement will drag the scale factor until a point is marked where the object will be scaled to that factor. Figure 2–177 and the following prompt are examples of an absolute value to perform the scaling operation of half the circle's normal size.

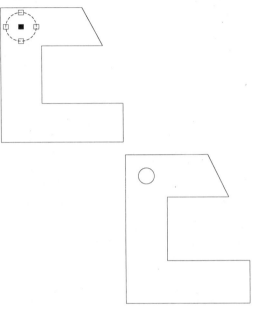

Command: *(Select the circle to enable grips, then select the warm grip at the center of the circle to make it hot. Press the Spacebar until the SCALE mode appears at the bottom of the prompt line or press the right mouse button to activate the grip cursor menu to pick Scale)*

SCALE
<Scale factor>/Base point/Copy/Undo/
    Reference/eXit:**0.50**
Command: *(Press ESC to remove object highlight)*
Command: *(Press ESC to remove the grips)*

Figure 2–177

Using the Grip—MOVE/COPY Mode

The multiple copy option of the MOVE mode is demonstrated by copying the circle shown in Figure 2–178 using polar coordinates at distances 2.50 and 5.00 both in the 270 direction. This multiple Copy option is actually disguised under the command options of object grips.

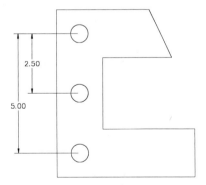

Figure 2–178

Select the circle, then the center grip of the circle to convert the warm grip into a hot grip. Use the Spacebar to scroll past the STRETCH mode to the MOVE mode. Issue a Copy inside of the MOVE mode to be placed in Multiple MOVE mode. See Figures 2–179A and 2–179B.

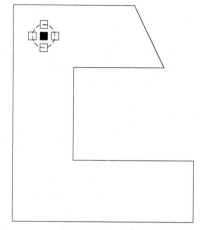

Figure 2–179A

Command: *(Select the circle to activate the grips at the center and quadrants; select the warm grip at the center of the circle to make it hot. Then press the Spacebar until the MOVE mode appears at the bottom of the prompt line or press the right mouse button to activate the grip cursor menu to pick Move)*
MOVE
<Move to point>/Base point/Copy/Undo/eXit: **C**
  (For Copy)
MOVE (multiple)
<Move to point>/Base point/Copy/Undo/eXit:
  @2.50<270
MOVE (multiple)
<Move to point>/Base point/Copy/Undo/eXit:
  @5.00<270
MOVE (multiple)
<Move to point>/Base point/Copy/Undo/eXit: **X**
  (To exit)
Command: *(Press ESC to remove object highlight)*
Command: *(Press ESC to remove the grips)*

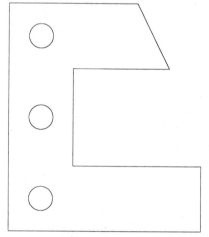

Figure 2–179B

Using the Grip—MIRROR Mode

Use the grip MIRROR mode to flip an object along a mirror line similar to the one used in the regular MIRROR command. Follow the prompts for the MIRROR option if an object needs to be mirrored but the original does not need to be saved. This performs the mirror, but does not produce a copy of the original. If the original object needs to be saved during the mirror operation, use the Copy option of MIRROR mode. This places you in multiple MIRROR mode. Locate a base point and a second point to perform the mirror. See Figure 2–180.

Command: *(Select the circle in Figure 2–180 to enable grips, then select the warm grip at the center of the circle to make it hot. Press the Spacebar until the MIRROR mode appears at the bottom of the prompt line or press the right mouse button to activate the grip cursor menu to pick Mirror)*
MIRROR
<Second point>/Base point/Copy/Undo/eXit: **C**
  (For Copy)
MIRROR (multiple)
<Second point>/Base point/Copy/Undo/eXit: **B**
  (For Base Point)
Base point: **Mid**
of *(Pick the midpoint at "A")*
MIRROR (multiple)
<Second point>/Base point/Copy/Undo/eXit:
  @1<90
MIRROR (multiple)
<Second point>/Base point/Copy/Undo/eXit: **X**
  (To Exit)
Command: *(Press ESC to remove object highlight)*
Command: *(Press ESC to remove the grips)*

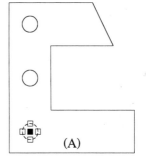

(A)

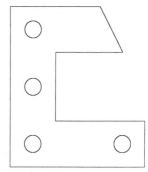

Figure 2–180

Using the Grip—ROTATE Mode

Numerous grips may be selected by window or crossing boxes. At the command prompt, pick a blank part of the screen; this should place you in Window/Crossing selection mode. Picking up or below and to the right of the previous point places you in Window selection mode; picking up or below and to the left of the previous point places you in Crossing selection mode. This method is used on all objects shown in Figure 2–181. Selecting the lower left grip and using the Spacebar to advance to the **ROTATE** option allows all objects to be rotated at a defined angle in relation to the previously selected grip.

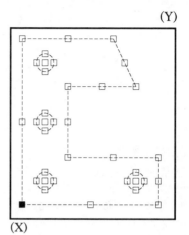

(Y)

(X)

Command: *(Pick near "X," then near "Y" to create a window selection set and enable all grips in all objects. Select the warm grip at the lower left corner of the object to make it hot. Then press the Spacebar until the ROTATE mode appears at the bottom of the prompt line or press the right mouse button to activate the grip cursor menu to pick Rotate)*
ROTATE
<Rotation angle>/Base point/Copy/Undo/ Reference/eXit: **30**
Command: *(Press ESC to remove object highlight)*
Command: *(Press ESC to remove the grips)*

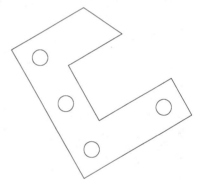

Figure 2–181

Using the Grip—Multiple ROTATE Mode

One problem of the regular ROTATE command is that although an object can be rotated, the original location can not be saved. You have to use points to mark the original location of an object before it is rotated. The ARRAY command has been used as a substitute to rotate and copy an object. Now with object grips, you may use ROTATE mode to rotate and copy an object without the use of reference points or the ARRAY command. Figure 2–182A is a line that needs to be rotated and copied at a 40-degree angle. With a positive angle, the direction of the rotation will be counterclockwise.

Selecting the line in Figure 2–182A enables grips located at the endpoints and midpoint of the line. Selecting the warm grip makes it hot. This hot grip also locates the vertex of the required angle. Press the Spacebar until the ROTATE mode is reached. Enter Multiple ROTATE mode by entering Copy when prompted in the command sequence below. Finally enter 40 to produce a copy of the original line segment at a 40-degree angle in the counterclockwise direction. (See Figure 182B.)

Command: *(Select line segment "A"; then select the warm grip at "B" to make it hot. Press the Spacebar until the ROTATE mode appears at the bottom of the prompt line or press the right mouse button to activate the grip cursor menu to pick Rotate)*
ROTATE
<Rotation angle>/Base point/Copy/Undo/Reference/eXit: **C** *(For Copy)*
ROTATE (multiple)
<Rotation angle>/Base point/Copy/Undo/Reference/eXit:**30**
<Rotation angle>/Base point/Copy/Undo/Reference/eXit: **X** *(To Exit)*
Command: *(Press ESC to remove object highlight)*
Command: *(Press ESC to remove the grips)*

The result is illustrated in Figure 2–182C.

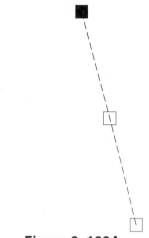

Figure 2–182A

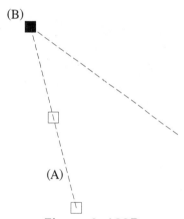

Figure 2–182B

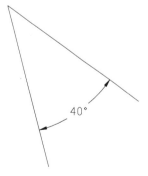

Figure 2–182C

Multiple Copy Mode and Offset Snap Locations for Rotations

All Multiple Copy modes inside of grips may be op-
erated in a snap location mode while holding down
the Shift key. Here is how it works. In Figure 2–183A,
the vertical centerline and circle are selected using
the grip pickbox. The objects highlight and the grips
appear. A multiple copy of the selected objects needs
to be made at an angle of 45 degrees. The grip RO-
TATE mode is used in Multiple Copy mode.

Command: *(Select centerline segment "A" and*
    circle "B"; then select the warm grip at the
    center of the circle to make it hot. Press the
    Spacebar until the ROTATE mode appears
    at the bottom of the prompt line)
ROTATE
<Rotation angle>/Base point/Copy/Undo/
    Reference/eXit: **C** *(To Copy)*
ROTATE (multiple)
<Rotation angle>/Base point/Copy/Undo/
    Reference/eXit: **B** *(For Base Point)*
Base point: **Cen**
of *(Select the circle at "C" to snap to the center*
    of the circle)
ROTATE (multiple)
<Rotation angle>/Base point/Copy/Undo/
    Reference/eXit: **45**

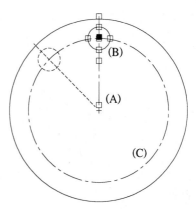

Figure 2–183A

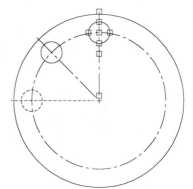

Figure 2–183B

Rather than enter another angle to rotate and copy the same objects, the Shift key is held down placing you in offset snap location mode. Moving the cursor snaps the selected objects. This keeps the value of the original angle of rotation, namely 45-degrees.

ROTATE (multiple)
<Rotation angle>/Base point/Copy/Undo/
   Reference/eXit: *(Hold down the Shift key and
   move the circle and centerline until it snaps
   to the next 45 degree position shown in
   Figure 2–183B)*
ROTATE (multiple)
<Rotation angle>/Base point/Copy/Undo/
   Reference/eXit: **X**

Command: *(Press ESC to remove object
   highlight)*

Command: *(Press ESC to remove the grips)*

The Rotate-Copy-Snap Location mode could allow you to create Figure 2–183C without the aid of the ARRAY command. Since all angle values are 45 degrees, continue holding down the Shift key to snap to the next 45-degree location and mark a point to place the next group of selected objects.

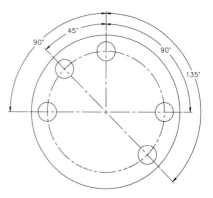

Figure 2–183C

Multiple Copy Mode and Offset Snap Locations for Moving

As with the previous example of using Offset Snap Locations for ROTATE Mode, these same snap locations apply to MOVE mode. Figure 2–184A shows two circles along with a common centerline. The circles and centerline are selected using the grip pickbox, which highlights these three objects and activates the grips. The intent is to move and copy the selected objects at two unit increments.

Command: *(Select the two circles and centerline to activate the grips; select the warm grip at the midpoint of the centerline to make it hot. Then press the Spacebar until the MOVE mode appears at the bottom of the prompt line or press the right mouse button to activate the grip cursor menu to pick Move)*
MOVE
<Move to point>/Base point/Copy/Undo/eXit: **C**
 (To Copy)
MOVE (multiple)
<Move to point>/Base point/Copy/Undo/eXit:
 @2.00<270

In Figure 2–184B, instead of remembering the previous distance and entering it to create another copy of the circles and centerline, hold down the Shift key and move the cursor down to see the selected objects snap to the previous distance.

MOVE (multiple)
<Move to point>/Base point/Copy/Undo/eXit:
 (Hold down the Shift key and move the cursor down to have the selected entities snap to another 2.00-unit distance)
MOVE (multiple)
<Move to point>/Base point/Copy/Undo/eXit: **X**
 (To exit)
Command: *(Press ESC to remove object highlight)*
Command: *(Press ESC to remove the grips)*

The completed hole layout is shown in Figure 2–184C using the Offset Snap Location method of object grips.

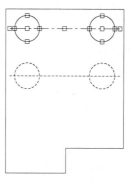

Figure 2–184A

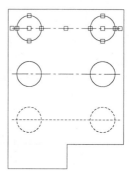

Figure 2–184B

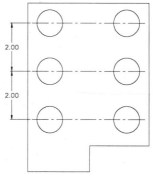

Figure 2–184C

Tutorial Exercise
Tile.Dwg

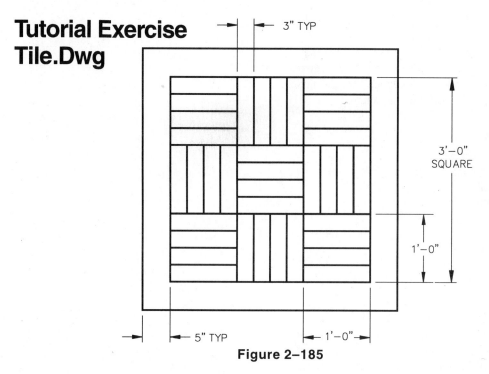

Figure 2–185

Purpose

This tutorial is designed to use the OFFSET and TRIM commands to complete the drawing of the floor tile shown in Figure 2–185.

System Settings

Use the UNITS command and change the units of MEASURE from decimal to architectural units. Keep the remaining default settings. Use the LIMITS command and change the limits of the drawing to (0,0) for the lower left corner and (10',8') for the upper right corner. Use the SNAP command and change the value from 1/2" to 4". (If the GRID command is set to 0, the snap setting of 4" will also change the grid spacing to 4").

Layers

Create the following layers with the format :

Name	Color	Linetype
Object	White	Continuous

Suggested Commands

The LINE command will be used to begin the Tile. The OFFSET command is used to copy selected line segments at a specified distance. The TRIM command is then used to clean up intersecting corners. The ERASE command can be used to delete objects from the drawing (Remember to use the OOPS command to bring back previously erased objects deleted by mistake.)

Whenever possible, substitute the appropriate command alias in place of the full AutoCAD command in each tutorial step. For example, use "Co" for the COPY command, "L" for the LINE command, and so on. The complete listing of all command aliases is located on pages 1–9 and 1–10.

Step #1

Begin this exercise by using the LINE command
and polar coordinate mode to draw a 3'0" square
shown in Figure 2–186. The direct distance mode
can also be used to construct the square.

 Command: **LINE**

From point: **12,12**
To point: **@3'<0**
To point: **@3'<90**
To point: **@3'<180**
To point: **C** *(To close the figure and exit the
    command)*

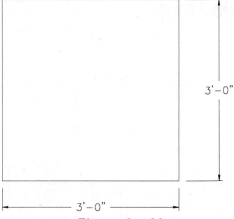

3'–0"

3'–0"

Figure 2–186

Step #2

Use the ARRAY command to copy the top line in a
rectangular pattern and have all lines spaced 3 units
away from each other. Select the top line as the object
to array and perform a rectangular array consisting
of twelve rows and one column. See Figure 2–187.
Since the top line selected will be copied straight
down, a negative distance must be entered to perform this operation. Another popular command that
could be used here is OFFSET. However since each
line must be offset separately, the ARRAY command is the more efficient command to use.

 Command: **ARRAY**

Select objects: *(Select the top horizontal line
    at "A")*
Select objects: *(Press* ENTER *to continue with this
    command)*
Rectangular or Polar array (<R>/P): **R**
Number of rows (---) <1>: **12**
Number of columns (|||) <1>: **1**
Unit cell or distance between rows (---): **-3**

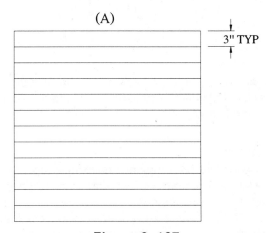

(A)

3" TYP

Figure 2–187

Step #3

Again use the ARRAY command but this time copy the right vertical line in a rectangular pattern and have all lines spaced 3 units away from each other. Select the right line as the object to array and perform a rectangular array consisting of one row and twelve columns (see Figure 2–188). Since the right line selected will be copied to the left, you must enter a negative distance to perform this operation.

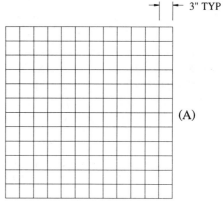

Figure 2–188

 Command: **ARRAY**

Select objects: *(Select the right vertical line at "A")*
Select objects: *(Press* ⸢ENTER⸣ *to continue with this command)*
Rectangular or Polar array (<R>/P): **R**
Number of rows (---) <1>: **1**
Number of columns (|||) <1>: **12**
Distance between columns (|||): **-3**

Step #4

Use the TRIM command. Select the two vertical dashed lines shown in Figure 2–189 as cutting edges; use Figure 2–190 as a guide for determining which lines to trim out.

 Command: **TRIM**

Select cutting edges: (Projmode = UCS, Edgemode = No extend)
Select objects: *(Select the two dashed lines shown in Figure 2–189)*
Select objects: *(Press* ⸢ENTER⸣ *to continue with this command)*
<Select object to trim>/Project/Edge/Undo: *(Continue to Step #5)*

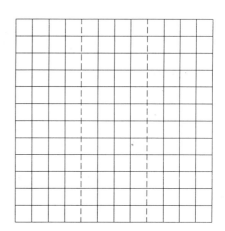

Figure 2–189

Step #5

For the last prompt of the TRIM command in Step #4, select all horizontal lines in the areas marked "A," "B," "C," and "D" in Figure 2–190. When finished selecting the objects to trim, press the ENTER key to exit the command.

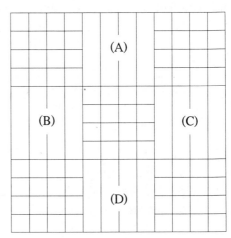

<Select object to trim>/Project/Edge/Undo:
    (Select all horizontal lines in area "A")
<Select object to trim>/Project/Edge/
    Undo:*(Select all horizontal lines in area "B")*
<Select object to trim>/Project/Edge/
    Undo:*(Select all horizontal lines in area "C")*
<Select object to trim>/Project/Edge/
    Undo:*(Select all horizontal lines in area "D")*
<Select object to trim>/Project/Edge/Undo:
    (Press ENTER *to exit this command)*

Figure 2–190

Step #6

Use the TRIM command and select the two horizontal dashed lines as cutting edges in Figure 2–191.

 Command:**TRIM**

Select cutting edges: (Projmode = UCS,
    Edgemode = No extend)
Select objects: *(Select the two dashed lines in
    Figure 2–191)*
Select objects: *(Press* ENTER *to continue with this
    command)*
<Select object to trim>/Project/Edge/Undo:
    (Continue to Step #7)

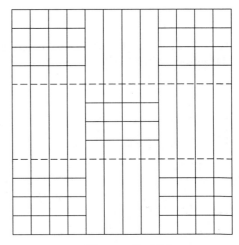

Figure 2–191

Step #7

For the last step of the TRIM command in Step #6, select all vertical lines in the areas marked "A," "B," "C," "D," and "E" in Figure 2–192. When finished selecting the objects to trim, press the ENTER key to exit the command.

<Select object to trim>/Project/Edge/Undo:
   (Select all vertical lines in area "A")
<Select object to trim>/Project/Edge/
   Undo:*(Select all vertical lines in area "B")*
<Select object to trim>/Project/Edge/
   Undo:*(Select all vertical lines in area "C")*
<Select object to trim>/Project/Edge/
   Undo:*(Select all vertical lines in area "D")*
<Select object to trim>/Project/Edge/
   Undo:*(Select all vertical lines in area "E")*
<Select object to trim>/Project/Edge/Undo:
   (Press [ENTER] to exit this command)

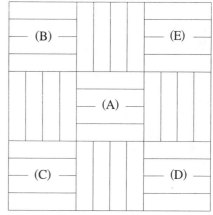

Figure 2–192

Step #8

Use the OFFSET command to offset the lines "A," "B," "C," and "D" five units in the directions shown in Figure 2–193.

 Command:**OFFSET**

Offset distance or Through <0'-1">:**5**
Select object to offset: *(Select the line at A)*
Side to offset? *(Pick a point above the line)*
Select object to offset: *(Select the line at B)*
Side to offset? *(Pick a point right of the line)*
Select object to offset: *(Select the line at C)*
Side to offset? *(Pick a point below the line)*
Select object to offset: *(Select the line at D)*
Side to offset? *(Pick a point left of the line)*
Select object to offset: *(Press [ENTER] to exit this
   command)*

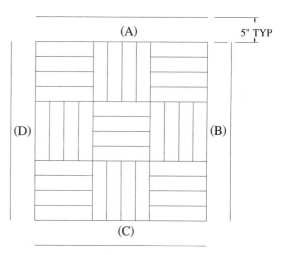

Figure 2–193

Step #9

Use the FILLET command set to a radius of 0 to place a corner at the intersection of lines "A" and "B." See Figure 2–194. The radius should already be set to 0 by default so simply pick the two lines and the corner is formed.

 Command:**FILLET**

(TRIM mode) Current fillet radius = 0'-0 1/2"
Polyline/Radius/Trim/<Select first object>:**R**
Enter fillet radius <0'-0 1/2">:**0**

 Command:**FILLET**

(TRIM mode) Current fillet radius = 0'-0"
Polyline/Radius/Trim/<Select first object>:
  (Select line "A")
Select second object: *(Select line "B")*

Repeat this procedure for the other three corners.

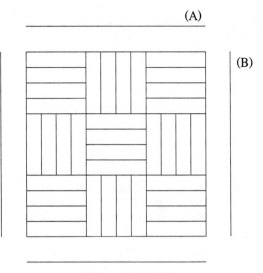

Figure 2–194

Step #10

The completed tile drawing is shown in Figure 2–195. Follow Step #11 to add more tiles in a rectangular pattern using the ARRAY command.

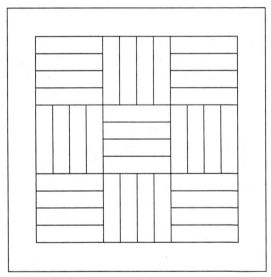

Figure 2–195

Step #11

As an alternate step, use the ARRAY command to copy the initial design in a rectangular pattern by row and column. Use the following prompts to perform this operation to construct a series of tiles as in Figure 2–196.

Figure 2–196

 Command: **ARRAY**

Select objects: *(Select the entire tile design from Step #10 using the All option)*
Select objects: *(Press ENTER to continue with this command)*
Rectangular or Polar array (<R>/P): **R**
Number of rows(---) <1>: **3**
Number of columns (III) <1>: **2**
Unit cell or distance between rows (---): **3'10**
Distance between columns (III): **3'10**

Tutorial Exercise
Inlay.Dwg

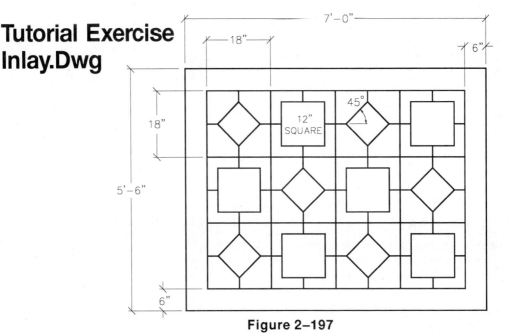

Figure 2–197

Purpose

This tutorial is designed to allow you to construct a drawing of the inlay shown in Figure 2–197 using the COPY command and the Multiple modifier.

System Settings

Use the UNITS command and change the units of measure from decimal to architectural units. Keep the remaining default settings. Use the LIMITS command and change the limits of the drawing to (0,0) for the lower left corner and (15'6,"9'6") for the upper right corner. Use the SNAP command and change the value from 1.0000 to 3". (If the Grid command is set to 0, the snap setting of 3" will also change the grid spacing to 3".)

Layers

Create the following layers with the format :

Name	Color	Linetype
Object	White	Continuous

Suggested Commands

Begin this tutorial by drawing a 6'0" x 4'6" rectangle using the LINE command. Offset the edges of the rectangle by a distance of 18". Then, using the 3" grid as a guide along with the Snap-on, draw the diamond and square shapes. Use the COPY command along with the Multiple modifier to copy the diamond and square shapes numerous times at the designated areas.

Whenever possible, substitute the appropriate command alias in place of the full AutoCAD command in each tutorial step. For example, use "Co" for the COPY command, "L" for the LINE command, and so on. The complete listing of all command aliases is located on pages 1–9 and 1–10.

Step #1

Begin this exercise by using the LINE command and polar coordinate mode to draw a rectangle 6'0" by 4'6" shown in Figure 2–198. The direct distance mode may also be used to construct the rectangle.

 Command: **LINE**

From point: **12,12**
To point: **@6'<0**
To point: **@4'6<90**
To point: **@6'<180**
To point: **C** *(To close the figure and exit the command)*

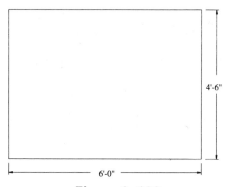

Figure 2–198

Step #2

Use the ARRAY command to copy the top line in a rectangular pattern and have all lines spaced 18 units away from each other. Select the top horizontal line at "A" as the object to array. Since this line will be copied straight down, a value of -18 for the spacing between rows will perform this operation. Repeat this command to copy the right vertical line at "B" 3 times to the left at a distance of 18 units. Enter a value of -18 units for the spacing in between columns since the copying is performed in the left direction as shown in Figure 2–199.

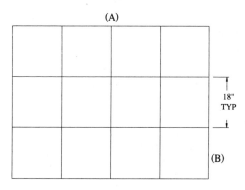

Figure 2–199

 Command: **ARRAY**

Select objects: *(Select the top horizontal line at "A")*
Select objects: *(Press* ENTER *to continue with this command)*
Rectangular or Polar array (<R>/P): **R**
Number of rows (---) <1>: **3**
Number of columns (|||) <1>: **1**
Unit cell or distance between rows (---): **-18**

Repeat the preceeding Rectangular ARRAY command above for the vertical line at "B." Use 1 for the number of rows, 4 for the number of columns, and -18 as the distance between columns.

Step #3

Begin drawing one 12" x 12" diamond figure in the position shown in Figure 2–200. Use the ZOOM command along with the Window option to magnify the area around the position of the diamond figure.

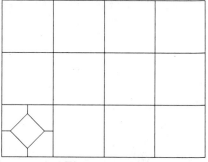

Figure 2–200

Step #4

Be sure the Grid and Snap values are set to 3"; the Snap should already be turned on. Then use the LINE command to draw the four lines shown in Figure 2–201. Use "A," "B," "C," and "D" as the starting points for the four lines.

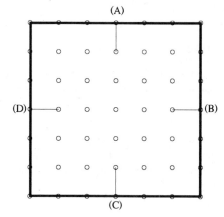

Figure 2–201

Step #5

Use the LINE command to draw the diamond shaped figure shown in Figure 2–202. Be sure Ortho mode is turned off before constructing the line segments.

 Command:**LINE**

From point: **Endp**
of *(Select the endpoint of the line at "A")*
To point: **Endp**
of *(Select the endpoint of the line at "B")*
To point: **Endp**
of *(Select the endpoint of the line at "C")*
To point: **Endp**
of *(Select the endpoint of the line at "D")*
To point: **C**

When finished, perform a ZOOM-Previous operation; turn the snap off using the F9 function key.

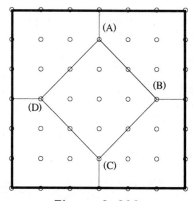

Figure 2–202

Step #6

To prepare for the following steps, use the Osnap Settings dialog box shown in Figure 2–203 and click on the Intersection mode for the current Running Osnap mode.

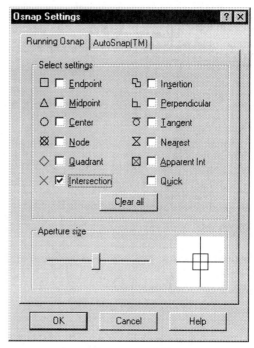

Figure 2–203

Step #7

Use the COPY command and the Multiple option to repeat the diamond-shaped pattern in Figure 2–204. The Osnap-Intersection mode should already be enabled.

 Command: **COPY**

Select objects: *(Select all dashed lines at the right)*

Select objects: *(Press* ENTER *to continue with this command)*

<Base point or displacement>/Multiple: **M**

Base point: *(Select the intersection at "A")*

Second point of displacement: *(Select the intersection at "B")*

Second point of displacement: *(Select the intersection at "C")*

Second point of displacement: *(Select the intersection at "D")*

Second point of displacement: *(Select the intersection at "* E *")*

Second point of displacement: *(Select the intersection at "F")*

Second point of displacement: *(Press* ENTER *to exit this command)*

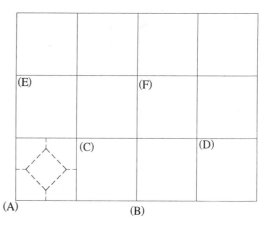

Figure 2–204

Step #8

The diamond pattern of the Inlay floor tile should be similar to Figure 2–205.

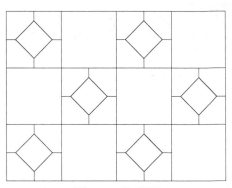

Figure 2–205

Step #9

Begin drawing one 12" x 12" square figure in the position shown in Figure 2–206. Use the ZOOM-WINDOW command to magnify the area around the position of the square figure.

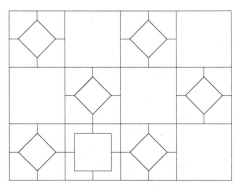

Figure 2–206

Step #10

Be sure the Grid and Snap values are set to 3" and turn the Snap back on using the F9 function key. Also, turn off running osnap mode by either double clicking on OSNAP in the status bar or pressing the F3 function key. Then use the LINE command to draw the four lines shown in Figure 2–207. Use "A," "B," "C," and "D" as the starting points for the four lines.

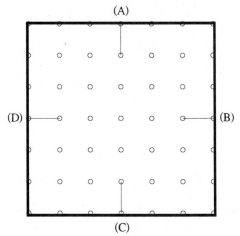

Figure 2–207

Step #11

Use the RECTANG command to construct the square-shaped object in figure 2–208. Be sure Snap mode is turned ON.

 Command: **RECTANG**

Chamfer/Elevation/Fillet/Thickness/Width/
  <First corner>: *(Pick a point at "A")*
Other corner: *(Pick a point at "B")*

When finished, perform a ZOOM-Previous operation; turn the Snap off using the F9 function key.

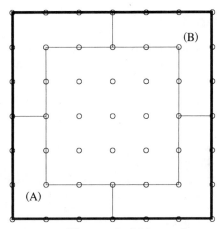

Figure 2–208

Step #12

Use the COPY command and the Multiple option to repeat the square shaped pattern in Figure 2–209. The Osnap-Intersection mode should already be enabled by pressing the F3 function key.

 Command: **COPY**

Select objects: *(Select all dashed lines in Figure 2–209)*
Select objects: *(Press [ENTER] to continue with this command)*
<Base point or displacement>/Multiple: **M**
Base point: *(Select the intersection at "A")*
Second point of displacement: *(Select the intersection at "B")*
Second point of displacement: *(Select the intersection at "C")*
Second point of displacement: *(Select the intersection at "D")*
Second point of displacement: *(Select the intersection at "E")*
Second point of displacement: *(Select the intersection at "F")*
Second point of displacement: *(Press [ENTER] to exit this command)*

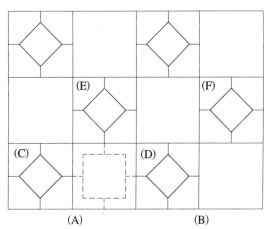

Figure 2–209

Step #13

The drawing of the inlay should appear similar to Figure 2–210.

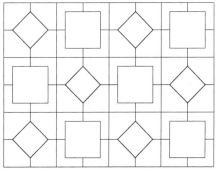

Figure 2–210

Step #14

Use the OFFSET command to copy the outline of the Inlay outward at a distance of six units as in Figure 2–211.

 Command:**OFFSET**

Offset distance or Through <0'-1">: **6**
Select object to offset: *(Select line "A")*
Side to offset? *(Pick a point above the line)*
Select object to offset: *(Select line "B")*
Side to offset? *(Pick a point left of the line)*
Select object to offset: *(Select line "C")*
Side to offset? *(Pick a point below the line)*
Select object to offset: *(Select line "D")*
Side to offset? *(Pick a point right of the line)*
Select object to offset: *(Press* ⌷ENTER⌷ *to exit this command)*

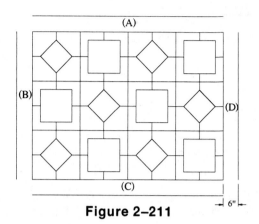

Figure 2–211

Step #15

Use the FILLET command set to a radius of "0" to place a corner at the intersection of lines "A" and "B." See Figure 2–212.

 Command:**FILLET**

(TRIM mode) Current fillet radius = 0'-0 1/2"
Polyline/Radius/Trim/<Select first object>:**R**
Enter fillet radius <0'-0 1/2">:**0**

 Command:**FILLET**

(TRIM mode) Current fillet radius = 0'-0"
Polyline/Radius/Trim/<Select first object>:
  (Select line "A" in Figure 2–212)
Select second object: *(Select line "B")*

Repeat this procedure for the other three corners.

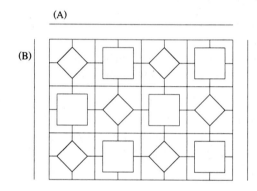

Figure 2–212

Step #16

The completed problem is shown in Figure 2–213. Dimensions may be added upon the request of your instructor.

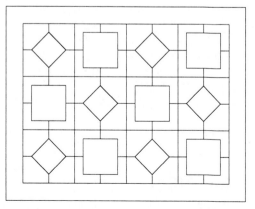

Figure 2–213

Tutorial Exercise
Clutch.Dwg

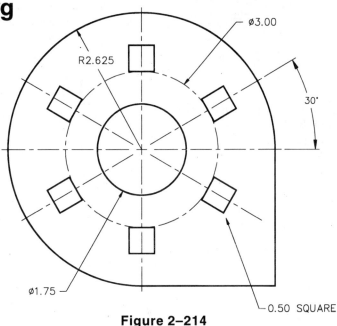

Figure 2-214

Purpose

This tutorial is designed to allow you to construct a one-view drawing of the clutch shown in Figure 2-214 using coordinates and the ARRAY command.

System Settings

Use the current default settings for the limits of this drawing, (0,0) for the lower left corner and (12,9) for the upper right corner. Use the GRID command and change the grid spacing from 1.0000 to 0.2500 units. The grid will be used only as a guide for constructing this object. Do not turn the SNAP or ORTHO commands on.

Layers

Create the following layers with the format :

Name	Color	Linetype
Object	White	Continuous
Center	Yellow	Center

Suggested Commands

Draw the basic shape of the object using the LINE and CIRCLE commands. Lay out a centerline circle, draw one square shape, and use ARRAY to create a multiple copy of the square in a circular pattern.

Whenever possible, substitute the appropriate command alias in place of the full AutoCAD command in each tutorial step. For example, use "Co" for the COPY command, "L" for the LINE command, and so on. The complete listing of all command aliases is located on pages 1-9 and 1-10.

Step #1

Check that the current layer is set to "Object." Begin drawing the clutch by placing a circle with the center at absolute coordinate (6.00,5.00) and radius of 2.625 units as shown in Figure 2–215. Place another circle using the same center point and a diameter of 1.75 units.

 Command: -**LAYER**

?/Make/Set/New/ON/OFF/Color/Ltype/Freeze/
   Thaw/LOck/Unlock: **S** *(To Set)*
New current layer <0>: **OBJECT**
?/Make/Set/New/ON/OFF/Color/Ltype/Freeze/
   Thaw/LOck/Unlock: *(Press* ENTER *to exit this command)*

 Command: **CIRCLE**

CIRCLE 3P/2P/TTR/<Center point>: **6.00,5.00**
Diameter/<Radius>: **2.625**

 Command: **CIRCLE**

CIRCLE 3P/2P/TTR/<Center point>: **@**
Diameter/<Radius> <2.6250>: **D**
Diameter <5.2500>: **1.75**

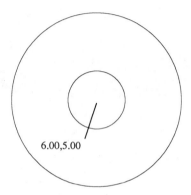

Figure 2–215

Step #2

Turn Ortho mode on and use direct distance mode to draw the lower right corner of the clutch. See Figure 2–216.

Command: **ORTHO**
ON/OFF <OFF>: **ON**

 Command: **LINE**

From point: **Qua**
of *(Pick the quadrant of the circle at "A")*
To point: *(Move the cursor to the right and enter a value of 2.625)*
To point: *(Move the cursor up and enter a value of 2.625)*
To point: *(Press* ENTER *to exit this command)*

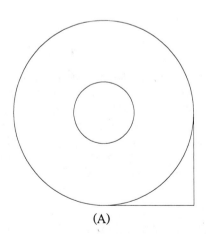

(A)

Figure 2–216

Step #3

Use the TRIM command to partially delete one-fourth of the circle. Use the horizontal and vertical lines shown in Figure 2–217 as the cutting edges and select the circle at "A" as the object to trim.

 Command:**TRIM**

Select cutting edges: (Projmode = UCS, Edgemode = No extend)
Select objects: *(Press [ENTER]; this will select all objects as cutting edges)*
<Select object to trim>/Project/Edge/Undo: *(Select the circle at "A")*
<Select object to trim>/Project/Edge/Undo: *(Press [ENTER] to exit this command)*

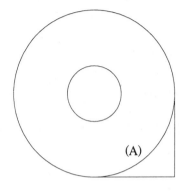

(A)

Figure 2–217

Step #4

Change to the CENTER layer. Set the DIMCEN variable to a value of -0.09 units. This will place a center mark at the center of the circle and extend centerlines just outside of the larger circle as shown in Figure 2–218. Use DIMCENTER to construct a centerline. Erase the bottom centerline segment. This line will be placed in a later step.

 Command: **-LAYER**

?/Make/Set/New/ON/OFF/Color/Ltype/Freeze/ Thaw/LOck/Unlock: **S** *(To Set)*
New current layer <OBJECT>: **CENTER**
?/Make/Set/New/ON/OFF/Color/Ltype/Freeze/ Thaw/LOck/Unlock: *(Press [ENTER] to exit this command)*

Command:**DIMCEN**
New value for DIMCEN <0.0900>: **-0.09**

 Command:**DIMCENTER**

Select arc or circle: *(Select the edge of the arc at "A")*

 Command:**ERASE**

Select objects: *(Pick the bottom centerline at "B")*
Select objects: *(Press [ENTER] to exit this command)*

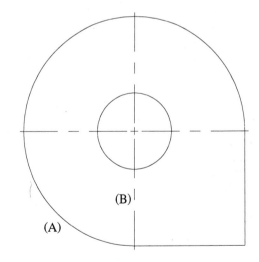

(B)

(A)

Figure 2–218

Step #5

While in the CENTER layer, construct the 3.00 diameter centerline circle shown in Figure 2–219.

 Command: **CIRCLE**

3P/2P/TTR/<Center point>:**6.00,5.00**
Diameter/<Radius><0.8750>:**D**
Diameter<1.7500>:**3.00**

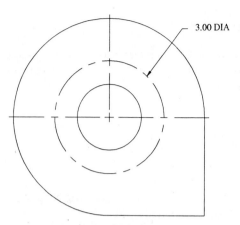

Figure 2–219

Step #6

Set the current layer back to OBJECT. Then, use the ZOOM command to magnify the upper portion of the clutch shown in Figure 2–220 for constructing a square in the next step.

 Command: **-LAYER**

?/Make/Set/New/ON/OFF/Color/Ltype/Freeze/
    Thaw/LOck/Unlock: **S** *(To Set)*
New current layer <CENTER>: **OBJECT**
?/Make/Set/New/ON/OFF/Color/Ltype/Freeze/
    Thaw/LOck/Unlock: *(Press* ⏎ *to exit this
    command)*

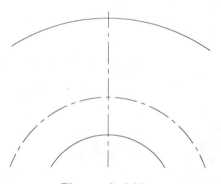

Figure 2–220

Step #7

Construct the 0.50 unit square using the RECTANG command with the assistance of the From and Intersection Osnap options.

 Command: **RECTANG**

Chamfer/Elevation/Fillet/Thickness/Width/
    <First corner>: **From**
Base point: **Int**
of *(Select the intersection at "A" in Figure
    2–221)*
<Offset>:**@0.25<180**
Other corner: **@0.50,0.50**

Perform a ZOOM-Previous operation to return to the original display.

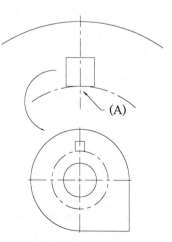

Figure 2–221

Step #8

Use the ARRAY command to copy the square and vertical centerline in a circular pattern six times. Use the following prompts to perform this operation and complete the clutch, as in Figure 2–222.

 Command: **ARRAY**

Select objects: *(Select the square)*
Select objects: *(Select the vertical centerline at "A")*
Select objects: *(Press* ⏎ *to continue with this command)*
Rectangular or Polar array (R/<P>): **P**
Base/<Specify center point of array>: **Int**
of *(Select the intersection at "B")*
Number of items: **6**
Angle to fill (+=ccw, -=cw) <360>: *(Press* ⏎ *to accept this default)*
Rotate objects as they are copied? <Y> *(Press* ⏎ *to perform the array operation)*

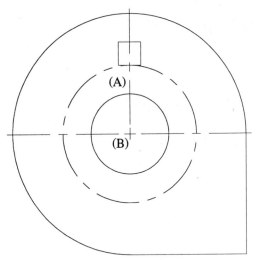

Figure 2–222

Step #9

The completed problem is in Figure 2–223. Dimensions may be added at a later date.

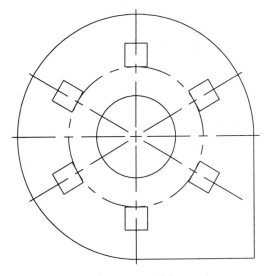

Figure 2–223

Tutorial Exercise
Angle.Dwg

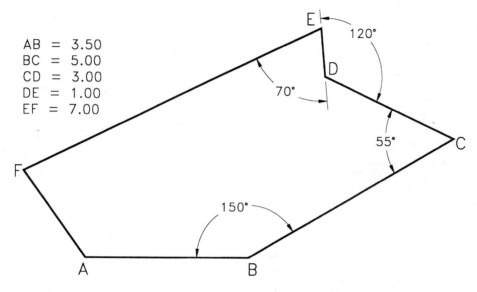

AB = 3.50
BC = 5.00
CD = 3.00
DE = 1.00
EF = 7.00

Figure 2–224

Purpose

This tutorial is designed to allow you to construct a one-view drawing of the angle shown in Figure 2–224 using the ARRAY and LENGTHEN commands.

System Settings

Use the current default settings for the limits of this drawing, (0,0) for the lower left corner and (12,9) for the upper right corner. Use the GRID command and change the grid spacing from 1.0000 to 0.2500 units. The grid will be used only as a guide for constructing this object.

Layers

Create the following layers with the format :

Name	Color	Linetype
Object	White	Continuous

Suggested Commands

Begin this drawing by constructing line AB. Then use the ARRAY command to copy and rotate line AB at an angle of 150 degrees in the clockwise direction. Once the line is copied, use the LENGTHEN command and modify the line to the proper length. Repeat this procedure for lines CD, DE, and EF. Complete the drawing by constructing a line segment from the endpoint at vertex "F" to the endpoint at vertex "A."

Whenever possible, substitute the appropriate command alias in place of the full AutoCAD command in each tutorial step. For example, use "Co" for the COPY command, "L" for the LINE command, and so on. The complete listing of all command aliases is located on pages 1–9 and 1–10.

Step #1

For the first line, turn Ortho mode on by pressing the F8 key or using the ORTHO command. Next, draw line AB using the polar coordinate or direct distance modes, as shown in Figure 2–225.

Command: **ORTHO**
ON/OFF <OFF>: **ON**

 Command: **LINE**

From point: **2,1**
To point: (*Move the cursor directly to the right of the last point and enter a value of 3.50; or use the polar coordinate mode and enter "@3.50<0")*
To point: (*Press* ENTER *to exit this command*)

Figure 2–225

Step #2

One technique of constructing the adjacent line at 150 degrees from line AB is to use the ARRAY command. Since the line needs to be arrayed in the clockwise direction, enter a value of -150 degrees for the angle to fill.

 Command: **ARRAY**

Select objects: (*Select line AB*)
Select objects: (*Press* ENTER *to continue with this command*)
Rectangular or Polar array (<R>/P): **P**
Base/<Specify center point of array>: **Endp**
of (*Pick the endpoint of the line at "B"*)
Number of items: **2**
Angle to fill (+=ccw, -=cw) <360>: **-150**
Rotate objects as they are copied? <Y> (*Press* ENTER *to perform the array operation*)

The result is shown in Figure 2–226.

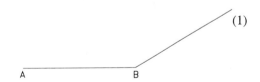

Figure 2–226

Step #3

The Array operation allowed line AB to be rotated and copied at the correct angle, namely -150 degrees. However the new line is the same length as line AB. Use the LENGTHEN command to increase the length of the new line to a distance of 5.00 units.

 Command: **LENGTHEN**

DElta/Percent/Total/DYnamic/<Select object>:**T**
  (For Total)
Angle/<Enter total length (1.0000)>:**5**
<Select object to change>/Undo: *(Pick the end of the line at "1")*
<Select object to change>/Undo: *(Press ⏎ to exit this command)*

The result is shown in Figure 2–227.

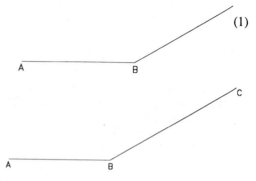

Figure 2–227

Step #4

Use the ARRAY command to rotate and copy line BC at an angle of -55 degrees, since the direction it is being copied is in the clockwise direction. Then, use the LENGTHEN command to reduce the length of the new line from 5.00 units to 3.00 units.

 Command: **ARRAY**

Select objects: *(Select line BC)*
Select objects: *(Press ⏎ to continue with this command)*
Rectangular or Polar array (R/<P>): **P**
Base/<Specify center point of array>: **Endp**
of *(Pick the endpoint of the line at "C")*
Number of items: **2**
Angle to fill (+=ccw, -=cw) <360>: **-55**
Rotate objects as they are copied? <Y> *(Press ⏎ to perform the array operation)*

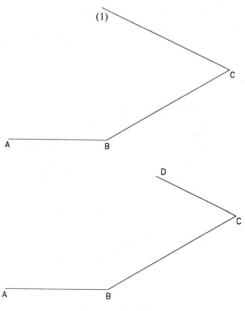

Figure 2–228

Command: **LENGTHEN**

DElta/Percent/Total/DYnamic/<Select object>:**T**
  (For Total)
Angle/<Enter total length (5.0000)>:**3**
<Select object to change>/Undo: *(Pick the end
  of the line at "1")*
<Select object to change>/Undo: *(Press ENTER to
  exit this command)*

The result is shown in Figure 2–228.

Step #5

Use the ARRAY command to rotate and copy line
CD at an angle of 120 degrees. A positive angle is
entered since it is being copied in the counterclockwise direction. Then, use the LENGTHEN command to reduce the length of the new line from 3.00
units to 1.00 unit.

 Command: **ARRAY**

Select objects: *(Select line CD)*
Select objects: *(Press ENTER to continue with this
  command)*
Rectangular or Polar array (R/<P>): **P**
Base/<Specify center point of array>: **Endp**
of *(Pick the endpoint of the line at "D")*
Number of items: **2**
Angle to fill (+=ccw, -=cw) <360>: **120**
Rotate objects as they are copied? <Y> *(Press
  ENTER to perform the array operation)*

 Command: **LENGTHEN**

DElta/Percent/Total/DYnamic/<Select object>:
  T *(For Total)*
Angle/<Enter total length (3.0000)>: **1**
<Select object to change>/Undo: *(Pick the end
  of the line at "1")*
<Select object to change>/Undo: *(Press ENTER to
  exit this command)*

The result is shown in Figure 2–229.

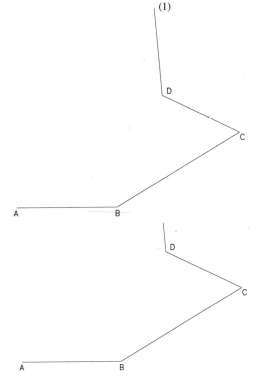

Figure 2–229

Step #6

Use the ARRAY command to rotate and copy line
DE at an angle of -70 degrees since the direction it
is being copied is again in the clockwise direction.
Then, use the LENGTHEN command to increase the
length of the new line from 1.00 unit to 7.00 units.

 Command:**ARRAY**

Select objects: *(Select line DE)*
Select objects: *(Press* ENTER *to continue with this
command)*
Rectangular or Polar array (R/<P>): **P**
Base/<Specify center point of array>: **Endp**
of *(Pick the endpoint of the line at "E")*
Number of items: **2**
Angle to fill (+=ccw, -=cw) <360>: **-70**
Rotate objects as they are copied? <Y> *(Press*
ENTER *to perform the array operation)*

 Command:**LENGTHEN**

DElta/Percent/Total/DYnamic/<Select object>:
T *(For Total)*
Angle/<Enter total length (1.0000)>: **7**
<Select object to change>/Undo: *(Pick the end
of the line at "1")*
<Select object to change>/Undo: *(Press* ENTER *to
exit this command)*

The result is shown in Figure 2–230.

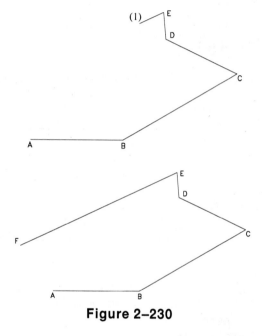

Figure 2–230

Step #7

Turn Ortho off. Connect endpoints "F" and "A" with
a line as shown in Figure 2–231A.

 Command: **LINE**

From point: **Endp**
of *(Pick the endpoint of the line at "F")*
To point: **Endp**
of *(Pick the endpoint of the line at "A")*
To point: *(Press* ⌐ENTER⌐ *to exit this command)*

The completed drawing is illustrated in Figure 2–231B.
You may add dimensions at a later date.

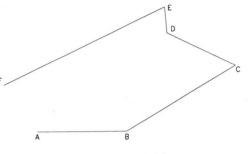

Figure 2–231A

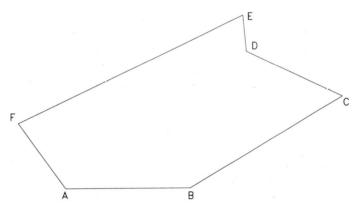

Figure 2–231B

Tutorial Exercise
Lug.Dwg

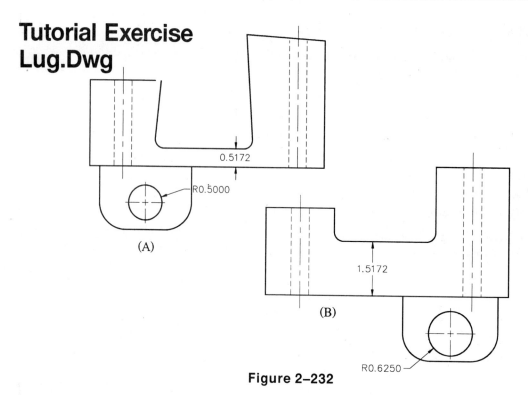

Figure 2–232

Purpose

This tutorial is designed to use object grips to edit the drawing of the lug shown in Figure 2–232 at "A" until it appears like the illustration at "B."

System Settings

Since this drawing is provided on diskette, open an existing drawing file called "Lug." Follow the steps in this tutorial for using object grips to edit and make changes to the Lug.

Layers

The following layers have already been created with the format:

Name	Color	Linetype
Object	White	Continuous
Center	Yellow	Center
Hidden	Red	Hidden
Dim	Yellow	Continuous
Defpoints	White	Continuous

Suggested Commands

Begin this tutorial by selecting the Tools area of the pull-down menu area. Next select "Grips..." and check the box "Enable Grips" to have grips present when an object is selected. Then use the STRETCH, MOVE, ROTATE, SCALE, and MIRROR modes of object grips to make changes to the existing drawing file.

Whenever possible, substitute the appropriate command alias in place of the full AutoCAD command in each tutorial step. For example, use "Co" for the COPY command, "L" for the LINE command, and so on. The complete listing of all command aliases is located on pages 1–9 and 1–10.

Step #1

Before beginning this tutorial, be sure that Object Grip mode is enabled by selecting "Tools" from the pull-down menu area. Next select "Grips...," which activates the Grips dialog box (see Figure 2–233). Use the dialog box to examine that grips are enabled when you place a check in the appropriate box.

Step #2

Begin by turning ORTHO off. Next, at the command prompt, use the grip cursor to select the inclined line "A" shown in Figure 2–234. Notice the appearance of the grips at the endpoints and midpoints of the line. Continue by going to Step #3.

Command: **ORTHO**
ON/OFF <ON>: **OFF**
Command: *(Select the inclined line "A")*

Step #3

While still at the command prompt, select the grip shown in Figure 2–234 at "A." This grip becomes the current base point for the following editing options. Use the STRETCH option to reposition the endpoint of the highlighted line to the endpoint of the horizontal line shown in Figure 2–235. When this operation is complete, press the ESC key twice at the command prompt to remove the object highlight and grips from the display screen. Continue by going to Step #4.

Command: *(Select the warm grip at the*
   endpoint of the line at "A" to make it hot)
STRETCH
<Stretch to point>/Base point/Copy/Undo/eXit:
   Endp
of *(Select the endpoint of the horizontal line at "B")*
Command: *(Press ESC to remove the object*
   highlight)
Command: *(Press ESC to remove the grips)*

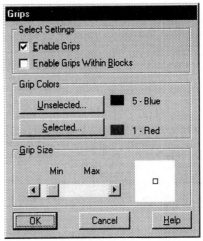

Figure 2–233

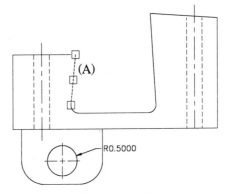

Figure 2–234

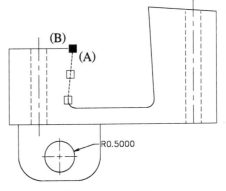

Figure 2–235

Step #4

At the command prompt, use the grip cursor to select the two inclined lines "A" and "B" shown in Figure 2–236. Notice the appearance of the grips at the endpoints and midpoints of the line. Continue by going to Step #5.

Command: *(Select the inclined lines "A" and "B")*

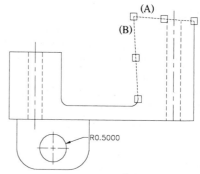

Figure 2–236

Step #5

While still at the command prompt, select the grip shown in Figure 2–237 at "A." This grip becomes the current base point for the following editing options. Use the STRETCH option in combination with Osnap-Tracking mode to reposition the corner of the highlighted lines to form a corner. When this operation is complete, press the ESC key twice at the command prompt to remove the object highlight and grips from the display screen. Continue by going to Step #6.

Command: *(Select the warm grip at the endpoint of the line at "A" to make it hot).*
** STRETCH **
<Stretch to point>/Base point/Copy/Undo/eXit:
 TK *(To enable Tracking)*
First tracking point: *(Select the grip at "B" in Figure 2–237)*
Next point (Press [ENTER] to end tracking): *(Select the grip at "C" in Figure 2–237)*
Next point (Press [ENTER] to end tracking): *(Press [ENTER])*
Command: *(Press ESC to remove the object highlight)*
Command: *(Press ESC to remove the grips)*

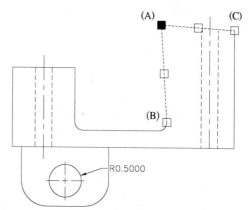

Figure 2–237

Step #6

At the command prompt, use the grip cursor to select the two vertical lines, one horizontal line, and both filleted corners shown in Figure 2–238. Notice the appearance of the grips at the endpoints and midpoints of the lines and arcs. Continue by going to Step #7.

Command: *(Select the two vertical lines, two arcs representing fillets, and the horizontal line)*

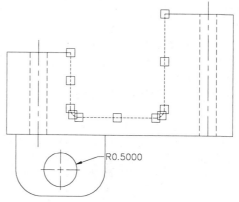

Figure 2–238

Step #7

While still at the command prompt, select the All Grips in Figure 2–239 by holding down the Shift key as the grips are selected. Lift off the Shift key and pick the selected grip at "A" again. This grip becomes the current base point for the following editing options. Use the STRETCH option in combination with a polar coordinate to stretch the objects a distance of one unit in the 90-degree direction. When this operation is complete, press the ESC key at the command prompt to remove the object highlight and grips from the display screen. Continue by going on to Step #8.

Command: *(Select all warm grips shown in Figure 2–239 to make them hot. Hold down the Shift key to make multiple warm grips hot. Lift off the Shift key and pick the grip at "A" again.)*
STRETCH
<Stretch to point>/Base point/Copy/Undo/eXit:
  @1.00<90
Command: *(Press ESC to remove the object highlight)*
Command: *(Press ESC to remove the grips)*

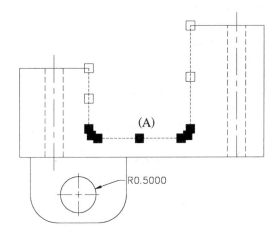

Figure 2–239

Step #8

At the command prompt, use the grip cursor to select the circle and dimension in Figure 2–240. Notice the appearance of the grips at the quadrants and center of the circle and center, starting point, and text location of the radius dimension. Continue by going on to Step #9.

Command: *(Select the circle and the radius dimension)*

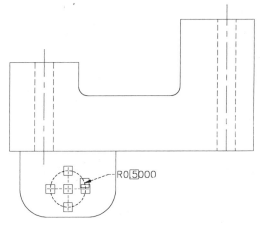

Figure 2–240

Step #9

While still at the command prompt, select the grip at the center of the circle shown in Figure 2–241. This grip becomes the current base point for the following editing options. Use the SCALE option to increase the size of the circle using the Reference option. Notice that this will also affect the value of the associative dimension. When this operation is complete, press the ESC key twice at the command prompt to remove the object highlight and grips from the display screen. Continue by going to Step #10.

Command: *(Select the warm grip at the center of the circle to make it hot. Press the Spacebar until the SCALE mode appears at the bottom of the prompt line)*
** SCALE **
<Scale factor>/Base point/Copy/Undo/ Reference/eXit: **R**
Reference length <1.0000>: **0.50**
** SCALE **
<New length>/Base point/Copy/Undo/ Reference/eXit: **0.625**
Command: *(Press ESC to remove the object highlight)*
Command: *(Press ESC to remove the grips)*

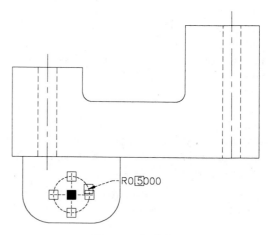

Figure 2–241

Step #10

At the command prompt, use the grip cursor to select the dimension shown in Figure 2–242. Notice the appearance of the grips at the center, starting point, and text location of the radius dimension. The dimension text will be relocated using grips. Continue by going to Step #11.

Command: *(Select the dimension in Figure 2–242)*

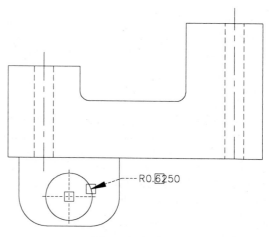

Figure 2–242

Step #11

Still at the command prompt, select the text grip shown in Figure 2–243. This grip becomes the current base point for the following editing options. Use the STRETCH option to reposition the dimension text to a better location. When this operation is complete, press the ESC key twice at the command prompt to remove the object highlight and grips from the display screen. Continue by going to Step #12.

Command: *(Select the warm grip at the text location to make it hot).*
** STRETCH **
\<Stretch to point>/Base point/Copy/Undo/eXit: *(Pick a convenient location on the display screen)*
Command: *(Press ESC to remove the object highlight)*
Command: *(Press ESC to remove the grips)*

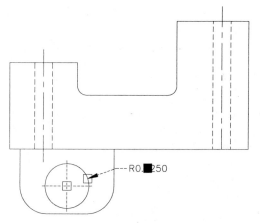

Figure 2–243

Step #12

At the command prompt, use the grip cursor to select the centerline in Figure 2–244. Notice the appearance of the grips at the endpoints and midpoints of the centerline. Continue by going to Step #13.

Command: *(Select the centerline shown in Figure 2–244)*

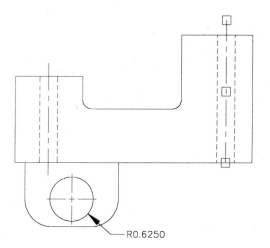

Figure 2–244

Step #13

While still at the command prompt, select the grip shown in Figure 2–245 at "A." This grip becomes the current base point for the following editing options. Use the STRETCH mode and Base option to extend the centerline from a new base point at "B" to the intersection at "C." When this operation is complete, press the ESC key twice at the command prompt to remove the object highlight and grips from the display screen. Continue by going to Step #14.

Command: *(Select the warm grip at the bottom of the centerline to make it hot).*
STRETCH
<Stretch to point>/Base point/Copy/Undo/eXit:
  B *(For Base)*
Base point: *(Select the grip at "B")*
<Stretch to point>/Base point/Copy/Undo/eXit:
  Int
of *(Select the intersection of the centerline and horizontal line at "C")*
Command: *(Press ESC to remove the object highlight)*
Command: *(Press ESC to remove the grips)*

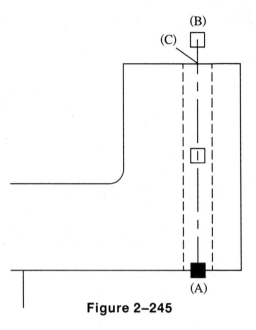

Figure 2–245

Step #14

At the command prompt, use the grip cursor to select the two hidden lines and centerline shown in Figure 2–246. Notice the appearance of the cold grips at the endpoints and midpoints of the lines. Continue by going to Step #15.

Command: *(Select the two hidden lines and the centerline)*

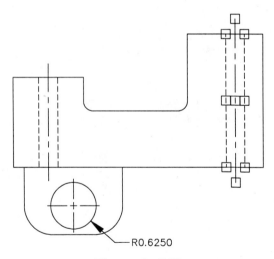

Figure 2–246

Step #15

While still at the command prompt, select the three grips at the midpoints of the center and hidden lines by holding down the SHIFT key as they are selected in Figure 2–247. These grips become the current base point for the following editing options. Lift off the Shift key and pick the selected middle grip again. Use the Move option to center the hidden and centerlines to the middle of the horizontal line. When this operation is complete, press the ESC key twice at the command prompt to remove the object highlight and grips from the display screen. Continue by going to Step #16.

Command: *(Use the Shift key to individually select the three warm grips at the middle of the three lines to make them hot. Lift off the Shift key and pick the middle grip again.Press the Spacebar until MOVE mode appears at the bottom of the prompt line)*
MOVE
<Move to point>/Base point/Copy/Undo/eXit: **B** *(For Base)*
Base point: **Int**
of *(Select the intersection of the lines at "A")*
<Move to point>/Base point/Copy/Undo/eXit: **Mid**
of *(Select the horizontal line at "B")*
Command: *(Press ESC to remove the object highlight)*
Command: *(Press ESC to remove the grips)*

Step #16

At the command prompt, use the grip cursor to select all of the objects shown in Figure 2–248. Notice the appearance of the grips at the various key points of these objects. Continue by going to Step #17.

Command: *(Select all objects shown in Figure 2–248. This can be easily accomplished using the Automatic Window mode and marking points at "X" and "Y")*

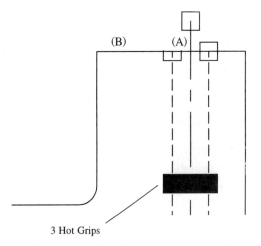

(B) (A)

3 Hot Grips

Figure 2–247

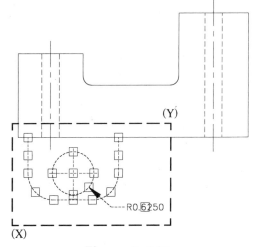

(Y)

R0.6250

(X)

Figure 2–248

Step #17

While still in the command prompt, pick the grip at "A" and press the Spacebar until MIRROR mode appears at the bottom of the prompt line. Mirror the selected objects using the midpoint of the line at "B" in Figure 2-249A.

Command: *(Pick the warm grip at "A" to make it hot. Press the Spacebar until MIRROR mode appears at the bottom of the prompt line)*
** MIRROR**
<Second point>/Base point/Copy/Undo/eXit: **B** *(For Base)*
New base point: **Mid**
of *(Select the midpoint of the horizontal line at "B")*
** MIRROR**
<Second point>/Base point/Copy/Undo/eXit: **@1<90**
Command: *(Press ESC to remove the object highlight)*
Command: *(Press ESC to remove the grips)*

The completed object is illustrated in Figure 2-249B.

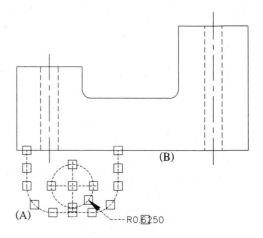

Figure 2-249A

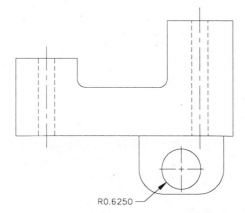

Figure 2-249B

Problems for Unit 2

Directions for Problems 2–1 through 2–13:
Construct each one-view drawing using the appropriate coordinate mode or direct distance mode. Utilize advanced commands such as ARRAY and MIRROR whenever possible.

Problem 2–1

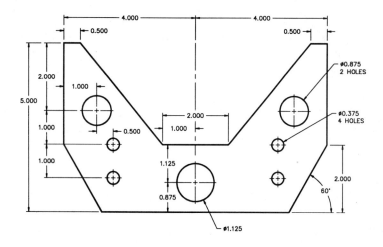

Problem 2–2

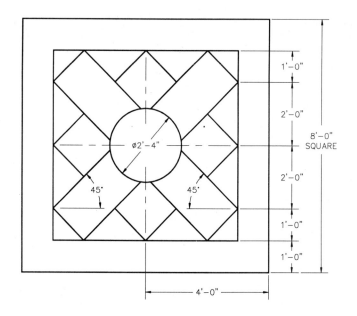

Problem 2–3

AB=2.85
BC=3.09
CD=1.93
DE=8.21
EF=5.53
FG=6.35

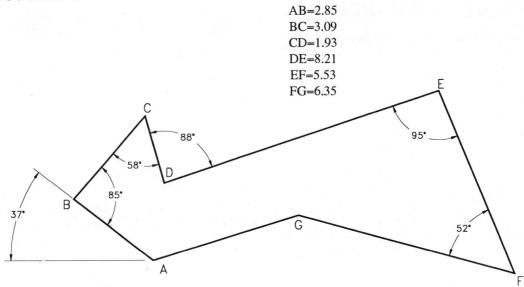

Problem 2–4

AB=8.37
BC=2.53
CD=8.01
DE=4.78
EF=7.30
FG=6.03
GH=4.10

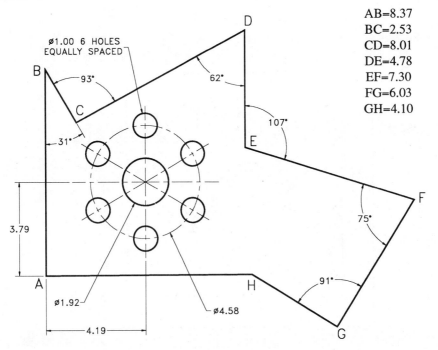

Problem 2–5

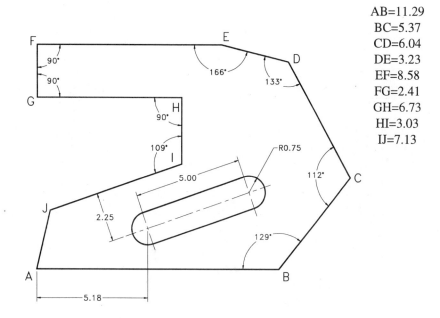

AB=11.29
BC=5.37
CD=6.04
DE=3.23
EF=8.58
FG=2.41
GH=6.73
HI=3.03
IJ=7.13

Problem 2–6

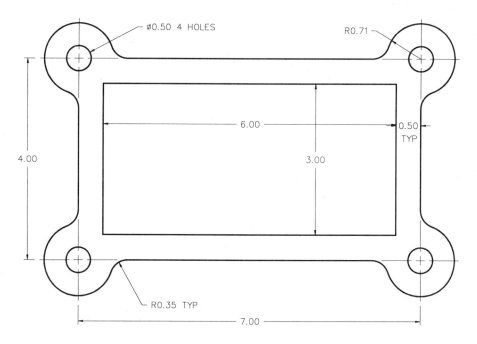

0.125 GASKET THICKNESS

Problem 2–7

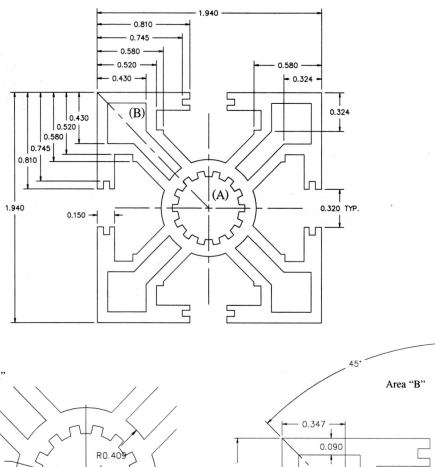

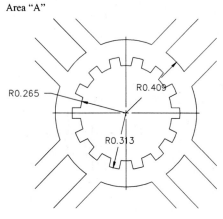

Area "A"

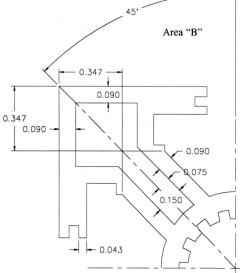

Area "B"

Problem 2–8

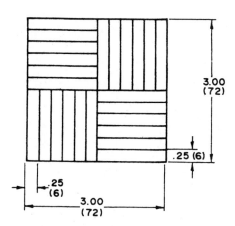

Problem 2–10

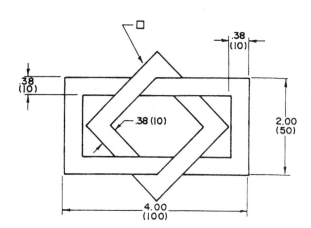

Problem 2–9

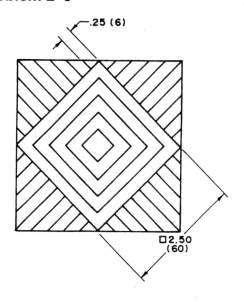

Problem 2–11

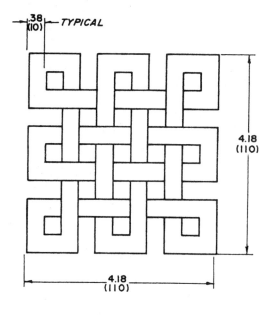

Problem 2–12

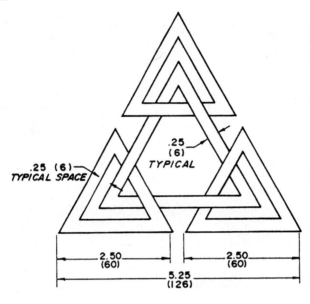

Problem 2–13

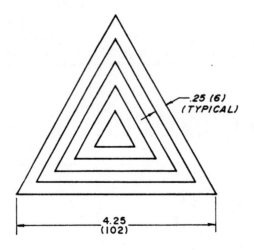

Directions for Problems 2–14 through 2–17:
    Construct one-view drawings of the following figures using a grid spacing of 0.25 units. Do not dimension these drawings unless otherwise specified by your instructor.

Problem 2–14

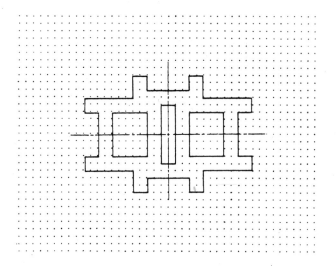

Problem 2–15

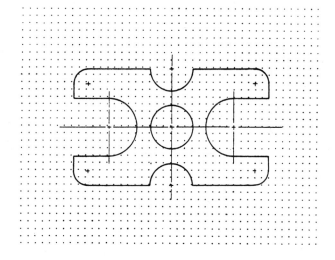

Problem 2-16

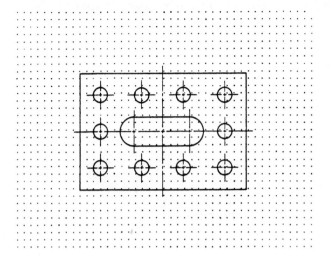

Problem 2-17

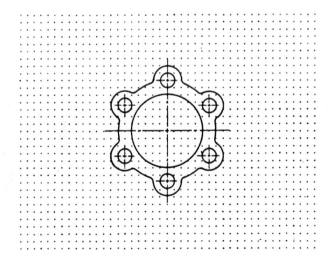

UNIT 3

Geometric Constructions

Many of the elements that go into the creation of a design revolve around geometric shapes. Applying and manipulating these shapes is the next step toward a successful design. Important manual drafting tools to assist in the construction of geometric shapes range from the T-square and drafting machine for drawing parallel and perpendicular lines, to the compass and dividers for drawing circles and arcs and for setting off distances. The computer offers the designer superior control and accuracy when dealing with geometric constructions. Much of the focus of this unit will be on how to manipulate the numerous object snap modes supplied by AutoCAD through construction examples and practical applications in the form of tutorial exercises. Numerous other problems are supplied at the end of this chapter to challenge the designer for a complete and correct solution.

Bisecting Lines and Arcs

Illustrated in Figure 3–1 is an arc intersected by a line. The purpose of this problem is to locate the midpoint of the line and the arc. The POINT command along with the Osnap-Midpoint option will be used to accomplish this. However, if the command is used to find the midpoint, it may not be visible. This is because the POINT command places a dot that is very difficult to see. The appearance of the point is controlled by the system variable PDMODE. There is also a system variable to control the size of the point called Pdsize. An alternate method of selecting different points is through the

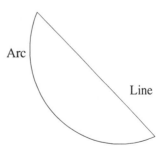

Figure 3–1

DDPTYPE dialog box which can be found in the pull-down menu under the headings Format > Point Style…

Command: **PDMODE**
New value for PDMODE <0>: **2**

Command: **PDSIZE**
New value for PDSIZE <0>: **.25**

Next, use the POINT command along with the Osnap-Midpoint option to locate the midpoint of the line and arc in Figure 3–2 using the following steps:

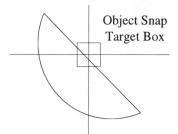

Object Snap
Target Box

Figure 3–2

 Command: **POINT**

Point: **Mid**
of *(Select the line or arc at any convenient location)*

The illustration in Figure 3–3 shows the midpoint locations of the line and arc. The current point is controlled by the system variable PDMODE. The style of point, namely the "plus," reflects the current Pdmode value of 2.

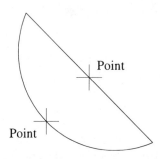

Point

Point

Figure 3–3

Bisecting Angles

Illustrated in Figure 3–4 are two lines forming an acute angle. (An acute angle measures any value less than 90 degrees.) The purpose of this problem is to bisect or divide the angle equally in two parts. In the past, construction lines or arcs were used to intersect with the endpoints of the legs of the angle. Then, a line was constructed to the midpoint of the construction object. Finally the construction object was erased. In this example, the XLINE command will be used to accomplish this operation. This method works if the two lines forming the angle are the exact same length or constructed with unequal lengths.

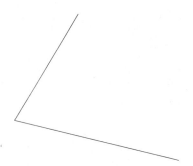

Figure 3–4

Use the XLINE command to perform the bisect operation:

 Command:**XLINE**

Hor/Ver/Ang/Bisect/Offset/<From point>:
  Bisect
Angle vertex point: **Endp**
of (Select the endpoint of the angle vertex at "X")
Angle start point: **Endp**
of (Select the endpoint of the angle at "Y")
Angle end point: **Endp**
of (Select the endpoint of the angle at "Z")
Angle end point: (Press ⌜ENTER⌝ to exit this
  command)

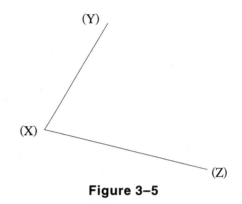

Figure 3–5

The illustration in Figure 3–6 shows the completed operation. Notice the xline continues in an infinite length in both directions.

 Command:**BREAK**

Select object: (Select the xline at "A")
Enter second point (or F for first point): **Endp**
of (Select the endpoint of the angle vertex at "B")

 Command:**BREAK**

Select object: (Select the object at "C")
Enter second point (or F for first point): (Select
  the object at "D")

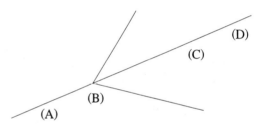

Figure 3–6

Once an xline is broken into two objects, it becomes a ray. The ray is an object beginning at a fixed point and continues infinitely in the opposite direction. Use the ERASE command to delete the two ray objects at "A" and "B" as shown in Figure 3–7.

 Command:**ERASE**

Select objects: (Select the ray at "A")
Select objects: (Select the ray at "B")
Select objects: (Press ⌜ENTER⌝ to execute this
  command)

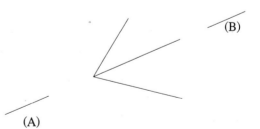

Figure 3–7

Dividing an Object Into Parts

Illustrated in Figure 3–8 is an inclined line. The purpose of this problem is to divide the line into an equal number of parts. This proved to be a tedious task using manual drafting methods but thanks to the AutoCAD DIVIDE command, this operation is much easier to perform. The DIVIDE command instructs the user to supply the number of divisions and performs the division by placing a point along the object to be divided. The size and shape are controlled by the system variables Pdsize and Pdmode, respectively. Be sure the Pdmode variable is set to a value that will produce a visible point. Otherwise, the results of the DIVIDE command will not be obvious.

Figure 3–8

Command: **PDMODE**
New value for pdmode <0>: **2**

Command: **PDSIZE**
New value for pdsize <0>: **.25**

Point modes and sizes may be set from the Point Style dialog box which can be found in the pull-down menu area under the headings of Format > Point Style...

Figure 3–9

Next, use the DIVIDE command, select the inclined line as the object to divide, enter a value for the number of segments, and the command divides the object by a series of points as shown in Figure 3–9. This command is located in the pull-down menu area under Draw > Point > Divide.

Command: **DIVIDE**
Select object to divide: *(Select the inclined line)*
<Number of segments>/Block: **9**

A practical application of the DIVIDE command may be in the area of screw threads where a number of threads per inch is needed to form the profile of the thread. See Figure 3–10.

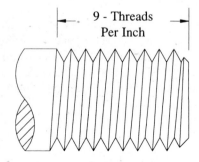

9 - Threads
Per Inch

Figure 3–10

Drawing Parallel Objects

Another fundamentally important operation regarding geometric constructions is the ability to construct objects parallel to each other. Illustrated in Figure 3–11 is an inclined line. The purpose of this problem is to create a matching object parallel to the original line at a set distance. The OFFSET command is used to accomplish this.

Figure 3–11

Use the OFFSET command, set the distance to offset, and select the object to offset and the side of the offset. The result will be similar to the illustration in Figure 3–12.

 Command: **OFFSET**

Offset distance or Through<Through>: **1.25**
Select object to offset: *(Select the line at "A")*
Side to offset? *(Select anywhere near "B")*
Select object to offset: *(Press* ENTER *to exit this command)*

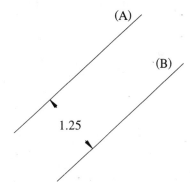

Figure 3–12

The OFFSET command will also produce concentric circles or arcs. Illustrated in Figure 3–13 is an arc. The purpose of this problem is to create an additional arc at a distance of 1.25 units away from the original arc.

Figure 3–13

Use the OFFSET command, set the distance to offset (which should already be set from the previous use of the command), select the object (this time the arc) to offset, and the side of the offset. The result will be similar to the illustration in Figure 3–14.

 Command: **OFFSET**

Offset distance or Through<Through>: **1.25**
Select object to offset: *(Select the arc at "A")*
Side to offset? *(Select anywhere near "B")*
Select object to offset: *(Press* ENTER *to exit this command)*

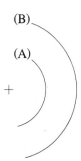

Figure 3–14

Constructing Hexagons

The Hexagon is an important geometric shape commonly used for such items as the plan view of a bolt or screw type of fastener. The illustrations in Figures 3–15 and 3–16 show two types of hexagons; one drawn in relation to its flat edges, and the other drawn in relation to its corners. The POLYGON command is used for drawing either example. Simply supply the following information: Number of the sides; Whether the figure is inscribed or circumscribed about a circle; and the radius of the circle. The polygon is drawn. One interesting characteristic of polygons is that they are constructed using a series of polylines, making the polygon one object. One other note concerning polygons is that the examples in Figures 3–15 through 3–18 illustrate constructing a hexagonal polygon. Using the POLYGON command, any size figure with up to 1,024 sides may be constructed; not just hexagons.

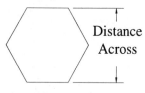

Figure 3–15

Illustrated in Figure 3–17 is an example of an inscribed hexagon, or a figure constructed inside of a circle. The POLYGON command is used to create this type of shape.

 Command:**POLYGON**

Number of sides:**6**
Edge/<Center of Polygon>: *(Select the center at "A")*
Inscribed in circle/Circumscribed about circle
  (I/C):**I**
Radius of circle: *(Enter a numerical value)*

Yet another use of the Polygon is illustrated in Figure 3–18 when the figure needs to be drawn outside of a circle. The Circumscribed option would be used for this example.

 Command:**POLYGON**

Number of sides:**6**
Edge/<Center of Polygon>: *(Select the center at "A")*
Inscribed in circle/Circumscribed about circle
  (I/C):**C**
Radius of circle: *(Enter a numerical value)*

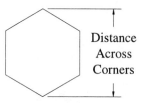

Figure 3–16

Inscribed Polygon

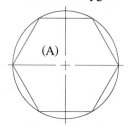

(A)

Figure 3–17

Circumscribed Polygon

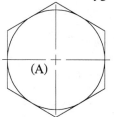

(A)

Figure 3–18

Constructing an Ellipse

The main parts of an ellipse are the major diameter or axis and minor diameter or axis illustrated in Figure 3–19. Numerous construction arcs and lines were needed to construct the ellipse using manual methods. The ELLIPSE command prompts the user for the center of the ellipse, the endpoint of one axis, and the endpoint of the other axis.

In the example in Figure 3–20, a polar coordinate can be used to identify both axes of the ellipse following the prompts below.

 Command:**ELLIPSE**

<Axis endpoint 1>/Center: **C**
Center of ellipse: *(Select a point at "A")*
Axis endpoint: **@4<0** *(Toward a point at "B")*
<Other axis distance>/Rotation: **@3<90**
    (Toward a point at "C")

The direct distance mode can also be used as a substitute for the polar coordinate mode using the following ELLIPSE command prompts:

Axis endpoint: *(With Ortho mode turned on, move the cursor to the right and enter the value **4**)*
<Other axis distance>/Rotation: *(Move the cursor directly up and enter the value **3**)*

The object in Figure 3–21 is an example of how to outline the view with an ellipse at "A" and how to use the OFFSET command to offset the ellipse in the direction inside of the object at "B."

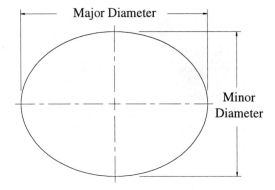

Figure 3–19

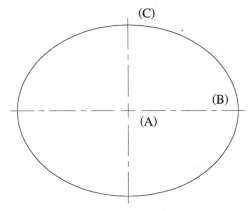

Figure 3–20

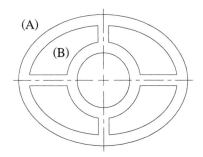

Figure 3–21

Tangent Arc to Two Lines

Illustrated in Figure 3–22 are two inclined lines. The purpose of this problem is to connect an arc tangent to the two lines at a specified radius. The Circle-TTR (Tangent-Tangent-Radius) command will be used here along with the TRIM command to clean up the excess geometry. To assist with this operation, the Osnap-Tangent mode is automatically activated when using the TTR option of the CIRCLE command.

First, use the CIRCLE-TTR command to construct an arc tangent to both lines.

 Command:**CIRCLE**
3P/2P/TTR/<Center point>:**TTR**
Enter Tangent spec: *(Select the line at "A")*
Enter second Tangent spec: *(Select the line at "B")*
Radius: *(Enter a desired radius value)*

Use the TRIM command to clean up the lines and arc. The completed result is illustrated in Figure 3–24. It is interesting to note that the FILLET command could have been used for this procedure. Not only will the curve be drawn, but this command will automatically trim the lines.

The object in Figure 3–25 is an example of a typical application where this procedure might be used.

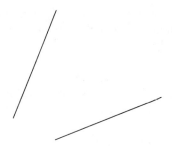

Figure 3–22

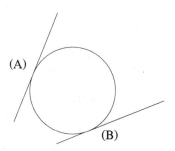

Figure 3–23

Figure 3–24

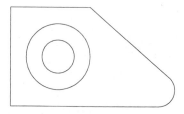

Figure 3–25

Tangent Arc to a Line and Arc

Illustrated in Figure 3–26 is an arc and an inclined line. The purpose of this problem is to connect an additional arc tangent to the original arc and line at a specified radius. The CIRCLE-TTR command will be used here along with the TRIM command to clean up the excess geometry.

First, use the CIRCLE-TTR command to construct an arc tangent to the arc and inclined line as shown in Figure 3–27.

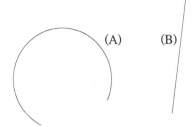

Figure 3–26

 Command: **CIRCLE**

3P/2P/TTR/<Center point>: **TTR**
Enter Tangent spec: *(Select the arc at "A")*
Enter second Tangent spec: *(Select the line at "B")*
Radius: *(Enter a desired radius value)*

Use the TRIM command to clean up the arc and line. The completed result is illustrated in Figure 3–28.

The object in Figure 3–29 is an example of a typical application where this procedure might be used.

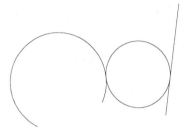

Figure 3–27

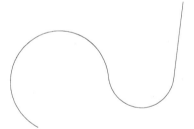

Figure 3–28

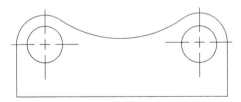

Figure 3–29

Tangent Arc to Two Arcs—Method #1

Illustrated in Figure 3–30 are two arcs. The purpose of this problem is to connect a third arc tangent to the original two at a specified radius. The CIRCLE-TTR command will be used here along with the TRIM command to clean up the excess geometry.

Use the CIRCLE-TTR command to construct an arc tangent to the two original arcs.

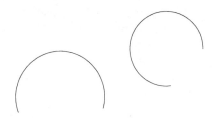

Figure 3–30

 Command:**CIRCLE**

3P/2P/TTR/<Center point>:**TTR**
Enter Tangent spec: *(Select the first arc at "A")*
Enter second Tangent spec: *(Select the second arc at "B")*
Radius: *(Enter a desired value)*

Use the TRIM command to clean up the two arcs using the circle as a cutting edge. The completed result is illustrated in Figure 3–32.

The object in Figure 3–33 is an example of a typical application where this procedure might be used.

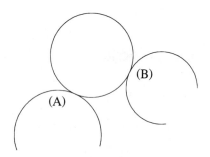

Figure 3–31

Figure 3–32

Figure 3–33

Tangent Arc to Two Arcs—Method #2

Illustrated in Figure 3–34 are two arcs. The purpose of this problem is to connect an additional arc tangent to both arcs and enclose both arcs at a specified radius. The CIRCLE-TTR command will be used here along with the TRIM command.

First, use the CIRCLE-TTR command to construct an arc tangent to both arcs and enclose both arcs.

Figure 3–34

 Command: **CIRCLE**

3P/2P/TTR/<Center point>: **TTR**
Enter Tangent spec: *(Select the arc at "A")*
Enter second Tangent spec: *(Select the arc at "B")*
Radius: *(Enter a desired radius value)*

Use the TRIM command to clean up all arcs. The completed result is illustrated in Figure 3–36.

The object in Figure 3–37 is an example of a typical application where this procedure might be used.

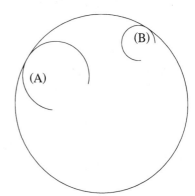

Figure 3–35

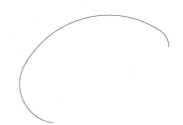

Figure 3–36

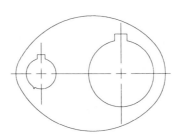

Figure 3–37

Tangent Arc to Two Arcs—Method #3

Illustrated in Figure 3–38 are two arcs. The purpose of this problem is to connect an additional arc tangent to one arc and enclose the other. The CIRCLE-TTR command will be used here along with the TRIM command to clean up unnecessary geometry.

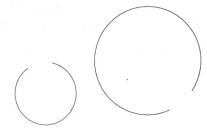

First, use the CIRCLE-TTR command to construct an arc tangent to the two arcs. Study the illustration in Figure 3–39 and the prompts below to understand the proper pick points for this operation.

Figure 3–38

 Command:**CIRCLE**

3P/2P/TTR/<Center point>:**TTR**
Enter Tangent spec: *(Select the arc at "A")*
Enter second Tangent spec: *(Select the line at "B")*
Radius: *(Enter a desired radius value)*

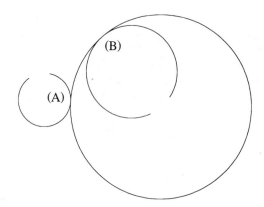

Use the TRIM command to clean up the arcs. The completed result is illustrated in Figure 3–40.

The object in Figure 3–41 is an example of a typical application where this procedure might be used.

Figure 3–39

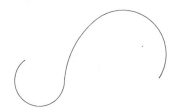

Figure 3–40

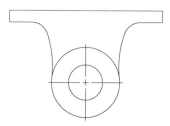

Figure 3–41

Perpendicular Construction Techniques

Illustrated in Figure 3–42 is an inclined line and
a point. The purpose of this problem is to construct a line from a point perpendicular to another line. The LINE command will be used for
this operation in addition to the Osnap-Node and
Osnap-Perpend options.

Use the LINE command and the Osnap-Node option to snap to the point illustrated in Figure 3–43.

 Command:**LINE**

From point: **Nod**
of *(Select the point at "A")*

Continue the LINE command and respond to the
prompt, "To point" by using the Osnap-Perpendicular option and selecting anywhere along the
inclined line as shown in Figure 3–44.

To point: **Per**
of *(Select the line at "A")*
To point: *(Press [ENTER] to exit this command)*

The completed solution is illustrated in Figure 3–45.

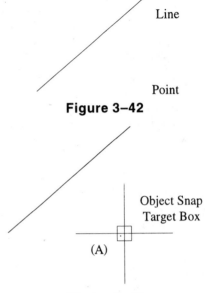

Line

Point

Figure 3–42

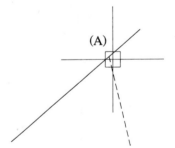

Object Snap
Target Box

(A)

Figure 3–43

(A)

Figure 3–44

Figure 3–45

Tangent Line to Two Arcs or Circles

Illustrated in Figure 3–46 are two circles. The purpose of this problem is to connect the two circles with two tangent lines. This can be accomplished using the LINE command and the Osnap-Tangent option.

Use the LINE command to connect two lines tangent to the circles. The following procedure is used for the first line. Use the same procedure for the second.

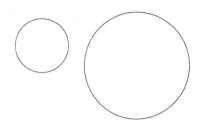

Figure 3–46

 Command: **LINE**

From point: **Tan**
to *(Select the circle near "A")*
To point: **Tan**
to *(Select the circle near "B")*
To point: *(Press* ENTER *to exit this command)*

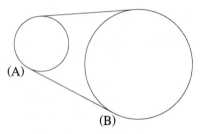

Figure 3–47

When using the Tangent option, the rubberband cursor is not present when drawing the beginning of the line. This is due to calculations required when identifying the second point.

Use the TRIM command to clean up the circles so the appearance of the object is similar to the illustration in Figure 3–48.

The object in Figure 3–49 is an example of a typical application where drawing lines tangent to circles might be used.

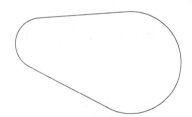

Figure 3–48

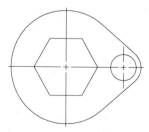

Figure 3–49

Quadrant Versus Tangent Osnap Option

Various examples have been given on previous pages concerning drawing lines tangent to two circles, two arcs, or any combination of the two. The object in Figure 3–50 illustrates the use of the Osnap-Tangent option when used along with the LINE command.

 Command: **LINE**

From point: **Tan**
to *(Select the arc at "A")*
To point: **Tan**
to *(Select the arc at "B")*
To point: *(Press ENTER to exit this command)*

Note that the angle of the line formed by points "A" and "B" is neither horizontal or vertical. The object in Figure 3–50 is a typical example of the capabilities of the Osnap option.

The object illustrated in Figure 3–51 is a modification of the drawing above with the inclined tangent lines changing to horizontal and vertical tangent lines. This example is to inform you that two Osnap options are available to perform tangencies, namely Osnap-Tangent and Osnap-Quadrant. However, it is up to you to evaluate under what conditions to use the Osnap options. In Figure 3–51, you can use the Osnap-Tangent or Osnap-Quadrant option to draw the lines tangent to the arcs. The Quadrant option could be used only since the lines to be drawn are perfectly horizontal or vertical. Usually it is impossible to know this ahead of time, and in this case, the Osnap-Tangent option should be used whenever possible.

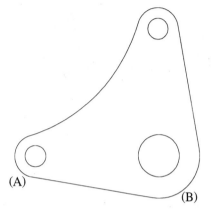

(A)

(B)

Figure 3–50

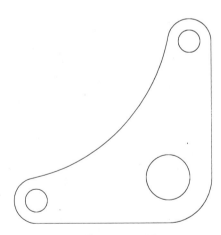

Figure 3–51

The slot in Figure 3–52 is an excellent example of when to use the Osnap-Quadrant option. Slots have two semicircles connected by two horizontal or vertical lines that enable you to use the Quadrant option. If the slot is positioned at an odd angle, simply construct it and use the ROTATE command to position it.

A slot is easily created using the FILLET command. If two lines are parallel, the FILLET command automatically calculates the distance between the two parallel lines and constructs an arc.

Figure 3–52

Ogee or Reverse Curve Construction

An ogee curve connects two parallel lines with a smooth flowing curve that reverses itself in symmetrical form. To begin constructing an ogee curve to line segments "AB" and "CD," first draw line "BC," which connects both parallel line segments. See Figure 3–53.

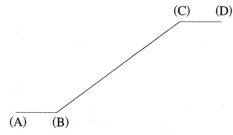

Figure 3–53

Use the DIVIDE command to divide line segment "BC" into four equal parts. Be sure to set a new point mode either through the PDMODE command or picking a new point from the Point Style dialog located in the pull-down menu area under the headings of Format > Point Style... Construct vertical lines from "B" and "C." Complete this step by constructing line segment "XY," which is perpendicular to line "BC" as shown in Figure 3–54. Do not worry about where line "XY" is located at this time.

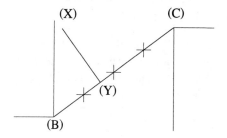

Figure 3–54

Move line "XY" to the location identified by the point in Figure 3–55. Complete this step by copying line "XY" to the location identified by point "Z" illustrated in Figure 3–55.

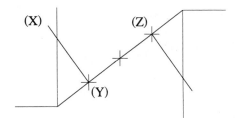

Figure 3–55

Construct two circles with centers located at points "X" and "Y" illustrated in Figure 3–56. Use the Osnap-Intersection mode to accurately locate the centers. Note: If an intersection is not found from the previous step, use the EXTEND command to find the intersection and continue with this step. The radii of both circles are equivalent to distances "XB" and "YC."

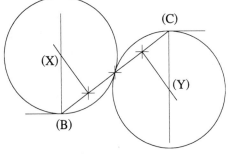

Figure 3–56

Use the TRIM command to trim away any excess arc segments to form the ogee curve as shown in Figure 3–57. This forms the frame of the ogee for the construction of objects such as the wrench illustrated in Figure 3–58.

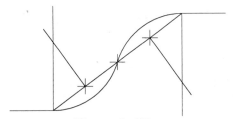

Figure 3–57

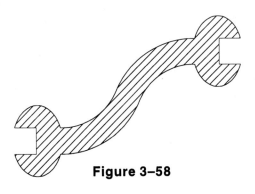

Figure 3–58

Tutorial Exercise
Pattern1.Dwg

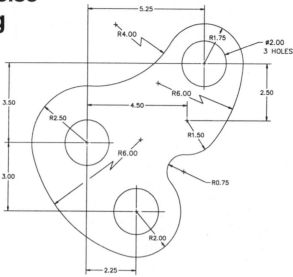

Figure 3–59

Purpose

This tutorial is designed to use geometric commands to construct a one-view drawing of Pattern1 such as the one in Figure 3–59. Refer to the following for special system settings and suggested command sequences.

System Settings

Begin a new drawing called "Pattern1." Use the UNITS command to change the number of decimal places past the zero from four to two. Keep the remaining default unit values. Using the LIMITS command, keep 0,0 for the lower left corner and change the upper right corner from 12,9 to 21.00,16.00.

Layers

Create the following layers with the format:

Name	Color	Linetype
Object	White	Continuous
Center	Yellow	Center
Dim	Yellow	Continuous

Suggested Commands

Begin constructing this object by first laying out four points which will be used as centers for circles. Use the CIRCLE-TTR command to construct tangent arcs to the circles already drawn. Use the TRIM command to clean up and partially delete circles to obtain the outline of the pattern. Then, add the 2.00 diameter holes followed by the center markers using the DIMCENTER command.

Whenever possible, substitute the appropriate command alias in place of the full AutoCAD command in each tutorial step. For example, use "Co" for the COPY command, "L" for the LINE command, and so on. The complete listing of all command aliases is located in Unit 01 on pages 1–9 and 1–10.

Step #1

Use the PDMODE command to change the point style to a value of 2. This will form a "plus sign" when using the POINT command. An alternate method of selecting different points is through the Point Style dialog box which can be found in the pull-down menu under the headings of Format > Point Style...

Locate one point at absolute coordinate 7.50,7.50. Then, use the COPY command and the dimensions in Figure 3–60 as a guide for duplicating the remaining points.

Command: **PDMODE**
New value for Pdmode <0>: **2**

 Command: **POINT**
Point: **7.50,7.50** (Locates the point at "A")

 Command: **COPY**
Select objects: **L** (This should select the point)
Select objects: (Press ⎚ENTER⎚ to continue)
<Base point or displacement>/Multiple: **M**
Base point: **Nod**
of (Select the point at "A")
Second point of displacement: **@2.25,-3.00**
   (Locates point "B")
Second point of displacement: **@4.50,1.00**
   (Locates point "C")
Second point of displacement: **@5.25,3.50**
   (Locates point "D")
Second point of displacement: (Press ⎚ENTER⎚ to exit this command)

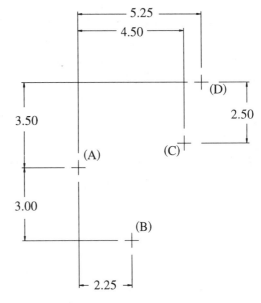

Figure 3–60

Step #2

Use the CIRCLE command to place four circles of
different sizes from points located at "A," "B," "C,"
and "D" as shown in Figure 3–61. When you have
completed drawing the four circles, use the ERASE
command to erase points "A," "B," "C," and "D."

 Command:**CIRCLE**

3P/2P/TTR/<Center point>:**Nod**
of *(Select the point at "A")*
Diameter/<Radius>: **2.50**

 Command:**CIRCLE**

3P/2P/TTR/<Center point>:**Nod**
of *(Select the point at "B")*
Diameter/<Radius>:**2.00**

 Command:**CIRCLE**

3P/2P/TTR/<Center point>:**Nod**
of *(Select the point at "C")*
Diameter/<Radius>:**1.50**

 Command:**CIRCLE**

3P/2P/TTR/<Center point>:**Nod**
of *(Select the point at "D")*
Diameter/<Radius>:**1.75**

 Command:**ERASE**

Select objects: *(Pick the four points labeled
 "A", "B", "C", and "D")*
Select objects: *(Press* ENTER *to execute this
 command)*

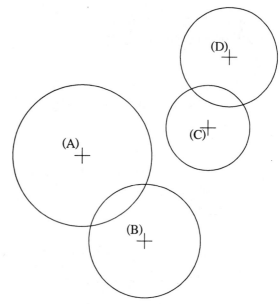

Figure 3–61

Step #3

Use the CIRCLE-TTR command to construct a 4.00-radius circle tangent to the two dashed circles in Figure 3–62. Then, use the TRIM command to trim away part of circle "C."

 Command: **CIRCLE**

3P/2P/TTR/<Center point>: **TTR**
Enter Tangent spec: *(Select the dashed circle at "A")*
Enter second Tangent spec: *(Select the dashed circle at "B")*
Radius: **4.00**

 Command: **TRIM**

Select cutting edges...
Select cutting edges: (Projmode = UCS, Edgemode = No extend)
Select objects: *(Select the two dashed circles shown in Figure 3–62)*
Select objects: *(Press* [ENTER] *to continue)*
<Select object to trim>/Project/Edge/Undo: *(Select the large circle at "C")*
<Select object to trim>/Project/Edge/Undo: *(Press* [ENTER] *to exit this command)*

A built-in undo is provided if a mistake is made during the trimming process.

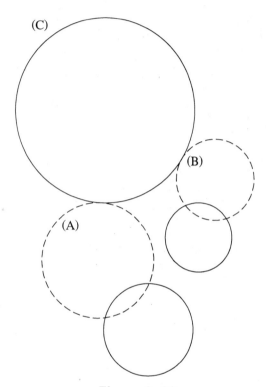

Figure 3–62

Step #4

Use the CIRCLE-TTR command to construct a 6.00-radius circle tangent to the two dashed circles in Figure 3–63. Then, use the TRIM command to trim away part of circle "C."

 Command:**CIRCLE**

3P/2P/TTR/<Center point>:**TTR**
Enter Tangent spec: *(Select the dashed circle at "A")*
Enter second Tangent spec: *(Select the dashed circle at "B")*
Radius:**6.00**

 Command:**TRIM**

Select cutting edges...
Select cutting edges: (Projmode = UCS, Edgemode = No extend)
Select objects: *(Select the two dashed circles at the right)*
Select objects: *(Press [ENTER] to continue)*
<Select object to trim>/Project/Edge/Undo: *(Select the large circle at "C")*
<Select object to trim>/Project/Edge/Undo: *(Press [ENTER] to exit this command)*

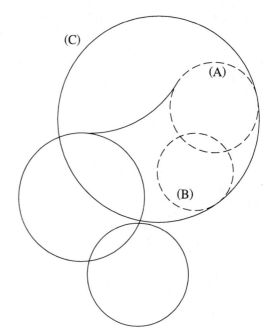

Figure 3–63

Step #5

Use the CIRCLE-TTR command to construct a 6.00-radius circle tangent to the two dashed circles at the right. Then, use the TRIM command to trim away part of circle "C." See Figure 3–64.

 Command:**CIRCLE**

3P/2P/TTR/<Center point>:**TTR**
Enter Tangent spec: *(Select the dashed circle at "A")*
Enter second Tangent spec: *(Select the dashed circle at "B")*
Radius:**6.00**

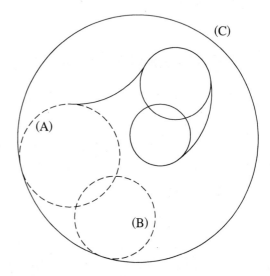

Figure 3–64

Command:**TRIM**

Select cutting edges...
Select cutting edges: (Projmode = UCS,
Edgemode = No extend)
Select objects: *(Select the two dashed circles
shown in Figure 3–64)*

Select objects: *(Press* [ENTER] *to continue)*
<Select object to trim>/Project/Edge/Undo:
(Select the large circle at "C")
<Select object to trim>/Project/Edge/Undo:
(Press [ENTER] *to exit this command)*

Step #6

Use the CIRCLE-TTR command to construct a 0.75-
radius circle tangent to the two dashed circles in
Figure 3–65. Then, use the TRIM command to trim
away part of circle "C."

 Command:**CIRCLE**

3P/2P/TTR/<Center point>:**TTR**
Enter Tangent spec: *(Select the dashed circle
at "A")*
Enter second Tangent spec: *(Select the
dashed circle at "B")*
Radius:**0.75**

 Command:**TRIM**

Select cutting edges...
Select cutting edges: (Projmode = UCS,
Edgemode = No extend)
Select objects: *(Select the two dashed circles
shown in Figure 3–65)*
Select objects: *(Press* [ENTER] *to continue)*
<Select object to trim>/Project/Edge/Undo:
(Select the circle at "C")
<Select object to trim>/Project/Edge/Undo:
(Press [ENTER] *to exit this command)*

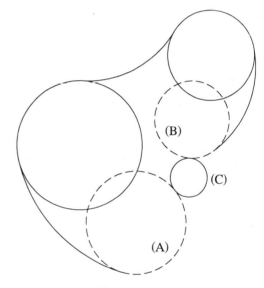

Figure 3–65

Step #7

Use the TRIM command, select all dashed arcs in Figure 3–66 as cutting edges, and trim away the circular segments to form the outline of the pattern1 drawing.

 Command: **TRIM**

Select cutting edges...
Select cutting edges: (Projmode = UCS, Edgemode = No extend)
Select objects: *(Select the four dashed arcs shown in Figure 3–66)*
Select objects: *(Press ENTER to continue)*
<Select object to trim>/Project/Edge/Undo: *(Select the circle at "A")*
<Select object to trim>/Project/Edge/Undo: *(Select the circle at "B")*
<Select object to trim>/Project/Edge/Undo: *(Select the circle at "C")*
<Select object to trim>/Project/Edge/Undo: *(Select the circle at "D")*
<Select object to trim>/Project/Edge/Undo: *(Press ENTER to exit this command)*

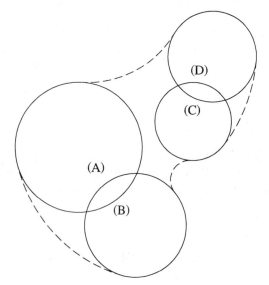

Figure 3–66

Step #8

Your drawing should be similar to the illustration in Figure 3–67.

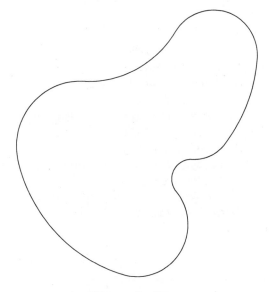

Figure 3–67

Step #9

Use the CIRCLE command to place a circle of 2.00-unit diameter at the center of arc "A". Then, use the COPY command to duplicate the circle at the center of arcs "B" and "C". See Figure 3–68.

 Command:**CIRCLE**

3P/2P/TTR/<Center point>:**Cen**
of *(Select the arc at "A")*
Diameter/<Radius>:**D**
Diameter:**2.00**

 Command:**COPY**

Select objects: **Last**
Select objects: *(Press ⸢ENTER⸣ to continue)*
<Base point or displacement>/Multiple:**M**
Base point: **Cen**
of *(Select the arc at "A")*
Second point of displacement: **Cen**
of *(Select the arc at "B")*
Second point of displacement: **Cen**
of *(Select the arc at "C")*
Second point of displacement: *(Press ⸢ENTER⸣ to exit this command)*

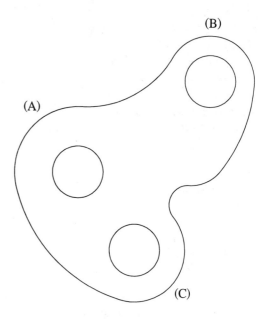

Figure 3–68

Step #10

Use the LAYER command to set the new current layer to Center. Change the variable Dimcen from a value of 0.09 to -0.12. This will place the center marker identifying the centers of circles and arcs when the DIMCENTER command is used. See Figure 3–69.

 Command -**LAYER**

?/Make/Set/New/ON/OFF/Color/Ltype/Freeze/
    Thaw/LOck/Unlock: **Set**
New current layer <0>: **Center**
?/Make/Set/New/ON/OFF/Color/Ltype/Freeze/
    Thaw/LOck/Unlock: *(Press* ENTER *to exit this command)*

Command: **DIMCEN**
Current value <0.09> New value: **-0.12**

 Command: **DIMCENTER**
Select arc or circle: *(Select the arc at "A")*

 Command: **DIMCENTER**
Select arc or circle: *(Select the arc at "B")*

 Command: **DIMCENTER**
Select arc or circle: *(Select the arc at "C")*

 Command: **DIMCENTER**
Select arc or circle: *(Select the arc at "D")*

Step #11

Dimensions may be added. Place them on the layer "Dim" as shown as Figure 3–70.

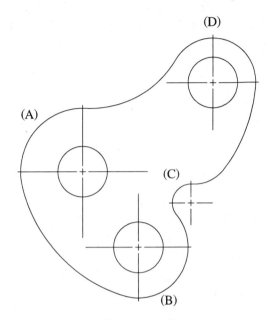

Figure 3–69

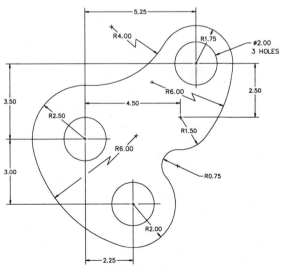

Figure 3–70

Tutorial Exercise
Gear-arm.Dwg

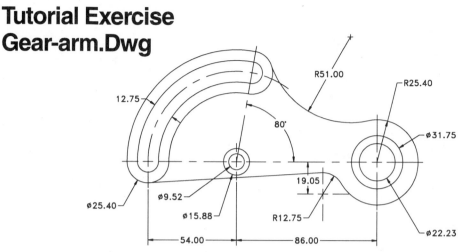

Figure 3–71

Purpose

This tutorial is designed to use geometric commands to construct a one-view drawing of the Gear-arm in metric format as illustrated in Figure 3–71.

System Settings

Begin a new drawing called "Gear-arm." Use the UNITS command to change the number of decimal places past the zero from four to two. Keep the remaining default unit values. Using the LIMITS command, keep 0,0 for the lower left corner and change the upper right corner from 12,9 to 265.00,200.00. Since a layer called "Center" must be created to display centerlines, use the LTSCALE command and change the default value of 1.00 to 25.40. This will make the long and short dashes of the centerlines appear on the display screen.

Layers

Create the following layers with the format:

Name	Color	Linetype
Object	White	Continuous
Center	Yellow	Center
Dim	Yellow	Continuous

Suggested Commands

The object consists of a combination of circles and arcs along with tangent lines and arcs. Use the POINT command to identify and lay out the centers of all circles for construction purposes. Use the ARC command to construct a series of arcs for the left side of the Gear-arm. The TRIM command will be used to trim circles, lines, and arcs to form the basic shape. Also, use the CIRCLE-TTR command for tangent arcs to existing geometry. Since this object is metric, commands such as LTSCALE and DIMSCALE need to be set to the metric-inch equivalent of 25.4 units. This value may be adjusted for better results.

Whever possible, substitute the appropriate command alias in place of the full AutoCAD command in each tutorial step. For example, use "Co" for the COPY command, "L" for the LINE command, and so on. The complete listing of all command aliases is located in Unit 01, on pages 1–9 and 1–10.

Step #1

Begin the gear-arm by drawing two circles of diameters 9.52 and 15.88 using the CIRCLE command and coordinate 112.00,90.00 as the center of both circles. See Figure 3–72.

 Command: **CIRCLE**

3P/2P/TTR/<Center point>:**112.00,90.00**
  (Point "A")
Diameter/<Radius>:**D**
Diameter:**9.52**

 Command: **CIRCLE**

3P/2P/TTR/<Center point>: **@** *(This identifies the last known point as the center of the circle)*
Diameter/<Radius>:**D**
Diameter:**15.88**

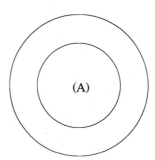

Figure 3–72

Step #2

Use the PDMODE command to change the point style to a value of 2. This will form a "plus" when using the POINT command. Use the PDSIZE command to change the point size to a value of 3 units. Use the POINT command and the Osnap-Center option to place a point at the center of the two circles. See Figure 3–73.

Command:**PDMODE**
New value for PDMODE <0>:**2**

Command:**PDSIZE**
New value for PDSIZE <0>:**3**

 Command:**POINT**

Point: **Cen**
of *(Select the large circle at "A")*

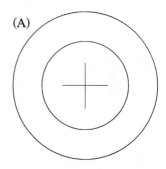

Figure 3–73

Step #3

Use the COPY command to duplicate the point using a polar coordinate distance of 86 units in the 0-degree direction. See Figure 3–74.

 Command:**COPY**

Select objects: Last *(This will select the point)*
Select objects: *(Press* ⏎ *to continue)*
<Base point or displacement>/Multiple: **Cen**
of *(Select the center of the large circle at "A")*
Second point of displacement: **@86<0**

 (A)

Figure 3–74

Step #4

Use the CIRCLE command to place three circles of different sizes from the same point at "A". See Figure 3–75.

 Command:**CIRCLE**

3P/2P/TTR/<Center point>:**Nod**
of *(Select the point at "A")*
Diameter/<Radius>:**25.40**

 Command:**CIRCLE**

3P/2P/TTR/<Center point>: **@** *(This identifies the last known point as the center of the circle)*
Diameter/<Radius>:**D**
Diameter:**31.75**

 Command:**CIRCLE**

3P/2P/TTR/<Center point>: **@** *(This identifies the last known point as the center of the circle)*
Diameter/<Radius>:**D**
Diameter:**22.23**

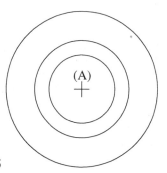
(A)

Figure 3–75

Step #5

Use the COPY command to duplicate point "A" using a polar coordinate distance of 54 units in the 180-degree direction. See Figure 3–76.

 Command:**COPY**

Select objects: *(Select the point at "A")*
Select objects: *(Press* ENTER *to continue)*
<Base point or displacement>/Multiple:**Cen**
of *(Select the center of the large circle at "A")*
Second point of displacement: **@54<180**

   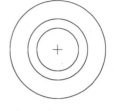

(A)

Figure 3–76

Step #6

Use the CIRCLE command to place two circles of different sizes at point "A" as shown in Figure 3–77. These circles will be converted to arcs in later steps.

 Command:**CIRCLE**

3P/2P/TTR/<Center point>:**Nod**
of *(Select the point at "A")*
Diameter/<Radius>:**D**
Diameter:**25.40**

 Command:**CIRCLE**

3P/2P/TTR/<Center point>: **@** *(This identifies the last known point as the center of the circle)*
Diameter/<Radius>:**D**
Diameter:**12.75**

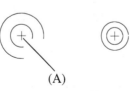   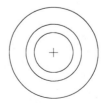

(A)

Figure 3–77

Step #7

Use the COPY command to duplicate point "A" using a polar coordinate distance of 54 units in the 80-degree direction as shown in Figure 3–78.

 Command:**COPY**

Select objects: *(Select the point at "A")*
Select objects: *(Press [ENTER] to continue)*
<Base point or displacement>/Multiple: **Cen**
of *(Select the center of the large circle at "A")*
Second point of displacement: **@54<80**

  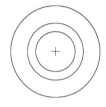

Figure 3–78

Step #8

Use the LINE command to draw a line using a polar coordinate distance of 70 and a direction of 80 degrees. Start the line at point "A". Then use the CIRCLE command to place two circles of different sizes at point "B" as in Figure 3–79. These circles will be converted to arcs in later steps.

 Command:**LINE**

From point: **Cen**
of *(Select the center of the large circle at "A")*
To point: **@70<80**
To point: *(Press [ENTER] to exit this command)*

 Command:**CIRCLE**

3P/2P/TTR/<Center point>:**Nod**
of *(Select the point at "B")*
Diameter/<Radius>:**D**
Diameter:**25.40**

 Command:**CIRCLE**

3P/2P/TTR/<Center point>: **@** *(This identifies the last known point as the center of the circle)*
Diameter/<Radius>:**D**
Diameter:**12.75**

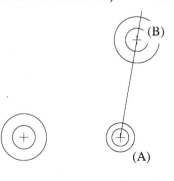 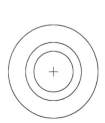

Figure 3–79

Step #9

Use the ARC command and draw an arc using point "A" as the center, point "B" as the start point, and point "C" as the endpoint as shown in Figure 3–80.

 Command: **ARC**

Center/<Start point>: **C**
Center: **Cen**
of *(Select the center of the circle at "A")*
Start point: **Int**
of *(Select the intersection of the line and circle at "B")*
Angle/Length of chord/<End point>: **Qua**
of *(Select the quadrant of the circle at "C")*

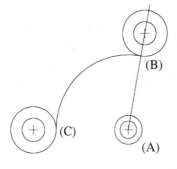
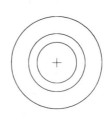

Figure 3–80

Step #10

Use the ARC command and draw an arc using point "A" as the center, point "B" as the start point, and point "C" as the endpoint as shown in Figure 3–81.

 Command: **ARC**
Center/<Start point>: **C**

Center: **Cen**
of *(Select the center of the circle at "A")*
Start point: **Int**
of *(Select the intersection of the line and circle at "B")*
Angle/Length of chord/<End point>: **Qua**
of *(Select the quadrant of the circle at "C")*

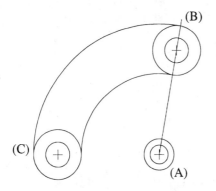

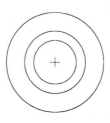

Figure 3–81

Step #11

Use the TRIM command, select the two dashed arcs in Figure 3–82 as cutting edges, and trim the two circles at points "C" and "D."

 Command:**TRIM**

Select cutting edge(s)...
Select cutting edges: (Projmode = UCS, Edgemode = No extend)
Select objects: *(Select the two dashed arcs "A" and "B")*

Select objects: *(Press ⏎ to continue)*
<Select object to trim>/Project/Edge/Undo: *(Select the circle at "C")*
<Select object to trim>/Project/Edge/Undo: *(Select the circle at "D")*
<Select object to trim>/Project/Edge/Undo: *(Press ⏎ to exit this command)*

A built-in undo is provided if a mistake is made by trimming the wrong object.

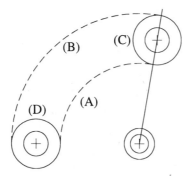
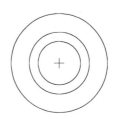

Figure 3–82

Step #12

Your drawing should be similar to the illustration in Figure 3–83.

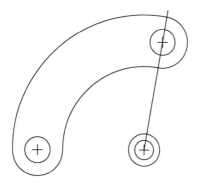
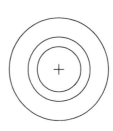

Figure 3–83

Step #13

Use the ARC command and draw an arc using point "A" as the center, point "B" as the start point, and point "C" as the endpoint as shown in Figure 3–84.

 Command: **ARC**

Center/<Start point>: **C**
Center: **Cen**
of *(Select the center of the circle at "A")*
Start point: **Int**
of *(Select the intersection of the line and circle at "B")*
Angle/Length of chord/<End point>: **Qua**
of *(Select the quadrant of the circle at "C")*

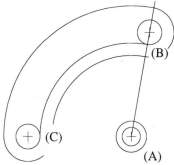

Figure 3–84

Step #14

Use the ARC command and draw an arc using point "A" as the center, point "B" as the start point, and point "C" as the endpoint as shown in Figure 3–85.

 Command: **ARC**

Center/<Start point>: **C**
Center: **Cen**
of *(Select the center of the circle at "A")*
Start point: **Int**
of *(Select the intersection of the line and circle at "B")*
Angle/Length of chord/<End point>: **Qua**
of *(Select the quadrant of the circle at "C")*

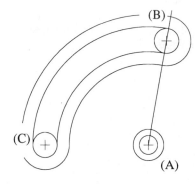

Figure 3–85

Step #15

Use the TRIM command, select the two dashed arcs at the right as cutting edges, and trim the two circles at points "C" and "D" as shown in Figure 3–86.

 Command:**TRIM**

Select cutting edge(s)...
Select cutting edges: (Projmode = UCS, Edgemode = No extend)
Select objects: *(Select the two dashed arcs "A" and "B")*
Select objects: *(Press* ⌷ENTER⌷ *to continue)*
<Select object to trim>/Project/Edge/Undo: *(Select the circle at "C")*

<Select object to trim>/Project/Edge/Undo: *(Select the circle at "D")*
<Select object to trim>/Project/Edge/Undo: *(Press* ⌷ENTER⌷ *to exit this command)*

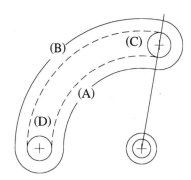

Figure 3–86

Step #16

Your drawing should be similar to the illustration in Figure 3–87. Always perform periodic screen redraws using the REDRAW command to clean up the display screen.

 Command:**REDRAW**

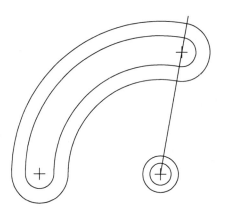

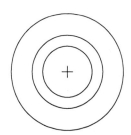

Figure 3–87

Step #17

Use the LINE command and draw a line from the quadrant of the small circle to the quadrant of the large circle in Figure 3–88. This line is used only for construction purposes.

 Command: **LINE**

From point: **Qua**
of *(Select the circle at "A")*
To point: **Qua**
of *(Select the circle at "B")*
To point: *(Press ENTER to exit this command)*

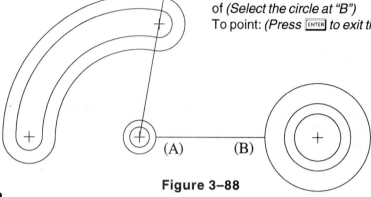

(A)   (B)

Figure 3–88

Step #18

Use the MOVE command to move the dashed line down the distance 19.05 units using the polar coordinate mode. Then, use the OFFSET command to offset the dashed circle the distance 12.75 units. The intersection of these two objects will be used to draw a 12.75 radius circle shown as that illustrated in Figure 3–89.

Select objects: *(Press ENTER to continue)*
Base point or displacement: **Endp**
of *(Select the endpoint of the line at "A")*
Second point of displacement: **@19.05<270**

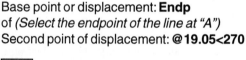 Command: **OFFSET**

Offset distance or Through<Through>: **12.75**
Select object to offset: *(Select the circle at "B")*
Side to offset? *(Select a blank part of the screen at "C")*
Select object to offset: *(Press ENTER to exit this command)*

 Command: **MOVE**

Select objects: *(Select the dashed line in Figure 3–89)*

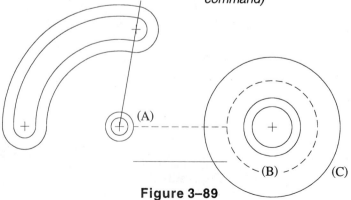

(A)

(B)   (C)

Figure 3–89

Step #19

Use the CIRCLE command to draw a circle with a radius of 12.75. Use the center of the circle as the intersection of the large dashed circle and the dashed horizontal line illustrated in Figure 3–90. Use the ERASE command to erase the dashed circle and dashed line.

 Command:**CIRCLE**

3P/2P/TTR/<Center point>:**Int**
of *(Select the intersection of the line and circle at "A")*
Diameter/<Radius>:**12.75**

 Command:**ERASE**

Select objects: *(Select the dashed circle and dashed line)*
Select objects: *(Press* ENTER *to execute this command)*

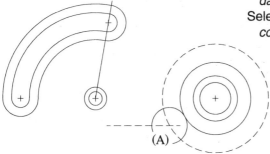

Figure 3–90

Step #20

Use the LINE command to draw a line from a point tangent to the arc at "A" to a point tangent to the circle at "B" as shown in Figure 3–91. Use the Osnap-Tangent option to accomplish this.

 Command:**LINE**

From point: **Tan**
to *(Select the arc at "A")*
To point: **Tan**
to *(Select the circle at "B")*
To point: *(Press* ENTER *to exit this command)*

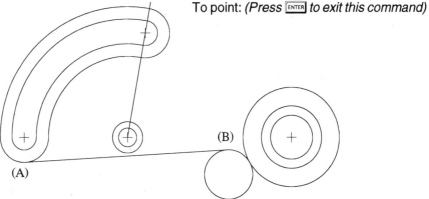

Figure 3–91

Step #21

Use the TRIM command, select the dashed line and dashed circle illustrated in Figure 3–92 as cutting edges, and trim the circle at "A".

 Command: **TRIM**

Select cutting edge(s)...

Select cutting edges: (Projmode = UCS, Edgemode = No extend)
Select objects: *(Select the dashed line and dashed circle in Figure 3–92)*
Select objects: *(Press* ENTER *to continue)*
<Select object to trim>/Project/Edge/Undo: *(Select the circle at "A")*
<Select object to trim>/Project/Edge/Undo: *(Press* ENTER *to exit this command)*

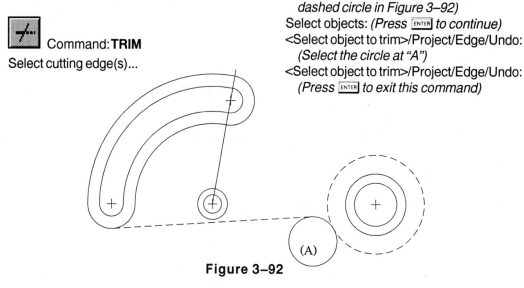

Figure 3–92

Step #22

Use the CIRCLE-TTR command to draw a circle tangent to the arc at "A" and tangent to the circle at "B" with a radius of 51 as in Figure 3–93. When using the TTR option in the CIRCLE command, the Osnap-Tangent option is automatically invoked.

Command: **CIRCLE**

3P/2P/TTR/<Center point>: **TTR**
Enter Tangent spec: *(Select the arc at "A")*
Enter second Tangent spec: *(Select the circle at "B")*
Radius: **51**

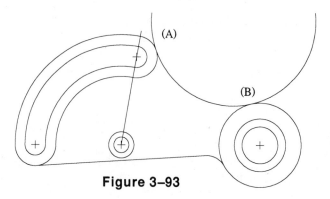

Figure 3–93

Step #23

Use the TRIM command, select the dashed arc and dashed circle illustrated in Figure 3–94 as cutting edges, and trim the large 51 radius circle at "A".

 Command:**TRIM**

Select cutting edge(s)...
Select cutting edges: (Projmode = UCS, Edgemode = No extend)

Select objects: *(Select the dashed arc and dashed circle in Figure 3–94)*
Select objects: *(Press* ENTER *to continue)*
<Select object to trim>/Project/Edge/Undo: *(Select the circle at "A")*
<Select object to trim>/Project/Edge/Undo: *(Press* ENTER *to exit this command)*

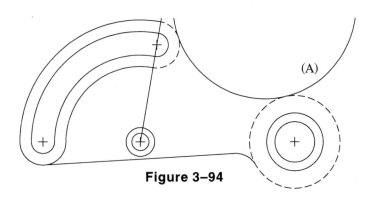

(A)

Figure 3–94

Step #24

Use the TRIM command, select the two dashed arcs illustrated in Figure 3–95 as cutting edges, and trim the circle at "C". Use the ERASE command to delete all four points used to construct the circles.

 Command:**TRIM**

Select cutting edge(s)...
Select cutting edges: (Projmode = UCS, Edgemode = No extend)
Select objects: *(Select the dashed arc "A" and dashed arc "B" in Figure 3–95)*

Select objects: *(Press* ENTER *to continue)*
<Select object to trim>/Project/Edge/Undo: *(Select the circle at "C")*
<Select object to trim>/Project/Edge/Undo: *(Press* ENTER *to exit this command)*

 Command:**ERASE**

Select objects: *(Select the four points illustrated in Figure 3–95)*
Select objects: *(Press* ENTER *to execute this command)*

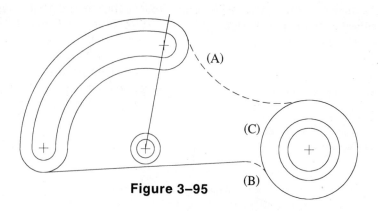

Figure 3–95

Step #25

Your display should appear similar to the illustration in Figure 3–96. Notice the absence of the points that were erased in the previous step. Standard center lines will be placed to mark the center of all circles in the next series of steps. Use the LAYER command to set the new current layer to Center. Change LTSCALE to 25.40.

Command:**LTSCALE**
New scale factor <1.0000>:**25.40**

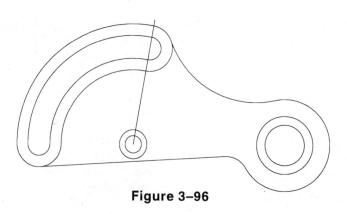

Figure 3–96

Step #26

To place center lines for all circles and arcs, two dimension variables need to be changed. The dimension variable, Dimscale, is set to a value of 1. Since this is a metric drawing, the Dimscale variable needs to be changed to a value of 25.4. This will increase all variables by this value which is necessary because we are drawing in metric units. Also, the Dimcen variable needs to be changed from a value of 0.09 to -0.09. The negative value will extend the center lines past the edge of the circle when using the DIMCENTER command. Erase the top of the center line at "E". See Figure 3–97.

 Command:**ERASE**

Select Objects: *(Select the top of the line at "E")*
Select Objects: *(Press ⌷ENTER⌷ to execute this command)*

Command:**DIMSCALE**
Current value <1> New value: **25.4**

Command:**DIMCEN**
Current value <0.09> New value: **-0.09**

 Command:**DIMCENTER**

Select arc or circle: *(Select the arc at "A")*
Repeat the above command to place a center marker at arcs "B", "C", and "D".

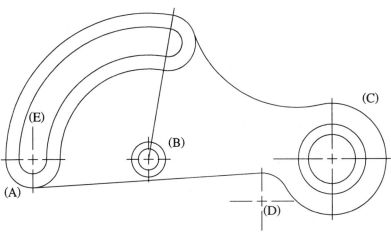

Figure 3–97

Step #27

Use the OFFSET command to offset the inside arc at "A" the distance 6.375 units. Indicate a point in the vicinity of "B" for the side to perform the offset as shown in Figure 3–98.

 Command: **OFFSET**

Offset distance or Through<Through>: **6.375**
Select object to offset: *(Select the arc at "A")*
Side to offset? *(Select a point in the vicinity of "B")*
Select object to offset: *(Press* ENTER *to exit this command)*

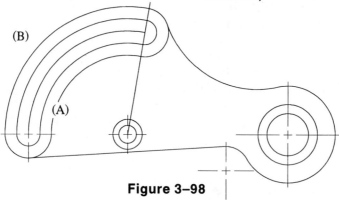

Figure 3–98

Step #28

Use the CHPROP command to change the arc at "A" and the line at "B" to the Center layer as shown in Figure 3–99. This layer should have been created earlier for the arc and line to change to the proper color and linetype.

 Command: **CHPROP**

Select objects: *(Select the middle arc at "A")*
Select objects: *(Select the line at "B")*
Select objects: *(Press* ENTER *to continue)*
Change what property(Color/LAyer/LType/
    Thickness): **LA**
New layer <0>: **Center**
Change what property(Color/LAyer/LType/
    Thickness): *(Press* ENTER *to exit this command)*

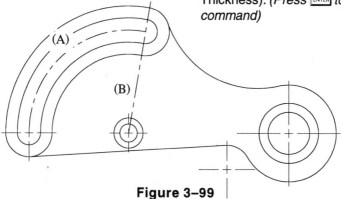

Figure 3–99

Step #29

Use the EXTEND command to extend the center line arc to intersect with the circular arc at "A"; see the illustration in Figure 3–100. Next, reset the dimension variable Dimcen from -0.09 to a new value of 0.09. This will change the center point to a plus without the center lines extending beyond the arc when using the DIMCENTER command. The finished object may be dimensioned as an optional step as illustrated in Figure 3–101.

 Command: **EXTEND**
Select boundary edge(s)...
Select boundary edges: (Projmode = UCS, Edgemode = No extend)
Select objects: *(Select the arc at "A")*
Select objects: *(Press* ENTER *to continue)*
<Select object to extend>/Project/Edge/Undo: *(Select the center line arc)*
Select object to extend: *(Press* ENTER *to exit this command)*

Command: **DIMCEN**
New value for DIMCEN <-0.09>: **0.09**

 Command: **DIMCENTER**
Select arc or circle: *(Select the arc at "B")*

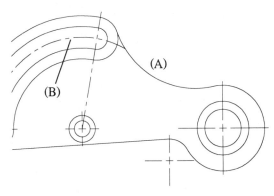

Figure 3–100

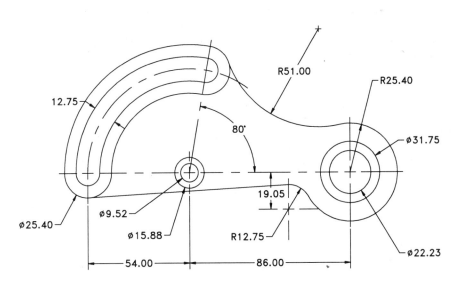

Figure 3–101

Problems for Unit 3

Directions for Problems 3-1 through 3-21:
  Construct these geometric construction figures using existing AutoCAD commands.

Problem 3-1

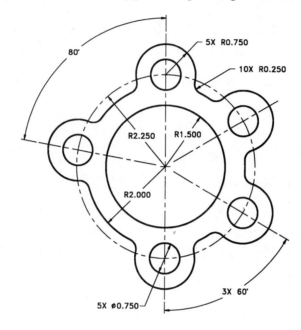

Problem 3-2

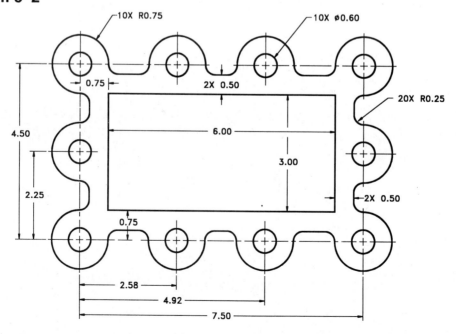

Problem 3–3

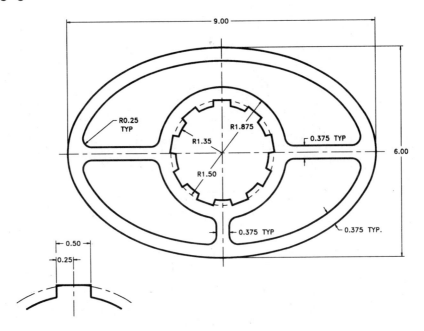

Problem 3–4

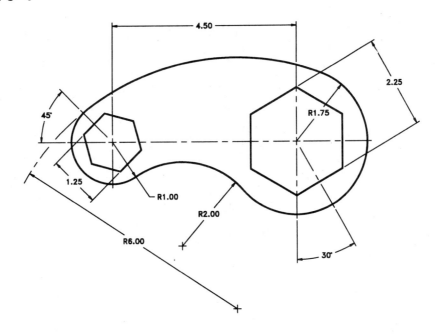

Problem 3–5

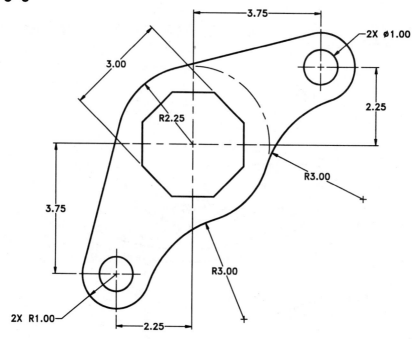

Problem 3–6

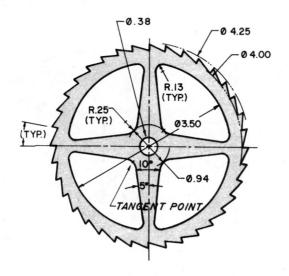

Problem 3–7

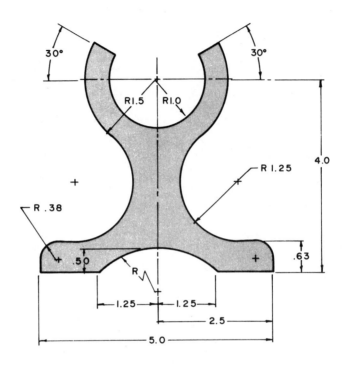

Problem 3–8

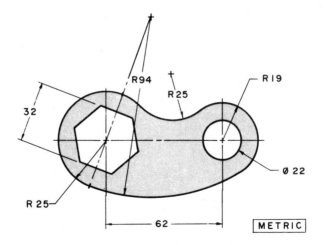

Problem 3–9

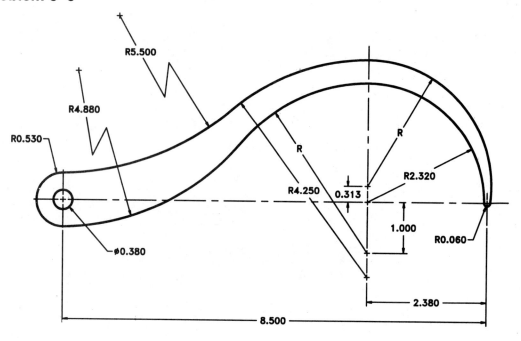

Problem 3–10

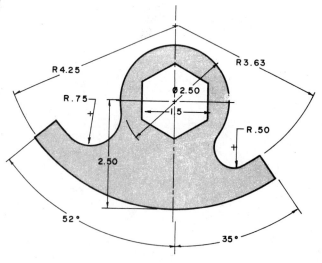

Problem 3–11

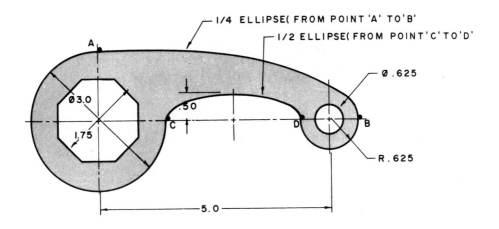

Problem 3–12

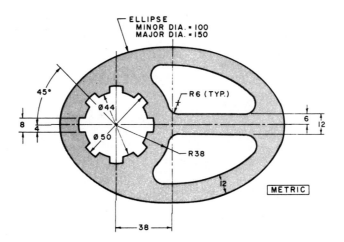

Problem 3–13

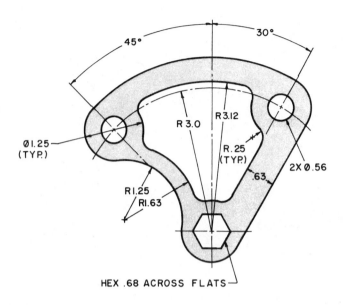

Problem 3–14

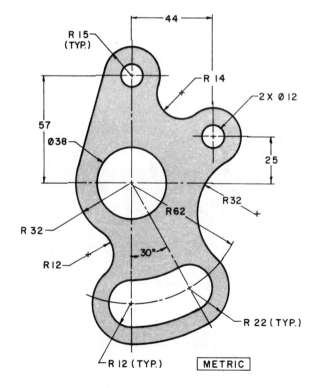

METRIC

Problem 3–15

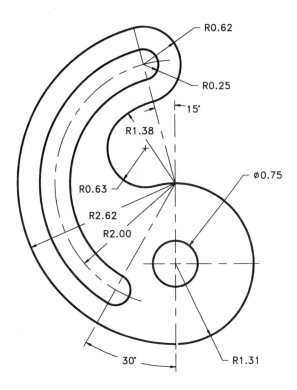

Problem 3–16

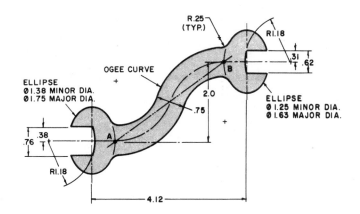

Problem 3–17

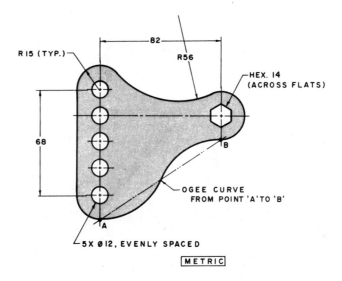

Problem 3–18

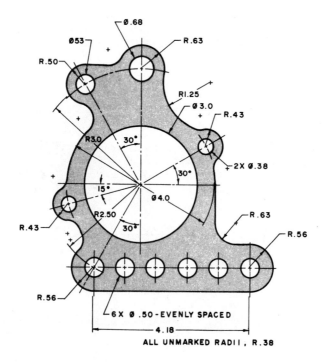

Problem 3–19

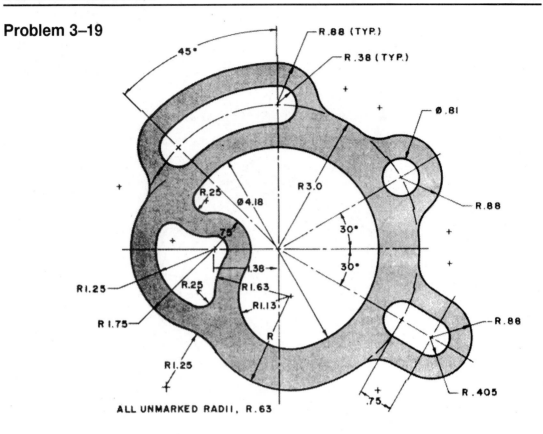

ALL UNMARKED RADII, R.63

Problem 3–20

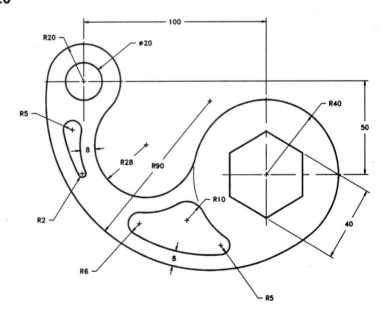

Problem 3–21

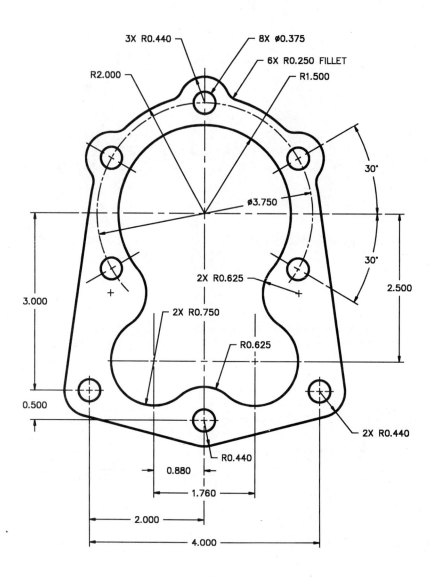

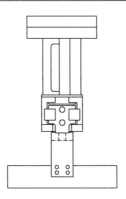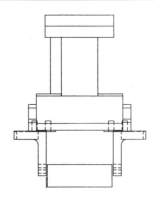

Shape Description/
Multiview Projection

Before any object is made in production, some type of drawing needs to be created. This is not just any drawing but rather an engineering drawing consisting of overall object sizes with various views of the object organized on the computer screen. This chapter introduces the topic of shape description or how many views are really needed to describe an object. The art of multiview projection includes methods of constructing one-view, two-view, and three-view drawings using AutoCAD commands. Linetypes are explained as a method of communicating hidden features located in different views of a drawing.

Shape Description

Before performing engineering drawings, you must first analyze the object being drawn. To do this describe the object by views or how an observer looks at the object. In Figure 4–1, the simple wedge, it is no surprise that this object can be viewed at almost any angle to get a better idea of its basic shape. However, to standardize how all objects are to be viewed and to limit confusion usually associated with complex multiview drawings, some standard method of determining how and where to view the object must be exercised.

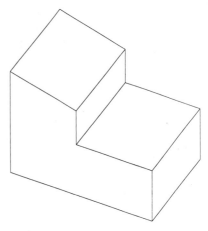

Figure 4–1

Even though the simple wedge is easy to understand because it is currently being displayed in picture or isometric form, it would be difficult to produce this object since it is unclear what the sizes of the front and top views are. In this way, the picture of the object is separated into six primary ways or directions to view an object, which are illustrated in Figure 4–2. The Front view begins the shape description followed by the Top view and Right Side view. Continuing on, the Left Side view, Back view, and Bottom view complete the primary ways to view an object.

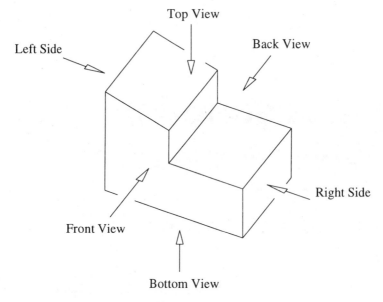

Figure 4–2

Now that the primary ways of viewing an object have been established, the views need to be organized to promote clarity and have the views reference themselves. Imagine the simple wedge positioned in a transparent glass box similar to the illustration in Figure 4–3. With the entire object at the center of the box, the sides of the box represent the ways to view the object. Images of the simple wedge are projected onto the glass surfaces of the box.

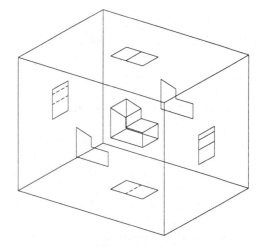

Figure 4–3

With the views projected onto the sides of the glass box, we must now prepare the views to be placed on a two-dimensional (2D) drawing screen. To accomplish this, the glass box, which is hinged, is unfolded as in Figure 4-4. All folds occur from the Front view, which remains stationary.

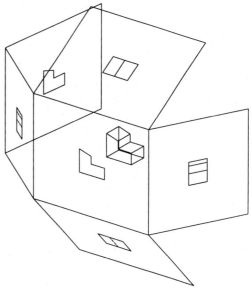

Figure 4-4

Figure 4-5 shows the views in their proper alignment to one another. However this illustration is still in a pictorial view. These views need to be placed flat before continuing.

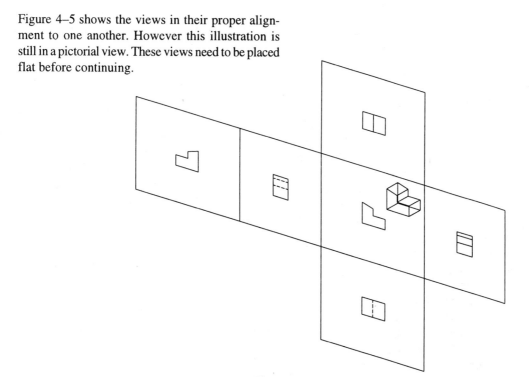

Figure 4-5

As the glass box is completely unfolded and laid flat, the result is illustrated in Figure 4–6. The Front view becomes the main view with other views being placed in relation to it. Above the Front view is the Top view. To the right of the Front is the Right Side view. To the left of the Front is the Left Side view followed by the Back view. Underneath the Front view is the Bottom view. This becomes the standard method of laying out the necessary views needed to describe an object. But are all views necessary? Upon closer inspection we find that, except for being a mirror image, the Front and Back views are identical. The Top and Bottom views appear similar as do the Right and Left Side views. One very important rule to follow in multiview objects is to select only those views that accurately describe the object and discard the remaining views.

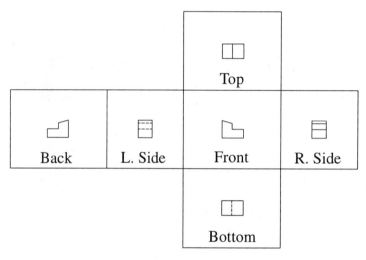

Figure 4–6

The complete multiview drawing of the simple wedge is illustrated in Figure 4–7. Only the Front, Top, and Right Side views are needed to describe this object. When laying out views remember that the Front view is usually the most important and holds the basic shape of the object being described. Directly above the Front view is the Top view, and to the right of the Front is the Right Side view. All three views are separated by a space of various sizes. This space is commonly called a dimension space because it becomes a good area to place dimensions describing the size of the object. The space also acts as a separator between views; without it, the views would touch on one another, making them difficult to read and interpret. The minimum distance of this space is usually 1.00 units.

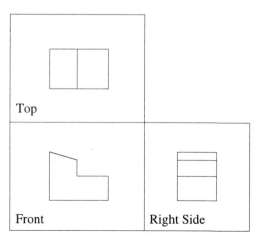

Figure 4–7

Relationships Between Views

Some very interesting and important relationships are set up when placing the three views in the configuration illustrated in Figure 4–8. Notice that since the Top view is directly above the Front view, both share the same Width dimension. The Front and Right Side views share the same Height. The relationship of Depth on the Top and Right Side views can be explained by constructing a 45-degree projector line at "A" and projecting the lines over and down or vice versa to get the Depth. Yet another principle illustrated by this example is that of projecting lines up, over, across, and down to create views. Editing commands such as ERASE and TRIM are then used to clean up unnecessary lines.

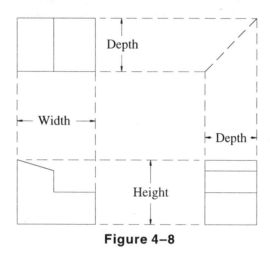

Figure 4–8

With the three views identified in Figure 4–9, the only other step, and it is an important one, is to annotate the drawing, or add dimensions to the views. With dimensions, the object drawn up in multiview projection can now be produced. Even though this has been a simple example, the methods of multiview projection work even for the most difficult and complex of objects.

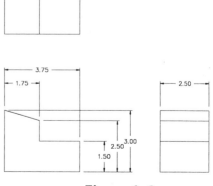

Figure 4–9

Linetypes and Conventions

At the heart of any engineering drawing is the ability to assign different types of lines to convey meaning to the drawing. When plotted out, all lines of a drawing are dark; border and title block lines are the thickest. Object lines outline visible features of a drawing and are made thick and dark, but not as thick as a border line. To identify features that are invisible in an adjacent view, a hidden line is used. This line is a series of dashes 0.12 units in length with a spacing of 0.06 units. Centerlines are used to identify the centers of circular features such as holes or to show that a hidden feature in one view is circular in another. The centerline consists of a series of long and short dashes. The short dash measures approximately 0.12 units whereas the long dash may vary from 0.75 to 1.50 units. A gap of 0.06 is placed in between dashes. Study the examples of these lines in Figure 4–10.

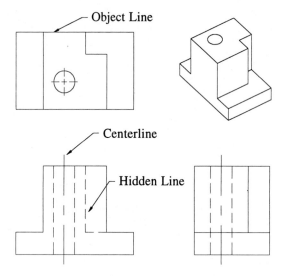

Figure 4–10

The object shown in Figure 4–11 is a good illustration of the use of phantom lines in a drawing. Phantom lines are expecially useful where motion is applied. The arm on the right is shown using standard object lines consisting of the continuous linetype. To show that the arm rotates about a center pivot, the identical arm is duplicated using the ARRAY command and all lines of this new element converted to phantom lines using the CHPROP, DDCHPROP, or DDMODIFY commands. Notice that the smaller segments are not shown as phantom lines due to their short sizes.

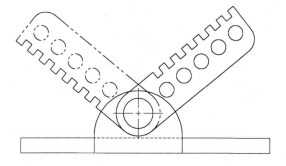

Figure 4–11

When using manual drafting techniques for drawing hidden and centerlines, the drafter first had to develop a skill of drawing a series of dashes as individual line segments along with spacing each dash equally. This was very tedious work, but with practice and experience, the method proved acceptable. Since CAD has come along, you have the option of assigning a linetype to the drawing. This means that if the current linetype is called HIDDEN, the series of dashes and spaces will automatically be drawn once a line, circle, arc, or any drawing entity is chosen. Illustrated in Figure 4–12 are the standard linetypes supplied with AutoCAD ever since the software was introduced in 1982.

```
DASHED,__ __ __ __ __ __ __ __ __ __ __ __ __ __ __
HIDDEN,_ _ _ _ _ _ _ _ _ _ _ _ _ _ _ _ _ _ _ _ _ _ _ _
CENTER,____ _ ____ _ ____ _ ____ _ ____ _ ____ _ ____
PHANTOM,____ __ __ ____ __ __ ____ __ __ ____ __ __
DOT,.................................................
DASHDOT,__ . __ . __ . __ . __ . __ . __ . __ . __ . __
BORDER,__ __ . __ __ . __ __ . __ __ . __ __ . __ __
DIVIDE,__ . . __ . . __ . . __ . . __ . . __ . . __
```

Figure 4–12

The linctypes illustrated in Figure 4–13 are some of the more commonly used linetype definitions supplied with AutoCAD. These linetypes may be viewed by using the LOAD option of either the Linetype or Layer control dialog boxes. The lines in Figure 4–13 are only a partial listing of the complete list. The CENTER linetype is identical to the standard linetype on the previous page. There also exist two more types of centerlines; CENTER2 is the same type of center linetype except that all dashes and spaces are half the size of the orginal. CENTERX2 has the original linetype doubled in size. All current linetypes of AutoCAD have three possible linetypesfrom which to choose.

```
BORDER,__ __ . __ __ . __ __ . __ __ . __ __ . __ __ . __ __
BORDER2,__ . __ . __ . __ . __ . __ . __ . __ . __ . __ . __ . __
BORDERX2,____ ____ . ____ ____ . ____ ____ . ____
CENTER,____ _ ____ _ ____ _ ____ _ ____ _ ____ _ ____
CENTER2,___ _ ___ _ ___ _ ___ _ ___ _ ___ _ ___ _ ___
CENTERX2,_____ __ _____ __ _____ __
DASHDOT,__ . __ . __ . __ . __ . __ . __ . __ . __ . __
DASHDOT2,__._._._._._._._._._._._._._._._._._._._
DASHDOTX2,____ . ____ . ____ . ____ . ____ . ____
DASHED,__ __ __ __ __ __ __ __ __ __ __ __ __
DASHED2,_ _ _ _ _ _ _ _ _ _ _ _ _ _ _ _ _ _ _ _ _ _ _
DASHEDX2,____ ____ ____ ____ ____ ____ ____
```

Figure 4–13

Use of linetypes in a drawing is crucial to the interpretation of the views and the final design before the object is actually made. Sometimes the linetype appears too long; in other cases the linetype does not appear at all even though using the LIST command on the object will show the proper layer and linetype. The LTSCALE command is used to manipulate the size of all linetypes loaded into a drawing. By default, all linetypes are assigned a scale factor of 1.00. This means that the actual dashes and/or spaces of the linetype are multiplied by this factor. The views illustrated in Figure 4–14 show linetypes that use the default value of 1.00 from the LTSCALE command.

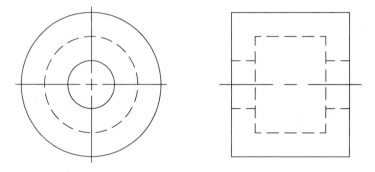

Figure 4–14

If a linetype appears too long, use the LTSCALE command and set a new value to less than 1.00. If a linetype appears too short, use the LTSCALE command and set a new value to greater than 1.00. The same views illustrated in Figure 4–15 show the effects of the LTSCALE command set to a new value of 0.75. Notice that the center in the right side view has one more series of dashes than the same object illustrated previously. The 0.75 multiplier that affects all dashes and spaces defined in the linetype.

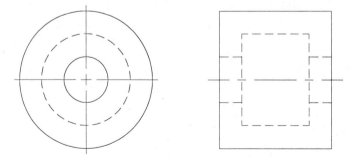

Figure 4–15

Figure 4–16 illustrates how using the LTSCALE command with a new value of 0.50 shortens the linetypes even more. Now even the center marks identifying the circles have been changed into centerlines. When using the LTSCALE command, the new value, whether larger or smaller that 1.00, affects all linetypes visible on the display screen.

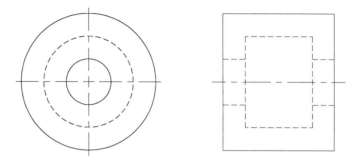

Figure 4–16

One-View Drawings

An important rule to remember concerning multiview drawings is to draw only enough views to accurately describe the object. In the drawing of the gasket in Figure 4–17, a front and side view are shown. However, the side view is so narrow that it is difficult to interpret the hidden lines drawn inside. A better approach would be to leave out the side view and construct a one-view drawing consisting of just the front view.

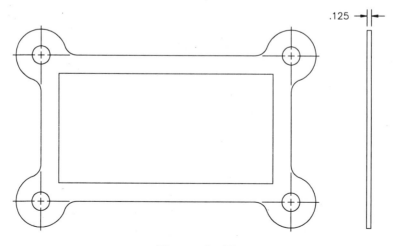

Figure 4–17

Begin the one-view drawing of the gasket by first laying out centerlines marking the centers of all circles and arcs as in Figure 4–18. A layer containing centerlines could be used to show all lines as centerlines.

.125 GASKET THICKNESS

Figure 4–18

Use the CIRCLE command to lay out all circles representing the bolt holes of the gasket in Figure 4–19. The OFFSET command could be used to form the large rectangle on the inside of the gasket. If lines of the rectangle extend past each other, use the FILLET command set to a value of "0." Selecting two lines of the rectangle will form a corner. Repeat this procedure for any other lines that do not form exact corners.

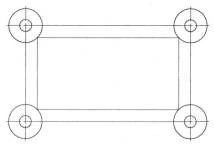

Figure 4–19

Use the TRIM command to begin forming the outside arcs of the gasket in Figure 4–20.

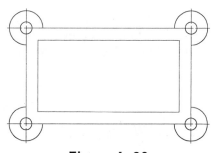

Figure 4–20

Use the FILLET command set to the desired radius to form a smooth transition from the arcs to the outer rectangle in Figure 4–21.

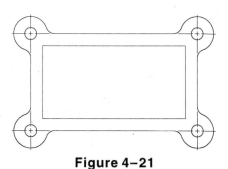

Figure 4–21

Two-View Drawings

Before attempting any drawing, determine how many views need to be drawn. A minimum number of views are needed to describe an object. Drawing extra views is not only time consuming, but may result in two identical views with mistakes in each view. The operator must interpret which is the correct set of views. The illustration in Figure 4–22 is a three-view multiview drawing of a coupler. The front view is identified by the circles and circular hidden circle. Except for their rotation angles, the top and right side views are identical. In this example or for other symmetrical objects, only two views are needed to accurately describe the object being drawn. The top view has been deleted to leave the front and right side views. The side view could have easily been deleted in favor of leaving the front and top views. This decision is up to the designer depending on sheet size and which views are best suited for the particular application.

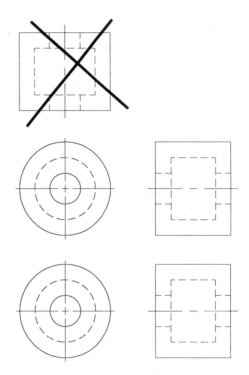

Figure 4–22

To illustrate how AutoCAD is used as the vehicle for creating a two-view engineering drawing, study the pictorial drawing illustrated in Figure 4–23 to get an idea of how the drawing will appear. Begin the two-view drawing by using the LINE command to lay out the front and side views. The width of the top view may be found by projecting lines up from the front since both views share the same width. Provide a space of 1.50 units in between views to act as a separator and allow for dimensions at a later time.

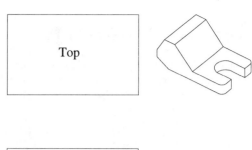

Figure 4–23

Begin adding visible details to the views such as circles, filleted corners, and angles as shown in Figure 4–24. Use various editing commands such as TRIM, EXTEND, and OFFSET to clean up unnecessary geometry.

Figure 4–24

From the front view, project corners up into the top view. These corners will form visible edges in the top view. Use the same projection technique to project features from the top view into the front view (Figure 4–25).

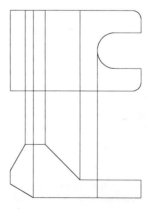

Figure 4–25

Use the TRIM command to delete any geometry that appears in the 1.50 dimension space. The views now must conform to engineering standards by determining which lines are visible and which are invisible as shown in Figure 4–26. The corner at "A" represents an area hidden in the top view. Use the CHPROP or DDCHPROP commands to convert the line in the top view from the continuous linetype to the hidden linetype. In the same manner, the slot visible in the top view is hidden in the front view. Again use CHPROP or DDCHPROP to convert the continuous line in the front view to the hidden linetype. Since the slot in the top view represents a circular feature, use the DIMCENTER command to place a center marker at the center of the semicircle. To show in the front view that the hidden line represents a circular feature, add one centerline consisting of one short dash and two short dashes. If the slot in the top view was square instead of circular, centerlines would not be necessary.

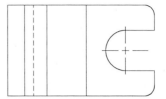

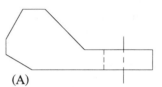

(A)

Figure 4–26

Use the spaces provided to properly add dimensions to the drawing as shown in Figure 4–27. Once the dimension spaces are filled with numbers, use outside areas to call out distances. Placing dimensions will be discussed in a later chapter.

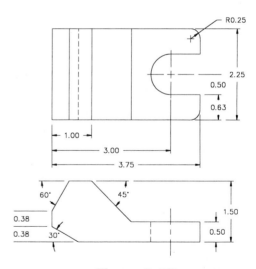

Figure 4–27

Three-View Drawings

If two views are not enough to describe an object, draw three views. This consists of front, top, and right side views. A three-view drawing of the guide block as illustrated in pictorial format in Figure 4–28, will be the focus of this segment. Notice the broken section exposing the spotfacing operation above a drill hole. Begin this drawing by laying out all views using overall dimensions of width, depth, and height. The LINE command along with OFFSET are popular commands used to accomplish this. Provide a space in between views to accommodate dimensions at a later time.

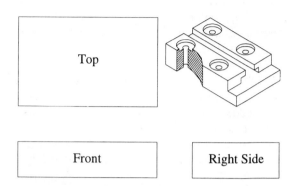

Figure 4–28

Begin drawing features in the views they appear visible in (Figure 4–29). Since the spotface holes appear above, draw these in the top view. The notch appears in the front view; draw it there. A slot is visible in the right side view and is drawn there.

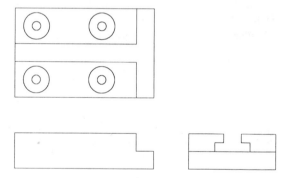

Figure 4–29

As in two-view drawings, all features are projected down from top to front view. To project depth measurements from top to right side views, construct a 45 degree line at "A." See Figure 4–30.

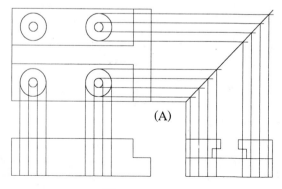

(A)

Figure 4–30

Use the 45-degree line to project the slot from the right side to the top view as shown in Figure 4–31. Project the height of the slot from the right side to the front view. Use the CHPROP or DDCHPROP command to convert continuous lines to hidden lines where features appear invisible such as the holes in the front and right side views.

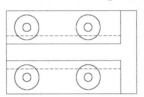

Figure 4–31

Use the CHPROP or DDCHPROP command to change the remaining lines from continuous to hidden. Erase any construction lines including the 45-degree projection line (Figure 4–32).

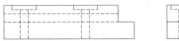

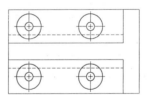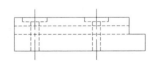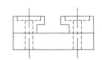

Figure 4–32

Begin adding centerlines to label circular features as shown in Figure 4–33. The DIMCENTER command is used where the circles are visible. Where features are hidden but represent circular features, the single centerline consisting of one short dash and two long dashes is used. In Figure 4–34, dimensions remain the final step in completing the engineering drawing before being checked and shipped off for production.

Figure 4–33

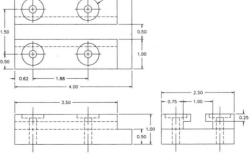

Figure 4–34

Fillets and Rounds

Numerous objects require highly finished and polished surfaces consisting of extremely sharp corners. Fillets and rounds represent the opposite case where corners are rounded off either for ornamental purposes or required by design. Generally a fillet consists of a rounded edge formed in the corner of an object as illustrated in Figure 4–35 at "A." A round is formed at an outside corner similar to "B." Fillets and rounds are primarily used where objects are cast or made from poured metal. The metal will form easier around a pattern that has rounded corners versus sharp corners which usually break away. Some drawings have so many fillets and rounds that a note is used to convey the size of all similar to "All Fillets and Rounds 0.125 Radius."

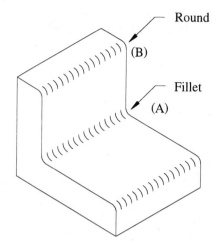

Figure 4–35

AutoCAD provides the FILLET command that allows the user to enter a radius followed by the selection of two lines. The result will be a fillet of the specified radius to the two lines selected. The two lines are also automatically trimmed leaving the radius drawn from the endpoint of one line to the endpoint of the other line. Illustrated in Figure 4–36 are examples of using the FILLET command. Because sometimes the command is used over and over again, the MULTIPLE automatically repeats the next command entered from the keyboard, in this case the FILLET command. Since MULTIPLE continually repeats the command, use the [Esc] key to cancel the command and return to the command prompt.

Command: **MULTIPLE**
FILLET *(Enter in this command)*
(TRIM mode) Current fillet radius = 0.5000
Polyline/Radius/Trim/<Select first object>:
    Radius
Enter fillet radius <0.0000>: **0.25**
Polyline/Radius/Trim/<Select first object>:
    (Select at "A" and "B")
Polyline/Radius/Trim/<Select first object>:
    (Select at "B" and "C")
Polyline/Radius/Trim/<Select first object>:
    (Select at "C" and "D")
Polyline/Radius/Trim/<Select first object>:
    (Strike the Esc key to cancel)

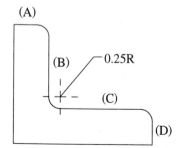

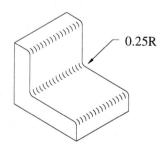

Figure 4–36

Yet another powerful feature of the FILLET command is to connect two lines at their intersections or corner the two lines. This is accomplished by setting the fillet radius to "0" and selecting the two lines. This is illustrated in Figure 4–37.

 Command:**FILLET**

(TRIM mode) Current fillet radius = 0.5000
Polyline/Radius/<Select two objects>:**R**
Enter fillet radius <0.2500>:**0**

 Command:**FILLET**

(TRIM mode) Current fillet radius = 0.5000
Polyline/Radius/<Select two objects>: *(Select at "A" and "B")*

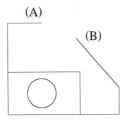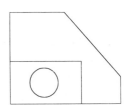

Figure 4–37

Slots are easily formed using the FILLET command. Regardless of the current radius setting, if two parallel lines are selected, the FILLET command reads the distance between the two lines and constructs an arc similar to Figure 4–38.

Figure 4–38

 Command:**FILLET**

(TRIM mode) Current fillet radius = 0.5000
Polyline/Radius/Trim/<Select first object>:
  (Select the line at "A")
Select second object: *(Select the line at "B")*

To complete the slot, repeat the above procedure for the ends of the line segments at "C" and "D" (Figure 4–39).

Figure 4–39

Chamfers

Chamfers represent yet another way to finish a sharp corner of an object. As fillets and rounds result from a pattern-making operation and remain unfinished, a chamfer is a machining operation that may even result in a polishing operation. Figure 4–40 is one example of an object that has been chamfered along its top edge.

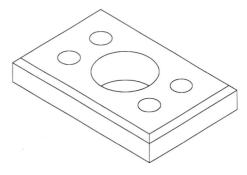

Figure 4–40

As with the FILLET command, AutoCAD also provides a CHAMFER command designed to draw an angle across a sharp corner given two chamfer distances. The most popular chamfer involves a 45 degree angle which is illustrated in Figure 4–41. Even though this command does not allow the user to specify an angle, the operator may control the angle by the distances entered. In the example in Figure 4–41, by specifying the same numeric value for both chamfer distances, a 45-degree chamfer will automatically be formed. As long as both distances are the same, a 45-degree chamfer will always be drawn. Study the illustration in Figure 4–41 and the following prompts:

 Command: **CHAMFER**

(TRIM mode) Current chamfer Dist1 = 0.5000, Dist2 = 0.5000
Polyline/Distance/Angle/Trim/Method/<Select first line>: **D**
Enter first chamfer distance <0.0000>: **0.15**
Enter second chamfer distance <0.1500>:
(Strike [ENTER] to accept)

Figure 4–42 is similar to the 45-degree chamfer shown in Figure 4–41 with the exception that the angle is different. This results from two different chamfer distances outlined in the following prompts. This type of edge is commonly called a bevel.

 Command: **CHAMFER**

(TRIM mode) Current chamfer Dist1 = 0.5000, Dist2 = 0.5000
Polyline/Distance/Angle/Trim/Method/<Select first line>: **D**
Enter first chamfer distance <0.0000>: **0.30**
Enter second chamfer distance <0.3000>: **0.15**

 Command: **CHAMFER**

(TRIM mode) Current chamfer Dist1 = 0.3000, Dist2 = 0.3000

 Command: **CHAMFER**

(TRIM mode) Current chamfer Dist1 = 0.1500, Dist2 = 0.1500
Polyline/Distance/Angle/Trim/Method/<Select first line>: (Select the line at "A")
Select second line: (Select the line at "B")

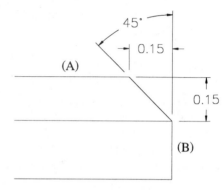

Figure 4–41

Polyline/Distance/Angle/Trim/Method/<Select first line>: (Select the line at "A")
Select second line: (Select the line at "B")

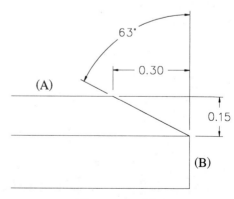

Figure 4–42

Runouts

Where flat surfaces become tangent to cylinders, there must be some method of accurately representing this using fillets. In the object illustrated in Figure 4–43, the front view shows two cylinders connected to each other by a tangent slab. The top view is complete; the front view has all geometry necessary to describe the object with the exception of the exact intersection of the slab with the cylinder.

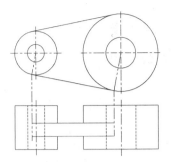

Figure 4–43

Figure 4–44 displays the correct method for finding intersections or runouts—areas where surfaces intersect others and blend in, disappear, or simply runout. A point of intersection is found at "A" in the top view with the intersecting slab and the cylinder. This actually forms a 90-degree angle with the line projected from the center of the cylinder and the angle made by the slab. A line is projected from "A" in the top view to intersect with the slab found in the front view.

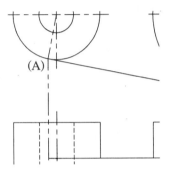

Figure 4–44

In Figure 4–45, fillets are drawn to represent the slab and cylinder intersections. This forms the runout.

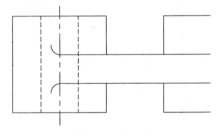

Figure 4–45

The resulting two-view drawing complete with runouts is illustrated in Figure 4–46.

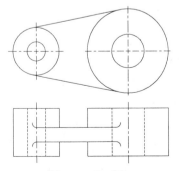

Figure 4–46

Tutorial Exercise
Shifter.Dwg

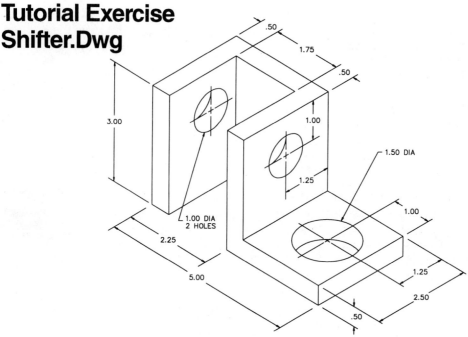

Figure 4–47

Purpose

This tutorial is designed to allow the user to construct a three-view drawing of the Shifter as shown in Figure 4–47.

System Settings

Use the UNITS command and change the number of decimal places from four to two. Keep the remaining default settings. Use the default settings for the screen limits (0,0 for the lower left corner and 12,9 for the upper right corner). The grid and snap do not need to be set to any certain values.

Layers

Create the following layers with the format:

Name	Color	Linetype
Object	Green	Continuous
Hid	Red	Hidden
Cen	Yellow	Center
Dim	Yellow	Continuous

Suggested Commands

The primary commands used during this tutorial are OFFSET and TRIM. The OFFSET command is used for laying out all views before using the TRIM command to clean up excess lines. Since different linetypes represent certain features of a drawing, the CHPROP OR DDCHPROP commands are used to convert to the desired linetype needed as set by the LAYER command. Once all visible details are identified in the primary views, project the visible features into the other views using the LINE command. A 45-degree inclined line is constructed to project lines from the top view to the right side view and vice versa.

Whenever possible, substitute the appropriate command alias in place of the full AutoCAD command in each tutorial step; for example, use "Co" for the COPY command, "L" for the LINE command, and so on. The complete listing of all command aliases is located in Unit 01 on pages 1–9 and 1–10.

Step #1

Begin the orthographic drawing of the Shifter by constructing a right angle consisting of one horizontal and one vertical line. The corner formed by the two lines will be used to orient the front view.

 Command: **LINE**

From point: **1,1**
To point: **@11<0**
To point: *(Press* ENTER *to exit this command)*

 Command: **LINE**

From point: **1,1**
To point: **@8<90**
To point: *(Press* ENTER *to exit this command)*

Figure 4–48

Step #2

Begin the layout of the primary views by using the OFFSET command to offset the vertical line at "A" the distance of 5.00 units, which represents the length of the Shifter (Figure 4–49).

 Command: **OFFSET**

Offset distance or Through <Through>:
5.00
Select object to offset: *(Select the vertical line at "A")*
Side to offset? *(Pick a point anywhere near "B")*
Select object to offset: *(Press* ENTER *to exit this command)*

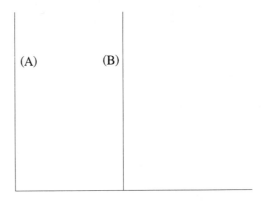

Figure 4–49

Step #3

Use the OFFSET command to offset the horizontal line at "A" the distance of 3.00 units, which represents the height of the Shifter (Figure 4–50).

 Command:**OFFSET**

Offset distance or Through <5.00>:**3.00**
Select object to offset: *(Select the horizontal line at "A")*
Side to offset? *(Pick a point anywhere near "B")*
Select object to offset: *(Press ⌜ENTER⌝ to exit this command)*

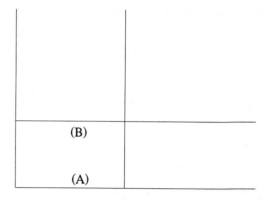

Figure 4–50

Step #4

Begin laying out dimension spaces that will act as separators between views and allow for the placement of dimensions once the Shifter is completed (Figure 4–51). A spacing of 1.50 units will be more than adequate for this purpose. Again use the OFFSET command to accomplish this.

 Command:**OFFSET**

Offset distance or Through <3.00>:**1.50**

Select object to offset: *(Select the vertical line at "A")*
Side to offset? *(Pick a point anywhere near "B")*
Select object to offset: *(Select the horizontal line at "C")*
Side to offset? *(Pick a point anywhere near "D")*
Select object to offset: *(Press ⌜ENTER⌝ to exit this command)*

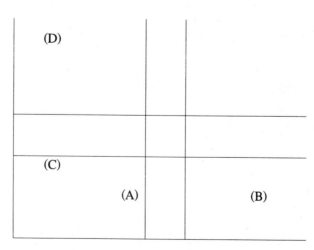

Figure 4–51

Step #5

Use the OFFSET command to lay out the depth of the Shifter at a distance of 2.50 units (Figure 4–52).

 Command: **OFFSET**

Offset distance or Through <1.50>:**2.50**
Select object to offset: *(Select the vertical line at "A")*
Side to offset? *(Pick a point anywhere near "B")*
Select object to offset: *(Select the horizontal line at "C")*
Side to offset? *(Pick a point anywhere near "D")*
Select object to offset: *(Press* ENTER *to exit this command)*

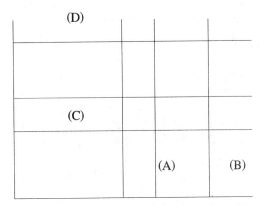

Figure 4–52

Step #6

Use the TRIM command to trim away excess construction lines used when laying out the primary views of the Shifter (Figure 4–53).

 Command: **TRIM**

Select cutting edges: (Projmode = UCS, Edgemode = No extend)
Select objects: *(Select the lines at "A" and "B")*
Select objects: *(Press* ENTER *to continue)*
<Select object to trim>/Project/Edge/Undo: *(Select the line at "C")*
<Select object to trim>/Project/Edge/Undo: *(Select the line at "D")*
<Select object to trim>/Project/Edge/Undo: *(Select the line at "E")*
<Select object to trim>/Project/Edge/Undo: *(Select the line at "F")*
<Select object to trim>/Project/Edge/Undo: *(Press* ENTER *to exit this command)*

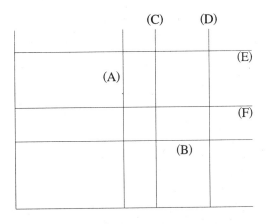

Figure 4–53

Step #7

Use the TRIM command again to complete trimming away excess construction lines used when laying out the primary views of the Shifter (Figure 4–54).

 Command:**TRIM**

Select cutting edges: (Projmode = UCS, Edgemode = No extend)
Select objects: *(Select the lines at "A" and "B")*
Select objects: *(Press ENTER to continue)*
<Select object to trim>/Project/Edge/Undo: *(Select the line at "C")*
<Select object to trim>/Project/Edge/Undo: *(Select the line at "D")*
<Select object to trim>/Project/Edge/Undo: *(Select the line at "E")*
<Select object to trim>/Project/Edge/Undo: *(Select the line at "F")*
<Select object to trim>/Project/Edge/Undo: *(Press ENTER to exit this command)*

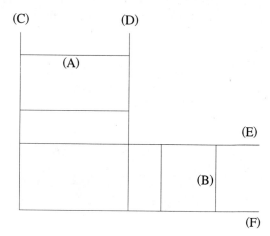

Figure 4–54

Step #8

Your display should appear similar to the illustration In Figure 4–55 with the layout of the front, top, and right side views. Begin adding details to all views through methods of projection. Use the OFFSET command to offset lines "A", "C", and "E" a distance of 0.50.

 Command:**OFFSET**

Offset distance or Through<2.50>:**0.50**
Select object to offset: *(Select the vertical line at "A")*
Side to offset? *(Pick a point anywhere near "B")*
Select object to offset: *(Select the horizontal line at "C")*
Side to offset? *(Pick a point anywhere near "D")*
Select object to offset: *(Select the horizontal line at "E")*
Side to offset? *(Pick a point anywhere near "F")*
Select object to offset: *(Press ENTER to exit this command)*

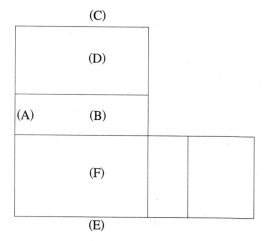

Figure 4–55

Step #9

Use the OFFSET command and offset the vertical
line at "A" the distance of 1.75 (Figure 4–56).

 Command: **OFFSET**

Offset distance or Through<0.50>: **1.75**
Select object to offset: *(Select the vertical line
   at "A")*
Side to offset? *(Pick a point anywhere near "B")*
Select object to offset: *(Press* ENTER *to exit this
   command)*

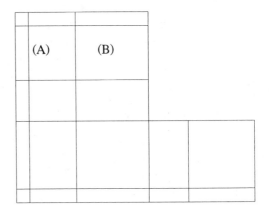

Figure 4–56

Step #10

Use the OFFSET command to offset the vertical
line at "A" a distance of 0.50 (Figure 4–57).

 Command: **OFFSET**

Offset distance or Through<1.75>: **0.50**
Select object to offset: *(Select the vertical line
   at "A")*
Side to offset? *(Pick a point anywhere near "B")*
Select object to offset: *(Press* ENTER *to exit this
   command)*

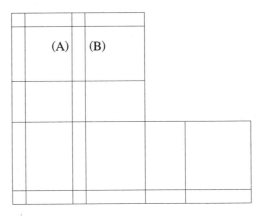

Figure 4–57

Step #11

Your display should appear similar to the illustration in Figure 4–58. Next, use the TRIM command to partially delete the line segments located in the spaces in between views.

 Command:**TRIM**

Select cutting edges: (Projmode = UCS, Edgemode = No extend)
Select objects: *(Select the lines at "A," "B," "C," and "D")*
Select objects: *(Press ⏎ to continue)*
<Select object to trim>/Project/Edge/Undo: *(Select the line at "E")*
<Select object to trim>/Project/Edge/Undo: *(Select the line at "F")*
<Select object to trim>/Project/Edge/Undo: *(Select the line at "G")*
<Select object to trim>/Project/Edge/Undo: *(Select the line at "H")*
<Select object to trim>/Project/Edge/Undo: *(Select the line at "I")*
<Select object to trim>/Project/Edge/Undo:

(Select the line at "J")
<Select object to trim>/Project/Edge/Undo: *(Select the line at "K")*
<Select object to trim>/Project/Edge/Undo: *(Select the line at "L")*
<Select object to trim>/Project/Edge/Undo: *(Press ⏎ to exit this command)*

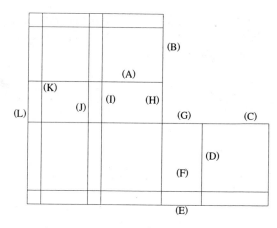

Figure 4–58

Step #12

Use the ZOOM-WINDOW command to magnify the top view similar to the illustration in Figure 4–59. Then use the TRIM command to trim the excess object labeled "B."

 Command:**TRIM**

Select cutting edges: (Projmode = UCS, Edgemode = No extend)
Select objects: *(Select the vertical line at "A")*
Select objects: *(Press ⏎ to continue)*
<Select object to trim>/Project/Edge/Undo: *(Select the line at "B")*

<Select object to trim>/Project/Edge/Undo: *(Press ⏎ to exit this command)*

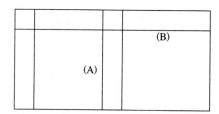

Figure 4–59

Step #13

Zoom back to the original display using the ZOOM-PREVIOUS command. Use the ZOOM-WINDOW command to magnify the display to show the front view illustrated in Figure 4–60. Use the TRIM command to clean-up the excess lines in the front view using the illustration in Figure 4–60 as a guide.

 Command:**TRIM**

Select cutting edges: (Projmode = UCS,
 Edgemode = No extend)
Select objects: *(Select the lines at "A" and "B")*
Select objects: *(Press* ENTER *to continue)*
<Select object to trim>/Project/Edge/Undo:
 (Select the line at "C")
<Select object to trim>/Project/Edge/Undo:
 (Select the line at "D")
<Select object to trim>/Project/Edge/Undo:
 (Select the line at "E")

<Select object to trim>/Project/Edge/Undo:
 (Select the short line at "F")
<Select object to trim>/Project/Edge/Undo:
 (Press ENTER *to exit this command)*

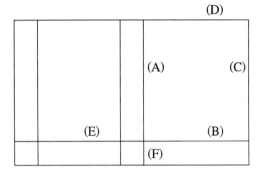

Figure 4–60

Step #14

Use the ZOOM-PREVIOUS command to zoom back to the previous screen display containing the three views. Use the FILLET command to create a corner by selecting lines "A", "B", and "C" illustrated in Figure 4–61.

 Command:**FILLET**

(TRIM mode) Current fillet radius = 0.5000
Polyline/Radius/Trim/<Select first object>:**R**
Enter fillet radius <0.5000>:**0.00**
Command: *(Press* ENTER *to replay the previous
 command)*

Command:**FILLET**

(TRIM mode) Current fillet radius = 0.0000
Polyline/Radius/Trim/<Select first object>:
 (Select line "A")
Select second object: *(Select line "B")*

Command:**FILLET**

(TRIM mode) Current fillet radius = 0.0000
Polyline/Radius/Trim/<Select first object>:
 (Select line "B")
Select second object: (Select line "C")

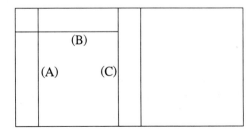

Figure 4–61

Step #15

Use the TRIM command to partially delete the horizontal line at "C" and form the upside-down "U" shape illustrated in Figure 4–62.

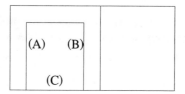

 Command: **TRIM**

Select cutting edges: (Projmode = UCS, Edgemode = No extend)
Select objects: *(Select the vertical lines at "A" and "B")*
Select objects: *(Press* [ENTER] *to continue)*
<Select object to trim>/Project/Edge/Undo: *(Select the horizontal line at "C")*
<Select object to trim>/Project/Edge/Undo: *(Press* [ENTER] *to exit this command)*

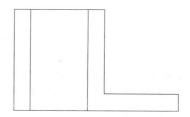

Figure 4–62

Step #16

Begin placing a circle in the top view representing the 1.50 diameter drill hole in Figure 4–63. Use the CIRCLE command and the Osnap-From and Osnap-Int options to set up a temporary point of reference illustrated in Figure 4–63.

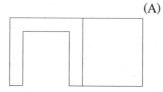

 Command: **CIRCLE**

3P/2P/TTR/<Center point>: **From**
Base point: **Int**
of *(Pick the corner intersection at "A")*
<Offset>: **@-1.00,-1.25**
Diameter/<Radius>: **D**
Diameter: **1.50**

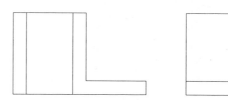

Figure 4–63

Step #17

Placing a circle in the right side view representing the 1.00 diameter drill hole. Use the CIRCLE command and the Osnap-From and Osnap-Int options to set up a temporary point of reference illustrated to the right at "A".

 Command:**CIRCLE**

3P/2P/TTR/<Center point>:**From**
Base point: **Int**
of *(Pick the corner intersection at "A")*
<Offset>:**@-1.25,-1.00**
Diameter/<Radius>:**D**
Diameter:**1.00**

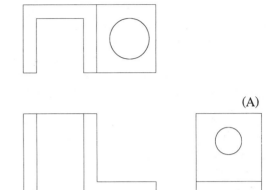

(A)

Figure 4–64

Step #18

Use the LINE command and draw projection lines from both circles into the front view as shown in Figure 4–65. Use the Osnap-Quadrant and Osnap-Perpendicular modes to accomplish this.

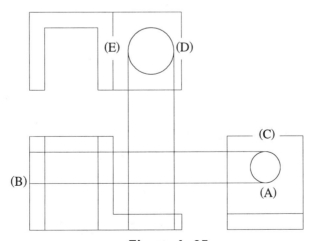 Command:**Line**

From point: **Qua**
of *(Select the quadrant of the circle at "A")*

To point: **Per**
to *(Select the line at "B" to obtain the perpendicular point)*
To point: *(Press* ENTER *to exit this command)*

Repeat the above procedure for the quadrants "C", "D", and "E".

Figure 4–65

Step #19

Place center marks at the centers of both circles as illustrated in Figure 4–66. Before proceeding with this operation, a system variable needs to be set to a certain value to achieve the desired results. Set the DIMCEN system variable to a value of -0.12. This will not only place the center mark when using the DIMCENTER command, but will also extend the centerline a short distance outside of both circles.

Command: **DIMCEN**
New value for DIMCEN <0.09>: **-0.12**

 Command: **DIMCENTER**

Select arc or circle: *(Select the circle at "A")*

 Command: **DIMCENTER**

Select the arc or circle: *(Select the circle at "B")*

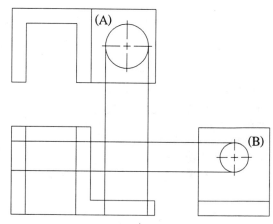

Figure 4–66

Step #20

Project two lines from the endpoints of both center marks using the OSNAP-ENDPOINT command. Turn ortho on to assist with this operation. The lines will be converted to centerlines at a later step. Draw the lines 0.50 units past the front view as illustrated in Figure 4–67.

 Command: **LINE**

From point: **Endp**
of *(Select the endpoint of the center mark in the top view at "A")*
To point: *(Mark a point just below the front view at "B")*
To point: *(Strike Enter to exit this command)*

 Command: **LINE**

From point: **Endp**
of *(Select the endpoint of the center mark in the side view at "C")*

To point: *(Mark a point to the left of the front view at "D")*
To point: *(Press* ENTER *to exit this command)*

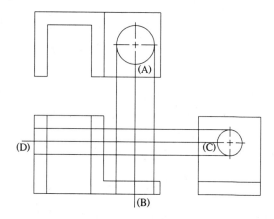

Figure 4–67

Step #21

Use the TRIM command and the illustration in Figure 4–68 to trim away excess lines.

 Command:**TRIM**

Select cutting edges: (Projmode = UCS, Edgemode = No extend)
Select objects: *(Select the lines at "A," "B," "C," and "D")*
Select objects: *(Press* ⌈ENTER⌉ *to continue)*
<Select object to trim>/Project/Edge/Undo: *(Select the line at "E")*

<Select object to trim>/Project/Edge/Undo: *(Select the line at "F")*
<Select object to trim>/Project/Edge/Undo: *(Select the line at "G")*
<Select object to trim>/Project/Edge/Undo: *(Select the line at "H")*
<Select object to trim>/Project/Edge/Undo: *(Select the line at "I")*
<Select object to trim>/Project/Edge/Undo: *(Select the line at "J")*
<Select object to trim>/Project/Edge/Undo: *(Press* ⌈ENTER⌉ *to exit this command)*

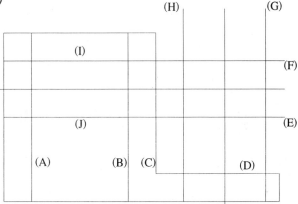

Figure 4–68

Step #22

Use the CHPROP or DDCHPROP command to change the six lines illustrated in Figure 4–69 from their current layer assignment of "0" to a new layer assignment of "Hid" which will change the lines from object to hidden lines.

 Command:**CHPROP**

Select objects: *(Select all six short lines labeled "A" to "F")*
Select objects: *(Press* ⌈ENTER⌉ *to continue this command)*
Change what property (Color/LAyer/LType/ ltScale/Thickness) ? **LA**
New layer <0>: **Hid**

Change what property (Color/LAyer/LType/ ltScale/Thickness) ? *(Press* ⌈ENTER⌉ *to exit this command)*

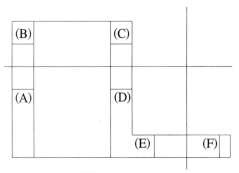

Figure 4–69

Step #23

Use the BREAK command to partially delete the lines illustrated in Figure 4–70 before converting them to centerlines. Remember, centerlines extend past the object lines when identifying hidden drill holes; it would be inappropriate to use the TRIM command for this step.

 Command: **BREAK**

Select object: *(Select the horizontal line approximately at "A")*
Enter second point (or F for first point): *(Select the line approximately at "B")*

 Command: **BREAK**

Select object: *(Select the vertical line approximately at "C")*
Enter second point (or F for first point): *(Select the line at approximately "D")*

 Command: **BREAK**

Select object: *(Select the horizontal line approximately at "E")*
Enter second point (or F for first point): **@**

Using the "@" symbol breaks the object at the exact location identified by the first point selected. Remember, the "@" symbol means "last point".

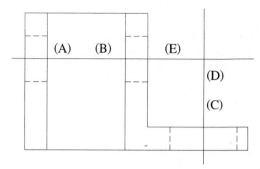

Figure 4–70

Step #24

The purpose of the "@" symbol in the previous step is to break a line into two segments without showing the break. The "@" symbol means "the last known point" which completes the BREAK command by satisfying the "Second point" prompt. To prove this, use the ERASE command to delete the segments no longer needed.

 Command: **ERASE**

Select objects: *(Carefully select the lines at "A" and "B")*
Select objects: *(Press ⏎ to perform the erase)*

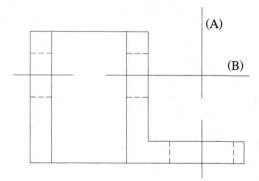

Figure 4–71

Step #25

Use the CHPROP or DDCHPROP command to change the three lines illustrated in Figure 4–72 from their current layer assignment of "0" to a new layer assignment of "Cen" which will change the lines from object to center lines.*

 Command: **CHPROP**

Select objects: *(Select all three short lines labeled "A" to "C")*
Select objects: *(Press* ENTER *to continue this command)*
Change what property (Color/LAyer/LType/ ltScale/Thickness) ? **LA**
New layer <0>: **Cen**

Change what property (Color/LAyer/LType/ ltScale/Thickness) ? *(Press* ENTER *to exit this command)*

*If the lines do not appear as centerlines. use the LTSCALE command and change the value from 1.00 to 0.50.

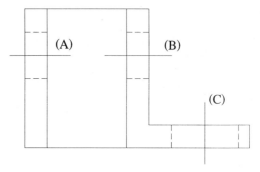

Figure 4–72

Step #26

A 45-degree angle needs to be constructed in order to begin projecting features from the top view to the right side view and then back again. This angle is formed by extending the bottom edge of the top view to intersect with the left edge of the side view as shown in Figure 4–73. Use the FILLET command to accomplish this.** Then draw the 45-degree line; the length of this line is not important.

 Command: **FILLET**

(TRIM mode) Current fillet radius = 0.0000
Polyline/Radius/Trim/<Select first object>:
  (Select line "A")
Select second object: *(Select line "B")*

 Command: **LINE**

From point: **Int**
of *(Select the intersection at "C")*
To point: **@4<45**
To point: *(Press* ENTER *to exit this command)*

**Check that the Fillet radius is currently set to 0.0000.

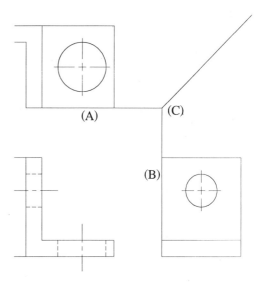

Figure 4–73

Step #27

Draw lines from points "A", "B", and "C" to intersect with the 45-degree angle projector as shown in Figure 4–74. Be sure Ortho mode is on to draw horizontal lines. Use the Osnap-Intersec, Endpoint, and Quadrant modes to assist in this operation.

 Command: **LINE**
From point: **Int**
of *(Select the intersection of the corner at "A")*
To point: *(Draw a horizontal line just past the 45-degree angle)*
To point: *(Press* ⟨ENTER⟩ *to exit this command)*

 Command: **LINE**
From point: **Endp**
of *(Select the endpoint of the center line at "B")*
To point: *(Draw a horizontal line just past the 45-degree angle)*
To point: *(Press* ⟨ENTER⟩ *to exit this command)*

 Command: **LINE**
From point: **Qua**
of *(Select the quadrant of the circle at "C")*
To point: *(Draw a horizontal line just past the 45-degree angle)*
To point: *(Press* ⟨ENTER⟩ *to exit this command)*

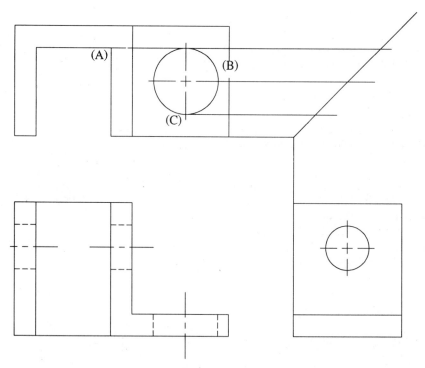

Figure 4–74

Step #28

Draw lines from the intersection of the 45-degree angle projection line to points in the right side view as shown in Figure 4–75. Use the following commands to accomplish this.

 Command: **LINE**

From point: **Int**
of *(Select the intersection of the angle at "A")*
To point: **Per**
to *(Select the bottom horizontal line of the right side view at "B")*
To point: *(Press* ENTER *to exit this command)*

 Command: **LINE**

From point: **Int**
of *(Select the intersection of the angle at "C")*
To point: **Per**
to *(Select the bottom line of the right side view at "B")*
To point: *(Press* ENTER *to exit this command)*

 Command: **LINE**

From point: **Int**
of *(Select the intersection of the angle at "D")*
To point: *(Select a point below the bottom of the side view at "E")*
To point: *(Press* ENTER *to exit this command)*

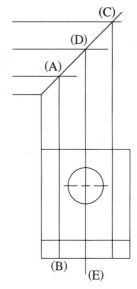

Figure 4–75

Step #29

Erase the three projection lines from the top view using the ERASE command as shown in Figure 4–76.

 Command: **ERASE**

Select objects: *(Select lines "A", "B", and "C")*
Select objects: *(Press* ENTER *to perform this command)*

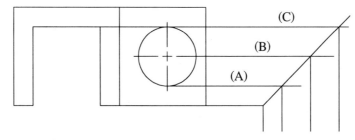

Figure 4–76

Step #30

Use the TRIM command to trim away unnecessary geometry using the illustration in Figure 4–77 as a guide. The hidden hole and slot will be formed in the side view through this operation.

 Command: **TRIM**

Select cutting edges: (Projmode = UCS, Edgemode = No extend)
Select objects: *(Select the horizontal line at "A")*
Select objects: *(Press [ENTER] to continue)*
<Select object to trim>/Project/Edge/Undo: *(Select the line at "B")*
<Select object to trim>/Project/Edge/Undo: *(Press [ENTER] to continue)*

Repeat this procedure for the line illustrated in Figure 4–77 using the horizontal line "C" as the cutting edge and the vertical line "D" as the line to trim.

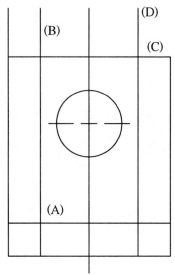

Figure 4–77

Step #31

Use the BREAK command to split the vertical line at "A" into two separate entities as shown in Figure 4–78. This is accomplished by typing @ in response to the prompt "Enter second point". This will split the line in two without the break being noticeable.

 Command: **BREAK**

Select object: *(Select the vertical line at "A")*
Enter second point (or F for first point): @

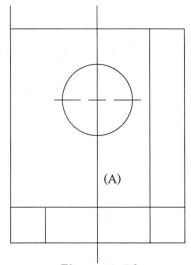

Figure 4–78

Step #32

Use the ERASE command to delete the top half of the line broken previously in Step #31. This will leave a short line segment that will be changed or converted into a centerline marking the center of the hidden hole, see Figure 4–79. Redraw the screen to refresh the vertical centerlines at the hole.

 Command: **ERASE**

Select objects: *(Select the vertical line at "A")*
Select objects: *(Press* ⏎ *to execute this command)*

 Command: **REDRAW**

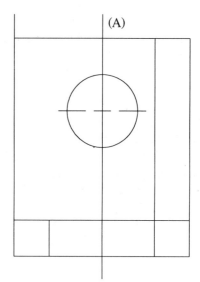

Figure 4–79

Step #33

Use the CHPROP or DDCHPROP command to convert the two vertical lines labeled "A" and "B" in Figure 4–80 from the object layer to the layer named "Hid." Do the same for the longer vertical line labeled "C" but change this line from the object layer to the layer named "Cen."

 Command: **CHPROP**

Select objects: *(Select the two vertical lines labeled "A" and "B")*
Select objects: *(Press* ⏎ *to continue)*
Change what property (Color/LAyer/LType/ ltScale/Thickness) ? **LA**
New layer <0>: **Hid**
Change what property (Color/LAyer/LType/ ltScale/Thickness) ? *(Press* ⏎ *to exit this command)*

 Command: **CHPROP**

Select objects: *(Select the vertical line labeled "C")*
Select objects: *(Press* ⏎ *to continue)*

Change what property (Color/LAyer/LType/ ltScale/Thickness) ? **LA**
New layer <0>: **Cen**
Change what property (Color/LAyer/LType/ ltScale/Thickness) ? *(Press* ⏎ *to exit this command)*

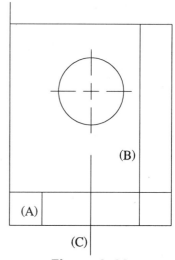

Figure 4–80

Step #34

Draw three lines from key features on the side view to intersect with the 45-degree angle projector as shown in Figure 4–81. Use Osnap options whenever possible. Ortho mode must be on.

 Command: **LINE**

From point: **Qua**
of *(Select the quadrant of the circle at "A")*
To point: *(Identify a point past the 45-degree angle)*
To point: *(Press* ENTER *to exit this command)*

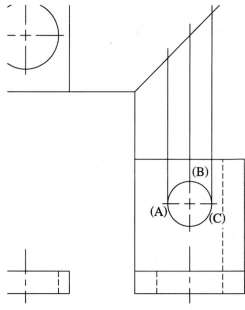

Figure 4–81

Repeat the above procedure for the other two projection lines. Use the Osnap-Endpoint option and begin the second projector line from the endpoint of the center marker at "B." Begin the third projector line from the quadrant of the circle at "C."

Step #35

Draw three lines from the intersection at the 45-degree projector across to the top view as shown in Figure 4–82. Draw the middle line longer since it will be converted into a centerline at a later step.

Repeat the above procedure for the other two lines. Do not use an Osnap option for the opposite end of the middle line but rather identify a point just to the left of the vertical line at "B."

 Command: **LINE**

From point: **Int**
of *(Select the intersection at "A")*
To point: **Per**
to *(Select the vertical line at "B")*
To point: *(Press* ENTER *to exit this command)*

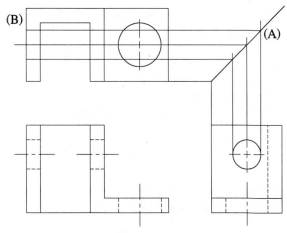

Figure 4–82

Step #36

Use the ERASE command to erase the three vertical projector lines from the side view as shown in
Figure 4–83.

Command:**ERASE**
Select objects: *(Select the vertical lines labeled
"A", "B", and "C")*
Select objects: *(Press* ENTER *to execute this
command)*

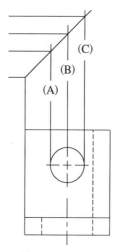

Figure 4–83

Step #37

Use the TRIM command to trim away excess lines
in the top view as illustrated in figure 4–84. Perform a redraw when completed.

 Command:**TRIM**

Select cutting edges: (Projmode = UCS,
Edgemode = No extend)
Select objects: *(Select the vertical lines labeled
"A", "B", and "C")*
Select objects: *(Press* ENTER *to continue)*
<Select object to trim>/Project/Edge/Undo:
(Select the line at "D")
<Select object to trim>/Project/Edge/Undo:
(Select the line at "E")
<Select object to trim>/Project/Edge/Undo:
(Select the line at "F")
<Select object to trim>/Project/Edge/Undo:
(Select the line at "G")
<Select object to trim>/Project/Edge/Undo:
(Press ENTER *to exit this command)*

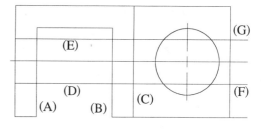

Figure 4–84

 Command:**REDRAW**

Step #38

Use the BREAK command to partially delete the horizontal line segment from "A" to "B" as shown in Figure 4–85. Use the BREAK command with the @ option to break the line into two segments at "C." Use the ERASE command to delete the trailing line segment at "D." Redraw the screen.

 Command: **BREAK**

Select object: *(Select the horizontal line at "A")*
Enter second point (or F for first): *(Select the line at "B")*

 Command: **BREAK**

Select object: *(Select the horizontal line at "C")*
Enter second point (or F for first): **@**

 Command: **ERASE**

Select objects: *(Select the horizontal line segment at "D")*
Select objects: *(Press* ENTER *to execute this command)*

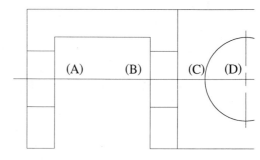

(A)   (B)   (C)   (D)

Figure 4–85

Step #39

Use the CHPROP or DDCHPROP commands to change the line segments labeled "A," "B," "C," and "D" to the new layer named "Hid," which will display hidden lines. See Figure 4–86. Change the line segments labeled "E" and "F" to the new layer named "Cen," which will display center lines.

 Command: **CHPROP**

Select objects: *(Select the four lines labeled "A" to "D")*
Select objects: *(Press* ENTER *to continue)*
Change what property (Color/LAyer/LType/ ltScale/Thickness) ? **LA**
New layer <0>: **Hid**
Change what property (Color/LAyer/LType/ ltScale/Thickness) ? *(Press* ENTER *to exit this command)*

 Command: **CHPROP**

Select objects: *(Select the two lines labeled "E" and "F")*
Select objects: *(Press* ENTER *to continue)*
Change what property (Color/LAyer/LType/ ltScale/Thickness) ? **LA**
New layer <0>: **Cen**
Change what property (Color/LAyer/LType/ ltScale/Thickness) ? *(Press* ENTER *to exit this command)*

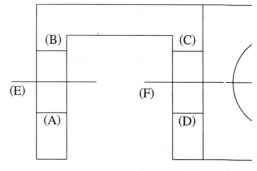

(B)   (C)

(E)   (F)

(A)   (D)

Figure 4–86

Step #40

Use the ERASE command to delete the 45-degree angle line. Use the FILLET command to create corners in the top view and right side view as shown in Figure 4–87. Check that the fillet radius is currently set to 0.0000.

 Command:**ERASE**

Select objects: *(Select the inclined line at "A")*
Select objects: *(Press* ⌷ENTER⌷ *to execute this command)*

 Command:**FILLET**

(TRIM mode) Current fillet radius = 0.0000
Polyline/Radius/Trim/<Select first object>:
(Select line "B")
Select second object: *(Select line "C")*

 Command:**FILLET**

(TRIM mode) Current fillet radius = 0.0000
Polyline/Radius/Trim/<Select first object>:
(Select line "D")
Select second object: *(Select line "E")*

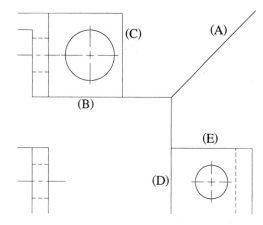

Figure 4–87

Step #41

Use the CHPROP or DDCHPROP command to change the center marks at "A" and "B" to the "Center" layer as shown in Figure 4–88. This completes this tutorial on performing a multiview projection drawing. The steps have been numerous in order to detail every command sequence. In reality, the process is much faster, especially since a few basic commands were used most of the time such as OFF-SET and TRIM. Use this tutorial as a guide in completing the many multiview drawing problems at the end of this unit.

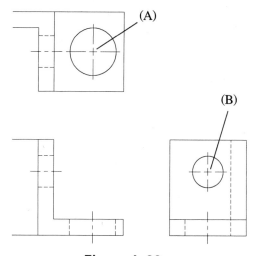

Figure 4–88

Step #42

In keeping with the layers created, all object lines are to be changed from the layer they were created on to the object layer using the CHPROP or DDCHPROP command. See Figure 4–89.

 Command: **CHPROP**

Select objects: *(Select all twenty-four object lines)*
Select objects: *(Press ENTER to continue)*

Change what property (Color/LAyer/LType/
  ltScale/Thickness) ? **LA**
New layer <0>: **Object**
Change what property (Color/LAyer/LType/
  Thickness)? *(Press ENTER to exit this
  command)*

The final process in completing a multiview drawing is to place dimensions to define size and locate features. This topic will be discussed in a later chapter.

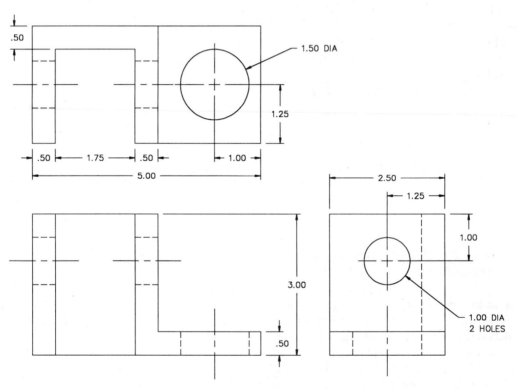

Figure 4–89

Tutorial Exercise
XYZ.Dwg

It is possible to perform multiview projections from one view to another without drawing construction lines and then trimming them to size. XYZ filters provide the means of accomplishing this along with a little practice. In the illustration in Figure 4–90, the problem is to draw a circle using the center point of the circle exactly at the center of the rectangle. To complicate the issue, no other command or setting can be used to perform this operation. First, construct the rectangle.

 Command:**LINE**
From point: **2,2** *(Point "A")*
To point: **@8<0** *(To Point "B")*
To point: **@6<90** *(To Point "C")*
To point: **@8<180** *(To Point "D")*
To point: **C** *(Back to Point "A" to close and exit the command)*

Use the CIRCLE command and begin with the normal "Center point" prompt sequence. Begin finding the center of the rectangle by first filtering out the midpoint of the X value at "A" followed by the Y value at "B" as shown in Figure 4–91. Answer the prompt for Z by entering a value of 0. Complete the CIRCLE command by supplying the radius value of 2.

 Command:**CIRCLE**
3P/2P/TTR/<Center point>:**.X**
of **Mid**
of *(Select the horizontal line at "A")*
(Need YZ):**.Y**
of **Mid**
of *(Select the vertical line at "B")*
(Need Z): **0**
Diameter/<Radius>:**2**

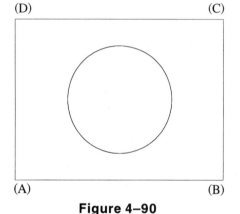

Figure 4–90

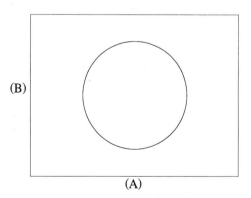

Figure 4–91

To review, filtering out the midpoint of the horizontal line at "A" saved the point to satisfy the center point of the circle as in Figure 4–92. However, a circle needs at least a second point; this is the reason for the prompt "Need YZ." A YZ point was needed to find the center point. Instead of selecting any point, the Y point was filtered out at the midpoint of the vertical line at "B." Now, the X and Y values were saved to satisfy the center of the circle. However, since AutoCAD now exists in a three-dimensional (3D) database, a prompt "Need Z" appears. Since the entire drawing is located at an elevation of 0, entering a value of 0 satisfies the centerpoint of the circle and marks a point. This may seem very tedious and too much trouble; with a little practice, however, filters become another drawing aid to work in your favor.

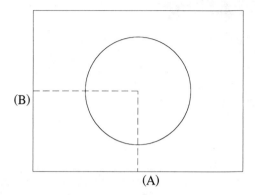

Figure 4–92

Using Tracking Mode

An alternate method of finding the center of a rectangle is through the use of tracking. Here, issuing tracking inside of an existing command prompts the user for a series of tracking points. Ortho mode is automatically turned On once tracking is activated. Study the following command sequence and illustrations in Figure 4–93.

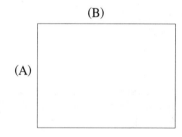

 Command:**CIRCLE**
3P/2P/TTR/<Center point>:**TK**
First tracking point: **Mid**
of *(Select the line near "A")*
Next point (Press ⌨ENTER to end tracking): **Mid**
of *(Select the line near "B")*
Next point (Press ⌨ENTER to end tracking): *(Press the ⌨ENTER key)*
Diameter/<Radius>: <Coords off> <Coords on>**1.50**

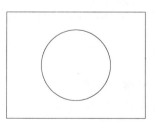

Figure 4–93

Tutorial Exercise
Gage.Dwg

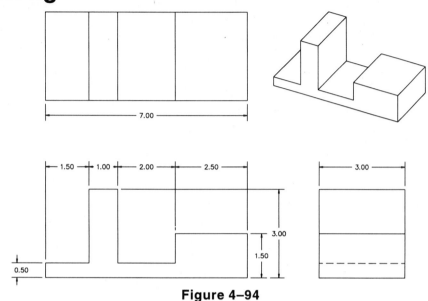

Figure 4–94

Purpose

This tutorial is designed to allow the user to construct a three-view drawing of the Gage with the aid of XYZ filters. See Figure 4–94.

System Settings

Begin a new drawing called "Gage." Use the UNITS command to change the number of decimal places past the zero from 4 to 2. Keep the remaining default unit values. Using the LIMITS command, keep 0,0 for the lower left corner and change the upper right corner from 12,9 to 15.50,9.50. Use the GRID command and change the grid spacing from 1.00 to 0.25 units. Do not turn the Snap or Ortho on.

Layers

Create the following layers with the format:

Name	Color	Linetype
Hidden	Red	Hidden

Suggested Commands

Begin this tutorial by laying out the three primary views using the LINE and OFFSET commands. Use the TRIM command to clean up any excess line segments. As an alternate method used for projection, use XYZ filters in combination with Osnap options to add features in other views.

Whenever possible, substitute the appropriate command alias in place of the full AutoCAD command in each tutorial step; for example, use "Co" for the COPY command, "L" for the LINE command, and so on. The complete listing of all command aliases is located in Unit 01 on pages 1–9 and 1–10.

Step #1

Begin the multiview drawing of the Gage by constructing the front view using absolute and polar coordinates as illustrated in Figure 4–95. Start the front view at coordinate 1.50,1.00.

The preceding coordinate entries could also be accomplished using the direct distance mode.

 Command: **LINE**

From point: **1.50,1.00**
To point: **@7.00<0**
To point: **@1.50<90**
To point: **@2.50<180**
To point: **@1.00<270**
To point: **@2.00<180**
To point: **@2.50<90**
To point: **@1.00<180**
To point: **@2.50<270**
To point: **@1.50<180**
To point: **C**

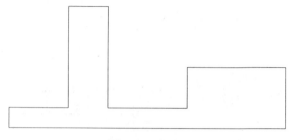

Figure 4–95

Step #2

Begin the construction of the top view by locating the lower left corner at coordinate 1.50,5.50 as shown in Figure 4–96.

 Command: **LINE**

From point: **1.50,5.50**
To point: **@7.00<0**
To point: **@3.00<90**
To point: **@7.00<180**
To point: **C**

Figure 4–96

Step #3

Begin the construction of the right side view by locating the lower left corner at coordinate 10.50,1.00 as show in Figure 4–97.

 Command: **LINE**

From point: **10.50,1.00**
To point: **@3.00<0**
To point: **@3.00<90**
To point: **@3.00<180**
To point: **C**

Figure 4–97

Step #4

Your display should appear similar to the illustration in Figure 4–98; with the placement of the top view above the front view and the right side view directly to the right of the front. Filtering methods will now be used to complete the missing lines in the top and right side views.

Figure 4–98

Step #5

Zoom into the drawing so the front and top views appear similar to the illustration in Figure 4–99. Follow the steps below for the proper use of XYZ filters when used in conjunction with the LINE command.

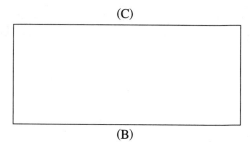

 Command: **LINE**

From point: **.X**
of **Int**
of *(Select the intersection of the lines at "A")*
(Need YZ): **Nea**
to *(Select anywhere along the horizontal line "B")*
To point: **Per**
to *(Select anywhere along the horizontal line "C")*
To point: *(Press ENTER to exit this command)*

The X value identified by selecting the intersection on the front view was saved for later use by projecting the value to the horizontal line of the top view and completing the LINE command with the Osnap-Perpendicular option.

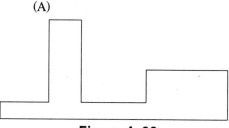

Figure 4–99

Step #6

Your display should appear similar to the illustration in Figure 4–100. Rather than use filters for the next line, simply duplicate the last line drawn using the COPY command.

 Command: **COPY**

Select objects: **L**
Select objects: *(Press ENTER to continue)*
<Base point or displacement>/Multiple: **Endp**
of *(Select the endpoint of the line at "A")*
Second point of displacement: **Endp**
of *(Select the endpoint of the line at "B")*

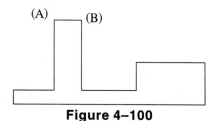

Figure 4–100

Step #7

After performing the COPY command, your display should appear similar to the illustration in Figure 4–101. Use XYZ filters to project the last object line in the top view from the front view.

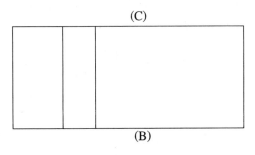

 Command: **LINE**

From point: **.X**
of **Int**
of *(Select the intersection of the lines at "A")*
(Need YZ): **Nea**
to *(Select anywhere along the horizontal line "B")*
To point: **Per**
to *(Select anywhere along the horizontal line "C")*
To point: *(Press* ENTER *to exit this command)*

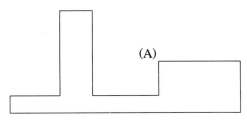

Figure 4–101

Step #8

The complete front and top views are illustrated in Figure 4–102. Use the same procedure with XYZ filters for completing the right side view.

Figure 4–102

Step #9

Begin using XYZ filters to add missing lines to the right side view. First, use the Zoom-Window option to magnify the area illustrated in Figure 4–103.

 Command: **ZOOM**

All/Center/Dynamic/Extents/Left/Previous/
   Vmax/Window/<Scale(X/XP)>:**W**
First corner: *(Select a point on the screen at "A")*
Second corner: *(Select a point on the screen at "B")*

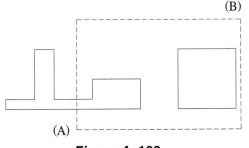

Figure 4–103

Step #10

Instead of filtering or saving the X coordinate, the same procedure can be used for saving the Y coordinate for later use. See Figure 4–104.

 Command: **LINE**
From point: **.Y**
of **Int**
of *(Select the intersection of the lines at "A")*
(Need XZ): **Nea**
to *(Select anywhere along the vertical line "B")*

To point: **Per**
to *(Select anywhere along the vertical line "C")*
To point: *(Press ⏎ to exit this command)*

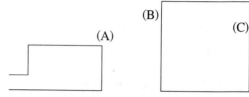

Figure 4–104

Step #11

Filter out the Y coordinate value of the intersecting area of the front view to project the coordinate to the right side view as in Figure 4–105.

 Command: **LINE**
From point: **.Y**
of **Int**
of *(Select the intersection of the lines at "A")*
(Need XZ): **Nea**
to *(Select anywhere along the vertical line "B")*

To point: **Per**
to *(Select anywhere along the vertical line "C")*
To point: *(Press ⏎ to exit this command)*

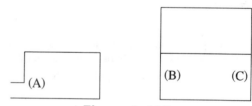

Figure 4–105

Step #12

Since the last line projected details a hidden surface, the object line in Figure 4–106 needs to be changed to a hidden line. The CHPROP command will be used to accomplish this only if a layer identifying hidden lines, such as Hidden, has already been created.

 Command: **CHPROP**
Select objects: **L**
Select objects: *(Press ⏎ to continue)*
Change what property(Color/LAyer/LType/
  Thickness)? **LA**

New layer<0>: **Hidden**
Change what property(Color/LAyer/LType/
  Thickness)? *(Press ⏎ to exit this command)*

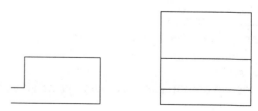

Figure 4–106

Step #13

The completed right side view is illustrated in Figure 4–107 complete with visible and invisible surfaces. Use the Zoom-Previous or Zoom-All options to demagnify your display and show all three views of the Gage.

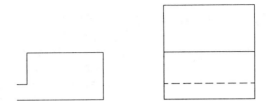

Figure 4–107

As a general rule of thumb, an X coordinate was filtered out and retrieved later for projecting lines from the front to the top view or vice versa. A Y coordinate was filtered out and retrieved later for projecting lines from the front to the right side view and vice versa. Filters may also be used for projecting such features as holes and slots between views. The key is to lock onto a significant part of an entity using one of the many options provided by the Osnap option. See Figure 4–108.

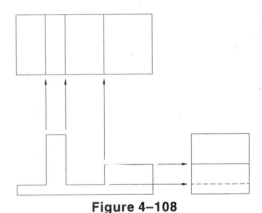

Figure 4–108

With the completed three-view drawing illustrated in Figure 4–109, the next step would be to add dimensions to the views for manufacturing purposes. This topic will be discussed in a later chapter.

Figure 4–109

Questions for Unit 4

Directions for Questions 4–1 through 4–4:
 Based on the isometric drawings, identify which set of orthographic projections correctly depict the object. Place your answer in the box provided below.

Question 4–1

Answer

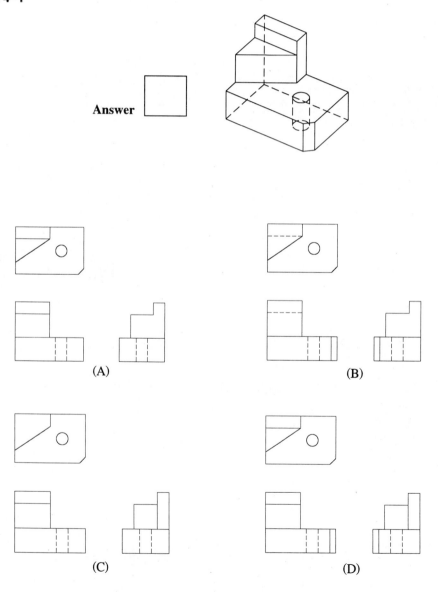

(A)

(B)

(C)

(D)

Question 4–2

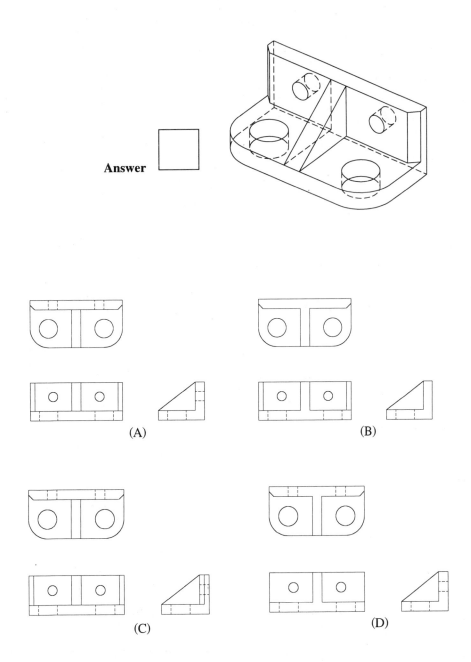

Answer

(A)

(B)

(C)

(D)

Question 4–3

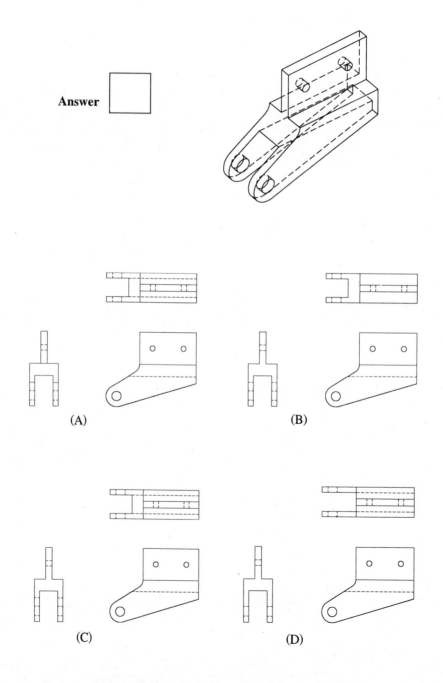

Answer ⬚

(A)

(B)

(C)

(D)

Question 4–4

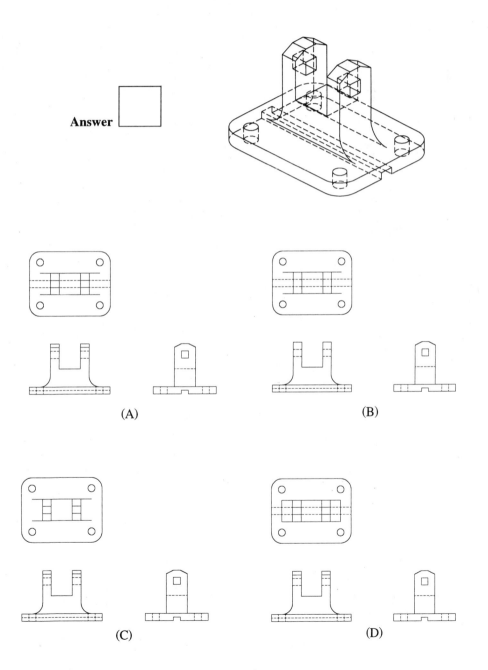

Answer [ ]

(A)

(B)

(C)

(D)

Problems for Unit 4

Directions for Problems 4–5 and 4–6:
Find the missing lines in these problems and sketch the correct solution.

Problem 4–5

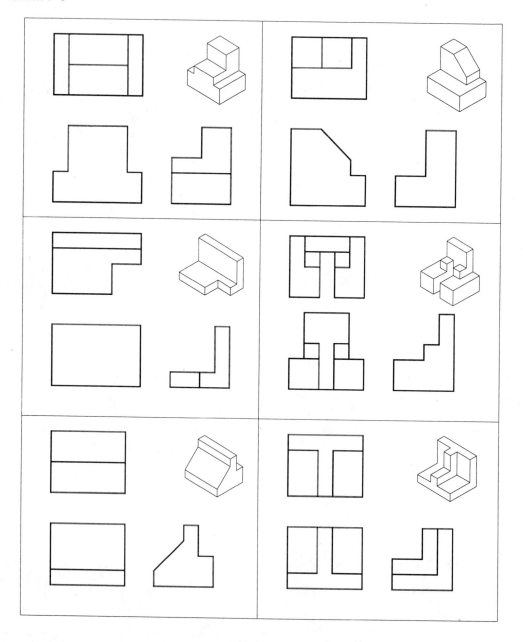

Problem 4–6

Directions for Problems 4–7 through 4–11:
Construct a multiview drawing by sketching the front, top, and right side views using the following grid as a guide.

Problem 4–7

Problem 4–8

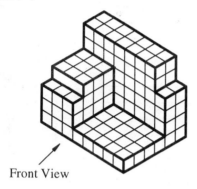

Front View

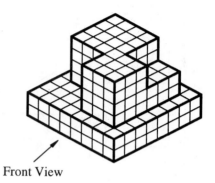

Front View

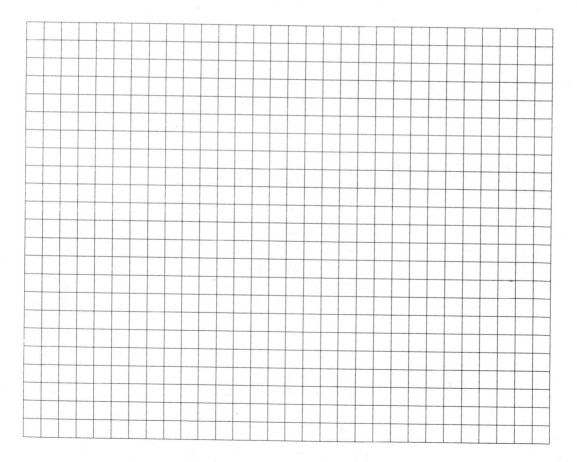

Problem 4–9

Problem 4–10

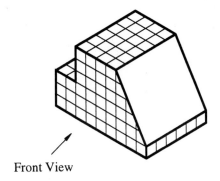

Front View

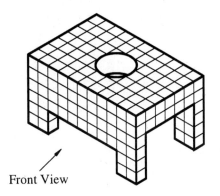

Front View

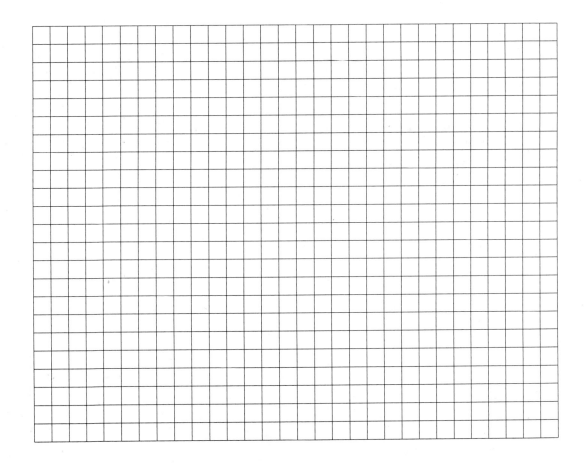

Problem 4–11

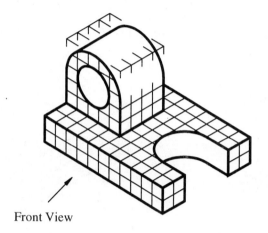

Front View

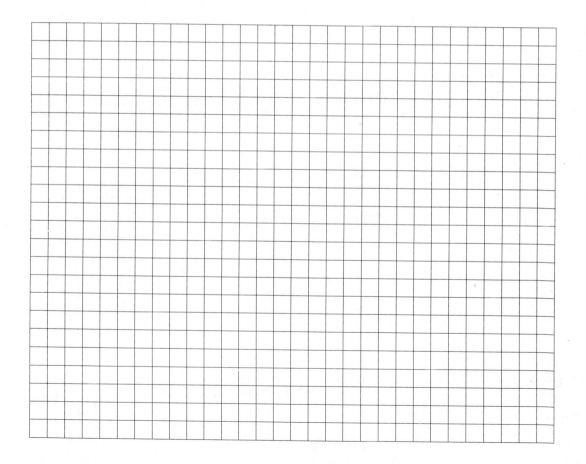

Directions for Problems 4–12 through 4–21:
   Construct a multiview drawing of each object. Be sure to construct only those views that accurately
   describe the object.

Problem 4–12

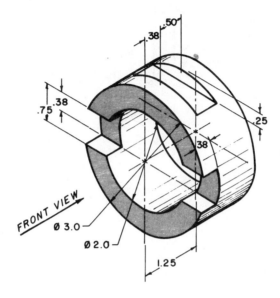

Problem 4–13

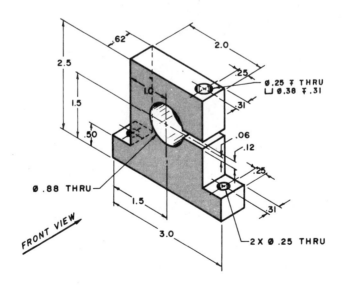

Problem 4–14

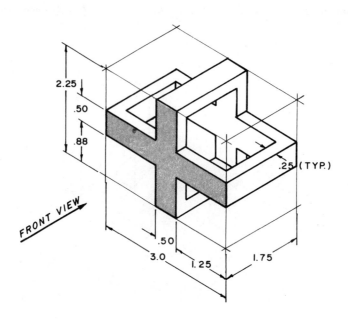

Problem 4–15

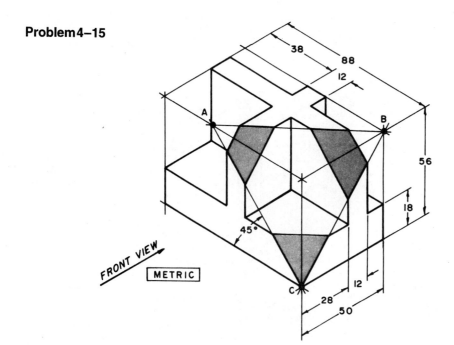

Problem 4–20

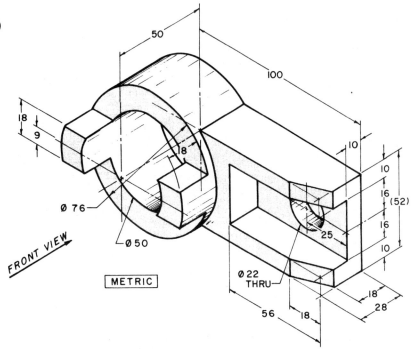

Problem 4–21

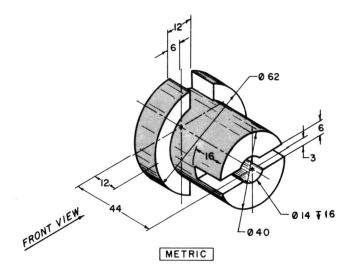

Directions for Problems 4–22 through 4–33
   Construct a 3-view drawing of each object.

Problem 4–22

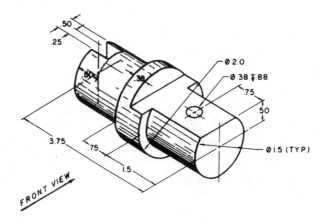

Problem 4–23

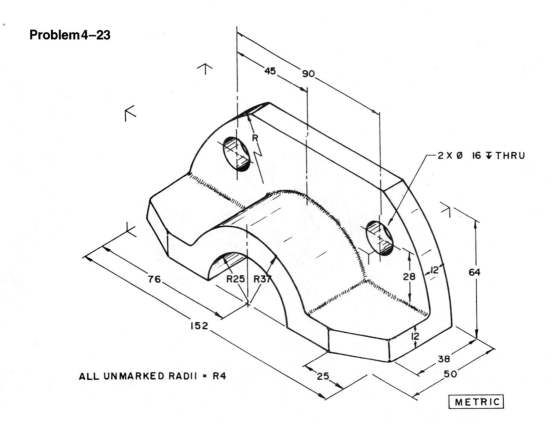

ALL UNMARKED RADII = R4

METRIC

Problem 4–24

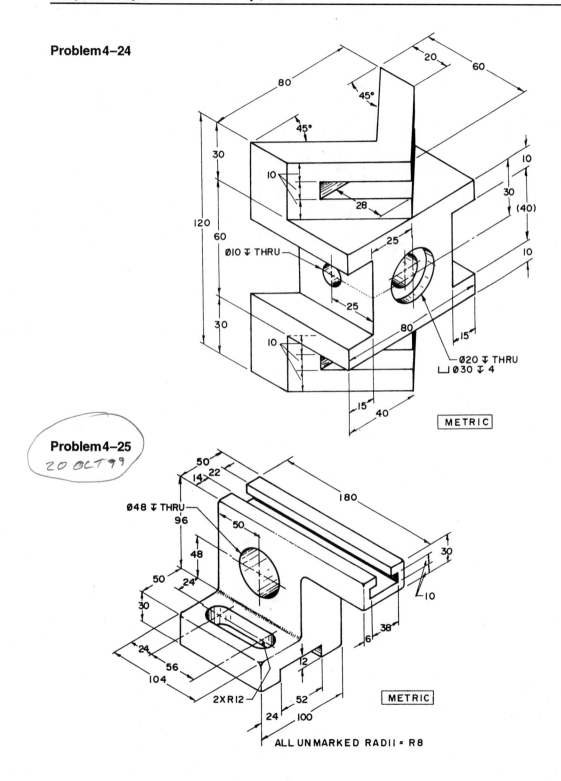

METRIC

Problem 4–25

20 OCT 99

ALL UNMARKED RADII = R8

METRIC

Problem 4–26

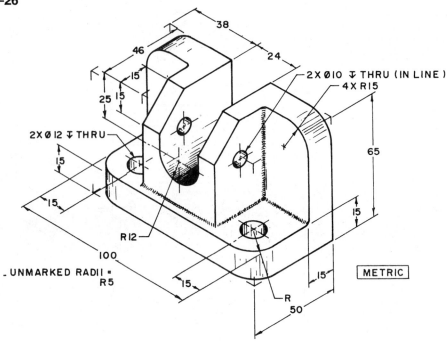

METRIC

2 X Ø10 ⊤ THRU (IN LINE)
4 X R15
2 X Ø12 ⊤ THRU
2 X Ø12 ⊤ THRU
R12
UNMARKED RADII =
R5
R
38
46
24
25
15
15
15
15
15
15
15
65
50
100

Problem 4–27

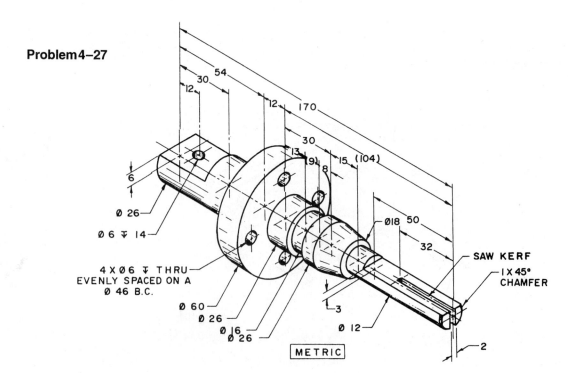

METRIC

54
30
12
12
170
30
13
9
8
15
(104)
6
Ø 26
Ø 6 ⊤ 14
4 X Ø6 ⊤ THRU
EVENLY SPACED ON A
Ø 46 B.C.
Ø 60
Ø 26
Ø 16
Ø 26
Ø 12
Ø18
50
32
3
2
SAW KERF
1 X 45°
CHAMFER

Problem 4–28

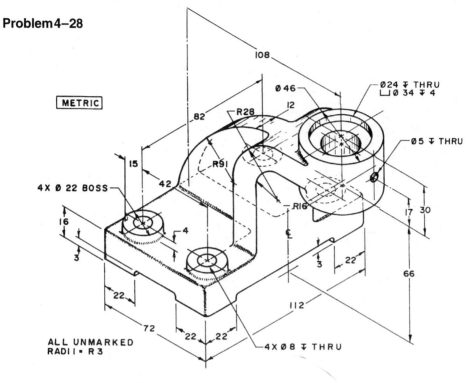

METRIC

4 X Ø 22 BOSS

ALL UNMARKED
RADII = R3

Ø 46

Ø24 ⌄ THRU
⌴ Ø 34 ⌄ 4

Ø5 ⌄ THRU

4 X Ø 8 ⌄ THRU

Problem 4–29

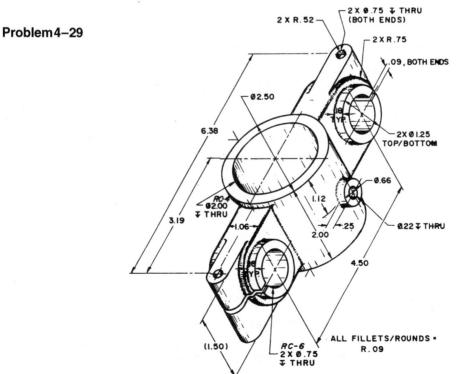

2 X Ø.75 ⌄ THRU
(BOTH ENDS)

2 X R.52

2 X R.75

.09, BOTH ENDS

Ø2.50

6.38

2 X Ø1.25
TOP/BOTTOM

Ø.66

RC-4
Ø2.00
⌄ THRU

3.19

1.06

1.12

2.00

.25

Ø.22 ⌄ THRU

4.50

(1.50)

RC-6
2 X Ø.75
⌄ THRU

ALL FILLETS/ROUNDS =
R.09

Problem 4–30

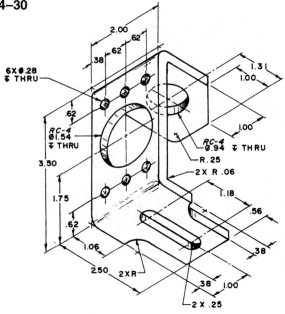

Problem 4–31

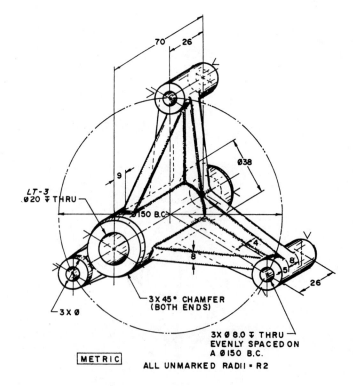

METRIC

ALL UNMARKED RADII = R2

Problem 4–32

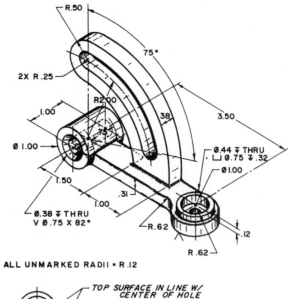

ALL UNMARKED RADII = R.12

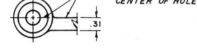

TOP SURFACE IN LINE W/
CENTER OF HOLE

Problem 4–33

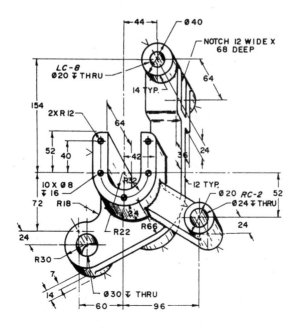

Directions for Problems 4–34 through 4–37
   Using the background grid as a guide, reproduce each problem on a CAD system. Use a grid spacing of 0.25 units for each problem.

Problem 4–34

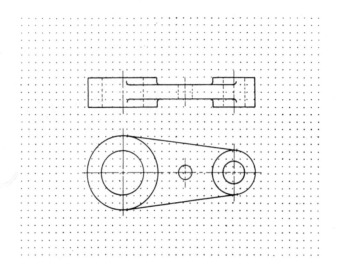

Problem 4–35

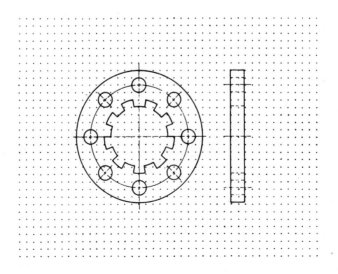

Problem 4-36

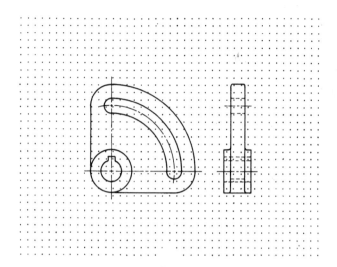

Problem 4-37

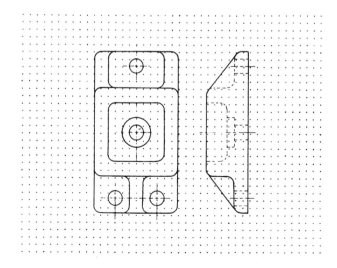

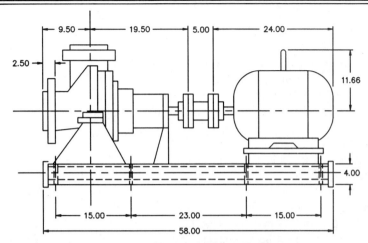

Dimensioning Techniques

Once views have been laid out, a design is not ready for the production line until numbers describing how wide, tall, or deep the object is are added to the drawing. However, these numbers must be added in a certain organized fashion; otherwise, the drawing becomes difficult to read. This may lead to confusion and the possible production of a part that is incorrect by the original design. This unit will focus on the basics of dimensioning and includes a few rules for proper dimensioning practices with numerous examples. Dimensioning techniques us-

ing AutoCAD will also be discussed in great detail including linear dimensioning, radius and diameter dimensioning, dimensioning angles, leader line usage, and a complete listing of all dimension settings with an explanation of their purpose. Special topics include dimensioning isometric drawings and ordinate or datum dimensioning. Three tutorials follow the main body of text; all three tutorials are complete regarding geometry. To complete them, dimensions need to be added or edited in some cases.

Dimension Basics

Before discussing the components of a dimension, remember that object lines (at "A") continue to be the thickest lines of a drawing with the exception of border or title blocks (See Figure 5–1). To promote contrasting lines, dimensions become visible, yet thin, lines. The heart of a dimension is the dimension line (at "B"), which is easily identified by the location of arrow terminators at both ends (at "D"). In mechanical cases, the dimension line is broken in the middle, which provides an excellent location for the dimension text (at "E"). For architectural applications, dimension text is usually placed

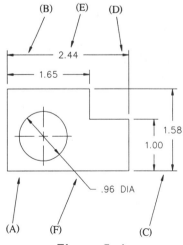

Figure 5–1

above an unbroken dimension line. The extremities of the dimension lines are limited by placing lines that act as stops for the arrow terminators. These lines, called extension lines (at "C"), begin close to the object without touching the object. Extension lines will be highlighted in greater detail in the pages that follow. For placing diameter and radius dimensions, a leader line consisting of an inclined line with a short horizontal shoulder is used (at "F"). Other applications of leader lines are for adding notes to drawings.

When you place dimensions in a drawing, provide a spacing of at least 0.38 units between the first dimension line and object being dimensioned (at "A" in Figure 5–2). If placing stacked or baseline dimensions, provide a minimum spacing of at least 0.25 units between the first and second dimension line at "B" or any other dimension lines placed thereafter. This will prevent dimensions from being placed too close to each other.

It is recommended that extensions never touch the object being dimensioned and begin approximately between 0.03 and 0.06 units away from the object at "A"(see Figure 5–3). As dimension lines are added, extension lines should extend no further than 0.12 beyond the arrow or any other terminator (at "B"). The height of dimension text is usually 0.125 units (at "C"). This value also applies to notes placed on objects with leader lines. Certain standards may require a taller lettering height. Become familiar with office practices that may deviate from these recommended values.

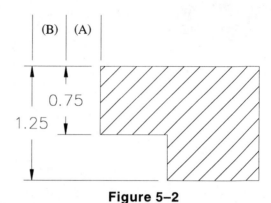

Figure 5–2

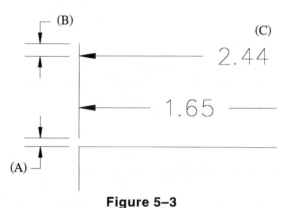

Figure 5–3

Placement of Dimensions

When placing multiple dimensions on one side of an object, place the shorter dimension closest to the object followed by the next larger dimension, as shown in Figure 5–4. When placing multiple horizontal and vertical dimensions involving extension lines that cross other extension lines, do not place gaps in the extension lines at their intersection points.

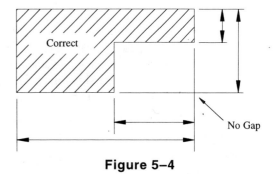

Figure 5–4

As it is acceptable for extension lines to cross each other, it is considered unacceptable practice for extension lines to cross dimension lines as in the example in Figure 5–5. The shorter dimension is placed closest to the object followed by the next larger dimension.

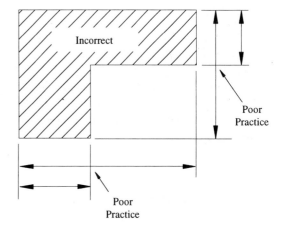

Figure 5–5

It is considered poor practice to dimension on the inside of an object when there is sufficient room to place dimensions on the outside. There may be exceptions to this rule, however. It is also considered poor practice to cross dimension lines since this may render the drawing confusing and possibly result in the inaccurate interpretation of the drawing. See Figure 5–6.

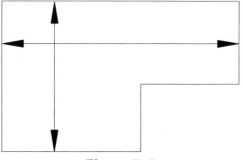

Figure 5–6

Placing Extension Lines

As two extension lines may intersect without providing a gap, so also may extension lines and object
lines intersect with each other without the need for a
gap in between them. This is the same rule practiced
when using centerlines that extend beyond the object without gapping as illustrated in Figure 5–7.

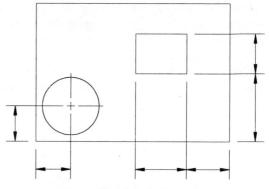

Figure 5–7

In the example in Figure 5–8, the gaps in the extension lines may appear acceptable; however, in
a very complex drawing, gaps in extension lines
would render a drawing confusing. Draw extension lines as continuous lines without providing
breaks in the lines.

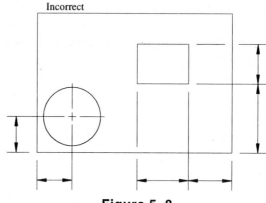

Figure 5–8

In the same manner, when centerlines are used as
extension lines for dimension purposes, no gap is
provided at the intersection of the centerline and
the object (as in Figure 5–9). As with extension lines,
the centerline should extend no further than 0.125
units past the arrow terminator.

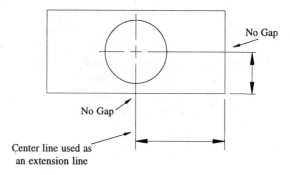

Figure 5–9

Grouping Dimensions

To promote ease of reading and interpretation, group dimensions whenever possible as shown in Figure 5–10. In addition to making the drawing and dimensions easier to read, this promotes good organizational skills and techniques. As in previous examples, always place the shorter dimensions closest to the drawing followed by any larger or overall dimensions.

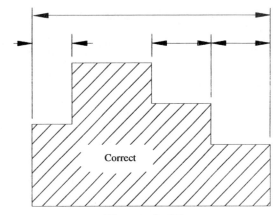

Figure 5–10

Avoid placing dimensions to an object line substituting for an extension line as in the illustration in Figure 5–11. The drawing is more difficult to follow with the dimensions being placed at different levels instead of being grouped. It must be noted, however, that there may be cases where even this practice of dimensioning is unavoidable.

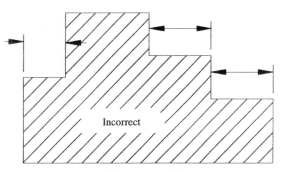

Figure 5–11

For tight spaces, arrange dimensions as in the illustration in Figure 5–12. Exercise extra care to follow proper dimension rules without sacrificing clarity.

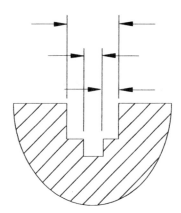

Figure 5–12

Dimensioning to Visible Features

In Figure 5–13, the object is dimensioned correctly, however, the problem is that hidden lines are used to dimension to. As there are always exceptions, try to avoid dimensioning to any hidden surfaces or features.

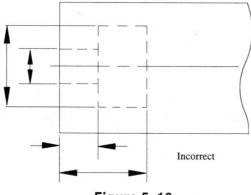

Figure 5–13

The object illustrated in Figure 5–14 is almost identical to that in Figure 5–13 with the exception that it has been converted into a full section. Surfaces that were previously hidden are now exposed. This example illustrates a better way to dimension details that were previously invisible.

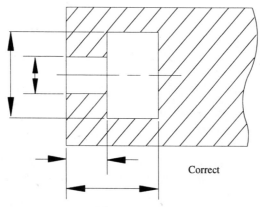

Figure 5–14

Dimensioning to Centerlines

Centerlines are used to identify the center of circular features as in "A" in Figure 5–15. You can also use centerlines to indicate an axis of symmetry as in "B." Here, the centerline consisting of a short dash flanked by two long dashes signifies that the feature is circular in shape and form. Centerlines may take the place of extension lines when placing dimensions in drawings.

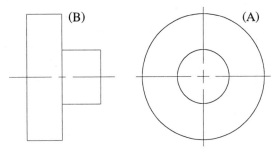

Figure 5–15

The illustration in Figure 5–16 represents the top view of a "U"-shaped object with two holes placed at the base of the "U." It also represents the correct way of utilizing centerlines as extension lines when dimensioning to holes. What makes this example correct is the rule of always dimensioning to visible features. The example in Figure 5–16 uses centerlines to dimension to holes that appear as circles. This is in direct contrast to Figure 5–17.

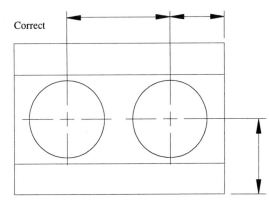

Figure 5–16

Figure 5–17 represents the front view of the "U"-shaped object. The hidden lines display the circular holes passing through the object along with centerlines. Centerlines are being used as extension lines for dimensioning purposes; however, it is considered poor practice to dimension to hidden features or surfaces. Always attempt to dimension to a view where the features are visible before dimensioning to hidden areas.

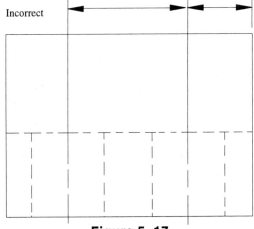

Figure 5–17

Arrowheads

Arrowheads are generally made three times as long as they are wide, or very long and narrow as shown in Figure 5–18. The actual size of an arrowhead would measure approximately 0.125 units in length.

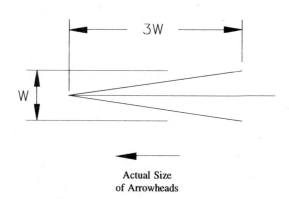

Actual Size
of Arrowheads

Figure 5–18

Dimension line terminators may take the form of shapes other than filled-in arrowheads (see Figure 5–19). The 45-degree slash or "tick" is a favorite dimension line terminator used by architects although they are sometimes seen in mechanical applications.

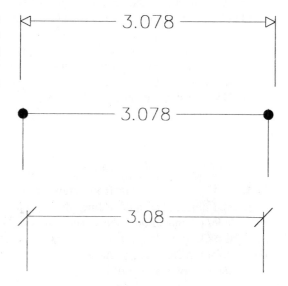

Figure 5–19

Dimensioning Systems

The Unidirectional System

When placing dimensions using AutoCAD, a typical example is illustrated in Figure 5–20. Here, all text items are right-reading or horizontal. This goes for all vertical, aligned, angular, and diameter dimensions. When dimension text can be read right-reading, this is the Unidirectional Dimensioning System. By default, all AutoCAD settings are set to dimension in the Unidirectional System.

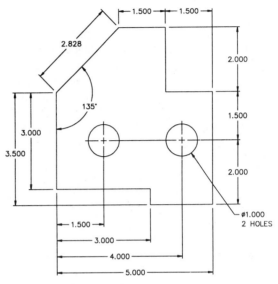

Figure 5–20

The Aligned System

The identical object from the previous example is illustrated in Figure 5–21. Notice that all horizontal dimensions have the text positioned in the horizontal direction as in Figure 5–20. However, vertical and aligned dimension text is rotated or aligned with the direction being dimensioned. This is the most notable feature of the Aligned Dimensioning System. Text along vertical dimensions is rotated in such a way that the drawing must be read from the right. Angular dimensions remain unaffected in the Aligned System; however, aligned dimension text is rotated parallel with the feature being dimensioned.

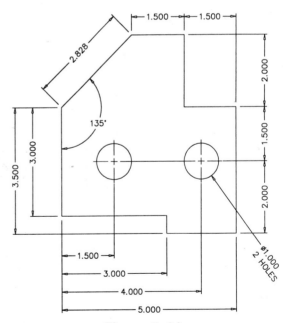

Figure 5–21

Repetitive Dimensions

Throughout this chapter, numerous methods of dimensioning have been discussed supported by many examples. Just as it was important in multiview projection to draw only those views that accurately described the object, so also is it important to dimension these views. In this manner, the actual production of the part may start. However, take care when placing dimensions; dimensioning takes planning to better place a dimension. The problem with the illustration in Figure 5–22 is that even though the views are correct and the dimensions call out the overall sizes of the object, there are too many cases where dimensions are duplicated. Once a feature has been dimensioned, such as the overall width of 3.75 units, this number does not need to be placed in the top view. This is the purpose of understand-

ing the relationship between views or what dimensions the views have in common with each other. Adding unnecessary dimensions also makes the drawing very busy and cluttered in addition to being very confusing to read. Compare the illustration in Figure 5–22 with the illustration in Figure 5–23, which shows just those dimensions needed to describe the size of the object. Do not be concerned that the top view has no dimensions; the designer should interpret the width of the top as 3.75 units from the front view, and the depth as 2.50 from the right side view.

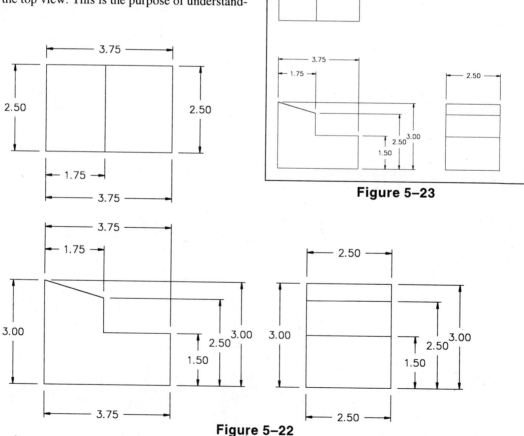

Figure 5–23

Figure 5–22

Dimension Toolbar Button Commands

Selection of dimensioning commands has been
made easier through the use of the Dimension
toolbar illustrated in Figure 5–24. Study this image and the corresponding command associated
with each button.

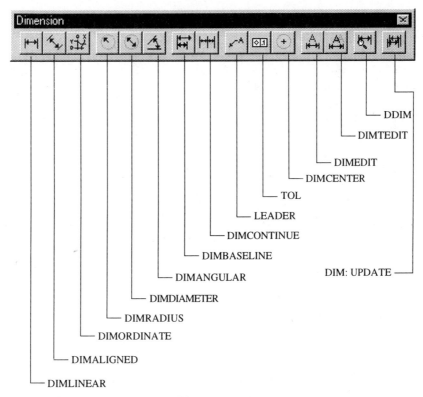

Figure 5–24

Linear Dimensions

Horizontal Linear Dimensions

The linear dimensioning mode generates either a horizontal or vertical dimension depending on the location of the dimension. The following prompts illustrate the generation of a horizon-

tal dimension using the DIMLINEAR command. Notice that identifying the dimension line location at "C" in Figure 5–25 automatically generates a horizontal dimension.

 Command:**DIMLINEAR**

First extension line origin or press ENTER to select:**Endp**
of *(Select the endpoint of the horizontal line at "A")*
Second extension line origin:**Endp**
of *(Select the other endpoint of the horizontal line at "B")*
Dimension line location (Mtext/Text/Angle/ Horizontal/Vertical/Rotated): *(Select a point near "C")*
Dimension text = 2.00

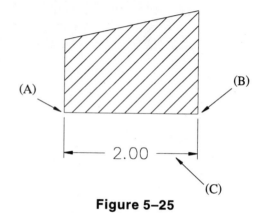

Figure 5–25

Vertical Linear Dimensions

The linear dimensioning command is also used to generate vertical dimensions. The following prompts illustrate the generation of a vertical dimension using the DIMLINEAR command. Notice that identifying the dimension line location at "C" in Figure 5–26 automatically generates a vertical dimension.

 Command:**DIMLINEAR**

First extension line origin or press ENTER to select:**Endp**
of *(Select the endpoint of the vertical line at "A")*
Second extension line origin:**Endp**
of *(Select the endpoint of the other vertical line at "B")*
Dimension line location (Mtext/Text/Angle/ Horizontal/Vertical/Rotated): *(Select a point near "C")*
Dimension text = 2.10

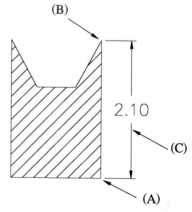

Figure 5–26

Aligned Dimensions

The aligned dimensioning mode generates a dimension line parallel to the distance specified by the location of two extension line origins as shown in Figure 5–27. The following prompts illustrate the generation of an aligned dimension using the DIMALIGNED command:

 Command:**DIMALIGNED**

First extension line origin or press ⌜ENTER⌟ to
   select:**Endp**
of *(Select the endpoint of the line at "A")*
Second extension line origin:**Endp**
of *(Select the endpoint of the line at "B")*
Dimension line location (Mtext/Text/Angle):
   (Select a point near "C")
Dimension text = 5.6569

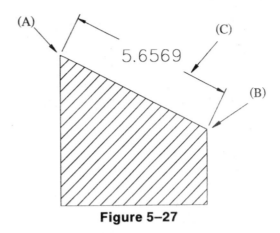

Figure 5–27

Rotated Linear Dimensions

This linear dimensioning mode generates a dimension line which is rotated at a specific angle (see Figure 5–28). The following prompts illustrate the generation of a rotated dimension using the DIMROTATE command.

 Command:**DIMLINEAR**

First extension line origin or press ⌜ENTER⌟ to
   select:**Endp**
of *(Select the endpoint of the line at "A")*
Second extension line origin:**Endp**
of *(Select the endpoint of the line at "B")*
Dimension line location (Mtext/Text/Angle/
   Horizontal/Vertical/Rotated):**R**
Dimension line angle <0>:**45**
Dimension line location (Mtext/Text/Angle/
   Horizontal/Vertical/Rotated): *(Select a point
   near "C")*
Dimension text = 2.8284

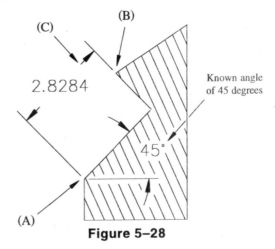

Figure 5–28

Continue Dimensions

The power of grouping dimensions for ease of reading has already been explained. The illustration in Figure 5–29 shows yet another feature while dimensioning in AutoCAD; namely, the practice of using Continue dimensions. With one dimension already placed, the DIMCONTINUE command is selected, which prompts you for the second extension line location. Picking the second extension line location strings the dimensions next to each other or continues the dimension.

 Command: **DIMLINEAR**

First extension line origin or press ENTER to select: **Endp**
of *(Select the endpoint at "A")*
Second extension line origin: **Endp**
of *(Select the endpoint at "B")*
Dimension line location (Mtext/Text/Angle/ Horizontal/Vertical/Rotated): *(Locate the 1.75 dimension)*
Dimension text = 1.75

 Command: **DIMCONTINUE**

Specify a second extension line origin or (<select>/Undo): **Endp**
of *(Select the endpoint at "C")*
Dimension text = 1.25
Specify a second extension line origin or (<select>/Undo): **Endp**
of *(Select the endpoint at "D")*
Dimension text = 1.50
Specify a second extension line origin or (<select>/Undo): **Endp**
of *(Select the endpoint at "E")*
Dimension text = 1.00
Specify a second extension line origin or (<select>/Undo): *(Press ENTER to exit this mode)*
Select continued dimension: *(Press ENTER to exit this command)*

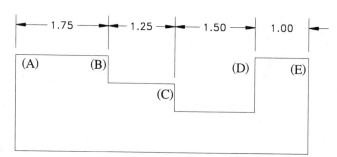

Figure 5–29

Baseline Dimensions

Yet another aid in grouping dimensions is using the DIMBASELINE command. Continue dimensions place dimensions next to each other; Baseline dimensions establish a base or starting point for the first dimension as shown in Figure 5–30. Any dimensions that follow in the DIMBASELINE command are calculated from the common base point already established. This is a very popular mode to use when one end of an object acts as a reference edge. As dimensions are placed using the DIMBASELINE command, an AutoCAD dimension setting controls the spacing of the dimensions away from each other. This value is set by default to 0.38 units and can be found in the Geometry dialog box of the Dimension Styles dialog box.

 Command:**DIMLINEAR**

First extension line origin or press ⌜ENTER⌝ to select: **Endp**
of *(Select the endpoint at "A")*
Second extension line origin: **Endp**
of *(Select the endpoint at "B")*

Dimension line location (Mtext/Text/Angle/ Horizontal/Vertical/Rotated): *(Locate the dimension)*
Dimension text = 1.75

 Command: **DIMBASELINE**

Specify a second extension line origin or (<select>/Undo):**Endp**
of *(Select the endpoint at "C")*
Dimension text = 3.00
Specify a second extension line origin or (<select>/Undo):**Endp**
of *(Select the endpoint at "D")*
Dimension text = 4.50
Specify a second extension line origin or (<select>/Undo):**Endp**
of *(Select the endpoint at "E")*
Dimension text = 5.50
Specify a second extension line origin or (<select>/Undo): *(Press ⌜ENTER⌝ to exit this mode)*
Select base dimension: *(Press ⌜ENTER⌝ to exit this command)*

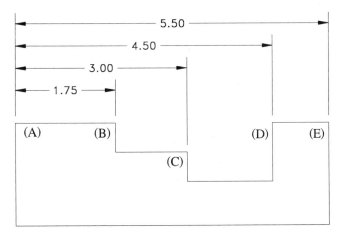

Figure 5–30

Radius and Diameter Dimensioning

Arcs and circles are to be dimensioned in the view where their true shape is visible. The mark in the center of the circle or arc indicate its center point as shown in Figure 5–31. You may place the dimension text either inside or outside of the circle; you may also use grips to aid in the dimension text location of a diameter or radius dimension. The prompts for Diameter and Radius dimensions are as follows:

 Command:**DIMDIAMETER**

Select arc or circle: *(Select the edge of the circle)*
Dimension text = 2.50
Dimension line location (Mtext/Text/Angle): *(Select the desired location of the leader line to display the diameter dimension).*

 Command:**DIMRADIUS**

Select arc or circle: *(Select the edge of the arc)*
Dimension text = 1.50
Dimension line location (Mtext/Text/Angle): *(Select the desired location of the leader line to display the radius dimension).*

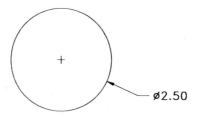

ø2.50

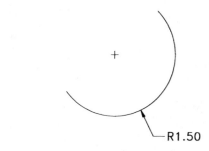

R1.50

Figure 5–31

Leader Lines

A leader line is a thin, solid line leading from a note or dimension ending with an arrowhead illustrated at "A" in Figure 5–32. The arrowhead should always terminate on an object line such as the edge of a hole or arc. A leader to a circle or arc should be radial; this means it is drawn so that if extended, it would pass through the center of the circle illustrated at "B." Leaders should cross as few object lines as possible and should never cross each other. The short horizontal shoulder of a leader should meet the dimension illustrated at "A." It is poor practice to underline the dimension with the horizontal shoulder illustrated at "C." Example "C" also illustrates a leader not lined up with the center or radial. This may affect the appearance of the leader. Again, check for the standard office practices to ensure this example is acceptable. Yet another function of a leader is to attach notes to a drawing illustrated at "D." Notice the two notes attached to the view have different terminators, arrows, and dots. It is considered good practice to adopt only one terminator for the duration of the drawing. The prompt sequence for the LEADER command is as follows:

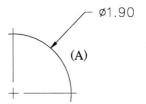

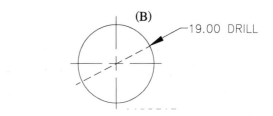

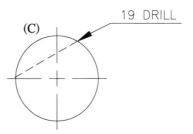

 Command: **LEADER**

From point: **Nea**
to (*Select the edge of the arc at "A"*)
To point: *(Locate the end of the leader)*
To point (Format/Annotation/
  Undo)<Annotation>: *(Press* ⌷ENTER⌷ *to accept this default)*
Annotation (or press ⌷ENTER⌷ for options): *(Press* ⌷ENTER⌷ *to accept this default)*
Tolerance/Copy/Block/None/<Mtext>: *(Press* ⌷ENTER⌷ *to display the Mtext dialog box. Enter the desired text to make up the leader note and click on the OK button to display the leader)*

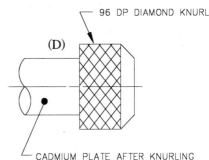

Figure 5–32

Dimensioning Angles

Dimensioning angles requires two lines forming the angle in addition to the location of the vertex of the angle along with the dimension arc location. Before going any further, understand where the curved arc for the angular dimension is derived from. In illustration "A" in Figure 5–33, the dimension arc is struck from an imaginary center or vertex of the arc. The following prompts are taken from the DIMANGULAR command:

 Command: **DIMANGULAR**

Select arc, circle, line, or press ENTER: *(Select line "A")*

Second line: *(Select line "B")*

Dimension arc line location (Mtext/Text/Angle): *(Select the location at "C")*

Dimension text = 53

Dimensioning Slots

For slots, first select the view where the slot is visible. Two methods of dimensioning the slot are illustrated in Figure 5–34. A slot may be called out by locating the center-to-center distance of the two semi-circles followed by a radius dimension to one of the semicircles; which radius dimension selected depends on the available room to dimension. A second method involves the same center-to-center distance followed by an overall distance designating the width of the slot. This dimension happens to be the same as the diameter of the semicircles. It is also considered good practice to place this dimension inside of the slot. A more complex example in Figure 5–35 involves slots formed by curves and angles. Here, the radius of the circular center arc is called out. Angles reference each other for accuracy. As in Figure 5–34, the overall width of the slot is dimensioned, which happens to be the diameter of the semicircles at opposite ends of the slot. Use the DIMALIGN command to place the 0.48 dimension. Use OSNAP-Nearest to identify the location at "A" and OSNAP-Perpendicular to identify the location at "B."

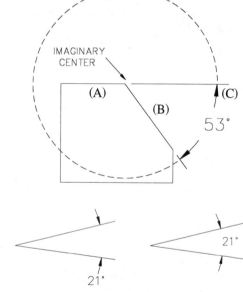

Figure 5–33

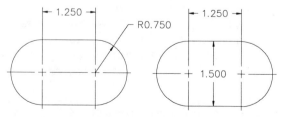

Figure 5–34

Ordinate Dimensioning

The plate in Figure 5–36 consists of numerous drill holes with a few slots in addition to numerous 90-degree angle cuts along the perimeter. This object is not considered difficult to draw or make since it mainly consists of drill holes. However, conventional dimensioning techniques make the plate appear complex since a dimension is required for the location of every hole and slot in both the X and Y directions. Add to that standard dimension components such as extension lines, dimension lines, and arrowheads and it is easy to get lost in the complexity of the dimensions even on this simple object.

A better dimensioning method to use is illustrated in Figure 5–37, called Ordinate or Datum dimensioning. Here, no dimension lines or arrowheads are drawn; instead, one extension line is constructed from the selected feature to a location specified by you. A dimension is added to identify this feature in either the X or Y directions. It is important to understand that all dimension calculations occur in relation to the current User Coordinate System (UCS), or the current 0,0 origin. In the example in Figure 5–37, with the 0,0 origin located in the lower left corner of the plate, all dimensions in the horizontal and vertical directions are calculated in relation to this 0,0 location. Holes and slots are called out using the DIMDIAMETER command. The following illustrates a typical ordinate dimensioning prompt sequence:

 Command: **DIMORDINATE**

Select feature: *(Select a feature using an Osnap option)*
Leader endpoint (Xdatum/Ydatum/Mtext/Text):
    (Locate an outside point)
Dimension text = Calculated value

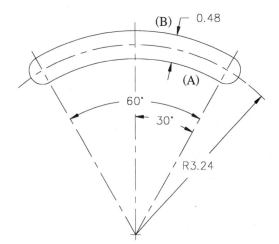

Figure 5–35

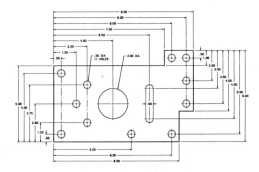

Figure 5–36

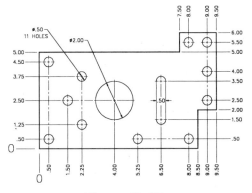

Figure 5–37

To illustrate how to place ordinate dimensions, see the example in Figure 5–38 and the prompt sequence below. Before placing any dimensions, a new UCS must be moved to a convenient location on the object using the UCS-Origin option. All ordinate dimensions will reference this new origin since it is located at coordinate 0,0. At the command prompt, type DIMORDINATE to enter ordinate dimensioning. Select the Quadrant of the arc at "A" as the feature. For the leader endpoint, pick a point at "B." Be sure Ortho mode is on. It is also helpful to snap to a convenient grid point for this and other dimensions along this direction. Follow the prompt sequence below:

 Command:**DIMORDINATE**

Select feature: *(Select the Quadrant of the arc at "A")*

Leader endpoint (Xdatum/Ydatum/Mtext/Text): *(Locate a point at "B")*

Dimension text = 1.50

With the previous example in Figure 5–38 highlighting horizontal ordinate dimensions, placing vertical ordinate dimensions is identical (see Figure 5–39). With the location of the UCS still located in the lower left corner of the object, select the feature at "A" using either the Endpoint or Quadrant modes. Pick a point at "B" in a convenient location on the drawing. Again, it is helpful if Ortho is on and a grid dot is snapped to.

 Command:**DIMORDINATE**

Select feature: *(Select the Quadrant of the arc at "A")*

Leader endpoint (Xdatum/Ydatum/Mtext/Text): *(Locate a point at "B")*

Dimension text = 3.00

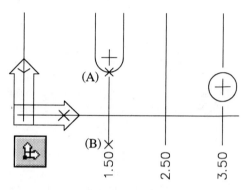

Figure 5–38

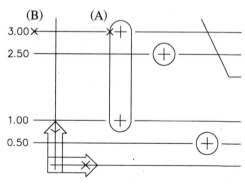

Figure 5–39

When spaces are tight to dimension to, two points not parallel to the X or Y axis will result in a "jog" being drawn (see Figure 5–40). It is still helpful to snap to a grid dot when performing this operation; however, be sure Ortho is off.

 Command:**DIMORDINATE**

Select feature: *(Select the Endpoint of the line or arc at "A")*

Leader endpoint (Xdatum/Ydatum/Mtext/Text): *(Locate a point at "B")*

Dimension text = 2.00

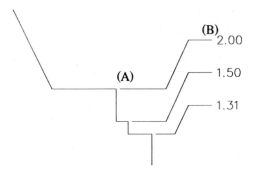

Figure 5–40

Ordinate dimensioning provides a neat and easy way of organizing dimensions for drawings where geometry leads to applications involving numerical control. Only two points are required to place the dimension that references the current location of the UCS. See Figure 5–41.

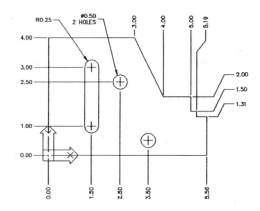

Figure 5–41

Dimensioning Isometric Drawings

In past versions of AutoCAD, it was possible to use the Aligned dimensioning mode to make the dimension line parallel with the surface being dimensioned. Arrowheads were also drawn parallel; however, extension lines were drawn perpendicular to the dimension lines and not at an isometric angle. The results of this type of dimensioning technique are illustrated in Figure 5–42. One of the only ways to simulate isometric dimensions was to turn off the extension lines, manually draw new extension lines at isometric angles, and move the dimension to a new location because of the position of the extension lines.

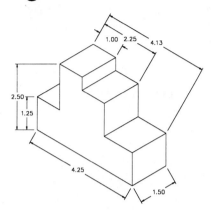

Figure 5–42

The DIMEDIT command and the Oblique option allows you to enter an obliquing angle which will rotate the extension lines and reposition the dimension line.

 Command: **DIMEDIT**

Dimension Edit (Home/New/Rotate/Oblique)
  <Home>: **Oblique**
Select objects: *(Select the 2.50 and 1.25 dimensions in Figure 5–43)*
Select objects: *(Press* ENTER *to continue)*
Enter obliquing angle (Press ENTER for none):
  150

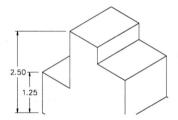

Figure 5–43

Both dimensions in Figure 5–44 were repositioned using the Oblique option of the DIMEDIT command. Notice that the extension lines and dimension line were affected. The DIMTEDIT command will allow text to be rotated at an angle.

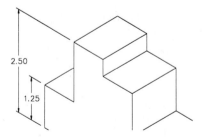

Figure 5–44

The Oblique option of the DIMEDIT command was used to rotate the dimension at "A" at an obliquing angle of 210 degrees (see Figure 5–45). An obliquing angle of -30 degrees was used to rotate the dimension at "B" and the dimensions at "C" required an obliquing angle of 90 degrees. This represents proper isometric dimensions except for the orientation of the text.

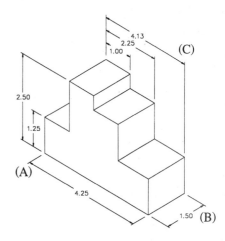

Figure 5–45

Tolerances

Interchangeability of parts requires replacement parts to fit in an assembly of an object no matter where the replacement part comes from. In the example in Figure 5–46, the "U"-shaped channel of 2.000 units in width is to accept a mating part of 1.995 units. Under normal situations, there is no problem with this drawing or callout. However, what if the production person cannot make the channel piece exactly at 2.000? What if he or she is close and the final product measures 1.997? Again, the mating part will have no problems fitting in the 1.997 slot. What if the mating part is not made exactly 1.995 units but is instead 1.997? You see the problem. As easy as it is to attach a dimension to a drawing, some thought needs to go into the possibility that based on the numbers, maybe the part cannot be easily made. Instead of locking dimensions in using one number, a range of numbers would allow the production person the flexibility to vary in any direction and still have the parts fit together. This is the purpose of converting some basic dimensions to tolerances.

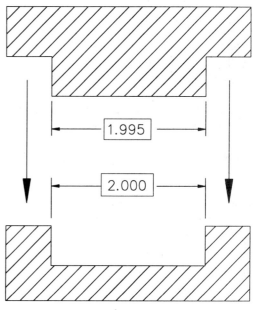

Figure 5–46

The example in Figure 5–47 shows the same two mating parts; this time, two sets of numbers for each part are assigned. For the lower part, the machinist must make the part anywhere in between 2.002 and 1.998, which creates a range of 0.004 units of variance. The upper mating part must be made within 1.997 and 1.993 units in order for the two to fit correctly. The range for the upper part is also 0.004 units. As we will soon see, no matter if upper or lower numbers are used, the parts will always fit together. If the bottom part is made to 2.002 and the top part is made to 1.993, the parts will fit. If the bottom part is made to 1.998 and the upper part is made to 1.997, the parts will fit. In any case or combination, if the dimensions are followed exactly as stated by the tolerances, the pieces will always fit. If the bottom part is made to 1.998 and the upper part is made to 1.999, the upper piece is rejected. The method of assigning upper and lower values to dimensions is called limit dimensioning. Here, the larger value in all cases is placed above the smaller value. This is also called a clearance fit since any combinations of numbers may be used and the parts will still fit together.

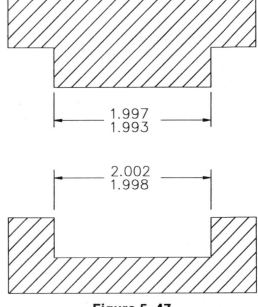

Figure 5–47

The object in Figure 5–48 has a different tolerance value assigned to it. The basic dimension is 2.000; in order for this part to be accepted, the width of this object may go as high as 2.002 units or as low as 1.999 units giving a range of 0.003 units by which the part may vary. This type of tolerance is called a plus/minus dimension with an upper limit of 0.002 difference from the lower limit of -0.001.

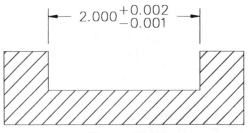

Figure 5–48

Illustrated in Figure 5–49 is yet another way to display tolerances. It is very similar to the plus/minus method except that both upper and lower limits are the same; for this reason, this method is called plus and minus dimensions. The basic dimension is still 2.000 with upper and lower tolerance limits of 0.002 units.

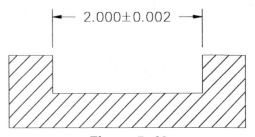

Figure 5–49

Geometric Dimensioning and Tolerancing (GD&T)

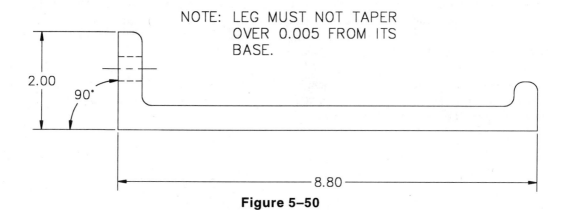

Figure 5–50

In the example of the object in Figure 5–50, approximately 1,000 items needed to be made based on the dimensions of the drawing. Also, once constructed, all objects needed to be tested to be sure the 90 degree surface did not deviate over 0.005 units from its base. Unfortunately the wrong base was selected as the reference for all dimensions. Which should have been the correct base feature? As all items were delivered, they were quickly returned since the long 8.80 surface drastically deviated from required 0.005 unit deviation or zone. This is one simple example that demonstrates the need for using geometric dimension and tolerancing techniques. First, this method deals with setting tolerances to critical characteristics of a part. Of course the function of the part must be totally un-

derstood by the designer in order to assign tolerances. The problem with the object in Figure 5–50 was the note which did not really specify which base feature to choose as a reference for dimensioning. Figure 5–51 is the same object complete with dimensioning and tolerancing symbols. The letter "-A-" inside of the rectangle identifies the datum or reference surface. The tolerance symbol at the end of the long edge tells the individual making this part that the long edge cannot deviate more than 0.005 units using surface "A" as a reference.

Using geometric tolerancing symbols insures more accurate parts with less error in interpreting the dimensions.

Figure 5–51

GD&T Symbols

Entering "TOL" at the command prompt brings up the first in a series of geometric dimensioning and tolerancing dialog boxes. The first dialog box to display is illustrated in Figure 5–52 and contains all major geometric dimensioning and tolerancing symbols. Choose the desired symbol by clicking the specific symbol. If the wrong symbol has been accidentally selected, click on the correct symbol. The next dialog box does not activate until the "OK" button is selected from the Symbol dialog box.

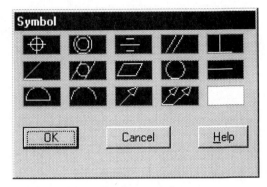

Figure 5–52

Illustrated in Figure 5–53 is a chart outlining all geometric tolerancing symbols supported in the Symbol dialog box. Alongside each symbol is the characteristic controlled by the symbol in addition to a tolerance type. Tolerances of Form such as Flat- ness and Straightness can be applied to surfaces without referencing a datum. On the other hand, tolerances of Orientation such as Angularity and Perpendicularity require datums as reference.

Symbol	Characteristic	Type of Tolerance
▱	Flatness	Form
—	Straightness	
◯	Roundness	
⌭	Cylindricity	
⌒	Profile of a Line	Profile
⌓	Profile of a Surface	
∠	Angularity	Orientation
⊥	Perpendicularity	
//	Parallelism	
⊕	Position	Location
◎	Concentricity	
⹀	Symmetry	
↗	Circular Runout	Runout
↗↗	Total Runout	

Figure 5–53

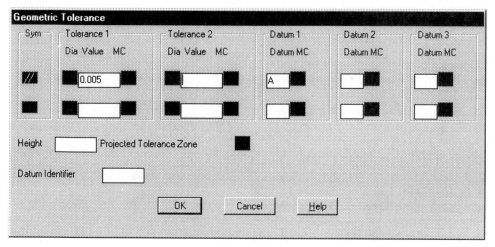

Figure 5–54

Once a symbol is picked and the Symbol dialog box exited, the Geometric Tolerance dialog box appears, illustrated in Figure 5–54. With the symbol placed inside of this dialog box, you now assign such items as tolerance values, maximum material condition modifiers, and datums. In the example in Figure 5–54, the tolerance of Parallelism is to be applied at a tolerance value of 0.005 units to Datum "A."

Additional GD&T Symbols

While in the main Geometric Tolerancing dialog box in Figure 5–54, clicking in the box under "MC" activates another dialog box illustrated in Figure 5–55. This dialog box is devoted to the concept of Material Condition and acts as a modifier to the main tolerance value.

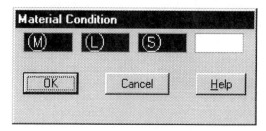

Figure 5–55

In brief, Maximum Material Condition refers to the condition of a characteristic such as a hole when the most material exists. Least Material Condition refers to the condition of a characteristic where the least material exists. Regardless of Feature Size indicates that the characteristic tolerance such as Flatness or Straightness must be maintained regardless of the actual produced size of the object.

Datums are represented by surfaces, points, lines, or planes and are used for referencing the features of an object. The symbols illustrated in Figure 5–56 are identified as "flags" used to identify a datum. The next series of figures illustrate how datums are used in combination with tolerances of Orientation.

The Feature Control Symbol

Once completing the desired information inside of the Geometric Tolerance dialog box, the result is illustrated in Figure 5–57 in the form of a Feature Control Symbol. This is the actual order that all symbols, values, modifiers, and datums are placed in. Follow the graphic and information in Figure 5–57 to view the contents of each area of the Feature control symbol box.

Ⓜ   Maximum Material Condition

Ⓛ   Least Material Condition

Ⓢ   Regardless of Feature Size

−A−   Primary Datum

−B−   Secondary Datum

−C−   Tertiary Datum

Figure 5–56

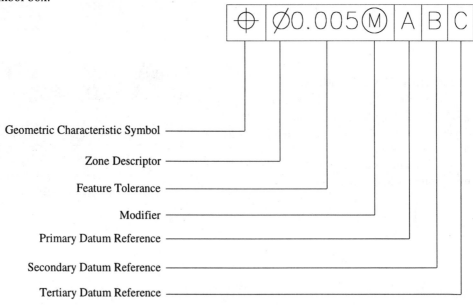

Geometric Characteristic Symbol

Zone Descriptor

Feature Tolerance

Modifier

Primary Datum Reference

Secondary Datum Reference

Tertiary Datum Reference

Figure 5–57

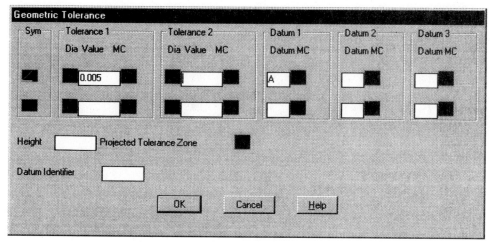

Figure 5–58

Tolerance of Angularity Example

Illustrated in the Geometric Tolerance dialog box in Figure 5–58 and 5–59 are examples of applying the tolerance of Angularity to a feature. First the symbol is identified along with the tolerance value. Since this tolerance requires a datum, the letter "A" identifies primary datum "A." The results are illustrated in Figure 5–59. Datum "A" identifies the surface of reference as the base of the object. The feature control symbol is applied to the angle and reads, "The surface dimensioned at 60 degrees cannot deviate more than 0.005 units from datum "A." The graphic in Figure 5–59 shows an exaggerated tolerance zone and is used for illustrative purposes only.

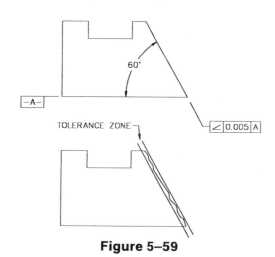

Figure 5–59

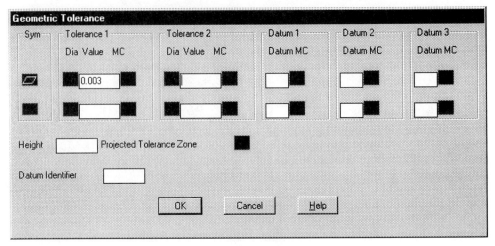

Figure 5–60

Tolerance of Flatness Example

Figure 5–60 is an example of applying a tolerance of Flatness to a surface. First the Geometric Tolerance dialog box has the Flatness symbol assigned in addition to a tolerance value of 0.003 units. The results are illustrated in Figure 5–61 with the feature control box being applied to the top surface. The tolerance box reads, "The surface must be flat with a 0.003 unit tolerance zone." The tolerance range is displayed in the second graphic and is exaggerated. The tolerance of Flatness does not usually require a datum for reference.

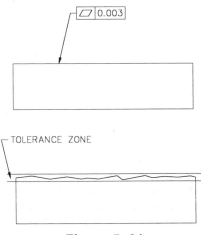

Figure 5–61

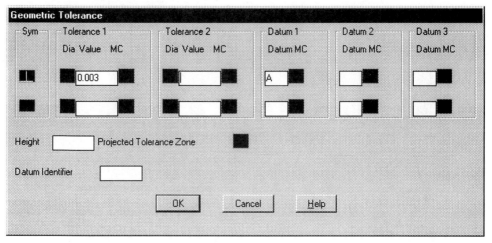

Figure 5–62

Tolerance of Perpendicularity Example

This next example illustrates a tolerance of Perpendicularity. The Geometric Tolerance dialog box reflects the perpendicularity symbol along with a tolerance value of 0.003 units. See Figure 5–62 This tolerance characteristic requires a datum to be most effective. In the example in Figure 5–63, the feature control box reads, "This surface must be perpendicular to datum 'A' within a tolerance zone of 0.003 units." The second graphic shows the tolerance zone and the amount of deviation that is acceptable. It is meant to be exaggerated and is used for illustrative purposes only.

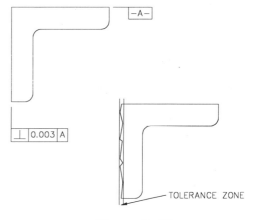

Figure 5–63

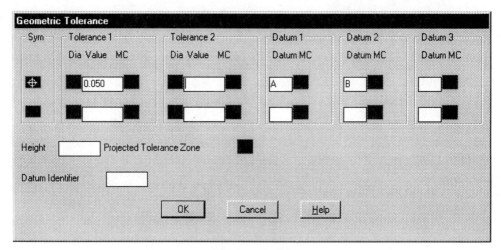

Figure 5–64

Tolerance of Position Example

This last example displays a tolerance of Position that is applied to a hole feature. Since a hole is centered from two edges, two datums are identified in the Geometric Tolerancing dialog box illustrated in Figure 5–64. The feature control box illustrated in Figure 5–65 reads, "The center of the hole must lie within a circular tolerance zone of 0.050 units in relation to datums 'A' and 'B.'" This is sometimes called a circular tolerance.

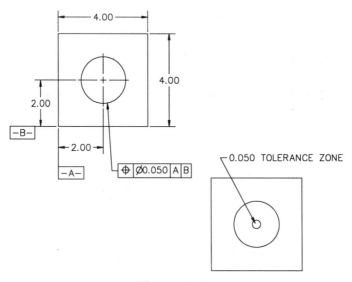

Figure 5–65

General Dimensioning Symbols

In today's global economy, the transfer of documents in the form of drawings is becoming a standard form of doing business. As drawings are shared with a subsidiary of an overseas company, two different forms of language may be needed to interpret the dimensions of the drawing. This in the past has led to confusion of interpreting a drawing and as a result, a system of dimensioning symbols has been developed. It is hoped that the recognition of a symbol will be easier to interpret than adding a note on the particular feature being dimensioned in a different language.

The illustration in Figure 5–66 shows some of the more popular dimensioning symbols in use today on drawings. Notice how the symbols are designed to make as clear and consistent an interpretation of the dimension as possible. As an example, the Deep or Depth symbol displays an arrow pointing down. This symbol is used to identify how far into a part the depth of a drill hole goes. Other symbols such as the Arc Length symbol identifies the length of an arc.

⌒	Arc Length
X.XX	Basic Dimension
▷	Conical Taper
⊔	Counterbore or Spotface
∨	Countersink
▽	Deep or Depth
∅	Diameter
X.XX	Dimension Not to Scale
2X	Number of Times—Places
R	Radius
(X.XX)	Reference Dimension
S∅	Spherical Diameter
SR	Spherical Radius
◁	Slope
□	Square

Figure 5–66

The size of all general dimensioning symbols is illustrated in Figure 5–67. All values are in relation to "h" or the relative height of the dimension numerals. As an example, with a dimension numeral height of 0.18, study the illustration in Figure 5–68 for finding the size of the counterbore dimension symbol. Since the height of the counterbore symbol is defined by "h," simply substitute the dimension numeral height of 0.18 for the height of the counterbore symbol. Since the width of the symbol is "2h," multiply 0.18 by 2 to obtain a value of 0.36. These symbols may be created and saved as blocks for insertion into drawings. They may also be part of an existing text font mapped to a particular key on the keyboard. Striking that key brings up the dimension symbol.

Applying the Basic Dimension Symbol

A basic dimension is identified by drawing a rectangular box around the dimension numeral (see Figure 5–69). The basic dimension represents the theoretical distance of the feature such as the dis-

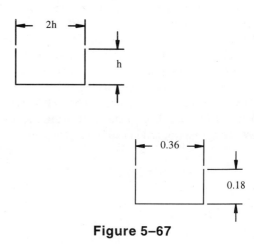

Figure 5–67

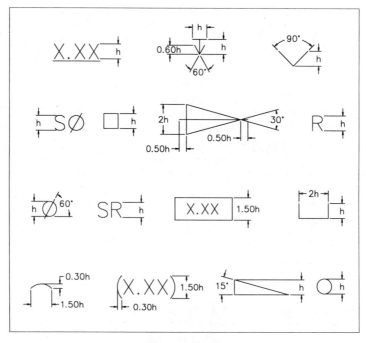

Figure 5–68

tance from the center hole to the left hole as 4.00 units with the distance from the left hole to the right hole as 8.00 units. It is a known fact that it is impossible to maintain tolerances to obtain an exact value of 4.00 or 8.00 units. A rectangle may be constructed; however, this may be a tedious task depending on the number of decimal places past the zero which determines the size of the rectangle. To constrain the rectangle around dimension text, go to the pull-down menu area and click on the following sequence: Format>Dimension Style...>Annotation. Locate the tolerance group and click in the edit box next to "Method:." Click on "Basic" to draw the rectangle when placing a dimension.

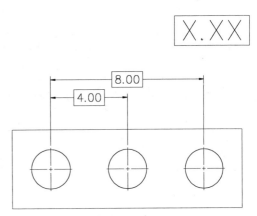

Figure 5-69

Applying the Reference Dimension Symbol

Illustrated in Figure 5–70 are a series of continue dimensions strung together in one line. The dimension numeral 3.00 units is enclosed in parentheses and is referred to as a reference dimension. Reference dimensions are not required drawing dimensions; they are placed for information purposes. The parentheses located on the keyboard are used along with the dimension numeral to create a reference dimension.

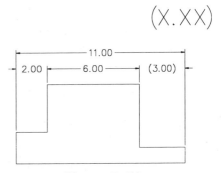

Figure 5-70

Applying the Not to Scale Dimension Symbol

At times it is necessary to place a dimension which is not to scale as in the example illustrated in Figure 5–71. To distinguish regular dimensions from a dimension which is not to scale, this dimension is identified in the drawing by drawing a line under the dimension. To perform this in an AutoCAD drawing, enter the following text string for the dimension value: %%u5.00%%u. The "%%u" toggles on underline text mode followed by the dimension text of 5.00 units. To toggle underline off, complete the text string with another "%%u."

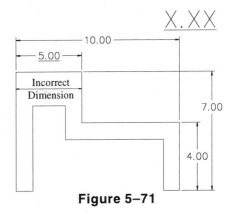

Figure 5-71

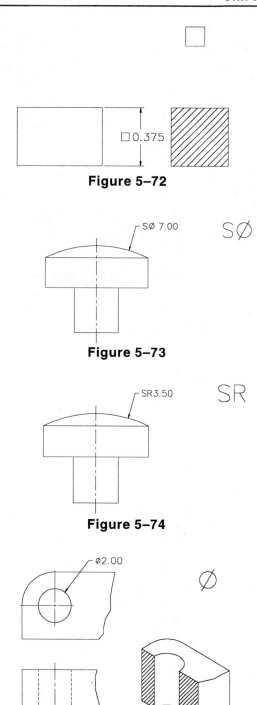

Applying the Square Dimension Symbol

Use this dimension symbol to identify the cross section of an object as a square. This will limit the need for two dimensions of the same value to be placed identifying a square. The square dimension symbol should be placed before the dimension numeral as in the illustration in Figure 5–72.

Figure 5–72

Applying the Spherical Diameter Dimension Symbol

The spherical illustrated in Figure 5–73 is in the form of a dome-shaped feature. Use the diameter symbol preceded by the letter "S" for spherical.

Figure 5–73

Applying the Spherical Radius Dimension Symbol

The illustration in Figure 5–74 is similar to the previous spherical example except the spherical is identified by a radius. Place the letter "S" in front of the "R" symbol.

Figure 5–74

Applying the Diameter Dimension Symbol

This symbol is represented by a circle with a diagonal line through it. This diameter symbol always precedes the dimension numeral with no space in between. This symbol is automatically placed before the dimension numeral when using the DIMDIAMETER command. All holes represented by the diameter symbol are understood to pass completely through a part as in the illustration in Figure 5–75.

Figure 5–75

Applying the Deep Dimension Symbol

When a hole does not pass completely through a part as in the illustration in Figure 5–76, the depth or deep dimension symbol is used. The diameter of the hole is first placed followed by the depth symbol and the distance the hole is to be drilled. Also in this figure is yet another method of identifying holes, although it is considered dated.

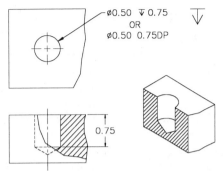

Figure 5–76

Applying the Deep Dimension Symbol

Illustrated in Figure 5–77 is yet another application of the depth dimension symbol. First the diameter of the hole is identified. Next, a counterdrill diameter with angle of the counterdrill is given. The depth of the counterdrill completes the dimension. This operation is used when a flat-head screw needs to be recessed below the surface of a part.

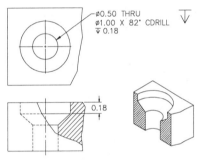

Figure 5–77

Applying the Counterbore Dimension Symbol

A counterbore is an enlarged portion of a previously drilled hole. The purpose of a counterbore is to receive the head of such screws as socket-head or fillister-head screws. The counterbore example in Figure 5–78 first identifies the diameter of the thru hole. On the next line comes the counterbore symbol and its diameter. The final specification is the depth of the counterbore identified by the Depth dimension symbol. Figure 5–78 shows another method of identifying counterbores, although it is considered dated.

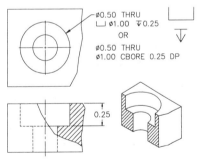

Figure 5–78

Applying the Spotface Dimension Symbol

A spotface is similar to the counterbore except that it is usually made quite shallower than the counterbore. The purpose of the spotface is to seat a washer from moving around along the surface of a part. The counterbore symbol is used to identify a spotface as in the example in Figure 5–79. The diameter of the thru hole is first given. On the next line, the counterbore symbol followed by diameter is given. Finally the depth of the counterbore is identified by the distance and the Depth dimension symbol. Figure 5–79 also shows another method of identifying spotfaces although it is considered dated.

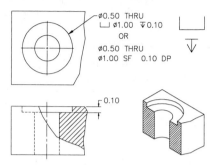

Figure 5–79

Applying the Slope Dimension Symbol

The slope dimension symbol applies to an inclined surface and the amount of rise in the surface given by two dimensions as in the illustration in Figure 5–80. This rise is indicated by the change in height per unit distance along a baseline. In the example in Figure 5–80, the slope symbol is placed followed by the change in height and the change in unit distance. This makes the slope dimension a ratio of the height and unit distance.

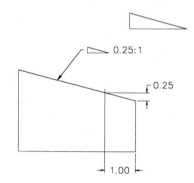

Figure 5–80

Applying the Countersink Dimension Symbol

A countersink is a V-shaped conical taper at one end of a hole. The purpose is to accept a flat-head screw and make it flush with the top surface of part. In the example in Figure 5–81, three holes of 0.50 diameter are first identified. On the next line, the countersink symbol followed by the number of degrees in the countersink is specified. Below this example is yet another method of identifying countersinks although it is considered dated.

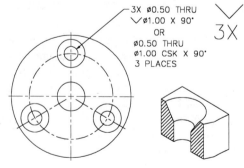

Figure 5–81

Applying the Arc Dimension Symbol

Illustrated in Figure 5–82 is the difference between dimensioning the chord of an arc and the actual distance of the arc. The 7.00 unit dimension in addition to defining the width of the part also specifies the distance of the chord of an arc. The 8.48 unit dimension is the distance of the arc and is specified by a small arc symbol above the distance. This is accomplished by drawing a small arc and moving it above the arc distance. Although AutoCAD does not dimension the length of an arc, it does allow you to use the DDMODIFY command to get the distance of an arc. This value is then entered at the dimension distance prompt.

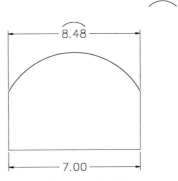

Figure 5–82

Applying the Origin Dimension Symbol

The origin dimension symbol is used to indicate the origin of the dimension. The small circle is substituted for an arrowhead as the termination of the dimension line as in Figure 5–83. To construct the open circle and filled arrowhead, go to the arrowheads group, click in the edit box next to the "1st" and pick the "dot blanked" terminator. Next, click in the edit box next to "2nd:" and pick the "closed filled" terminator.

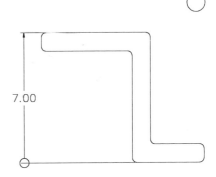

Figure 5–83

Applying the Conical Taper Dimension Symbol

Similar to the slope dimension symbol, the conical taper symbol is used when the amount of taper per unit of length is desired. The conical taper symbol is placed before the value of the taper. The value of the taper is based on a ratio of 1.00 units in length to a Delta diameter illustrated in Figure 5–84. The Delta diameter is based on the large shaft diameter minus the diameter taken at the area where the 1.00 unit length is located. This value becomes the ratio to 1.00 unit of length.

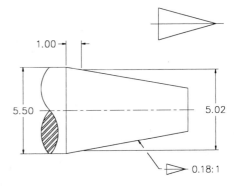

Figure 5–84

Location of Uniformly Spaced Features

Illustrated in Figure 5–85 is an example of a plate consisting of 8 holes equally spaced at 45-degree angles about a 7.30 diameter bolt hole circle. Rather than dimension each individual angle, the dimension symbol "X" is used for the number of times a feature is repeated. Notice in Figure 5–85 that "8X" is used to call out the number of times the angle and diameter of the holes repeat.

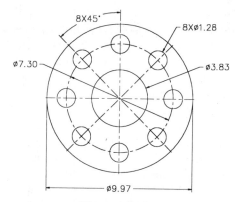

Figure 5–85

Location of Uniformly Spaced Features

The example in Figure 5–86 is similar to the previous illustration. Five 72-degree angles are used to locate five slots each with a width of 1.50 units.

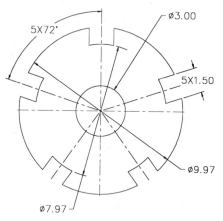

Figure 5–86

Location of Uniformly Spaced Features

The example in Figure 5–87 shows a different application to the location of features that are uniformly spaced. Six holes of 1.00 diameter are located along a bolt circle of 7.30 diameter. All six holes are spaced at 30-degree angles. Since the holes are not laid out in a full circle, only five angles are required to locate the six holes. The symbol "5X" is used to identify the number of angles while the symbol "6X" is used to identify the number of holes.

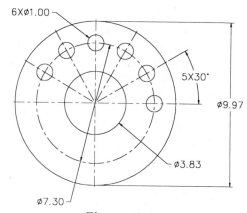

Figure 5–87

Character Mapping for Dimension Symbols

A text font has been supplied with the enclosed diskette; this file is called "ANSI_SYM.SHX." Rather than create the symbols as individual drawings, save them as Wblocks, insert them into a drawing, and move the symbols into position. This special text font contains the following dimension symbols: Square, Depth, Diameter, Conical Taper, Countersink, Counterbore, and Slope. The following regular keyboard characters have been replaced by the dimension symbols: ~, !, @, $, ^, &, and *. This means to bring up the Square dimension symbol, simply hold down the shift key and type the exclamation point from the keyboard. The Square dimension symbol will appear on the screen. The ANSI_SYM.SHX font was derived from the original SIMPLEX.SHP font by replacing the seven

characters previously identified with the new definitions containing the dimension symbols. Assign this font to a new text style and begin dimensioning with these symbols. See the following breakdown and Figure 5–88 for matching the dimension symbols with certain keys based on the ANSI_SYM.SHX font:

Shift + ~ = Square
Shift + ! = Depth
Shift + @ = Diameter
Shift + $ = Conical Taper
Shift + ^ = Countersink
Shift + & = Counterbore
Shift + * = Slope

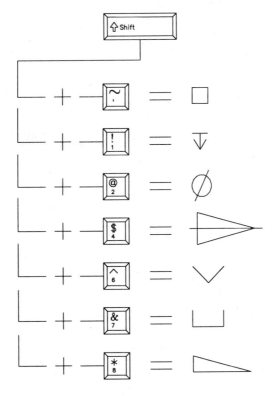

Figure 5–88

Generating Dimension Symbols

Illustrated in Figure 5–89 is a typical counterbore operation and the correct dimension layout. The LEADER command was used to begin the diameter dimension. The counterbore and depth symbols were generated using the MTEXT command. In any case, the first step to using the dimension symbols on the previous page is to first create a new text style of which any name may be used. For the purposes of this illustration, a new text style will be created called "ANSI" using the supplied text font "ANSI_SYM.SHX." Follow Figure 5–90 for the creation of this new text style.

Issue the STYLE command from the command prompt and the dialog box in Figure 5–90 appears. Click on the New button and enter in the new style name of "ANSI." In the Font Name edit box, search for and select the Font Name "ANSI_SYM.SHX." This file holds the geometry symbols just explained. Click on the Apply button; then click on the Close button. This closes the dialog box and makes "ANSI" the current text style.

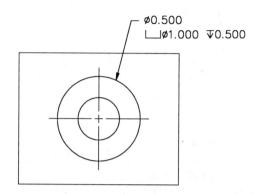

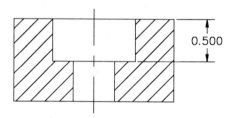

Figure 5–89

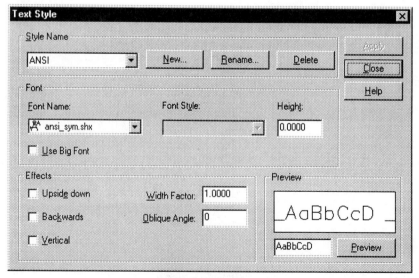

Figure 5–90

Begin placing a leader identifying the drill diameter and counterbore diameter and depth. Once the leader is drawn, the dimension text along with special control characters is entered in through the use of the Mtext dialog box as shown in Figure 5–91.

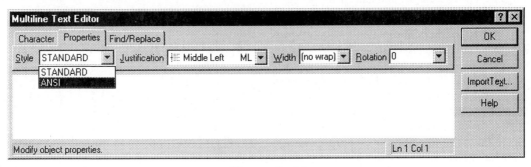

Figure 5–91

 Command: **LEADER**

From point: **Nea**
of *(Select a point along the edge of the circle at "A" below in Figure 5–93)*
To point: *(Pick a point at a convenient for the end of the leader)*
To point (Format/Annotation/

Undo)<Annotation>: *(Press* [ENTER] *to accept this default)*
Annotation (or press [ENTER] for options): *(Press* [ENTER] *for the list of options)*
Tolerance/Copy/Block/None/<Mtext>:*(Press* [ENTER] *to display the Mtext dialog box)*

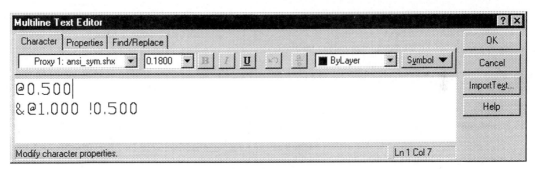

Figure 5–92

Click on the Properties folder and be sure ANSI is the current text style in Figure 5–91. When completed, click on the Character folder to begin entering in the leader note. See Figure 5–92.

Enter in the leader note in the Multiline Text Editor dialog box in Figure 5–92 and click on the OK button when finished. The leader complete with note is shown in Figure 5–93.

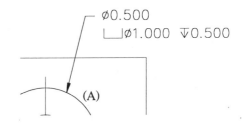

Figure 5–93

Grips and Dimensions

Grips have a tremendous amount of influence on dimensions. Grips allow dimensions to be positioned to better locations; grips also allow the dimension text to be located at a better position along the dimension line. In the example in Figure 5–94, notice the various unacceptable dimension placements. The 2.88 horizontal dimension lies almost on top of another extension line. To relocate this dimension to a better position, click on the dimension. Notice the grips appear and the entire dimension highlights. Now click on the grip near "A" in the illustration in Figure 5–94. (This grip is located at the left end of the dimension line). As the grip turns red, the Stretch mode is currently active. Stretch the dimension above the extension line but below the 3.50 dimension. The same results can be accomplished with the 4.50 horizontal dimension as it is stretched closer to the 3.50 dimension.

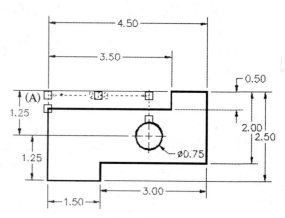

Figure 5–94

Notice the two vertical dimensions in Figure 5–95 on the left side of the object do not line up with each other; this would be considered poor practice. Pick both dimensions and notice the appearance of the grips in addition to both dimensions being highlighted. Click on the upper grip at "A" of the 1.25 dimension. As this grip turns red and places you in Stretch mode, select the grip at "B" of the opposite dimension. The result will be that both dimensions now line up with each other. The same can be accomplished with the 1.50 and 3.00 horizontal dimensions.

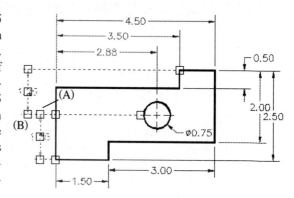

Figure 5–95

In the illustration in Figure 5–96, the 2.50 dimension text is too close to the 2.00 vertical dimension on the right side of the object. Click on the 2.50 dimension; the grips appear and the dimension highlights. Click in the grip representing the text location at "A." As this grip turns red, stretch the dimension text to a better location.

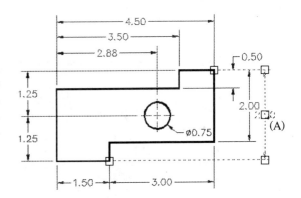

Figure 5–96

Using grips to control the placement of diameter and radius text is very easy to use. In the illustration in Figure 5–97, click on the diameter dimension; the grips appear in addition to the diameter dimension being highlighted. Click on the grip that locates the dimension text at "A." As this grip turns red, relocate the diameter dimension text to a better location using the stretch mode.

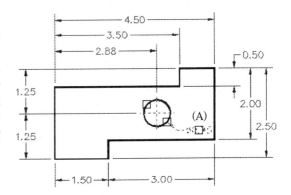

Figure 5–97

The completed object that has had the dimensions edited using grips is displayed in Figure 5–98.

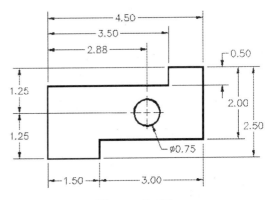

Figure 5–98

The Dimension Styles Dialog Box

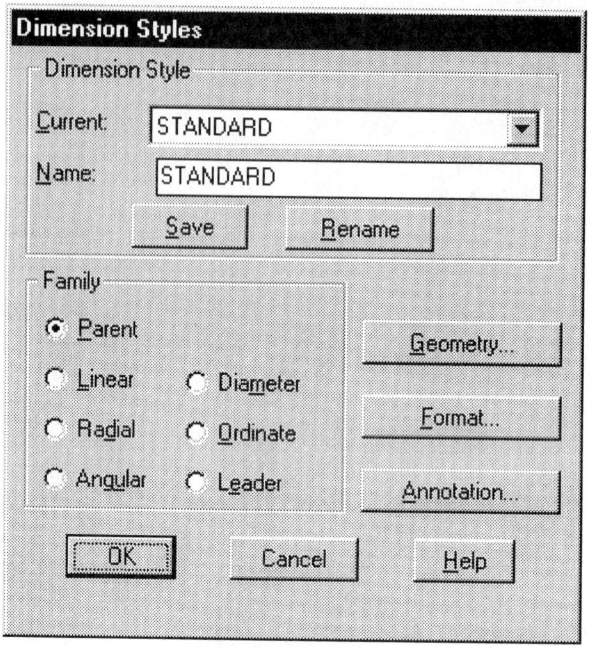

Figure 5–99

The concept of creating a dimension style has not changed very much in Release 14. It is still used to group a series of dimension settings under a unique name to determine the appearance of dimensions. The dialog box illustrated in Figure 5–99 is the Dimension Styles dialog box, which is used to create a dimension style much the same way as previous versions of AutoCAD. However, instead of picking numerous other dialog boxes to change dimension settings, three buttons are added to this main dialog box to control settings. This dialog box is activated by clicking on the Dimension Styles button or by entering in the DDIM command from the keyboard.

The "Geometry…" button activates a dialog box controlling dimension line, extension line, arrowhead, and center mark settings.

The "Format…" button activates a dialog box controlling placement of dimension text, horizontal justification and vertical justification of dimension text.

The "Annotation…" button activates a dialog box controlling primary units, alternate units, tolerance, and text settings.

Another new feature of the Dimension Styles dialog box is in the form of Dimension Style Families. Using families allows for variables to be assigned to a particular dimension type, such as linear dimensions and diameter dimensions, without having to set up an entirely separate dimension style for each type of dimension placed. Dimension style families will be covered in greater detail later in this unit.

The Geometry Dialog Box

Use the Geometry dialog box to control variable settings dealing with the dimension line, extension line, arrowheads, and center markers (Figure 5–100). Variables affecting the dimension line include the suppressing of either the first or second dimension line, the increment that the dimension lines will be spaced from each other in the case of a baseline dimension, the extension distance of the dimension line which is popular in architectural applications, and the color the dimension line will be drawn in. Settings affecting the extension line include the suppressing of either the first or second dimension line, the distance the extension line extends past the arrow, the dis-

tance the extension line is placed away from the object, and the color the extension line will be drawn in. Settings affecting arrowheads include the form of the first and second arrowheads including the size of the arrow. Settings affecting the center marker include the ability to place a center mark, adding centerlines to the center mark, not having any center mark placed, and the size of the center mark. Picking the graphic of the center mark also scrolls through all three types of center marks. The very important overall dimension scaling value is also present in this dialog box. These various settings will be covered in the next series of pages that follow.

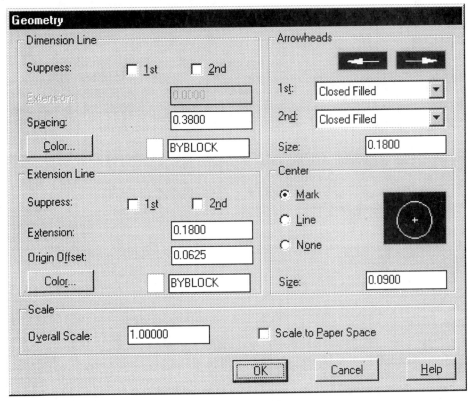

Figure 5–100

Overall Scale Setting

This setting acts as a multiplier and globally affects all current dimension settings that are specified by sizes or distances. This means if the current overall scale value is 1.00, other settings such as dimension text height will remain unchanged. If the overall scale value changes to 2.00, the dimension text height will be doubled in size. See Figure 5–101.

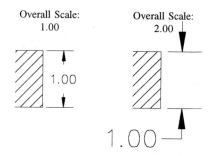

Figure 5–101

Dimension Line Settings

Use the areas of the dialog box in Figure 5–102 to control the visibility and spacing of the dimension line.

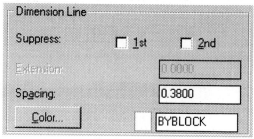

Figure 5–102

By default, dimension line suppression is turned off. To turn on suppression of dimension lines, place a check in the appropriate box next to Suppress in the Dimension Line area of the Geometry dialog box. This operation turns off the display of dimension lines similar to the illustration in Figure 5–103.

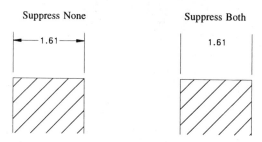

Figure 5–103

Clicking on the Color button of the Dimension Line area of the Geometry dialog box allows you to assign a different color to the dimension line as shown in Figure 5–104. The color of the current arrowhead terminator will also be affected by the dimension line color. A practical use of this operation is to assign color to the dimension line as a means of controlling line quality by assigning different pen weights to the appropriate color.

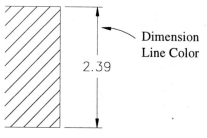

Figure 5–104

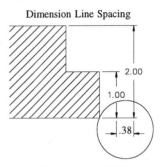

The Spacing setting in the Dimension Line area of the Geometry dialog box controls the spacing of multiple dimensions as they are placed away from each other similar to the illustration in Figure 5–105. This value affects the DIMBASELINE command.

Figure 5–105

Arrowhead Settings

Use the Arrowheads area of the Geometry dialog box to control the type of arrowhead terminator displayed in the drawing (shown in Figure 5–106). This dialog box also controls the size of the arrowhead.

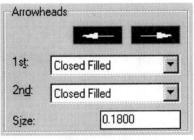

Figure 5–106

Clicking on the 1st edit box displays a number of arrowhead terminators many of which have been added to the current version of AutoCAD. Choose the desired terminator from the list in Figure 5–107. When choosing the desired arrowhead from the 1st edit box, the 2nd edit automatically updates to the selection made in the 1st. If you choose an arrowhead from the 2nd edit box, both 1st and 2nd arrowheads may be different at opposite ends of the dimension line; this is desired of some applications.

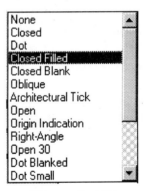

Figure 5–107

Illustrated in Figure 5–108 is the complete set of arrowheads along with their names. The last arrow type is a User Arrow. This allows you to define their own custom arrowhead.

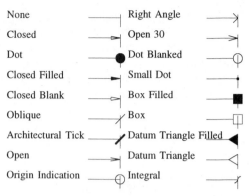

Figure 5–108

Use the Size setting of the Arrowheads area of the Geometry dialog box to control the size of the arrowhead terminator. Arrow types that appear filled in are controlled by the FILL command. With Fill turned on, all arrowheads are filled in as in Figure 5–109. With Fill turned off, only the outline of the arrow is displayed.

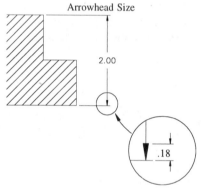

Figure 5–109

Extension Line Settings

The Extension Line area of the Geometry dialog box controls the display of extension lines, how far the extension extends past the arrowhead, and how far the extension line begins from the object. See Figure 5–110.

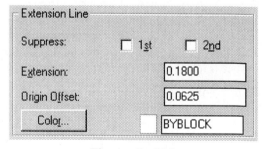

Figure 5–110

Suppress 1st and 2nd control the display of extension lines. They are useful when dimensioning to
an object line and to avoid placing the extension
line on top of the object line. Placing a check in the
1st box of Suppress turns off the 1st extension line.
The same is true when placing a check in the 2nd
box of Suppress; the second extension line is turned
off. Study the many examples illustrated in Figure
5–111 to get a better idea on how suppression of
extension lines operate.

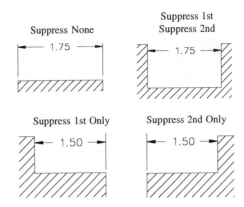

Figure 5–111

The Extension setting of the Extension Line area controls how far the extension extends past the arrowhead or dimension line as shown in Figure 5–112.
By default, a value of 0.18 is assigned to this setting.

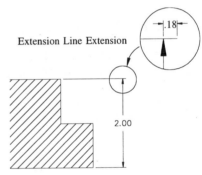

Figure 5–112

The Origin Offset setting of the Extension Line area
controls how far away from the object the extension line will start. See Figure 5–113. By default, a
value of 0.06 is assigned to this setting.

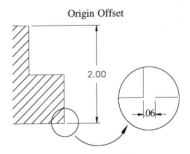

Figure 5–113

As with dimension lines, extension lines may be set to a different color to control line quality by clicking on the Color button and assigning a new color to the extension lines. See Figure 5–114.

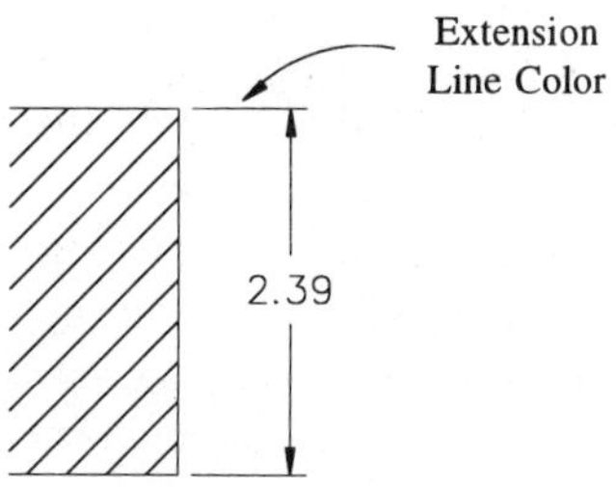

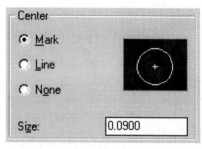

Figure 5–114

Center Mark Settings

The Center area of the Geometry dialog box allows you to control the type of center marker used when identifying the centers of circles and arcs as in Figure 5–115. Settings can either be made by clicking in the appropriate radio button as in Figure 5–115, or by clicking directly on the image of the center marker. Clicking on this image scrolls you through different types of center markers. The size of the center marker is also controlled in this area.

Figure 5–115

The three types of center marks are illustrated in Figure 5–116. The Mark option places a "plus" in the center of the circle. The Line option places the "plus" and extends a line past the edge of the circle. The None option is void of any center marks.

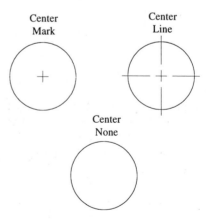

Figure 5–116

The Format Dialog Box

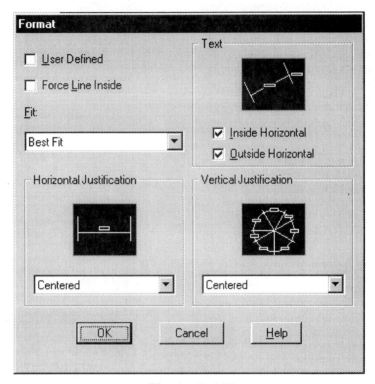

Figure 5–117

Use the Format dialog box to control variable settings dealing with the placement of dimension text, horizontal text justification and vertical text justification, and the fitting of text between extension lines as shown in Figure 5–117. Settings affecting the placement of text include whether text placed inside or outside will be horizontal or will align parallel to the dimension line. Settings affecting the vertical placement of text include whether text is centered along the dimension line, placed above the dimension line, or placed outside of the dimension line. A Japan International Standard has been implemented in this dialog box to address dimension globalization issues. Settings affecting the horizontal justification of text include whether dimension text is centered along the dimension line, placed next to the left extension line, or placed next to the right extension line. Settings affecting how text will fit between extension lines include text to take on a best fit. For radius and diameter dimensions, other best fit options include whether the text is placed outside of the diameter, and whether to include a center mark, whether the text is placed outside of the diameter but whether to force the interior dimension line with no center mark, or force the text inside of the diameter along with the dimension line with no center mark. These various settings will be covered in the following pages.

User Defined Settings

The User Defined and Force Line Inside areas of the Format dialog box control how the dimension text will be placed in the drawing and what control you have over this location. Clicking in the User Defined box allows you to control dimension text instead of placing text in a default location (see Figure 5–118). For example, if you can define the location of dimension text in a diameter dimension, this permits you to position the text anywhere in the drawing; by default the software places this text in the center of the circle.

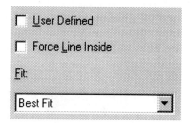

Figure 5–118

By default, text that does not fit inside of its extension lines will be placed on the outside of the extension lines. Placing a check in the box labeled "Force Line Inside" forces the dimension line to be drawn in between extension lines. See Figure 5–119.

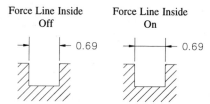

Figure 5–119

Six fit modes are available to better fit the dimension text or arrowheads in a space depending on the size of the space and the mode being used. All modes are explained in the following illustrations.

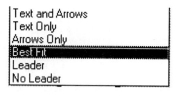

Figure 5–120

Best Fit Mode:

The best fit mode in Figure 5–120 calculates the dimension text height and arrowhead size and fits either the text or arrowhead or both depending on the space being dimensioned. The radius dimension allows the text to be placed outside of the feature being dimensioned when using this fit mode as in Figure 5–121.

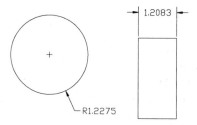

Figure 5–121

Text and Arrows Mode:

Use this fit mode to force text and arrows to be placed inside of a radial or linear dimension. This is true for the radius dimension in Figure 5–122 but not for the linear dimension. If the text and arrows cannot be placed inside due to a tight space, they are placed outside of the extension lines.

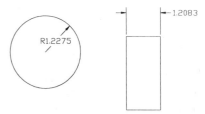

Figure 5–122

Text Only Mode:

This fit mode forces the dimension text to be placed in between the extension lines or inside of a circle as illustrated in Figure 5–123.

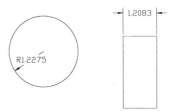

Figure 5–123

Arrows Only Mode:

Use this fit mode to force only the arrow inside of the extension lines leaving the text outside of the extension lines. This is evident in both the radius and linear dimensions in Figure 5–124. If the circle is large enough, the dimension text and arrow are placed inside of the circle.

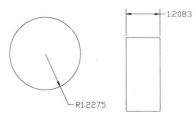

Figure 5–124

Leader Mode:

This fit mode mainly affects radius and diameter dimensions. The results may not be evident in the illustration in Figure 5–125. However when trying this mode, it is very important to select the desired first point for a radius or diameter dimension. Instead of freely moving the dimension text around to a better location, the first point anchors the leader. The dimension text may only be brought further from or closer to the circle or arc.

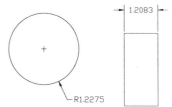

Figure 5–125

No Leader Mode:

This fit mode behaves similar to the one above. The first point anchors the dimension in place. This mode allows you to place the dimension text away from the circle or arc. However in doing so, a leader is not constructed to the dimension text in the illustration in Figure 5–126.

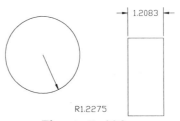

Figure 5–126

Text Settings

Use the Text area of the Format dialog box to control the alignment of text (see Figure 5–127). Dimension text can either be placed horizontally or parallel to the edge being dimensioned, or a combination of both. Two additional controls exist for text placement; you have control if the dimension text can fit inside of the extension lines; you also have control on the dimension text if it is placed outside of the extension lines. The appearance of the check in the box means the feature is on. Clicking inside of the image of the text will scroll you through all settings controlled by this area.

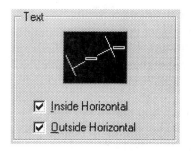

Figure 5–127

In the example in Figure 5–128, if both Inside Horizontal and Outside Horizontal boxes are checked, all dimension text is read horizontally. This includes text placed inside of the extension lines as well as text placed outside of the extension lines.

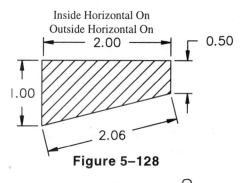

Figure 5–128

In the example in Figure 5–129, both Inside Horizontal and Outside Horizontal boxes are turned off; the dialog box would show the absence of any checks. The results show how all text is read parallel to the edge being dimensioned. Not only will vertical dimensions align the text vertically, but the 2.06 dimension used to dimension the incline is also parallel to the edge.

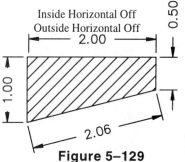

Figure 5–129

The example in Figure 5–130 shows how all text that fits inside of the extension lines will be read parallel to the edge being dimensioned while the text placed outside of the extension lines (the 0.50 dimension) remains horizontal since Outside Horizontal is turned on.

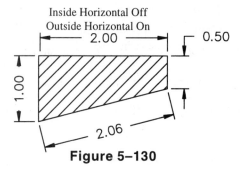

Figure 5–130

Horizontal Justification Settings

At times, dimension text needs to be better located in the horizontal direction. This is the purpose of the Horizontal Justification area of the Format dialog box (see Figure 5–131). This is a means of locating the dimension text so it can be read with better clarity. In addition to choosing settings from the drop down edit box, the image icon may be clicked on which will scroll you through all horizontal justification settings.

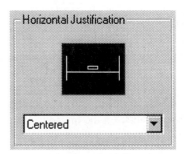

Figure 5–131

Illustrated in Figure 5–132 are the five modes of justifying text horizontally. By default, the horizontal text justification is centered in the dimension.

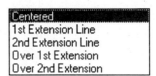

Figure 5–132

Clicking on the Centered option of the Horizontal Justification area displays dimension text illustrated in Figure 5–133. This is the default setting since it is the most commonly used text justification mode used in dimensioning.

Clicking on the 1st Extension Line option of the Horizontal Justification area displays the dimension text illustrated in Figure 5–134 where the dimension text slides over close to the first extension line. Use this option to position the text out of the way of other dimensions. Notice the corresponding option to have the text favor the 2nd extension line.

Clicking on the Over 1st Extension option of the Horizontal Justification area displays the dimension text parallel to and over the first extension line (see Figure 5–135). Notice the corresponding option to control text over the 2nd extension line.

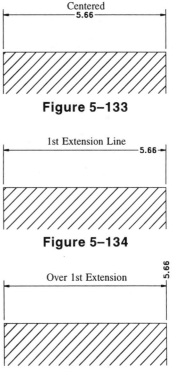

Figure 5–133

Figure 5–134

Figure 5–135

Vertical Justification Settings

Use the Vertical Justification area of the Format dialog box to control the vertical justification of dimension text as in Figure 5–136. Vertical justification modes may be set by clicking in the drop down edit box. Modes may also be scrolled through by clicking directly on the image icon.

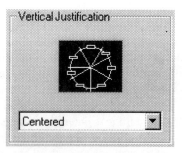

Figure 5–136

By default, dimension text is centered vertically in the dimension line. Other modes include justifying vertically above the dimension line, justifying vertically outside of the dimension line, and using the JIS (Japan International Standard) for placing text vertically. See Figure 5–137.

Figure 5–137

Illustrated in Figure 5–138 is the result of setting the Vertical Justification to Centered. Here, the text is centered vertically; the dimension line will automatically be broken to accept the dimension text.

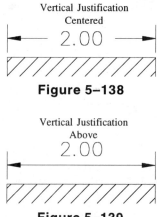

Figure 5–138

Illustrated in Figure 5–139 is the result of setting the Vertical Justification to Above. Here the text is placed directly above the dimension line that remains unbroken. This mode is very popular with architectural applications.

Figure 5–139

The Annotation Dialog Box

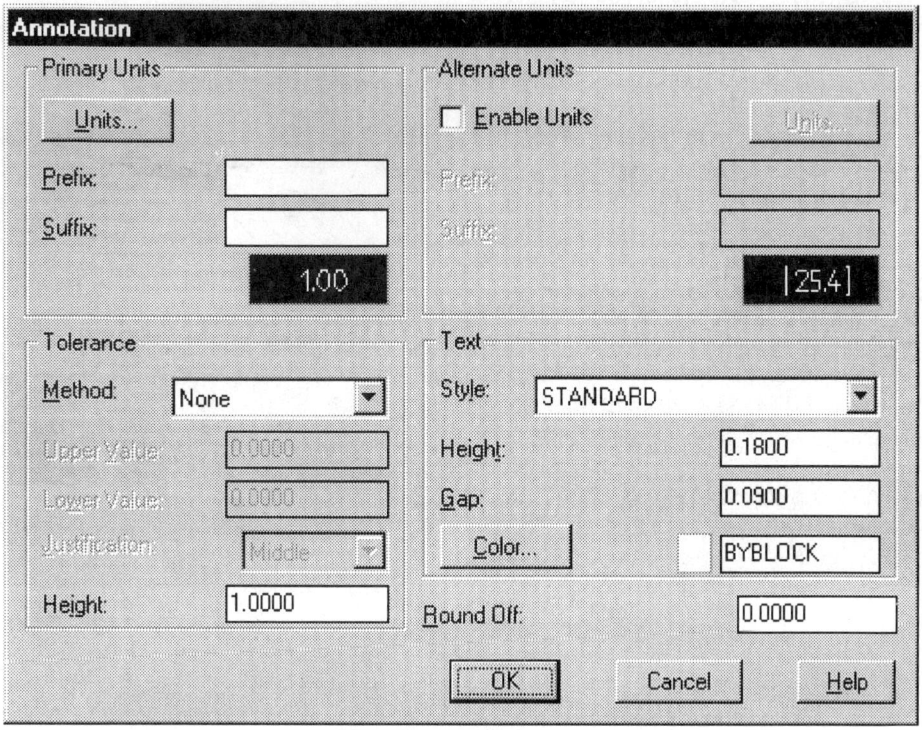

Figure 5–140

Use the Annotation dialog box to control variable settings dealing with the primary units, alternate units, tolerance method, and general text settings (see Figure 5–140). Settings affecting the primary units include the type of units the dimensions will be constructed in (decimal, engineering, architectural, etc.), and if the dimension text requires a prefix or suffix. Settings affecting alternate units include whether alternate units are enabled or disabled, units for alternate dimensions, and if the alternate dimension text requires a prefix or suffix. Settings affecting tolerances include an option for no tolerances, symmetrical tolerances, tolerances of deviation, limit dimensions, and basic dimensions. Other controls for tolerances include the setting of upper and lower limit values, the justification of the tolerance in relation to the main text, and the height of the tolerance. Settings affecting the dimension text include the current text style governing the dimension text, the height of dimension text, the gap formed on either side with the dimension text and the dimension line, and the color of the dimension text.

Use a Round Off value to round off all dimension distances to the nearest unit based on the round off value. With a round off value of 0.0000, the dimension text reflects the actual distance being dimensioned as in the illustration in Figure 5–141.

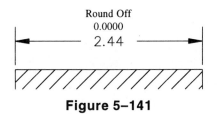

Figure 5–141

With a round off value set to 0.2500, the dimension text reflects the next 0.2500 increment, namely 2.50. See Figure 5–142.

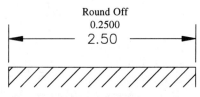

Figure 5–142

Primary Units Settings

Clicking on the Units button of the Primary Units area of the Annotation dialog box (shown in Figure 5–143) displays the dialog box similar to the illustration in Figure 5–144. Be sure to use this dialog box to place dimensions in their proper units.

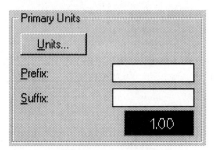

Figure 5–143

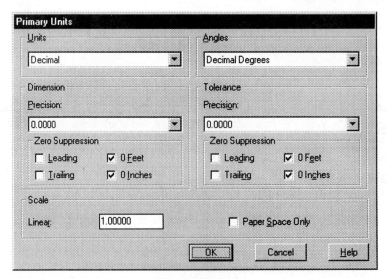

Figure 5–144

The units of this dialog box are similar to the DDUNITS command with the exception of a few additional dimension unit modes; namely Architectural (stacked) and Fractional (stacked). See Figure 5–145. Use these additional modes to layout fractions in standard numerator and denominator mode similar to the following:

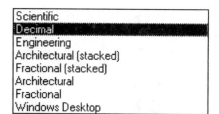

Figure 5–145

Architectural: 9 1/2

Architectural (stacked): $9\frac{1}{2}$

Use the Suffix mode of the Primary Units area to control the placement of a character string immediately after the dimension value. In the illustration in Figure 5–146, the letters "IN" have been added to the dimension signifying "inches."

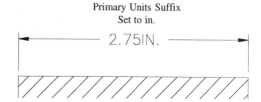

Figure 5–146

The Linear value setting of the Primary Units dialog box acts as a multiplier for all linear dimension distances including radius and diameter dimensions. As a dimension distance is calculated, the current value set in the Linear edit box is multiplied by the dimension to arrive at a new dimension value. In the illustration in Figure 5–147, and with a Linear Scale value of 1.00, the dimension distances are taken at their default values.

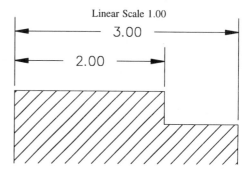

Figure 5–147

In the illustration in Figure 5–148, the Linear Scale value has been changed to 2.00 units. This means every dimension distance will be multiplied by 2.00; the results are the previous 3.00 and 2.00 dimensions being changed to 6.00 and 4.00. In a similar fashion, having a Linear Scale value set to 0.50 will cut all dimension values in half.

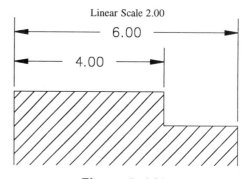

Figure 5–148

Alternate Units Settings

Use the Alternate Units area of the Annotation dialog box to enable or disable alternate units. See Figure 5–149. These units will be added to the existing dimension units currently calculated. Clicking on the Units button of the Alternate Units area displays a dialog box similar to that displayed in Figure 5–144 when using the DDUNITS command.

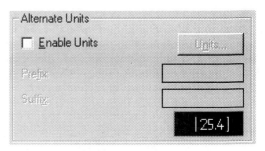

Figure 5–149

The illustration in Figure 5–150 shows the effect of turning Off Alternate Units. In this example, a single dimension value is calculated and placed in between dimension and extension lines. Figure 5–151 shows the effects of Alternate Units being enabled. Next to the calculated dimension value, the alternate value is placed in brackets. This value depends on the current setting of the Alternate Units Factor. With this factor set to 25.40, it is used as a multiplier for all calculated dimension values. Yet another feature of Alternate Units is the setting of the number of decimal places the alternate units will be set to (see Figure 5–152). This value is set when clicking on the Units button of the Alternate Units area of the Annotation dialog box.

Enable Alternate Units	Off
Alternate Units Factor	25.40
Alternate Units Decimal Places	0

Figure 5–150

Enable Alternate Units	On
Alternate Units Factor	25.40
Alternate Units Decimal Places	0

Figure 5–151

Enable Alternate Units	On
Alternate Units Factor	25.40
Alternate Units Decimal Places	2

Figure 5–152

Tolerance Settings

Various edit boxes are available in the Tolerance area of the Format dialog box shown in Figure 5–153. Depending on the type of tolerance being constructed, an Upper Value and Lower Value may be set to call out the current tolerance variance. The Tolerance Justification allows you to determine where the tolerance will be drawn in relation to the location of the body text. The Tolerance Height controls the text size of the tolerance. Usually this value is smaller than the body text size.

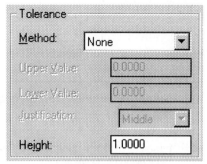

Figure 5–153

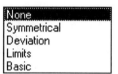

Figure 5–154

The five tolerance types are identified in the drop down edit box illustrated in Figure 5–154 and the graphic illustrated in Figure 5–155. No tolerance uses the calculated dimension value without applying any tolerances. The Symmetrical tolerance uses the same value set in the Upper and Lower Value. The deviation tolerance will have a value set in the Upper Value and an entirely different value set in the Lower Value. The Limits tolerance will use the Upper and Lower values and place the results with the larger limit dimension placed above the smaller limit dimension. The basic dimension does not add any tolerance value; instead a box is drawn around the dimension value.

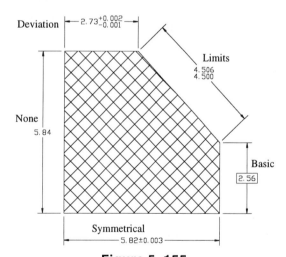

Figure 5–155

The Tolerance Height setting affects just the height of the tolerance text. This value acts as a multiplier; it affects the current dimension text height setting. In the illustration in Figure 5–156, a tolerance height setting of 1.00 has no affect on the height of the tolerance text; in fact the tolerance text is the same height as the main dimension text. Figure 5–157 shows the effects of setting the tolerance height to a value of 0.70 units. Here, the tolerance height is noticeably smaller than the main dimension text height. It is good practice to have the main dimension text value set to a higher value than the tolerance heights for greater emphasis of the main dimension.

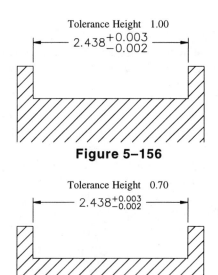

Figure 5–156

Figure 5–157

Text Settings

Use the Text area of the Annotation dialog box to make changes to various settings such as the Dimension Text style, Dimension Text Height, Dimension Text Gap, and Color of the dimension text. See Figure 5–158.

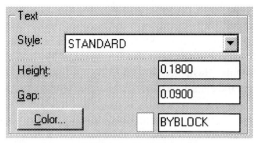

Figure 5–158

Use the Dimension text height settting to control the size of the dimension text illustrated in Figure 5–159. By default, a value of 0.18 is assigned to this setting.

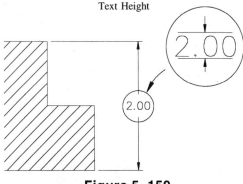

Figure 5–159

When placing dimensions, a gap is established between the inside ends of the dimension lines and the dimension text. This gap can be controlled by entering different values depending on the desired results. Study Figures 5–160 and 5–161 that have different text gap settings.

A Color button exists in the Text area of the Annotation dialog box. Use this to assign a different color exclusively to the dimension text (see Figure 5–162). This can prove very beneficial especially when assigning a medium line weight to the dimension text and a thin line weight to the dimension and extension lines.

Miscellaneous Dimensioning Settings

DIMASO—The Associative Dimensioning Control Setting

Use this dimension setting to turn associative dimensioning on or off; it must be entered in from the keyboard. If on, all objects that make up the dimension will be considered one object. If off, all dimension components such as arrowheads, extension lines, dimension lines, and dimension text will be considered single objects (see Figures 5–163 and 5–164). This is the same effect as using the Explode command on an associative dimension. With DIMASO set to off, commands that normally affect dimensions that are associative will have no effect. In fact, grips have little or no effect on a dimension considered nonassociative. It is considered poor practice to turn this variable off or to explode dimensions that are associative. The prompt for this dimension setting is:

Command:**DIMASO**
New value for DIMASO<On>: *(Enter off or keep the default)*

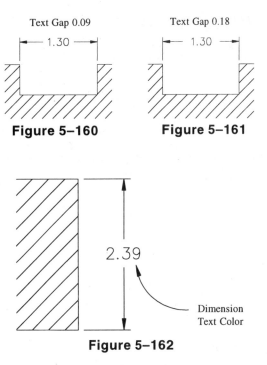

Text Gap 0.09 — 1.30 —

Text Gap 0.18 — 1.30 —

Figure 5–160   **Figure 5–161**

2.39

Dimension
Text Color

Figure 5–162

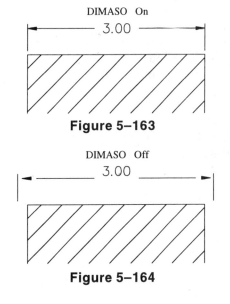

DIMASO On
3.00

Figure 5–163

DIMASO Off
3.00

Figure 5–164

Tutorial Exercise
Dimex.Dwg

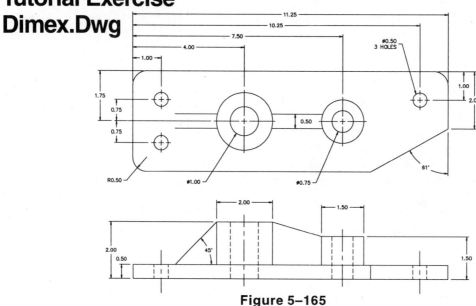

Figure 5–165

Purpose

The purpose of this tutorial is to draw the two view object illustrated In Figure 5–165 and place dimensions on the drawing. This drawing is available on diskette under the file name "Dimex.Dwg."

System Settings

Either copy "Dimex.Dwg" from the diskette provided or begin a new drawing called "Dimex." Units and Limits of the drawing should be already set. Use the GRID command and change the grid spacing from 1.00 units to 0.25 units to aid in the placement of dimension lines. Do not turn the Snap or Ortho modes on.

Layers

The drawing file "Dimex.Dwg" has the following layers already created for this tutorial.

Name	Color	Linetype
Object	Magenta	Continuous
Hidden	Red	Hidden
Center	Yellow	Center
Dim	Yellow	Continuous

Suggested Commands

The following dimension commands will be used: DIMLINEAR, DIMCONTINUE, DIMBASELINE, DIMCENTER, DIMRADIUS, DIMDIAMETER, DIMANGULAR, and LEADER. All dimension commands may be picked from the Dimension Toolbox, pull-down menu, or entered from the keyboard. Use the ZOOM command to get a closer look at details and features that are being dimensioned.

Whenever possible, substitue the appropriate command alias in place of the full AutoCAD command in eacy tutorial step; for example, use "Co" for the COPY command, "L" for the LINE command, and so on. The complete listing of all command aliases is located on pages 1–9 and 1–10.

Step #1

To prepare for the dimensioning of the drawing, issue the DDIM command and create two dimension styles. Immediately click on the Save button to update the STANDARD dimension style. For the first style, enter the name "MECHANICAL" in the "Name:" edit box of the Dimension Styles dialog box and click on the Save button. Notice a message appears at the bottom of the dialog box stating "Created MECHANICAL from STANDARD." See Figure 5–166.

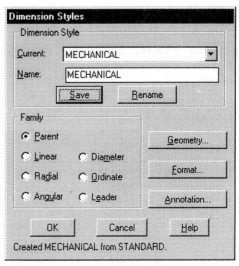

Figure 5–166

Step #2

Click on the Geometry... button and the Geometry dialog box will appear as in Figure 5–167. Make the following changes to this dialog box: Change the size of the arrowheads from 0.18 to a new value of 0.12; Change the size of the Extension Line Extension from a value of 0.18 to a new value of 0.07; Change the Extension Line Origin Offset from a value of 0.06 to a new value of 0.12. Finally, click on the Line radio button of the Cen-

ter area to draw full centerlines that mark the center of circles (this may already be set).

When completed with making these changes, click on the OK button at the bottom of the Geometry dialog box; this returns to the main Dimension Styles dialog box. Click on the Save button to save these changes to the main "MECHANICAL" dimension style.

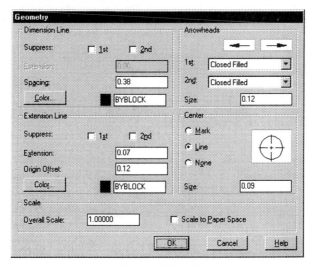

Figure 5–167

Step #3

While in the main Dimension Styles dialog box, click on the Annotation... button and change the Text Height from a value of 0.18 to a new value of 0.12. The dimension precision for decimal units should already be set to two places. When finished with this operation, click on the OK button in Figure 5–168 and the main Dimension Styles dialog box, click on the Save button to save this change to the "MECHANICAL" dimension style.

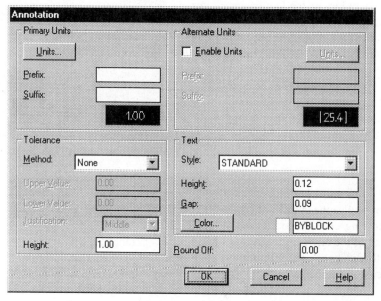

Figure 5–168

Step #4

As in Step #1, create a new dimension style. For this style, enter the name "EXT-OFF" in the "Name:" edit box of the Dimension Styles dialog box and click on the Save button. Notice a message appears at the bottom of the dialog box stating "Created EXT-OFF from MECHANICAL." (See Figure 5–169). The name "EXT-OFF" is short for "Extension Lines Off" and is designed to turn off both extension lines when placing dimensions inside of the drawing. It is good practice to place dimensions outside of the drawing views; however, there are times when a dimension must be place inside of the drawing for clarity.

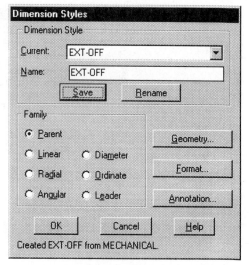

Figure 5–169

Step #5

Click on the Geometry... button and the Geometry dialog box will appear as in Figure 5–170. Place checks inside of the two Suppress: boxes under the category of Extension Line. This will turn off the extension lines when placing a dimension inside of one of the views.

When you have completed making these changes, click on the OK button at the bottom of the Geometry dialog box; this returns to the main Dimension Styles dialog box. Click on the Save button to save these changes to the main "EXT-OFF" dimension style.

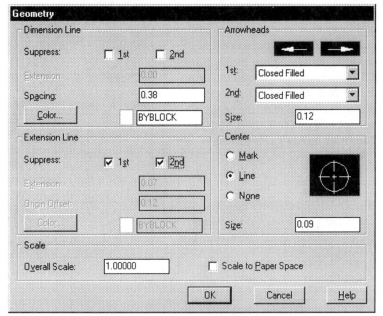

Figure 5–170

Step #6

While inside of the Dimension Styles dialog box, click in the Current edit box and make "MECHANI-CAL" the current dimension style, as shown in Figure 5–171.

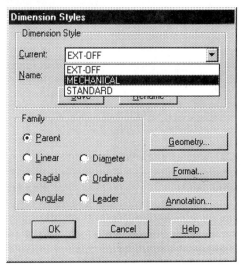

Figure 5–171

Step #7

While in the MECHANICAL Dimension Style
(Figure 5–172), click on the "Radial" button in the
"Family" area of the dialog box. You will now set
a series of dimension variables that will be used
whenever placing a radius dimension. Click on the
"Format..." button and the dialog box illustrated
in Figure 5–173 appears. Place a check in the box
labeled "User Defined." This will allow you to have
better control in placing the radius dimension. Also,
place a check in the box labeled "Force Line Inside." This will construct a line from the center of
the radius to the edge of the radius dimension. Finally, click in the "Fit:" edit box and activate the
"Arrows Only" item. This will place an arrow inside of the radius. When completed with this operation, click on the OK button; this will return
you to the main Dimension Styles dialog box. Before
continuing, click on the Save button to save these
changes to all radial dimensions to the "MECHANI-
CAL" dimension style.

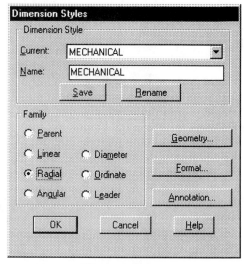

Figure 5–172

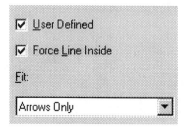

Figure 5–173

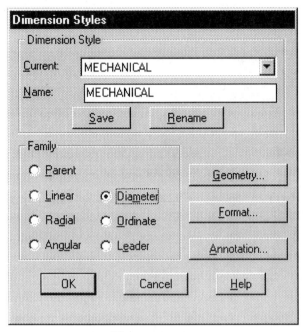

Figure 5–174

Step #8

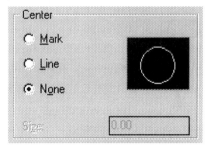

Figure 5–175

Click on the Diameter button in the main Dimension Styles dialog box under the "Family" area. The next series of settings will only affect diameter dimensions as they are placed in the drawing (Figure 5–174). Click on the "Geometry..." button and click in the "None" button that deals with type of Center marker in Figure 5–175. Since the center marks will be placed by the DIMCENTER command, they do not need to be on when a diameter dimension is placed. When completed with this setting, click on the OK button. While inside the Dimension Styles dialog box, click on the "Format..." button. Be sure there is a check next to "User Defined;" also, set the type of Fit to include Text and Arrows in Figure 5–176. When completed, click on the OK button, click on the "Save" button in the Dimension Styles dialog box, and click on the OK button to exit the Dimension Styles dialog box and return to the drawing editor.

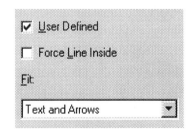

Figure 5–176

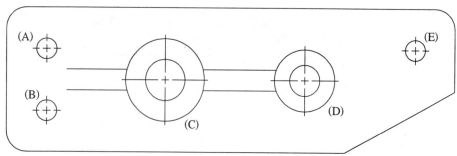

Step #9

Figure 5–177

Make the "CENTER" layer current. Begin placing center markers on this layer to identify the centers of all circular features in the Top view. The type of center marker has already been determined by saving it to the "MECHANICAL" dimension style. Use the DIMCENTER command to place these center markers.

Command:**-LAYER**

?/Make/Set/New/ON/OFF/Color/Ltype/Freeze/
   Thaw/LOck/Unlock:**Set**

New current layer <DIM>: **Center**
?/Make/Set/New/ON/OFF/Color/Ltype/Freeze/
   Thaw/LOck/Unlock: *(Press the* ⌤ *key to exit this command)*

   Command:**DIMCENTER**

Select arc or circle: *(Select the edge of circle "A")*

Repeat the above procedure for circles "B", "C", "D", and "E" as shown in Figure 5–177

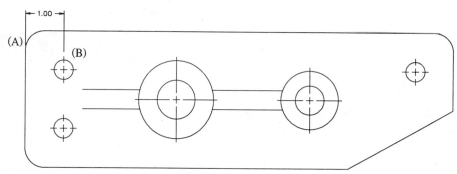

Figure 5–178

Step #10

Change the current layer from "CENTER" to "DIM." This layer will hold all dimensioning information. Next, place a horizontal dimension using the DIMLINEAR command. See Figure 5–178.

Command:**-LAYER**

?/Make/Set/New/ON/OFF/Color/Ltype/Freeze/
   Thaw/LOck/Unlock:**Set**

New current layer <CENTER>: **Dim**
?/Make/Set/New/ON/OFF/Color/Ltype/Freeze/
   Thaw/LOck/Unlock: *(Press the* ⌤ *key to exit this command)*

   Command:**DIMLINEAR**

First extension line origin or press ⌤ to
   select: **Endp**

of *(Select the endpoint of the line at "A")*
Second extension line origin: **Endp**
of *(Select the endpoint of the centerline at "B")*
Dimension line location (Mtext/Text/Angle/
   Horizontal/Vertical/Rotated): *(Locate the
   1.00 dimension)*
Dimension text=1.00

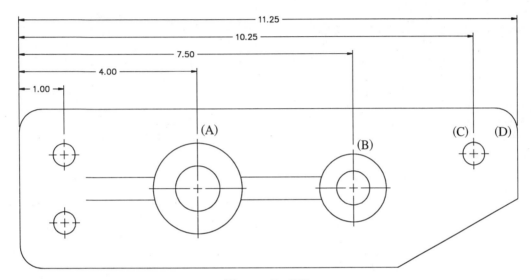

Figure 5-179

Step #11

Use the DIMBASELINE command to create a string of dimensions that reference the same starting point; namely the left side of the object, as shown in Figure 5-179.

 Command: **DIMBASELINE**

Specify a second extension line origin or
   (Undo/<Select>): **Endp**
of *(Select the endpoint of the centerline at "A")*
Dimension text = 4.00
Specify a second extension line origin or
   (Undo/<Select>): **Endp**
of *(Select the endpoint of the centerline at "B")*

Dimension text = 7.50
Specify a second extension line origin or
   (Undo/<Select>): **Endp**
of *(Select the endpoint of the centerline at "C")*
Dimension text = 10.25
Specify a second extension line origin or
   (Undo/<Select>): **Endp**
of *(Select the endpoint of the line at "D")*
Dimension text = 11.25
Specify a second extension line origin or
   (Undo/<Select>): *(Press* ENTER *to exit this
   option)*
Select base dimension: *(Press* ENTER *to exit this
   command).*

Step #12

Use the ZOOM command to get a closer look at the left side of the top view illustrated in Figure 5–180. Then use the DIMLINEAR command to place the 0.75 vertical dimension.

 Command:**DIMLINEAR**

First extension line origin or press ENTER to
    select: **Endp**
of *(Select the endpoint of the centerline at "A")*
Second extension line origin: **Endp**
of *(Select the endpoint of the centerline at "B")*
Dimension line location (Mtext/Text/Angle/
    Horizontal/Vertical/Rotated): *(Select a point
    at "C")*
Dimension text = 0.75

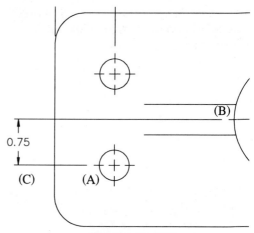

Figure 5–180

Step #13

Use the DIMCONTINUE command to place the next dimension in line with the previous dimension (Figure 5–181). When placing this type of dimension, the placement of the dimension line is remembered from the previous dimension.

 Command: **DIMCONTINUE**

Specify a second extension line origin or
    (Undo/<Select>):**Endp**
of *(Select the endpoint of the centerline at "A")*
Dimension text = 0.75
Specify a second extension line origin or
    (Undo/<Select>): *(Press ENTER to exit this
    mode)*
Select continued dimension: *(Press ENTER to exit
    this command)*

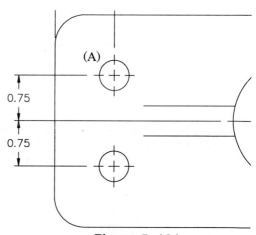

Figure 5–181

Step #14

Use the DIMLINEAR command to place the 1.75 vertical dimension illustrated in Figure 5–182.

 Command: **DIMLINEAR**

First extension line origin or press [ENTER] to select: **Endp**
of *(Select the endpoint of the extension line at "A")*
Second extension line origin: **Endp**
of *(Select the endpoint of the line at "B")*
Dimension line location (Mtext/Text/Angle/
   Horizontal/Vertical/Rotated): *(Select a point near "C")*
Dimension text = 1.75

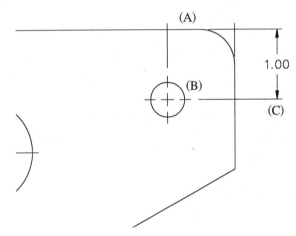

Figure 5–182

Step #15

Use the ZOOM command to get a closer look at the right side of the top view as illustrated in Figure 5–183 (or use the PAN command to keep the same zoom percentage and slide the current view to the left thus exposing the right side). Now place the 1.00 vertical dimension using the DIMLINEAR command.

 Command: **DIMLINEAR**

First extension line origin or press [ENTER] to select: **Endp**
of *(Select the endpoint of the line at "A")*
Second extension line origin: **Endp**
of *(Select the endpoint of the center line at "B")*
Dimension line location (Mtext/Text/Angle/
   Horizontal/Vertical/Rotated): *(Select a point near "C")*
Dimension text = 1.00

Figure 5–183

Step #16

Use the DIMBASELINE command to place the 2.00 dimension illustrated in Figure 5–184.

 Command: **DIMBASELINE**

Specify a second extension line origin or
   (Undo/<Select>): **Endp**
of *(Select the endpoint of the line at "A")*
Dimension text = 2.00
Specify a second extension line origin or
   (Undo/<Select>): *(Press* ENTER *to exit this
   mode)*
Select base dimension: *(Press* ENTER *to exit this
   command)*

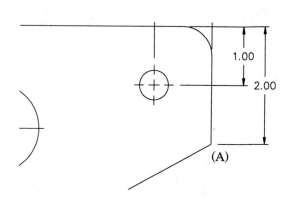

Figure 5–184

Step #17

Use the DIMANGULAR command to place the 61-degree dimension illustrated in Figure 5–185.

 Command: **DIMANGULAR**

Select arc, circle, line, or press ENTER: *(Select
   the line at "A")*
Second line: *(Select the line at "B")*
Dimension arc line location (Mtext/Text/Angle):
   (Select the location near "C")
Dimension text = 61

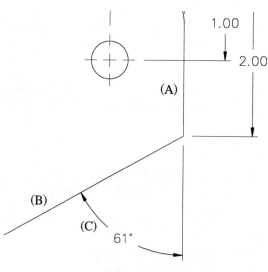

Figure 5–185

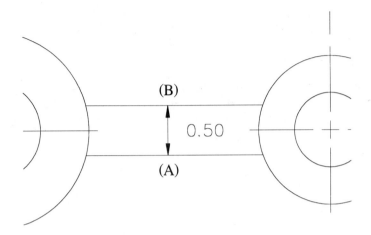

Figure 5–186

Step #18

Use the scroll bars to pan over and display the illustration in Figure 5–186. This area will be used to place the vertical 0.50 dimension which will callout the thickness of the web. Because this dimension will be placed inside of the web, the EXT-OFF dimension style will be used to control the display of the extension lines.

Click on the Dimension pull-down menu and pick Style...; when the Dimension Styles dialog box appears, click in the "Current" edit box and make "EXT-OFF" the new current dimension style. Click on the OK button to dismiss this dialog box.

Now place the 0.50 dimension using this dimension style.

 Command:**DIMLINEAR**

First extension line origin or press ENTER to
   select: **Nea**
to *(Select the line at "A")*
Second extension line origin: **Per**
to *(Select the line at "B")*
Dimension line location (Mtext/Text/Angle/
   Horizontal/Vertical/Rotated): *(Locate the
   dimension at "A")*
Dimension text = 0.50

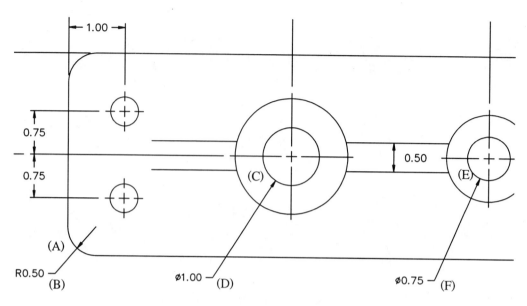

Figure 5–187

Step #19

Chate back to the "MECHANICAL" dimension style, then place the radius and diameter dimensions illustrated in Figure 5–187. The settings for both diameter and radius dimensions were already set in Steps #7 and #8. The DIMRADIUS command is used for placing the radius dimension; the DIMDIAMETER command is used for placing diameter dimensions.

 Command:**DIMRADIUS**

Select arc or circle: *(Select the arc at "A")*
Dimension text = 0.50
Dimension line location (Mtext/Text/Angle):
   (Locate the radius dimension at "B")

 Command:**DIMDIAMETER**

Select arc or circle: *(Select the circle at "C")*
Dimension text = 1.00
Dimension line location (Mtext/Text/Angle):
   (Locate the diameter dimension at "D")

 Command:**DIMDIAMETER**

Select arc or circle: *(Select the circle at "E")*
Dimension text = 0.75
Dimension line location (Mtext/Text/Angle):
   (Locate the diameter dimension at "F")

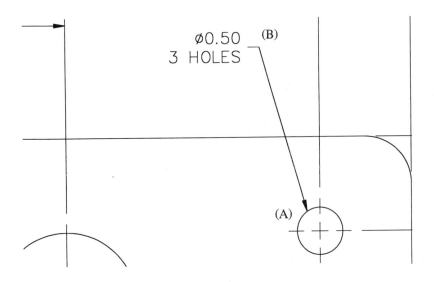

∅0.50   (B)
3 HOLES

(A)

Figure 5–188

Step #20

Call-out the three 0.50 holes by placing a leader illustrated in Figure 5–188. To insure that the text size for the leader is the same size as the other dimension text, set the Textsize system variable to a value of 0.12.

Command:**TEXTSIZE**
New value for TEXTSIZE <0.20>: **0.12**

  Command:**LEADER**

From point: **Nea**
to *(Select the circle at "A")*

To point: *(Locate the leader at "B")*
To point (Format/Annotation/
   Undo)<Annotation>: *(Press* ENTER *to continue)*
Annotation (or press ENTER for options): *(Press*
   ENTER*)*
Tolerance/Copy/Block/None/<Mtext>: *(Press*
   ENTER *to display the Mtext dialog box and enter*
   in the following):
%%c0.50
3 HOLES

Click on the OK button to place the leader annotation.

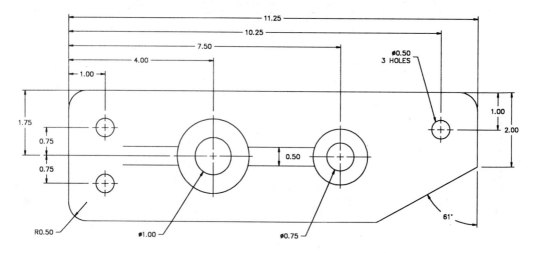

Figure 5–189

The completed top view including dimensions is illustrated in Figure 5–189. Use this example to check that all dimensions have been placed and all features (holes, fillets, etc.) have been properly identified.

The front view in Figure 5–190 will now be the focus for the next series of dimensioning steps. Again use the ZOOM command whenever dimensioning to smaller features.

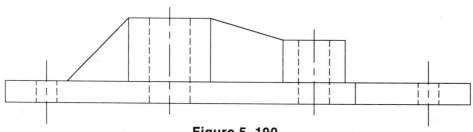

Figure 5–190

Step #21

Place the horizontal dimension illustrated in Figure 5–191 using the DIMLINEAR command.

 Command: **DIMLINEAR**

First extension line origin or press [ENTER] to select: *(Press [ENTER] to select a line to dimension)*

Select object to dimension: *(Select the line at "A")*

Dimension line location (Mtext/Text/Angle/Horizontal/Vertical/Rotated): *(Locate the dimension at "B")*

Dimension text = 2.00

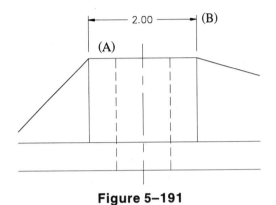

Figure 5–191

Step #22

Place the horizontal and vertical dimensions illustrated in Figure 5–192 using the DIMLINEAR command.

 Command: **DIMLINEAR**

First extension line origin or press [ENTER] to select: *(Press [ENTER] to select a line to dimension)*

Select object to dimension: *(Select the line at "A")*

Dimension line location (Mtext/Text/Angle/Horizontal/Vertical/Rotated): *(Locate the dimension near "B")*

Dimension text = 1.50

 Command: **DIMLINEAR**

First extension line origin or press [ENTER] to select: **Endp**

of *(Select the endpoint of the line at "C")*

Second extension line origin: **Int**

of *(Select the endpoint of the line at "D")*

Dimension line location (Mtext/Text/Angle/Horizontal/Vertical/Rotated): *(Locate the dimension near "E")*

Dimension text = 1.50

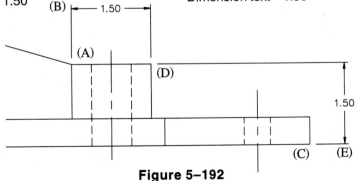

Figure 5–192

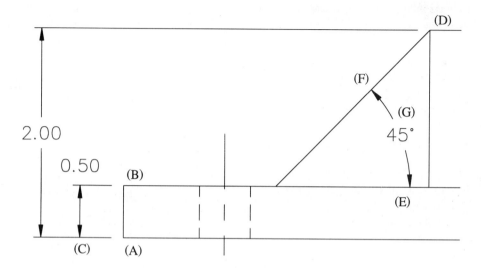

Figure 5–193

Step #23

Complete the dimensioning by placing the 0.50 vertical dimension, the 2.00 baseline dimension, and the 45-degree angular dimension as shown in Figure 5–193.

 Command: **DIMLINEAR**

First extension line origin or press [ENTER] to select: **Endp**
of *(Select the endpoint of the line at "A")*
Second extension line origin: **Endp**
of *(Select the endpoint of the line at "B")*
Dimension line location (Mtext/Text/Angle/Horizontal/Vertical/Rotated): *(Locate the dimension at "C")*
Dimension text = 0.50

Locate the dimension text to a better location through the use of grips.

 Command: **DIMBASELINE**

Specify a second extension line origin or (Undo/<Select>): **Endp**
of *(Select the endpoint of the line at "D")*
Dimension text = 2.00
Specify a second extension line origin or (Undo/<Select>): *(Press [ENTER] to exit this prompt)*
Select base dimension: *(Press [ENTER] to exit this command)*

 Command: **DIMANGULAR**

Select arc, circle, line, or press [ENTER]: *(Select the line at "E")*
Second line: *(Select the line at "F")*
Dimension arc line location (Mtext/Text/Angle): *(Locate the angular dimension at "G")*
Dimension text = 45

Once you have completed all dimensioning steps,
your drawing should appear similar to the illustration in Figure 5–194.

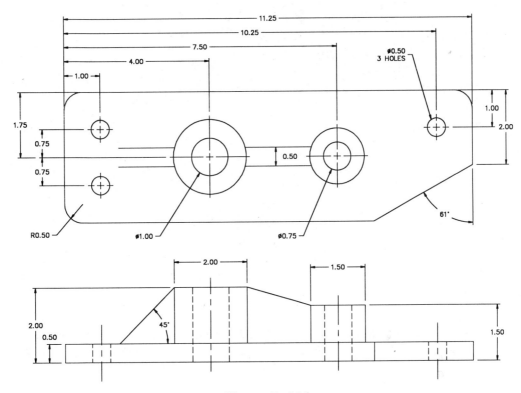

Figure 5–194

Tutorial Exercise
Tblk-iso.Dwg

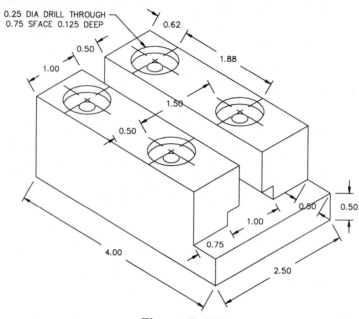

0.25 DIA DRILL THROUGH
0.75 SFACE 0.125 DEEP

0.62

0.50

1.88

1.00

1.50

0.50

0.50

0.80   0.50

1.00

4.00

0.75

2.50

Figure 5–195

Purpose

The purpose of this tutorial is to convert dimensions on an isometric drawing to oblique dimensions.

System Settings

The drawing in Figure 5–195 is already constructed and dimensioned. Enter the AutoCAD Drawing Editor and open the drawing file "TBLK-ISO." Follow the steps in this tutorial for converting the dimensions to proper isometric mode.

Layers

All layers have already been created:

Name	Color	Linetype
Center	Yellow	Center
Defpoints	White	Continuous
Dim	Yellow	Continuous
Object	White	Continuous

The layer "Defpoints" is automatically created when placing a dimension. It stands for "Definition Points" used for associative dimensioning practices.

Suggested Commands:

The DIMEDIT command along with the Oblique option will be used to complete this tutorial exercise.

Whenever possible, substitute the appropriate command alias in place of the full AutoCAD command in each tutorial step; for example, use "co" for the COPY command, "L" for the LINE command, and so on. The complete listing of all command aliases is located on pages 1–9 and 1–10.

Step #1

Use the Oblique option of the DIMEDIT command to rotate the selected dimensions by 150 degrees as shown in Figure 5–196.

 Command: **DIMEDIT**

Dimension Edit (Home/New/Rotate/Oblique)
  <Home>: **Oblique**
Select objects: *(Select dimensions "A" through "G")*
Select objects: *(Press* ENTER *to continue)*
Enter obliquing angle *(Press* ENTER *for none):* **150**

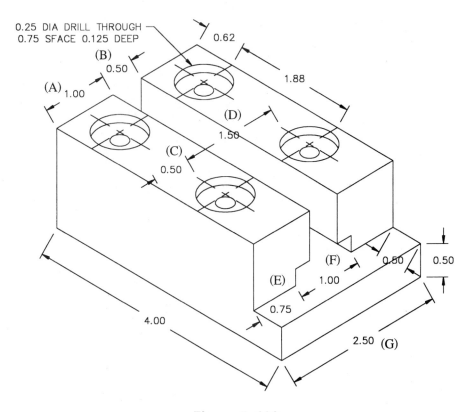

Figure 5–196

Step #2

Use the Oblique option of the DIMEDIT command
to rotate the selected dimensions by 210 degrees as
in Figure 5–197.

 Command: **DIMEDIT**

Dimension Edit (Home/New/Rotate/Oblique)
   <Home>: **Oblique**
Select objects: *(Select dimensions "A" through "D")*
Select objects: *(Press* ENTER *to continue)*
Enter obliquing angle (press ENTER for none): **210**

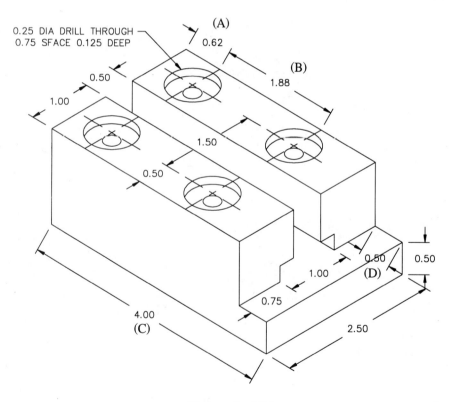

Figure 5–197

Step #3

Place two vertical dimensions using the DIMLINEAR
command as in Figure 5–198.

 Command: **DIMLINEAR**

First extension line origin or press ENTER to
   select: **Endp**
of *(Select the endpoint at "A")*
Second extension line origin: **Endp**
of *(Select the endpoint at "B")*
Dimension line location (Mtext/Text/Angle/
   Horizontal/Vertical/Rotated): *(Locate the
   dimension at a convenient distance)*
Dimension text = 1.00

 Command: **DIMLINEAR**

First extension line origin or press ENTER to
   select: **Endp**
of *(Select the endpoint at "C")*
Second extension line origin: **Endp**
of *(Select the endpoint at "D")*
Dimension line location (Mtext/Text/Angle/
   Horizontal/Vertical/Rotated): *(Locate the
   dimension at a convenient distance)*
Dimension text = 0.75

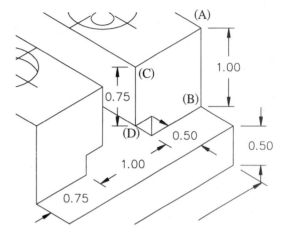

Figure 5–198

Step #4

Use the Oblique option of the DIMEDIT command
to rotate the selected dimensions by 30 degrees as
in Figure 5–199.

 Command: **DIMEDIT**

Dimension Edit (Home/New/Rotate/Oblique)
   <Home>: **Oblique**
Select objects: *(Select dimensions "A" through
   "C")*
Select objects: *(Press ENTER to continue)*
Enter obliquing angle (press ENTER for none): **30**

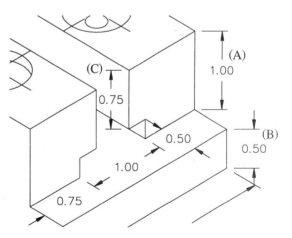

Figure 5–199

Step #5

Enter the Dimension Styles dialog box through the command button or pull-down menu area. Click on the "Geometry…" button and click in the two check boxes to suppress both extension lines located in the 0.50 dimension illustrated in Figure 5–200. This is to prevent the overlapping of object and dimension lines. When completed with this operation, click on the OK button to return to the main Dimension Styles dialog box. Click on the OK button to return to the drawing editor. Now update the 0.50 dimension to the latest dimension style change using the DIMSTYLE command and the Apply option.

Command: **DIMSTYLE**
dimension style: STANDARD
dimension style overrides:
    DIMASZ 0.12
    DIMDEC 2
    DIMEXE 0.07
    DIMEXO 0.10
    DIMFIT 5
    DIMSCALE 0.00
    DIMSE1 On
    DIMSE2 On
    DIMTDEC 2
    DIMTIX On
    DIMTXT 0.10

Dimension Style Edit (Save/Restore/STatus/
    Variables/Apply/?) <Restore>: **Apply**
Select objects: *(Select the 0.50 dimension at "A")*
Select objects: *(Press* [ENTER] *to apply the change in the dimension variable and exit the command)*

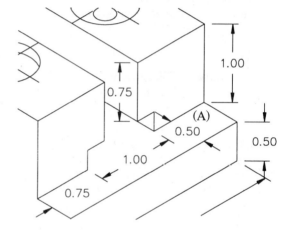

Figure 5–200

Once you have completed all dimensioning steps, your drawing should appear similar to the illustration in Figure 5–201.

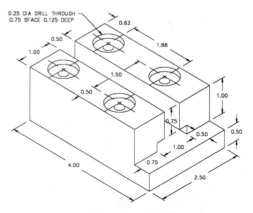

Figure 5–201

Tutorial Exercise
Bas-plat.Dwg

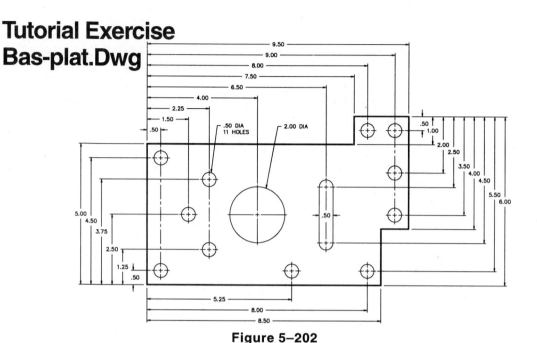

Figure 5–202

Purpose

The purpose of this tutorial is to convert the drawing of the Bas-plat (Base plate) from the conventional dimensioning style to the ordinate dimensioning style.

System Settings

The drawing in Figure 5–202 is already constructed and dimensioned up to a certain point. Enter AutoCAD and edit an existing drawing called "BAS-PLAT." Follow the steps in this tutorial for converting the dimensions to proper ordinate dimensions.

Layers

All layers have already been created:

Name	Color	Linetype
Cen	Yellow	Center
Defpoints	White	Continuous
Dim	Yellow	Continuous
Object	White	Continuous

Suggested Commands

Use the DIMORDINATE command for placing ordinate dimensions throughout this tutorial.

Whenever possible, substitue the appropriate command alias in place of the full AutoCAD command in eacy tutorial step; for example, use "Co" for the COPY command, "L" for the LINE command, and so on. The complete listing of all command aliases is located on pages 1–9 and 1–10.

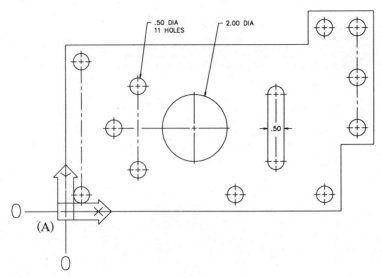

Figure 5-203

Step #1

All ordinate dimensions make reference to the current 0,0 location identified by the position of the user coordinate system (see Figure 5–203). Since this icon is located in the lower left corner of the display screen by default, the coordinate system must be moved to a point on the object where all ordinate dimensions will be referenced from. First use the UCS command to define a new coordinate system with the origin at the lower left corner of the object. Then use the UCSICON command to force the icon to display at the new origin.

 Command:**UCS**
Origin/ZAxis/3point/OBject/View/X/Y/Z/Prev/
    Restore/Save/Del/?/<World>:**Origin**
Origin point <0,0,0>:**Int**
of *(Select the intersection at "A")*

Command:**UCSICON**
ON/OFF/All/Noorgin/ORigin <ON>:**OR**

Step #2

Use the ZOOM-CENTER command to magnify the screen similar to the illustration in Figure 5–204.

 Command:**ZOOM**
All/Center/Dynamic/Extents/Left/Previous/
    Vmax/Window/<Scale(X/XP)>:**C**
Center point: 0,0
Magnification or Height <11.23>: **4**

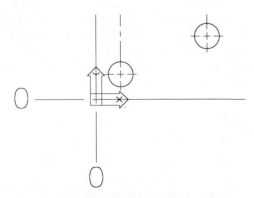

Figure 5-204

Step #3

Before continuing with this next step, be sure that the grid is turned on. Use the F7 function key to do this. Next, begin to place the first ordinate dimension using the DIMORDINATE command, or choose the DIMORDINATE command button from the Dimension Toolbox. Use the Osnap-Quadrant mode to select the circle as a feature. With the Snap and Ortho modes turned on, locate a point two grid dots below the object to identify the leader endpoint. AutoCAD will automatically determine if the dimension is Xdatum or Ydatum. See Figure 5–205.

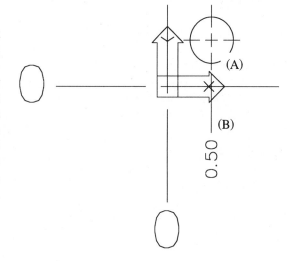

 Command: **DIMORDINATE**

Select feature: **Qua**
of *(Select the quadrant of the circle at "A")*
Leader endpoint (Xdatum/Ydatum/Mtext/Text):
   (Pick a point two grid dots below the edge of the object at "B")
Dimension text = 0.50

Figure 5–205

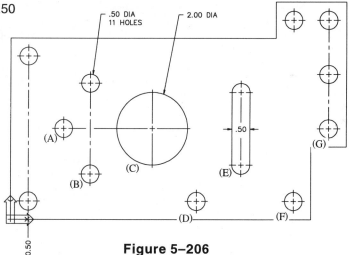

Figure 5–206

Step #4

Perform a Zoom-Previous operation to display the overall drawing of the bas-plat similar to the illustration in Figure 5–206. Repeat the procedure in Step #3 to place ordinate dimensions at locations "A" through "G." Be sure Ortho mode is on and that the leader location is two grid dots below the bottom edge of the object. The Osnap-Quadrant mode should be used on each circle and arc to satisfy the prompt, "Select feature." Use the Osnap Settings dialog box to set Quadrant as the current running Object Snap mode.

Step #5

Continue placing ordinate dimensions similar to the procedure used in Step #3. Use the Osnap-Endpoint mode to select features at "A" and "B" in the illustration in Figure 5–207. Have Ortho mode on and identify the leader endpoint two grid dots below the bottom edge of the object.

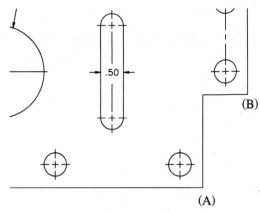

Figure 5–207

Step #6

The UCSICON command is used to turn off the display of the icon while still keeping the 0,0 origin at the lower left corner of the object. After turning off the icon, your display should appear similar to Figure 5–208.

Command:**UCSICON**
ON/OFF/All/Noorgin/ORigin <ON>:**Off**

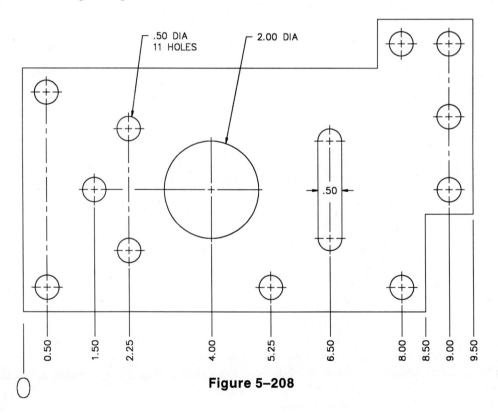

Figure 5–208

Step #7

With Snap on and Ortho on, begin placing the first vertical ordinate dimension using the DIMORDINATE command. The procedure and prompts are identical to that of placing a horizontal ordinate dimension. Follow Figure 5–209 and the following prompts for performing this operation.

 Command: **DIMORDINATE**

Select feature: **Qua**
of *(Select the quadrant of the circle at "A")*
Leader endpoint (Xdatum/Ydatum/Mtext/Text):
 *(Pick a point two grid dots to the left of the
 object at "B")*
Dimension text = 0.50

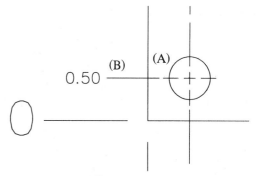

Figure 5–209

Step #8

Follow the procedure in the previous step to complete the vertical ordinate dimensions along this edge of the object as in Figure 5–210. Use the Osnap-Quadrant mode for "A" through "D" and Osnap-Endpoint for "E." Again have Ortho on and Snap on. For the leader endpoint, count two grid dots to the left of the object and place the dimensions.

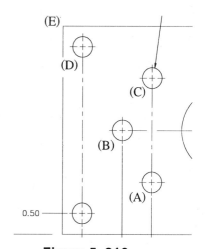

Figure 5–210

Step #9

Your display should appear similar to the illustration in Figure 5–211.

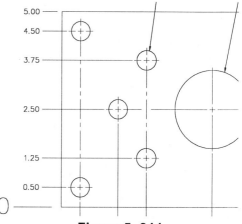

Figure 5–211

Step #10

Magnify the right portion of the object using the
ZOOM-WINDOW command. Use ordinate dimensions and a combination of Osnap-Endpoint and
Quadrant modes to place vertical ordinate dimensions from "A" to "I" as shown in Figure 5–212.

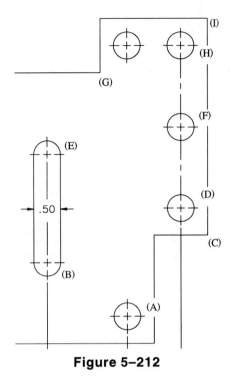

Figure 5–212

Step #11

Your display should appear similar to the illustration in Figure 5–213.

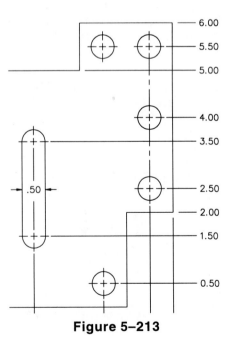

Figure 5–213

Step #12

Complete the dimensioning by placing two horizontal ordinate dimensions at the locations illustrated in Figure 5–214. As in the past, have Ortho on, Snap on, and use a combination of the Osnap-Quadrant and Endpoint modes to select the features. Place the leader endpoint two grid dots above the top line of the object. Use the Zoom-All option to return the entire object to your display. See Figure 5–215.

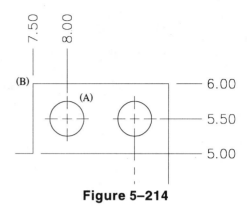

Figure 5–214

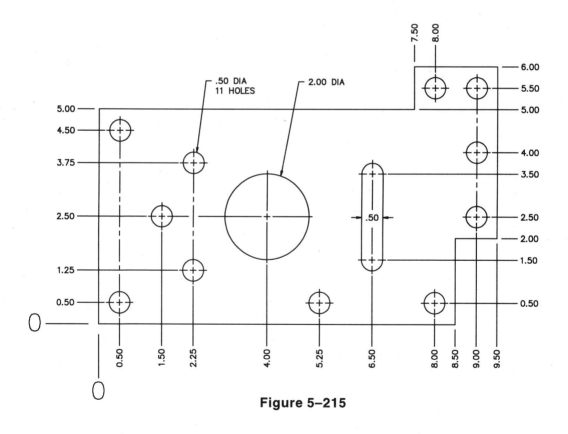

Figure 5–215

Problems for Unit 5

Problems 5–1 through 5–19

    1. Use the grid and spacing of 0.50 units to determine all dimensions.
    2. Reproduce the views shown, and fully dimension the drawings.

Problem 5–1

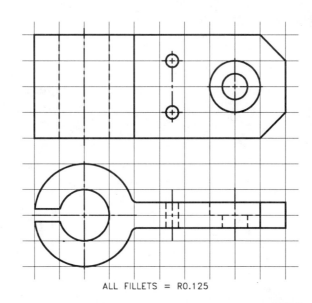

ALL FILLETS = R0.125

Problem 5–2

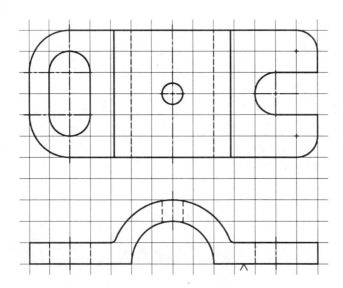

Problem 5–3

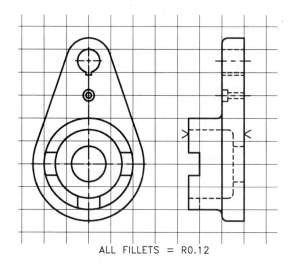

ALL FILLETS = R0.12

Problem 5–4

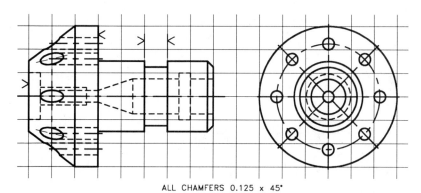

ALL CHAMFERS 0.125 x 45°

Problem 5–5

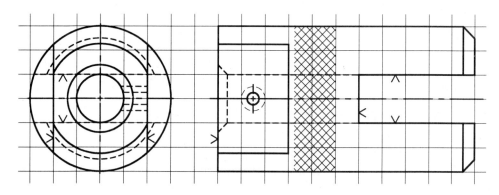

Problem 5–6

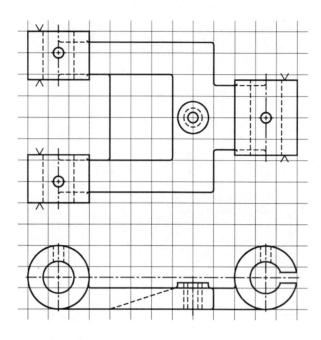

Problem 5–7

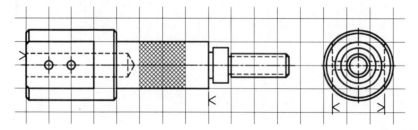

Problem 5–8

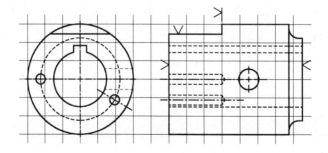

Problem 5–9

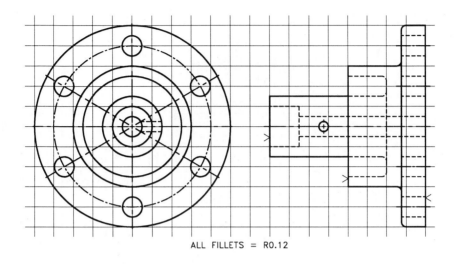

ALL FILLETS = R0.12

Problem 5–10

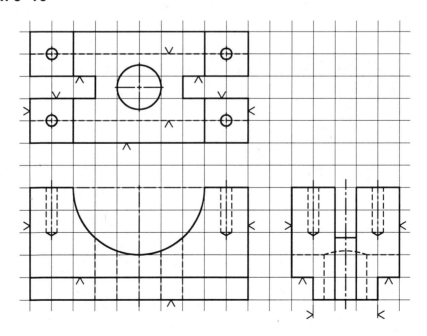

Problem 5–11

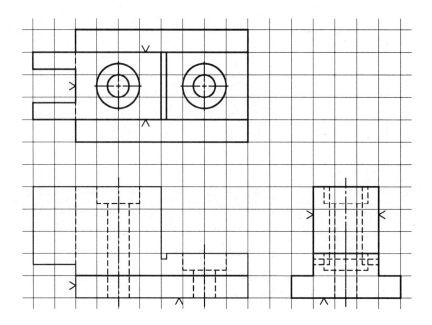

Problem 5–12

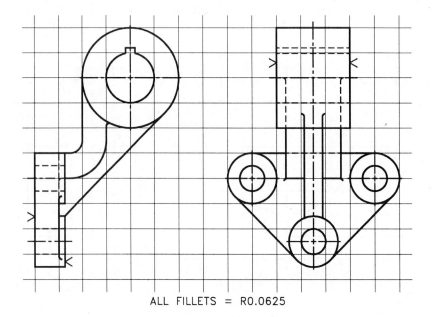

ALL FILLETS = R0.0625

Problem 5–13

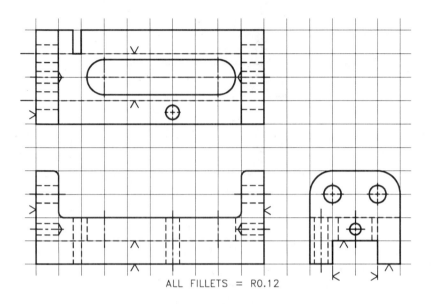

ALL FILLETS = R0.12

Problem 5–14

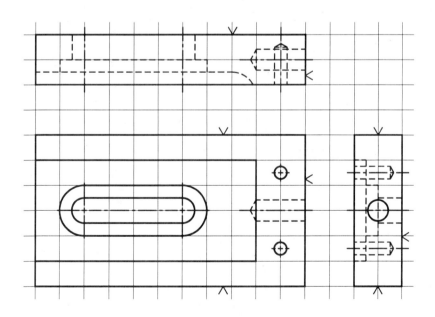

Problem 5–15

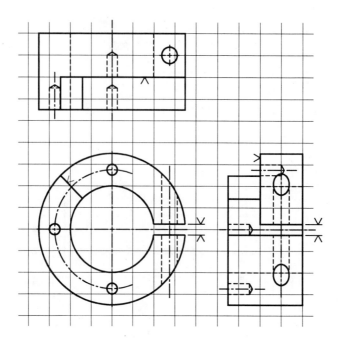

Problem 5–16

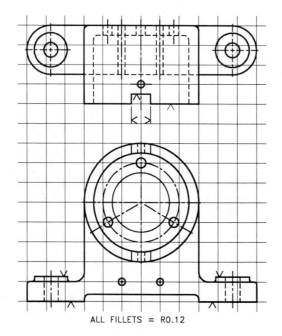

ALL FILLETS = R0.12

Problem 5–17

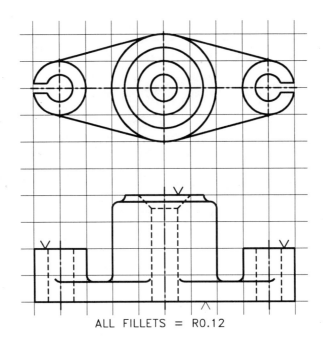

ALL FILLETS = R0.12

Problem 5–18

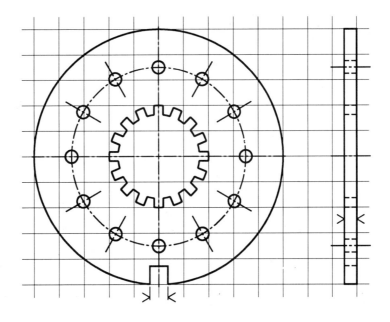

Problem 5–19

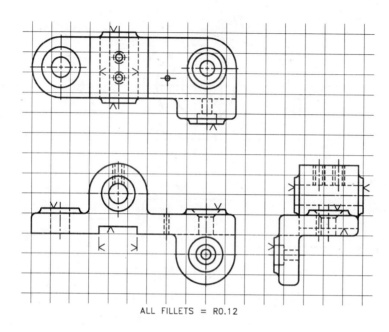

ALL FILLETS = R0.12

Problems 5–20 through 5–25

1. Convert the isometric drawings provided into orthographic drawings, showing as many views as necessary to communicate the design.
2. Fully dimension your drawings.

Problem 5–20

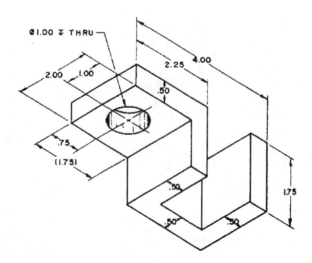

Problem 5–21

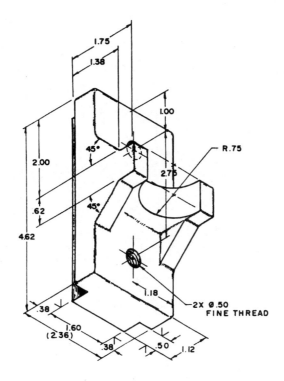

Problem 5–22

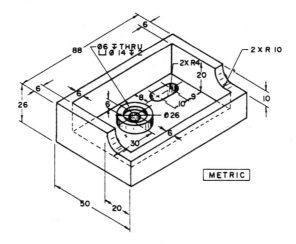

Problem 5–23

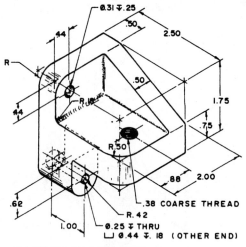

.31 ⊤ .25
.44
.50
2.50
R
.50
1.75
R.18
.44
.75
R.50
1.75
Ø.38 COARSE THREAD
.88   2.00
.62
R.42
1.00   Ø.25 ⊤ THRU
⊔ Ø.44 ⊤ .18 (OTHER END)

ALL UNMARKED RADII = R.09

Problem 5–24

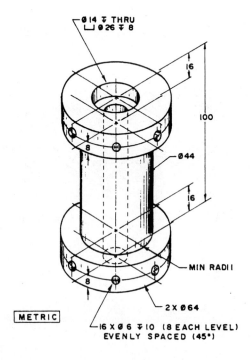

Ø14 ⊤ THRU
⊔ Ø26 ⊤ 8
16
100
8
Ø44
16
MIN RADII
8
2 X Ø64

METRIC

16 X Ø6 ⊤ 10 (8 EACH LEVEL)
EVENLY SPACED (45°)

Problem 5–25

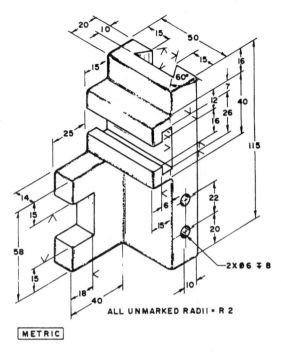

ALL UNMARKED RADII = R 2

METRIC

Directions for Problem 5–26
Use ordinate dimensioning techniques to dimension this drawing.

Problem 5–26

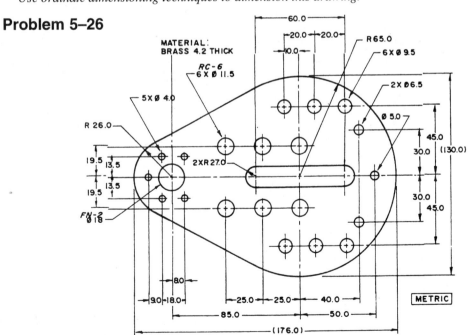

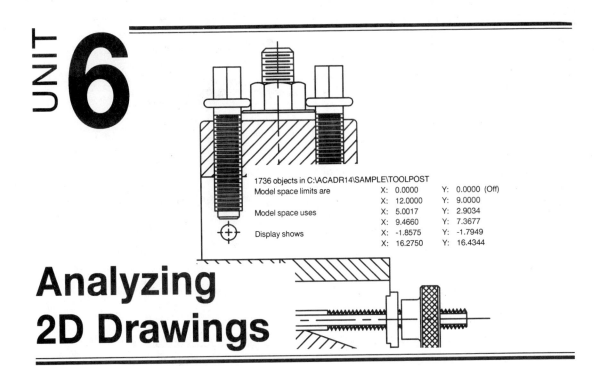

UNIT 6

1736 objects in C:\ACADR14\SAMPLE\TOOLPOST

	X:	Y:
Model space limits are	0.0000	0.0000 (Off)
	12.0000	9.0000
Model space uses	5.0017	2.9034
	9.4660	7.3677
Display shows	-1.8575	-1.7949
	16.2750	16.4344

Analyzing 2D Drawings

Completed drawings are usually plotted out and checked with scales for accuracy. Depending on the thickness of the pen used to perform the plot and the scale used, a range of accuracy or tolerance is assigned. A proper computer-aided design system is equipped with a series of commands to calculate distances and angles of selected objects. Surface areas may be performed on complex geometric shapes.

The following pages highlight all Inquiry commands and how they are used to display useful information on an object or group of objects. The DDMODIFY command is also explained in great detail on what type of object control it supplies to the user.

Use the information in this unit to become more familiar with all Inquiry commands and the DDMODIFY command.

Using Inquiry Commands

AutoCAD's Inquiry commands may be selected from the Inquiry Toolbar, may be keyed in at the keyboard, or may be selected from the pull-down menu bar or side bar screen menus as illustrated in Figures 6–1 and 6–2. The following is a list of the Inquiry commands with a short description of each:

AREA—used to calculate the surface area given a series of points or by selecting a polyline or circle. Multiple objects may be added or subtracted to calculate the area with holes and cutouts.

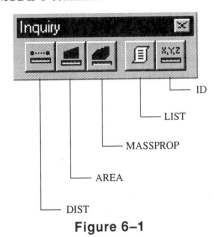

Figure 6–1

DIST—calculates the distance between two points. Also provides the delta X, Y, Z coordinate values, the angle in the X-Y plane, and the angle from the X-Y plane.

HELP—provides online help for any command. May be entered at the keyboard or selected from the pull-down menu bar.

ID—displays the X, Y, Z absolute coordinate of a selected point.

LIST—displays key information depending on the object selected.

STATUS—displays important information on the current drawing.

TIME—displays the time spent in the drawing editor.

The following pages give a detailed description of how these commands are used to interact on different AutoCAD objects.

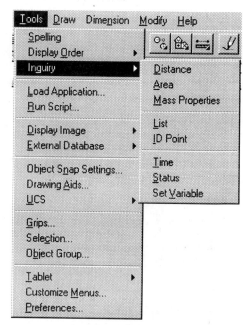

Figure 6–2

Finding the Area of an Enclosed Shape

The AREA command is used to calculate the area through the selection of a series of points. Select the endpoints of all vertices of the image in Figure 6–3 with the OSNAP-Endpoint option. Once the first point is selected along with the remaining points in either a clockwise or counterclockwise pattern, the command prompt "Next point:" is followed by the [ENTER] key to calculate the area of the shape. Along with the area is a calculation for the perimeter. Use Figures 6–3 and 6–4 to better understand of the prompt sequence used for finding the area by identifying a series of points.

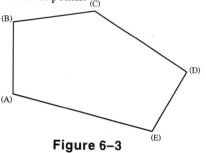

Figure 6–3

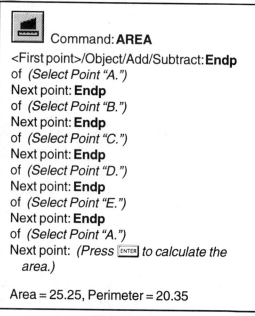

Command: **AREA**

<First point>/Object/Add/Subtract: **Endp**
of *(Select Point "A.")*
Next point: **Endp**
of *(Select Point "B.")*
Next point: **Endp**
of *(Select Point "C.")*
Next point: **Endp**
of *(Select Point "D.")*
Next point: **Endp**
of *(Select Point "E.")*
Next point: **Endp**
of *(Select Point "A.")*
Next point: *(Press* [ENTER] *to calculate the area.)*

Area = 25.25, Perimeter = 20.35

Figure 6–4

Finding the Area of an Enclosed Polyline or Circle

Previously, an example was given on finding the area of an enclosed shape using the AREA command and identifying the corners and intersections of the enclosed area by a series of points. For a complex area, this could be a very tedious operation. As a result, the AREA command has a built-in Object option that will calculate the area and perimeter on a polyline and the area and circumference on a circle. Study Figures 6–5 and 6–6 for these operations.

Finding the area of a polyline can only be accomplished if one of the following conditions are satisfied:

– The shape must have already been constructed using the PLINE command.

– The shape must have already been converted into a polyline using the PEDIT command if originally constructed out of individual objects.

 Command: **AREA**

<First point>/Object/Add/Subtract: **Object**
Select objects: *(Select image "A.")*

Area = 24.88, Perimeter = 19.51

 Command: **AREA**

<First point>/Object/Add/Subtract: **Object**
Select objects: *(Select circle "B.")*

Area = 7.07, Circumference = 9.42

Figure 6–5

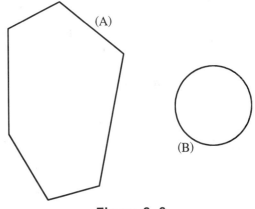

Figure 6–6

Finding the Area of a Surface by Subtraction

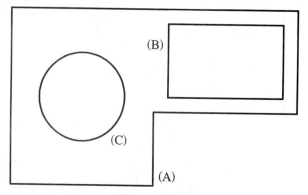

Figure 6–7

The steps you use to calculate the total surface area are: (1) calculate the area of the outline; and (2) subtract the objects inside of the outline. All individual objects, with the exception of circles, must first be converted into polylines using the PEDIT command. Next, the overall area is found and added to the database using the Add mode of the AREA command. Add mode is exited and the inner objects are removed using the Subtract mode of the AREA command. Remember, all objects must be in the form of a circle or polyline. This means the inner shape at "B" in Figure 6–7 must also be converted into a polyline using the PEDIT command before calculating the area. Care must be taken when selecting the objects to subtract. If an object is selected twice, it is subtracted twice and may yield an inaccurate area in the final calculation.

For the image in Figure 6–7, the total area with the circle and rectangle removed is 30.4314. See Figure 6–8.

 Command: **AREA**
<First point>/Object/Add/Subtract: **Add**
<First point>/Object/Subtract: **Object**
(ADD mode) Select objects:
(Select image "A.")
Area = 47.5000, Perimeter = 32.0000
Total area = 47.5000
(ADD mode) Select objects: *(Press [ENTER])*
<First point>/Object/Subtract: **Subtract**
<First point>/Object/Add: **Object**
(SUBTRACT mode) Select objects:
(Select shape "B.")
Area = 10.0000, Perimeter = 13.0000
Total area = 37.5000
(SUBTRACT mode) Select objects:
(Select shape "C.")
Area = 7.0686, Circumference = 9.4248
Total area = 30.4314
(SUBTRACT mode) Select objects: *(Press [ENTER])*
<First point>/Object/Add: *(Press [ENTER] to exit.)*

Figure 6–8

Using the DIST (Distance) Command

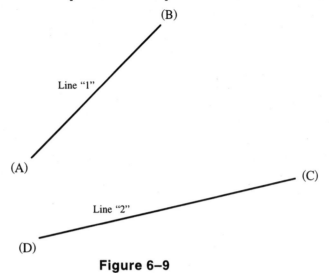

Figure 6–9

The DIST command calculates the linear distance between two points on an object whether it be the distance of a line, the distance between two points, or the distance from the quadrant of one circle to the quadrant of another circle (see Figures 6–9 and 6–10). The following information is also supplied when using the DIST command: the angle in the X-Y plane; the angle from the X-Y plane; the delta X,Y,Z coordinate values. The angle in the X-Y plane is given in the current angular mode set by the Units control dialog box. The delta X,Y,Z coordinate is a relative coordinate value taken from the first point identified by the DIST command to the second point.

 Command: **DIST**
First point: **Endp**
of *(Select the endpoint of the line at "A.")*
Second point: **Endp**
of *(Select the endpoint of the line at "B.")*
Distance=6.36, Angle in XY Plane=45.0000,
   Angle from XY Plane=0.0000
Delta X=4.50, Delta Y=4.50, Delta Z=0.00

 Command: **DIST**
First point: **Endp**
of *(Select the endpoint of the line at "C.")*
Second point: **Endp**
of *(Select the endpoint of the line at "D.")*
Distance=9.14, Angle in XY
   Plane=192.7500, Angle from XY
   Plane=0.0000
Delta X=-8.91, Delta Y=-2.02, Delta Z=0.00

Figure 6–10

Interpretation of Angles Using the DIST command

Previously, it was noted that the DIST command yields information regarding distance, delta X, Y coordinate values, and angle information. Of particular interest is the angle in the X-Y plane formed between two points. In Figure 6–11, picking the endpoint of the line segment at "A" as the first point followed by the endpoint of the line segment at "B" as the second point displays an angle of 42 degrees. This angle is formed from an imaginary horizontal line drawn from the endpoint of the line segment at "A" in the zero direction.

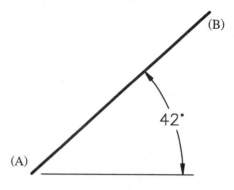

Figure 6–11

Take care when using the DIST command to find an angle on an identical line segment illustrated in Figure 6–12 as with the example in Figure 6–11. However, notice that the two points for identifying the angle are selected differently. Using the DIST command, the endpoint of the line segment at "B" is selected as the first point followed by the endpoint of the segment at "A" for the second point. A new angle in the X-Y plane of 222 degrees is formed. In Figure 6–12, the angle is calculated by constructing a horizontal line from the endpoint at "B," the new first point of the DIST command. This horizontal line is also drawn in the zero direction. Notice the relationship of the line segment to the horizontal baseline. Be careful identifying the endpoints of line segments when extracting angle information.

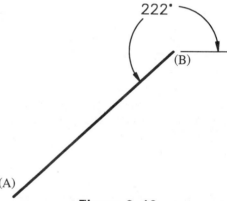

Figure 6–12

Using the ID (Identify) Command

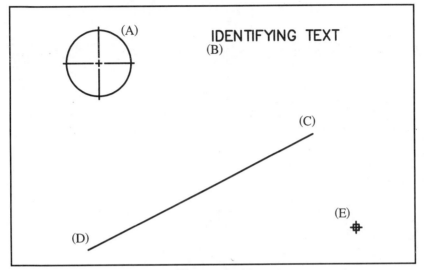

Figure 6–13

The ID command is probably one of the more straightforward of the Inquiry commands. ID stands for "Identify" and allows you to obtain the current absolute coordinate listing of a point along or near an object.

In Figure 6–13, the coordinate value of the center of the circle at "A" was found by using ID and the OSNAP-Center mode; the coordinate value of the starting point of text string "B" was found using ID and the OSNAP-Insert mode; the coordinate value of the endpoint of line segment "C" was found using ID and the OSNAP-Endpoint mode; the coordinate value of the midpoint of line segment at "CD" was found by using ID and the OSNAP-Midpoint mode; and the coordinate value of the current position of point "E" was found by using ID and the OSNAP-Node mode.

 Command: **ID**

Point: **Cen**
of *(Select the circle "A.")*
X = 2.00 Y = 7.00 Z = 0.00

 Command: **ID**

Point: **Ins**
of *(Select the text at "B.")*
X = 5.54 Y = 7.67 Z = 0.00

 Command: **ID**

Point: **Endp**
of *(Select the line at "C.")*
X = 8.63 Y = 4.83 Z = 0.00

 Command: **ID**

Point: **Mid**
of *(Select line "CD.")*
X = 1.63 Y = 1.33 Z = 0.00

 Command: **ID**

Point: **Nod**
of *(Select the point at "E.")*
X = 9.98 Y = 1.98 Z = 0.00

Using the DDMODIFY Command

Modify Line

Properties

| Color... | ■ BYLAYER | Handle: | 20 |

Layer... 0 Thickness: 0.0000

Linetype... BYLAYER Linetype Scale: 1.0000

From Point

Pick Point <

X: 4.065527954
Y: 2.551675964
Z: 0.0000

To Point

Pick Point <

X: 7.917080739
Y: 3.506983242
Z: 0.0000

Delta XYZ:

X: 3.8516
Y: 0.9553
Z: 0.0000

Length: 3.9683
Angle: 14

OK   Cancel   Help

Figure 6–14

The DDMODIFY command allows you to select an object and display its properties in a dialog box on the screen. Choose this command by selecting "Properties..." from the "Modify" section of the pull-down menu area. Entering DDMODIFY at the command prompt is yet another way to access this command. A typical Modify Object dialog box consists of the following:

Properties:
Color
Linetype
Layer
Thickness

Depending upon the object selected, the Modify Object dialog box in Figure 6–14 also displays important information on the object. Since this information is different for each type of object, it will be discussed in the next series of pages. Below is a brief description of the additional dialog boxes directly related to DDMODIFY, namely Color..., Linetype..., Layer..., and Thickness.

Changing the Color of an Object

Under the "Properties" section of the Modify Object dialog box in Figure 6–15 is the Color option. Use this property to change the color of the object selected. Choosing the color property brings up an additional dialog box displaying the current number of colors supported by the monitor configured to use AutoCAD. Picking a different color changes the selected object to that color. This is a quick way to change color. However, changing colors this way may affect the original color of an object set by the Layer and Linetype Properties dialog box.

Changing the Layer of an Object

The "Properties" section of the Modify Object dialog box in Figure 6–15 also has an option for modifying the layer of a selected object. By default, only layer 0 is created when entering a new drawing. As layers are created, they will appear in the "Select Layer" dialog box of the DDMODIFY command.

Changing the Linetype of an Object

The "Properties" section of the Modify Object dialog box in Figure 6–15 also has an option for modifying the linetype of a selected object. Choosing the linetype property brings up an additional dialog box displaying the current linetypes loaded into the drawing. By default, only the continuous linetype displays in this dialog box. As linetypes are loaded, they will appear in this dialog box. As with color, picking a different linetype changes the selected object to that linetype.

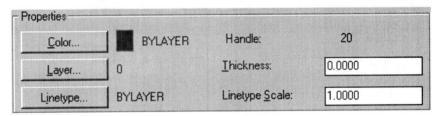

Figure 6–15

Listing the Properties of an Arc

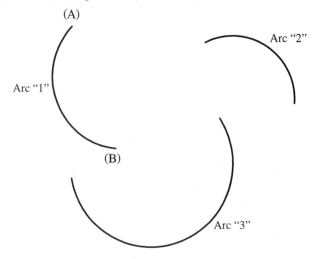

Figure 6–17

This segment of the Inquiry commands begins a study of the LIST command and all of the properties displayed on the particular object listed. Using the LIST command on an arc provides you with the following information: the name of the object being listed (Arc); the layer the object was drawn on; whether the object occupies model or paper space; the object handle; the center point of the arc; the radius of the arc; the starting angle of the arc in degrees; the ending angle of the arc in degrees; and the length of the arc segment.

The starting angle and ending angle of an arc are determined by the original construction of the arc in the counterclockwise direction. In the illustration in Figure 6–17, listing "Arc 1" displays a starting angle at "A" of 136 degrees. Moving in a counterclockwise direction, the end angle at "B" measures 265 degrees. See Figure 6–16.

```
ARC        Layer: 0
           Space: Model space
           Handle = 528
center point, X= 4.2065 Y= 6.3587 Z= 0.0000
radius 2.3677
start angle 136
end angle 265
length 5.3259
ARC   Layer: 0
           Space: Model space
           Handle = 1
center point, X= 7.9286 Y= 5.6429 Z= 0.0000
radius 2.0763
start angle 356
end angle 117
Length 4.3672
ARC   Layer: 0
           Space: Model space
           Handle = 3
center point, X= 5.2000 Y= 3.5000 Z= 0.0000
radius 2.7459
start angle 190
end angle 33
length 9.7106
```

Figure 6–16

Using DDMODIFY on an Arc

Modify Arc

Properties

Color...	■ BYLAYER	Handle:	508
Layer...	0	Thickness:	0.0000
Linetype...	BYLAYER	Linetype Scale:	1.0000

Center

Pick Point <   Radius: 2.367719664   Arc Length: 5.3259

X: 4.206521739   Start Angle: 136.1160103

Y: 6.358695652   End Angle: 264.9960800

Z: 0.0000   Total Angle: 129

OK   Cancel   Help

Figure 6–18

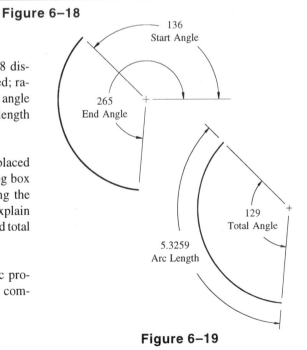

The DDMODIFY dialog box in Figure 6–18 displays the following: center of the arc selected; radius of the arc; starting angle of the arc; ending angle of the arc; total included angle of the arc; and length of the arc.

All of these parameters of the arc that are placed inside of a box may be changed in this dialog box and the object will update itself after exiting the dialog box. The illustrations in Figure 6–19 explain the starting angle, ending angle, total angle, and total length of the arc.

Using the DDMODIFY command on an arc provides a more powerful listing of the object compared to the LIST command in Figure 6–16.

Figure 6–19

Listing the Properties of an Arc Converted Into a Lightweight Polyline

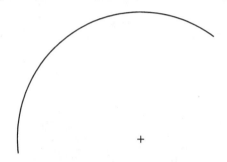

Figure 6–21

As shown in figure 6–20 using the LIST command on the polyarc in Figure 6–21 lists the absolute coordinate values of each vertex of the lightweight polyline along with the following information: the name of the object being listed (polyline); the layer the polyarc was drawn on; whether the polyarc occupies model or paper space; the specific vertex being listed in absolute coordinates; the center point of the polyarc segment; the radius of the polyarc; the starting angle of the polyarc; and the ending angle of the polyarc.

Once the last screen is displayed, the LIST command calculates the area occupied by a closed or open polyarc. If the polyarc is open, the total length of the polyarc segment is given. See Figure 6–20.

Using the EXPLODE command on a polyarc separates the object into individual arc segments.

```
LWPOLYLINE      Layer: 0
                Space: Model space
                Handle = 24
Open
Constant width 0.0000

at point X=9.0000 Y=6.0000 Z=0.0000
bulge 0.6510
center X=6.7746 Y=2.9221 Z=0.0000
radius 3.7981
start angle 0.9448
end angle 3.2530
at point X=3.0000 Y=2.5000 Z=0.0000

area 11.3095
length 8.7668
```

Figure 6–20

Using DDMODIFY on a Polyarc

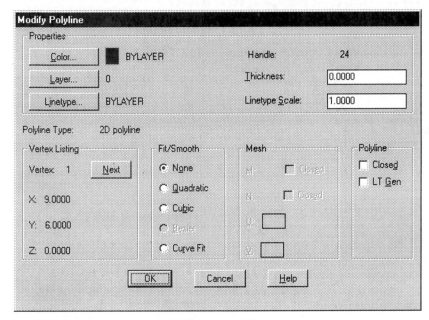

Figure 6–22

Using the DDMODIFY command on a polyarc lists the usual object properties such as color, linetype, layer name, and object thickness (see Figure 6–22). This command also lists the following information: the *X,Y,Z* coordinates of each polyarc vertex; the type of curve fitted to the polyarc; whether the polyarc is closed or open; and whether to generate linetype scaling per vertex.

The coordinates of all lightweight polyline vertices are identified in Figure 6–22 in the Modify Polyline dialog box. Selecting Next lists the next set of co-ordinates. This may not be as important a function on a polyarc as on a multiple vertex polyline since the polyarc only has two vertices: one at the beginning of the polyarc and the other at the end of the polyarc (see Figure 6–23). Also, the methods of fitting a curve have no effect on a polyarc.

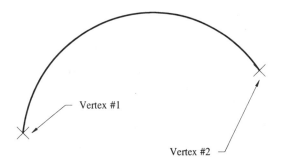

Figure 6–23

Listing the Properties of an Associative Dimension

Using the LIST command on the diameter dimension in Figure 6–25 lists the following information: the specific object being listed (Dimension); the layer the dimension was drawn on; whether the dimension occupies model or paper space; the first defining point of the leader; the second defining point of the leader; the text position (in absolute coordinates); the default text; and the current dimension style; (see Figure 6–24).

Using the LIST command on the linear dimension in Figure 6–25 lists the following information: the specific object being listed (Dimension); the layer the dimension was drawn on; whether the dimension occupies model or paper space; the first extension line defining point; the second extension defining point; the text position (in absolute coordinates); the default text; and the current dimension style. See Figure 6–24.

Dimension"1"
DIMENSION   Layer: 0
            Space: Model space
            Handle = 83
type: diameter
defining point: X=6.41 Y=3.57 Z=0.00
defining point: X=5.59 Y=5.65 Z=0.00
leader length   2.83
default text position: X=3.91 Y=8.27 Z=0.00
default text
dimension style: STANDARD

Dimension"2"
DIMENSION   Layer: 0
    Space: Model space
    Handle = 82
type: horizontal
1st extension defining point:X=2.00 Y=2.11 Z=0.00
2nd extension defining point:X=6.00 Y=3.37 Z=0.00
dimension line defining point:X=6.00 Y=1.25 Z=0.00
default text position: X=4.00 Y=1.25 Z=0.00
default text
dimension style: STANDARD

Figure 6–24

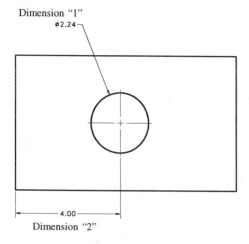

Figure 6–25

Using DDMODIFY on an Associative Dimension

Using the DDMODIFY command on a linear dimension lists the following information: the dimension type; the dimension text; and the current dimension style (see Figure 6–26).

The selected dimension can also be changed to a different dimension style by clicking the down arrow next to Style. Also present in this dialog box are three additional buttons enabling an individual to change various dimension settings. Clicking on the Geometry... button displays the Geometry sub-dialog box of the DDIM command. Clicking on For-mat... displays the Format sub-dialog box of the DDIM command. Clicking on the Annotation... button displays the Annotation sub-dialog box of the DDIM command.

The button labeled Full editor allows the dimension text to be changed or added to an existing dimension. Clicking on the Full editor button displays the Mtext editor.

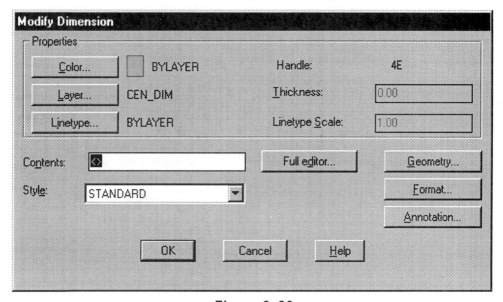

Figure 6–26

Listing the Properties of a Block

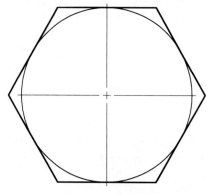

Figure 6–27

Using the LIST command on a block provides you with the following information: the name of the object being listed (Block Reference); the layer the block was inserted on; whether the block occupies model or paper space; the name of the block which in this case is "HEX"; the insertion point of the block in *X,Y,Z* coordinates; the *X,Y,Z* scale factors of the block; and the rotation angle of the block. See Figures 6–27 and 6–28.

```
BLOCK REFERENCE Layer: 0
                       Space: Model space
                       Handle = 61
    HEX
    at point, X=7.38 Y=4.49 Z=0.00
    X scale factor    1.00
    Y scale factor    1.00
    rotation angle    0
    Z scale factor    1.00
```

Figure 6–28

::A Using DDMODIFY on a Block

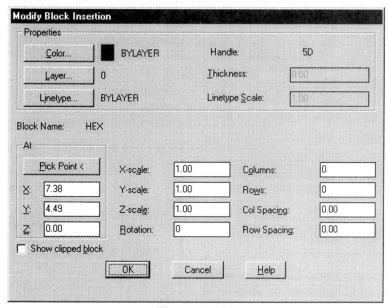

Figure 6–29

Control of blocks is enhanced through the use of the DDMODIFY command which displays the Modify Block Insertion dialog box in Figure 6–29. Using this dialog box allows you to dynamically change each of the following items: the insertion point of the block; the X, Y, Z scale values of the block; and the rotation angle of the block.

Using DDMODIFY to control multiple block insertions is also shown in Figure 6–29. With the insertion point of the block at the center of the Hex, two columns and three rows are specified. Spacings between columns and rows are 2.50 units. This creates the arrangement of hex shapes in Figure 6–30.

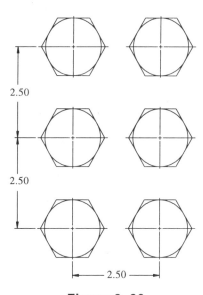

Figure 6–30

Listing the Properties of a Block with Attributes

TRANSITION PIECE

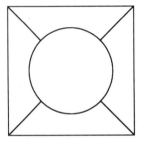

Figure 6–32

Using the LIST command on the block with attribute information in Figure 6–32 provides the following: the name of the object being listed (Block Reference); the layer the block was inserted on; whether the block occupies model or paper space; the name of the block which in this case is "DUCT"; the insertion point of the block in X, Y, Z coordinates; the X, Y, and Z scale factors of the block; and the rotation angle of the block. See Figure 6–31.

In addition to the block information supplied in Figure 6–31, the following attribute information is also supplied: the name of the object being listed (Attribute); the layer the attribute was inserted on; the style of text for the attribute; the font file of the attribute text; the justification of the attribute text; the height of the attribute text; the attribute value; the attribute tag; and the rotation angle, width factor, and obliquing angle of the attribute text.

```
BLOCK REFERENCE  Layer: 0
                 Space: Model space
                 Handle = 2F
     DUCT
     at point, X=6.00 Y=4.50 Z=0.00
     X scale factor 1.00
     Y scale factor 1.00
     rotation angle 0
     Z scale factor 1.00

ATTRIBUTE            Layer: 0
                    Space: Model space
                    Handle = 30
     Style=STANDARD Font file=SIMPLEX
     center point,X=5.99 Y=6.23 Z=0.00
     height 0.20
     value TRANSITION PIECE
     tag PART_NAME
     rotation angle 0
     width scale factor 1.00
     obliquing angle 0
     flags normal
     generation normal
```

Figure 6–31

Using DDMODIFY on a Block with Attributes

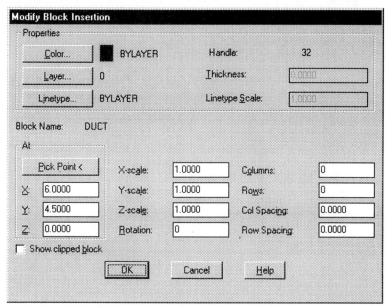

Figure 6–33

Control of blocks with attributes is enhanced similar to blocks without attributes through the use of the DDMODIFY command which displays the dialog box in Figure 6–33. Using this dialog box allows you to dynamically change each of the following items: the insertion point of the block; the *X,Y,Z* scale values of the block; and the rotation angle of the block.

This dialog box also controls multiple inserts of a block by allowing you to change each of the following items: the number of columns; the number of rows; a value for the spacing in between columns; and a value for the spacing in between rows.

With the insertion point of the block at the center of the duct, two columns and two rows are specified. Spacing between rows is 5.00 units while spacing between columns is 4.00 units. The results are shown in Figure 6–34. Notice that since the attribute information is part of the block, it is inserted along with the block at multiple intervals.

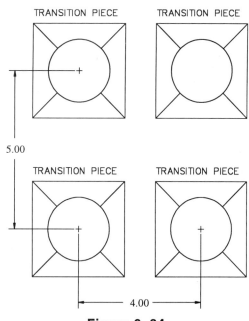

Figure 6–34

Listing the Properties of a Circle

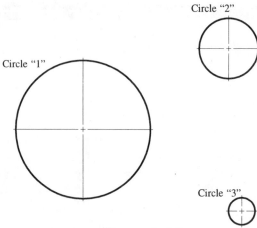

Figure 6–36

Using the LIST command on a circle (Figure 6–35) displays the following information: the name of the object being listed (Circle); the layer the circle was drawn on; whether the circle occupies model or paper space; the center point of the circle; the radius of the circle; the circumference of the circle; and the area of the circle.

As with all circle properties such as center point, radius, radius, circumference, and area, the number of decimal places of accuracy is set by the Units Control dialog box. See also Figure 6–36.

```
                                                Circle "1"
CIRCLE            Layer: 0
                  Space: Model space
                  Handle = 1
center point, X=4.5000 Y=4.5000 Z=0.0000
radius            2.5495
circumference     16.0190
area                          20.4204
                                                Circle "2"
CIRCLE            Layer: 0
                  Space: Model space
                  Handle = 2
center point, X=10.0000 Y=7.5000 Z=0.0000
radius            1.1180
circumference     7.0248
area                          3.9270
                                                Circle "3"
CIRCLE            Layer: 0
                  Space: Model space
                  Handle = 3
center point, X=10.5000 Y=1.5000 Z=0.0000
radius            0.5000
circumference     3.1416
area                          0.7854
```

Figure 6–35

Using DDMODIFY on a Circle

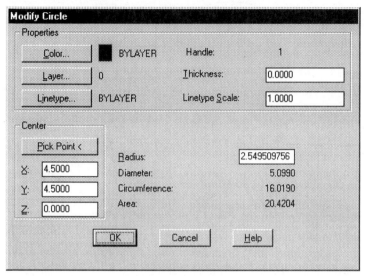

Figure 6–37

Using the DDMODIFY command on a circle lists the identical properties to change as with such objects as lines and arcs (see Figure 6–37). The Color, Linetype, and Layer name buttons all bring up additional dialog boxes to make it easier to change the properties of the object listed. In addition to these properties, the following additional parameters are listed of the circle: center point of the circle; radius of the circle; diameter of the circle; circumference of the circle; and area of the circle.

A new center point may be selected by clicking on the Pick Point< button or by entering a new coordinate value in the appropriate X, Y, and/or Z boxes as in Figure 6–38A. Also, the radius of the circle may be changed by editing the value in the radius box.

The diameter, circumference, and area values are all listed but may not be changed in this dialog box. Notice that when changing the circle radius, the diameter, circumference, and area values update themselves based on the new radius, as shown in Figure 6–38B.

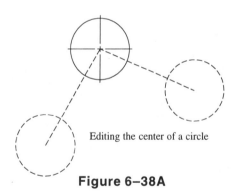

Editing the center of a circle

Figure 6–38A

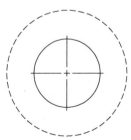

Editing the radius of a circle

Figure 6–38B

Listing the Properties of an Ellipse

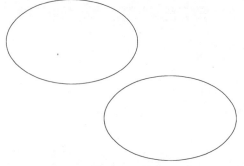

Figure 6–40

Two types of ellipse objects may be constructed inside of Release 14. Illustrated in Figure 6–39A is the information supplied on a true ellipse. Area information is not supplied, although the circumference, center, and major and minor axes of the ellipse are given.

```
ELLIPSE        Layer: 0
               Space: Model space
               Handle = 4F
Circumference: 20.69
Center: X=11.43 , Y=13.00 , Z=0.00
Major Axis: X=4.00 , Y=0.00, Z=0.00
Minor Axis: X=0.00 , Y=2.50 , Z=0.00
Radius Ratio: 0.63
```

Figure 6–39A

The true ellipse generated by the ELLIPSE command in Release 14 may be redrawn as an ellipse consisting of a series of polyarcs joined together into one single lightweight polyline (see Figure 6–40). This is accomplished by first setting the system variable PELLIPSE to a value of 1 or on. After turning on this variable and entering the ELLIPSE command, the ellipse is generated as the lightweight polyline object. Using the LIST command on this object displays each vertex of the polyline. Area and perimeter information on the ellipse is given after scrolling through all vertices. See Figure 6–39B.

```
LWPOLYLINE   Layer: 0
             Space: Model space
             Handle = 22
Closed
Constant width 0.0000

at point X=15.3112 Y=3.4478 Z=0.0000
bulge            0.1474
center X=13.7049 Y=3.4478 Z=0.0000
radius           1.6063
start angle      6.2832
end angle        0.5853
at point X=15.0438 Y=4.3352 Z=0.0000
bulge            0.1071
center X=12.5608 Y=2.6896 Z=0.0000
radius           2.9787
start angle      0.5853
end angle        1.0122
```

Figure 6–39B

Using DDMODIFY on an Ellipse

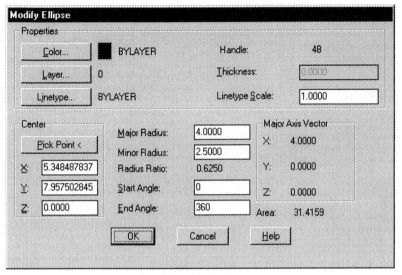

Figure 6–41

Using the DDMODIFY command on an ellipse lists the usual object properties such as color, linetype, layer name, and object thickness (see Figure 6–41). Since the object is considered a true ellipse, you have the ability to change the center point of the ellipse. You can make additional changes to the major radius, minor radius, start angle, and end angle. The area of the ellipse is also listed.

Illustrated in Figure 6–42A are the effects of Object Snap modes on a true ellipse. Snap points exist at the four quadrants of the ellipse in addition to the center of the ellipse.

Illustrated in Figure 6–42B are the effects of Object Snap modes on an ellipse generated as a series of polyarcs when the system variable PELLIPSE is turned on. Snap points exist at every vertice of the lightweight polyline.

True Ellipse Object Snap Points

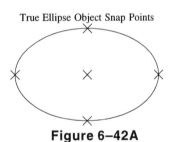

Figure 6–42A

Polyline Ellipse Object Snap Points

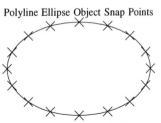

Figure 6–42B

Listing the Properties of a Hatch Pattern

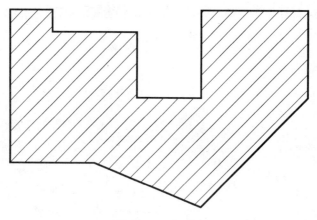

Figure 6–44

Using the LIST command on a hatch pattern (Figure 6–44) displays the following information: the name of the object being listed, (HATCH); the layer the hatch pattern was drawn on; whether the hatch pattern occupies model or paper space; the name of the hatch pattern being used; the scale of the hatch pattern; the angle of the hatch pattern; and if the hatch pattern is associative or not. See Figure 6–43.

```
HATCH          Layer: 0
               Space: Model space
               Handle=65
Hatch pattern ANSI31
Hatch scale (or space for User pattern) 2.0000
Hatch angle 0
Associative
```

Figure 6–43

Using DDMODIFY on an Hatch Pattern

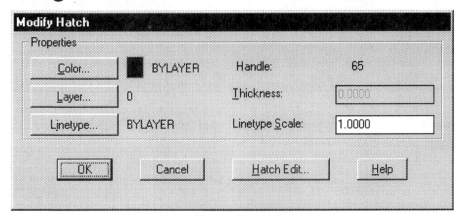

Figure 6–45

Using the DDMODIFY command on a hatch pattern allows you to make changes to such items as the color of the hatch pattern, the layer on which the hatch pattern was drawn, and the linetype of the hatch pattern (see Figure 6–45).

A special button is present that allows you to edit the current hatch pattern without erasing the current pattern and to construct it again to the correct parameters.

From the illustration of the hatch pattern in Figure 6–44, it has been determined that the hatch pattern scale needs to be increased to a new value of 3.0 units. Also, the angle of the hatching needs to be changed to 135 degrees. Since the hatch pattern is already designed to be drawn at a 45-degree angle, rotating the current pattern by 90 degrees will result in the new pattern being drawn at 135 degrees. When clicking on the "Hatch Edit..." button, the Hatchedit dialog box appears in Figure 6–46. Notice the hatch scale has been changed to 3.0000 units and the hatch angle changed to 90 degrees.

When you exit the dialog box and return to the drawing, the result is illustrated in Figure 6–47.

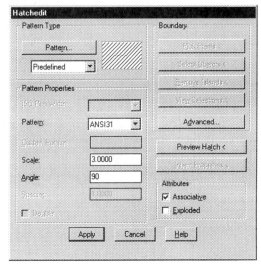

Figure 6–46

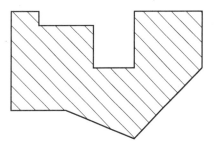

Figure 6–47

Listing the Properties of a Line

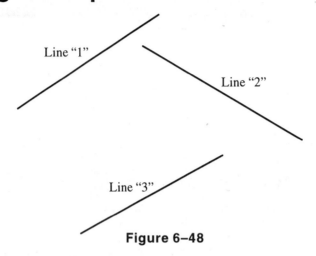

Line "1"

Line "2"

Line "3"

Figure 6–48

Using the LIST command on a line lists the following information: the specific specific object being listed (LINE); the layer the line segment was drawn on; whether the line occupies model or paper space; the starting endpoint of the line segment; the ending endpoint of the line segment; the length of the line segment; the angle of the line segment in the X-Y plane; and the relative coordinate value from the starting point to the ending point of the line segment.

Study Figures 6–48 and 6–49 to isolate the information on the individual line objects. The number of decimal places for the beginning and ending points of the line in addition to the length of the line are all governed by the Units Control dialog box.

                                   Line "1"
LINE   Layer: 0
               Space: Model space
               Handle = 22
from point, X=1.00 Y=5.00 Z=0.00
to point, X=5.50 Y=8.00 Z=0.00
Length=5.41, Angle in XY Plane=34
Delta X=4.50, Delta Y=3.00, Delta Z=0.00

                                   Line "2"
LINE   Layer: 0
               Space: Model space
               Handle = 23
from point, X=5.00 Y=7.00 Z=0.00
to point, X=10.00 Y=4.00 Z=0.00
Length=5.83, Angle in XY Plane=329
Delta X=5.00, Delta Y=-3.00, Delta Z=0.00

                                   Line "3"
LINE   Layer: 0
               Space: Model space
               Handle = 24
from point, X=3.00 Y=1.00 Z=0.00
to point, X=7.50 Y=3.50 Z=0.00
Length=5.15, Angle in XY Plane=29
Delta X=4.50, Delta Y=2.50, Delta Z=0.00

Figure 6–49

Using DDMODIFY on a Line

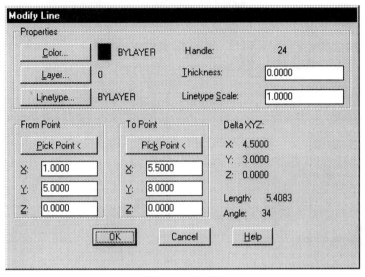

Figure 6–50

Using the DDMODIFY command on a line segment as in Figure 6–50 lists the following information: the X, Y, Z coordinates of the starting of the line; the X, Y, Z coordinates of the end of the line; the delta X, Y, Z coordinate value of the line; the total length of the line segment; and the angle the line segment makes in the X-Y plane.

The X, Y, Z values in the dialog box above may be changed to affect the beginning or end of the line segment. When any of these values change, the delta X, Y, Z, length, and angle values update themselves to the new values of the line segment.

Figure 6–51A shows how the length and angle are calculated. Angles will always be calculated in the counterclockwise direction based on the current setting in the Units control dialog box.

The Delta XY value in Figure 6–51B shows the horizontal and vertical distances from the beginning of the line segment to the end of the line segment.

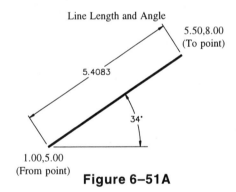

Line Length and Angle

Figure 6–51A

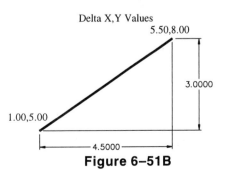

Delta X,Y Values

Figure 6–51B

Listing the Properties of a Point

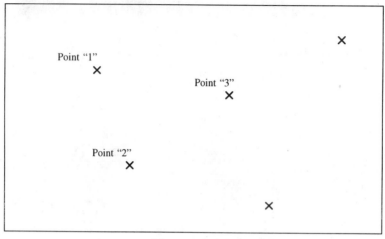

Figure 6–53

Using the LIST command on a point provides you with the following information: the name of the object being listed (POINT); the layer the point was drawn on; whether the point occupies model or paper space; and the coordinates of the point in *X,Y,Z* values See Figure 6–52.

As with all points, the size and appearance is controlled by the system variables PDSIZE and PDMODE. In Figure 6–53, a PDMODE value of 3 has been specified displaying all points in the appearance of an X.

Study the typical screens in Figure 6–52 and see how they relate to the points in the illustration in Figure 6–53.

	Point "1"
POINT	Layer: 0
	Space: Model space
	Handle = 22
at point, X=2.5482 Y=6.9289 Z=0.0000	

	Point "2"
POINT	Layer: 0
	Space: Model space
	Handle = 24
at point, X=3.5675 Y=4.1313 Z=0.0000	

	Point "3"
POINT	Layer: 0
	Space: Model space
	Handle = 23
at point, X=6.5819 Y=6.2133 Z=0.0000	

Figure 6–52

Using DDMODIFY on a Point

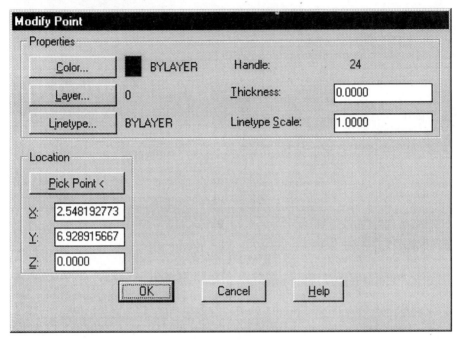

Figure 6–54

Using the DDMODIFY command on a point lists the usual object properties such as color, linetype, layer name, and thickness.

This command also lists the following information: the *X, Y, Z* coordinates of the point; and the current handle assigned to the point.

Notice in Figure 6–54 the appearance of the *X, Y, Z* coordinate values located in edit boxes. A new value may be entered in one of these boxes which will change the location of the point.

If the absolute coordinates of a point to move are not known, the Pick Point< button may be used to locate the point with the current pointing device such as a digitizing puck or mouse. Object Snap modes are usually used to locate the point on an existing object.

Listing the Properties of a Lightweight Polyline

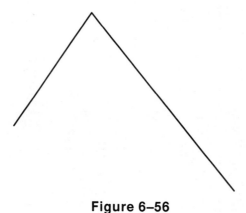

Figure 6–56

Using the LIST command on the lightweight polyline in Figure 6–56 lists the absolute coordinate values of each vertex of the polyline along with the following information: the name of the object being listed (LWPOLYLINE); the layer the polyline was drawn on; whether the lightweight polyline occupies model or paper space; the specific vertex of the lightweight polyline being listed in absolute coordinates; and the absolute coordinate value at the vertex. See Figure 6–55.

Using the EXPLODE command on a lightweight polyline separates the object into individual line segments.

The LIST command also calculates the area occupied by a closed or open lightweight polyline. For an open lightweight polyline, the total length of the polyline is given. If the lightweight polyline was closed, the perimeter of the object would be calculated.

```
LWPOLYLINE   Layer: 0
             Space: Model space
             Handle = 29
   Open
   Constant width 0.0000

   at point X=3.0000 Y=3.5000 Z=0.0000
   at point X=5.5000 Y=7.0000 Z=0.0000
   at point X=10.0000 Y=1.5000 Z=0.0000

   area 14.7500
   length 11.4075
```

Figure 6–55

Using DDMODIFY on a Lightweight Polyline

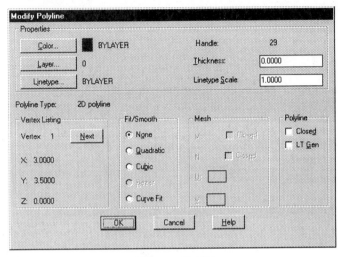

Figure 6–57

Using the DDMODIFY command on a lightweight polyline (Figure 6–57) lists the usual object properties along with the following information: the X, Y, Z coordinates of each vertex; the type of curve fitted to the polyline; whether the lightweight polyline is closed or open; and whether to generate linetype scaling per vertex.

By default, a polyline has normal curve generation which means it is absent of any curves. Selecting Quadratic, Cubic, and Fit Curve displays the results similar to the illustration in Figure 6–58.

Lightweight polylines have been designed to improve drawing performance especially if the drawing is large. When converting the lightweight polyline into a spline, the object is also converted into a 2D polyline object from Release 13. This operation could affect drawing efficiency.

The LT Gen option stands for linetype generation. If LT Gen is unselected in the DDMODIFY dialog box, or is set to off, the linetype is applied to each individual vertex. In Figure 6–59, with the

LTSCALE command set to 1.10, one leg of the polyline has a single centerline while the other leg has three centerlines. Checking the LT Gen box above in DDMODIFY turns LT Gen on with the results at the far right. Here, the first leg of the polyline has two centerlines that continue into the second leg.

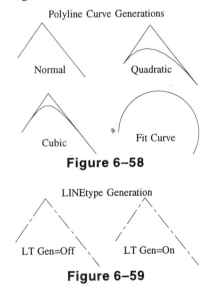

Figure 6–58

Figure 6–59

Listing the Properties of a Polygon

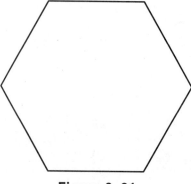

Figure 6-61

Using the LIST command on the polygon object (Figure 6-61) lists the absolute coordinate values of each vertex of the polygon along with the following information: the name of the object being listed (LWPOLYLINE); the layer the polygon was drawn on; whether the polygon occupies model or paper space; the specific vertex of the polygon being listed in absolute coordinates; and the absolute coordinate value at the vertex. See also Figure 6-60.

The LIST command also calculates the area and perimeter of the polygon since it is closed.

```
LWPOLYLINE        Layer: 0
                  Space: Model space
                  Handle = 24
Closed
Constant width 0.0000

at point X=8.4783 Y=4.9325 Z=0.0000
at point X=6.9783 Y=7.5306 Z=0.0000
at point X=3.9783 Y=7.5306 Z=0.0000
at point X=2.4783 Y=4.9325 Z=0.0000
at point X=3.9783 Y=2.3345 Z=0.0000
at point X=6.9783 Y=2.3345 Z=0.0000

area 23.3827
length 18.0000
```

Figure 6-60

Using DDMODIFY on a Polygon

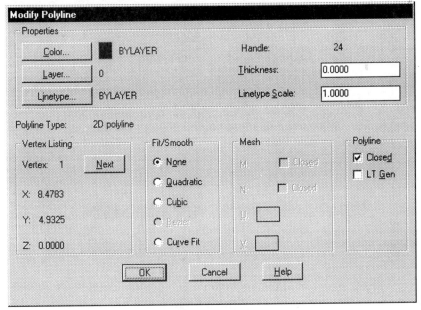

Figure 6–62

Using the DDMODIFY command on a polygon lists the usual object properties such as color, linetype, layer name, and object thickness (Figure 6–62). This command also lists the following additional information: the X,Y,Z coordinates of each polygon vertex; the type of curve fitted to the polygon; whether the polygon is closed or open; and whether to generate linetype scaling per vertex.

In Figure 6–63 are examples of an inscribed polygon (inside of a circle) and a circumscribed polygon (around a circle).

If LT Gen is unselected in the DDMODIFY dialog box, or is set to off, the linetype is applied to each individual vertex. In Figure 6–64, with the LTSCALE command set to 1.10, each leg of the polyline has a single centerline. Checking the LT Gen box in Figure 6–62 in DDMODIFY turns LT Gen on with the results in Figure 6–64 (right) where the linetype is distributed throughout the entire polygon.

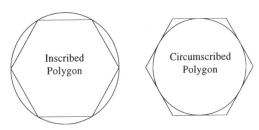

Figure 6–63

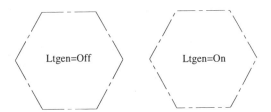

Figure 6–64

▤ Listing the Properties of a Ray

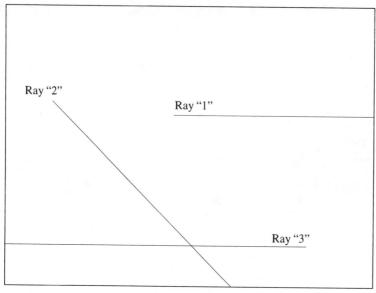

Figure 6–66

Using the LIST command on a ray (Figure 6–66) lists the name of the object being listed (RAY); the layer the ray was drawn on; and whether the ray occupies model or paper space. See also Figure 6–65.

The base point of the ray is given in absolute coordinates. This point identifies the beginning or start of the ray before it is drawn into infinity. The unit direction of the ray defines a point signifying the direction the ray is drawn in absolute coordinates. The angle the ray forms in the XY plane is also given.

In Figure 6–66, the ray identified at "1" begins at 10.7062,10.5078. Notice that the ray is drawn directly to the right in an infinite direction at an angle of 0 degrees.

Ray "1"

RAY   Layer: 0
         Space: Model space
         Handle = 30
base point, X=10.7062 Y=10.5078 Z=0.0000
unit direction, X=1.0000 Y=0.0000 Z=0.0000
angle in XY plane=0

Ray "2"

RAY   Layer: 0
         Space: Model space
         Handle = 32
base point, X=3.0350 Y=11.3261 Z=0.0000
unit direction, X=0.7071 Y=-0.7071 Z=0.0000
angle in XY plane=315

Ray "3"

RAY   Layer: 0
         Space: Model space
         Handle = 31
base point, X=19.0477 Y=2.5107 Z=0.0000
unit direction, X=-1.0000 Y=0.0000 Z=0.0000
angle in XY plane=180

Figure 6–65

Using DDMODIFY on a Ray

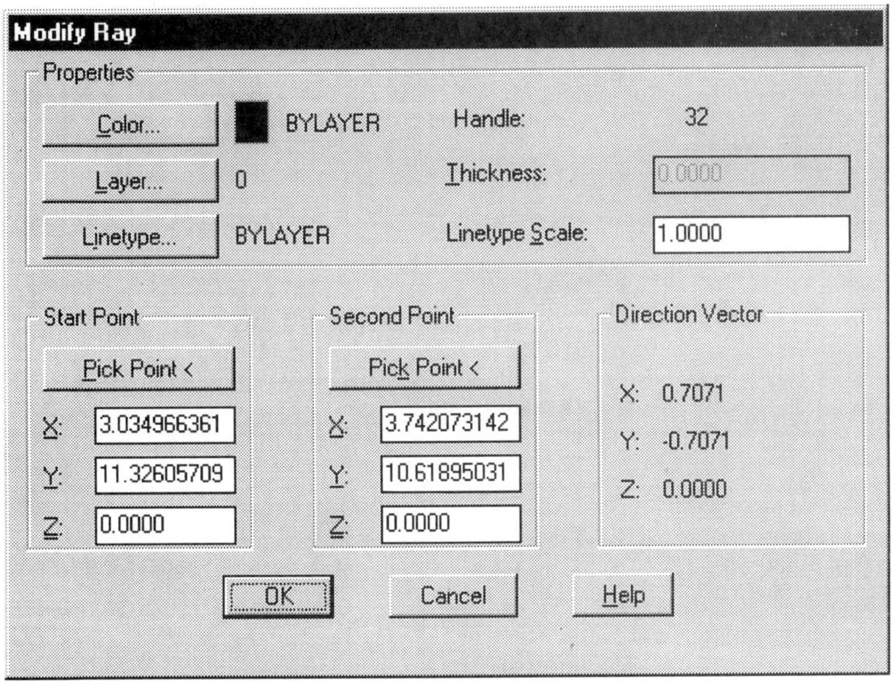

Figure 6–67

Using the DDMODIFY command on a ray lists the usual object properties such as color, linetype, layer name, and object thickness (see Figure 6–67).

This command also lists the following additional information: the starting and second points of the ray in addition to the current direction vector. All of these parameters are given in absolute coordinates.

In Figure 6–66, Ray "2" appears in Figure 6–67 in the dialog box. The start point of the ray is 3.0350,11.3261. The second point determines the direction vector calculation; since the current direction vector for this ray is 0.7071,-0.7071, it is drawn in a -45-degree angle off to the right and down.

Changing X and Y values for the start and second points will update the ray to the new start and direction vector.

Listing the Properties of a Multiline

Figure 6–69

Using the LIST command on an Mline (Figure 6–69) lists the name of the object being listed (MLINE); the layer the Mline was drawn on; whether the Mline occupies model or paper space; and the justification, scale, and style of the Mline. Vertice information in absolute coordinates identifies each segment of the Mline. See Figure 6–68.

In Figure 6–70, the DDMODIFY dialog box displays information such as the Mline style; however, this dialog box prevents the modification of the Mline at its individual vertice locations.

```
MLINE            Layer: 0
                 Space: Model space
                 Handle = 25

Justification = Top, Scale = 1.00,
Style = FDN-WALLS
Vertex 0: X=-103.4876 Y=20.7017 Z=0.0000
Vertex 1: X=-49.1857 Y=20.7017 Z=0.0000
Vertex 2: X=-49.1857 Y=-23.0983 Z=0.0000
Vertex 3: X=-13.6275 Y=-23.0983 Z=0.0000
```

Figure 6–68

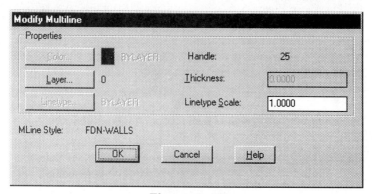

Figure 6–70

Listing the Properties of a Region

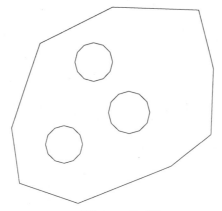

Figure 6–72

As shown in Figure 6–71, using the LIST command on a region lists the name of the object being listed (REGION); the layer the region was drawn on; whether the region occupies model or paper space; the area and perimeter of the region; and the bounding box of the region given in absolute coordinates. See also Figure 6–72.

In Figure 6–73, the DDMODIFY dialog box displays information such as the current linetype scale; however, this dialog box prevents the modification of the region.

REGION	Layer: 0
	Space: Model space
	Handle = 128
Area: 132.9654	
Perimeter: 71.9731	
Bounding Box:	
Lower Bound X=28.0067, Y=8.1002, Z=0.0000	
Upper Bound X=42.4143, Y=21.7554, Z=0.0000	

Figure 6–71

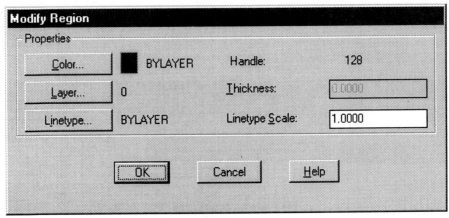

Figure 6–73

Listing the Properties of Mtext

CHAMFER
Text "1"

FILLET
Text "2"

Figure 6–75

As shown in Figure 6–74, using the LIST command on mtext displays the following information: the name of the object being listed (MTEXT); the layer the mtext was drawn on; whether the mtext occupies model or paper space; the location or starting point for the mtext; the width of the mtext box; the normal vector on how the mtext is being viewed; the rotation angle for the mtext; the current text style of mtext; the attachment or justification mode for the mtext object; the flow direction control for the mtext object; and the contents of the mtext object. See also Figure 6–75.

In Figure 6–74, the width scale factor, obliquing angle, and generation of text are all controlled by the STYLE command when the text font was originally associated with a text style.

	Text "1"
MTEXT	Layer: 0
	Space: Model space
	Handle = 2D
Location:	X=10.7798 Y=3.3793
Z=0.0000	
Width:	7.4813
Normal:	X=0.0000 Y=0.0000 Z=1.0000
Rotation:	0
Text style:	STANDARD
Text height:	1.0000
Attachment:	MiddleCenter
Flow direction:	ByStyle
Contents:	CHAMFER

	Text "2"
MTEXT	Layer: 0
	Space: Model space
	Handle = 2B
Location:	X=7.8996 Y=1.8236 Z=0.0000
Width:	7.0750
Normal:	X=0.0000 Y=0.0000 Z=1.0000
Rotation:	0
Text style:	ITALICC
Text height:	1.0000
Attachment:	TopLeft
Flow direction:	ByStyle
Contents:	FILLET

Figure 6–74

Using DDMODIFY on Mtext

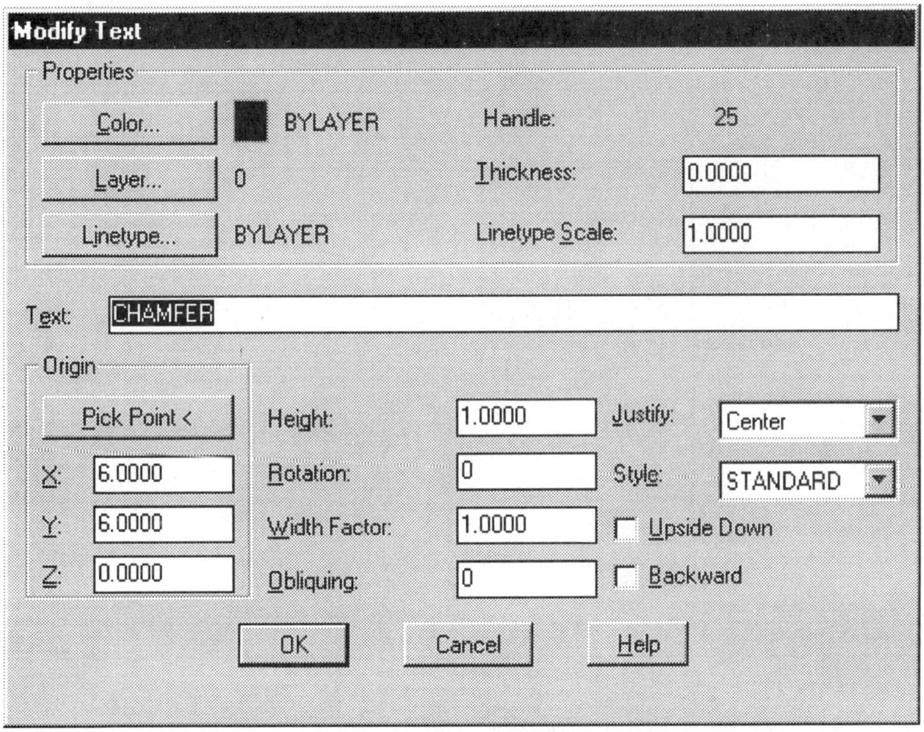

Figure 6–76

Using the DDMODIFY command on a mtext object provides for superior control of text (see Figure 6–76). All listings appearing in a box may be dynamically changed to affect the final form of the mtext object. In addition to the usual object properties that may be changed such as color, linetype, layer name, and thickness, the following may be changed using this dialog box. You may edit the actual text in a way similar to the DDEDIT command; you may select a new text origin point; enter a new text height; enter a new rotation angle for the text; enter a new width factor for the text; enter an obliquing angle to make text inclined; select a new justification position for the text; and select a new text style.

Keep in mind that all of these changes only apply to the text object selected and does not globally affect all text objects.

Clicking in the Style: edit box displays a pull-down menu consisting of all text styles currently defined in the drawing. Use this area to select a different text style for the text object just picked.

Selecting the Justify: edit box displays a pull-down menu consisting of all valid text justification positions, and it allows you to scroll up or down to select a new justification position.

Listing the Properties of an Xline

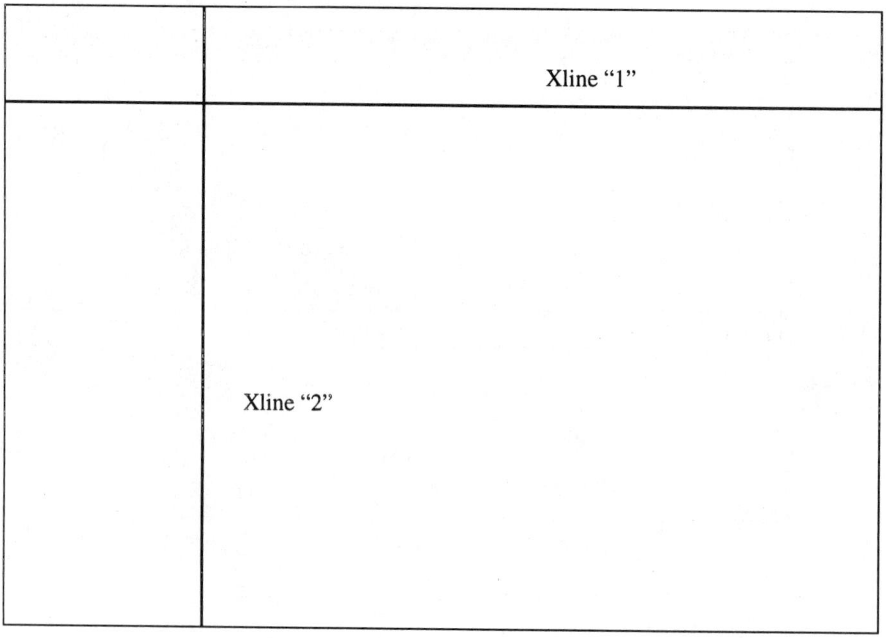

Figure 6–78

Using the LIST command on an Xline lists the name of the object being listed (XLINE); the layer the Xline was drawn on; and whether the Xline occupies model or paper space (see Figure 6–77).

The base point of the Xline is given in absolute coordinates. This point identifies the beginning or start of the Xline before it is drawn into infinity. The unit direction of the Xline defines a point signifying the direction the Xline is drawn in absolute coordinates. The angle the Xline forms in the X,Y plane is also given.

In Figure 6–78, the Xline identified at "2" begins at 5.7534,7.3089. Notice that even as the Xline is drawn straight up and down in infinite directions, an angle of 90 degrees is calculated in the X,Y plane.

```
                    Xline "1"
XLINE               Layer: 0
                    Space: Model space
                    Handle = 28
base point, X=10.2593 Y=11.2145 Z=0.0000
unit direction, X=1.0000 Y=0.0000 Z=0.0000
angle in XY plane = 0

                    Xline "2"
XLINE               Layer: 0
                    Space: Model space
                    Handle = 2A
base point, X=5.7534 Y=7.3089 Z=0.0000
unit direction, X=0.0000 Y=1.0000 Z=0.0000
angle in XY plane = 90
```

Figure 6–77

Using DDMODIFY on an Xline

Modify Xline

Properties

Color... ▉ BYLAYER   Handle: 28

Layer... 0   Thickness: 0.0000

Linetype... BYLAYER   Linetype Scale: 1.0000

Root Point
Pick Point <
X: 10.25930350
Y: 11.21447032
Z: 0.0000

Second Point
Pick Point <
X: 11.25930350
Y: 11.21447032
Z: 0.0000

Direction Vector
X: 1.0000
Y: 0.0000
Z: 0.0000

OK   Cancel   Help

Figure 6–79

Using the DDMODIFY command on an Xline lists the usual object properties such as color, linetype, layer name, and object thickness. See Figure 6–79.

This command also lists the following additional information: the root and second points of the Xline in addition to the current direction vector. All of these parameters are given in absolute coordinates.

In Figure 6–78, Xline "1" appears in Figure 6–79 in the dialog box. The root or starting point of the

Xline is 10.2593,11.2145. Notice the second point is identical to the root point. Since this Xline was started with a point and forced to be drawn horizontal, both root and second points share the same absolute coordinate values. As the Xline is drawn infinitely in two directions from the root point, it still is identified with the direction vector of 1,0,0.

As with the ray, changing X and Y values for the root point will update the Xline to the new root location.

Using the STATUS Command

```
Command: STATUS
31 objects in C:\ACADR13\WIN\3-15
Model space limits are    X: 0.0000      Y: 0.0000 (Off)
                          X: 12.0000     Y: 9.0000
Model space uses          X: 5.0017      Y: 2.9034
                          X: 9.4660      Y: 7.3677
Display shows             X: -1.8575     Y: -1.7949
                          X: 16.2750     Y: 16.4344
Insertion base is         X: 0.0000      Y: 0.0000  Z: 0.0000
Snap resolution is        X: 1.0000      Y: 1.0000
Grid spacing is           X: 0.0000      Y: 0.0000

Current space:            Model space
Current layer:            0
Current color:            BYLAYER -- 7 (white)
Current linetype:         BYLAYER -- CONTINUOUS
Current elevation:        0.0000 thickness: 0.0000
Fill on Grid off Ortho off Qtext off Snap off Tablet off
Object snap modes:        None
Free disk (dwg+temp=C:):143032320 bytes
Free physical memory: 0.2 Mbytes (out of 13.6M).
Free swap file space: 29.8 Mbytes (out of 50.2M).
Virtual address space: 25.8 Mbytes.
```

Figure 6–80

Once inside of a large drawing, sometimes it becomes difficult to keep track of various settings that have been changed from their default values to different values required by the drawing. To obtain a listing of these important settings contained in the current drawing file, use the STATUS command. Once the STATUS command is invoked, the graphics screen changes to a text screen displaying all of the information shown in Figure 6–80. The following information is supplied by the STATUS command:

- The number of objects contained in the current drawing file.
- The current Model space limits set by the LIMITS command.
- The area used by Model space.
- The coordinates of the current drawing display.
- The current insertion base point of the drawing.

- The current snap resolution value.
- The current grid spacing.
- The current space occupied by the drawing (Model or Paper).
- The current layer of the drawing.
- The current color.
- The current linetype.
- The current elevation and thickness of the drawing.
- Whether the following modes are on or off: Fill, Grid, Ortho, Qtext, Snap, Tablet
- The current running Object Snap modes currently activated.

The remaining lines appearing after the Object snap modes deal with the operating system of the central processing unit of the computer used to generate the current drawing.

Using the TIME Command

```
Current time:              Friday, June 23, 1995 at 10:54:34:360 PM
Times for this drawing:
Created:                   Sunday, January 29, 1995 at 4:24:51:480 PM
Last updated:              Sunday, January 29, 1995 at 5:05:50:330 PM
Total editing time:        0 days 00:23:53.780
Elapsed timer (on):        0 days 00:23:53.780
Next automatic save in:    <no modifications yet>
```

Figure 6–81

The TIME command (Figure 6–81) provides the operator with the following information: the current date and time; the date and time the drawing was created; the last time the drawing was updated; the total time spent editing the drawing so far; the total time spent while in AutoCAD, not necessarily in a particular drawing; and the current automatic save time interval.

Current Time

Displays the current date and time.

Created

This date and time value is set whenever using the NEW command for creating a new drawing file. This value is also set to the current date and time whenever a drawing is saved under a different name using the SAVE or SAVEAS commands.

Last Updated

This data consists of the date and time the current drawing was last updated. This value updates itself whenever using the SAVE command or the END command.

Total Editing Time

This represents the total time spent editing the drawing. The timer is always updating itself and cannot be reset to a new or different value.

Elapsed Timer

This timer runs while AutoCAD is in operation and you can turn it on or off or reset it.

Next Automatic Save In

This timer displays when the next automatic save will occur. This value is controlled by the system variable SAVETIME. If this system variable is set to zero, the automatic save utility is disabled. If the timer is set to a nonzero value, the timer displays when the next automatic save will take place. The increment for automatic saving is in minutes.

Tutorial Exercise
Extrude.Dwg

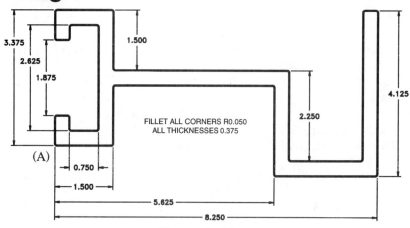

Figure 6–82

Purpose

This tutorial is designed to show you the various methods to construct the extruded pattern in Figure 6–82. The surface area of the extrusion will also be found using the AREA command.

System Settings

Keep the default drawing limits at 0.0000,0.0000 for the lower left corner and 12.0000,9.0000 for the upper right corner. Use the UNITS command and change the number of decimal places past the zero from four units to three units.

Layers

No special layers need to be created for this drawing although it is always good practice to create and draw on a separate layer for all object lines.

Suggested Commands

Begin drawing the extrusion with point "A" in Figure 6–82 at absolute coordinate 2.000,3.000.

Use either of the following methods to construct the extrusion:
1. Using a series of absolute, relative, and polar coordinates to construct the profile of the extrusion.
2. Construct a few lines; then use the OFFSET command followed by the TRIM command to construct the extrusion profile.

The FILLET command is used to create the 0.050 radius rounds at all corners of the extrusion. Before calculating the area of the extrusion, convert and join all objects into one single polyline. This will allow the AREA command to be used in a more productive way. Do not dimension this drawing.

Whenever possible, substitute the appropriate command alias in place of the full AutoCAD command in each tutorial step; for example, use "Co" for the COPY command, "L" for the LINE command, and so on. The complete listing of all command aliases is located on pages 1–9 and 1–10.

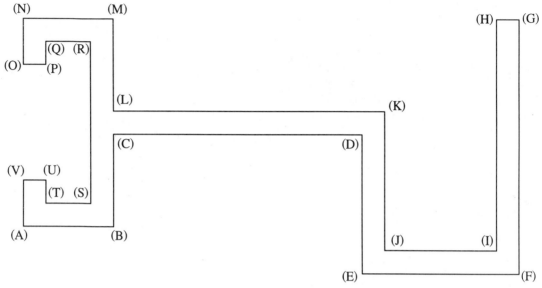

Figure 6–83

Step #1

One method of constructing the extrusion in Figure 6–83 is to use the measurements on the previous page to calculate a series of polar coordinate system distances. The direct distance mode could also be used to accomplish this step.

 Command: **LINE**
From point: **2.000,3.000** *(Starting at "A")*
To point: **@1.500<0** *(To "B")*
To point: **@1.500<90** *(To "C")*
To point: **@4.125<0** *(To "D")*
To point: **@2.250<270** *(To "E")*
To point: **@2.625<0** *(To "F")*
To point: **@4.125<90** *(To "G")*
To point: **@0.375<180** *(To "H")*
To point: **@3.750<270** *(To "I")*
To point: **@1.875<180** *(To "J")*
To point: **@2.250<90** *(To "K")*

To point: **@4.500<180** *(To "L")*
To point: **@1.500<90** *(To "M")*
To point: **@1.500<180** *(To "N")*
To point: **@0.750<270** *(To "O")*
To point: **@0.375<0** *(To "P")*
To point: **@0.375<90** *(To "Q")*
To point: **@0.750<0** *(To "R")*
To point: **@2.625<270** *(To "S")*
To point: **@0.750<180** *(To "T")*
To point: **@0.375<90** *(To "U")*
To point: **@0.375<180** *(To "V")*
To point: **Close** *(Back to "A")*

Step #2

All corners come together at 90-degree intersections. From the original dimensions of the extrusion, a note calls out that all corners be rounded off with a 0.050 radius. This is easily accomplished using the FILLET command. However, since so many corners need to be filleted, the risk is high of forgetting to fillet one or more corners. All corners may be filleted at one time only if the entire extrusion consists of one polyline. First, use the PEDIT command to perform this conversion followed by the FILLET command.

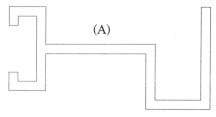

Figure 6–84

 Command:**PEDIT**

Select polyline: *(Select the object labeled "A" in Figure 6–84)*
Object selected is not a polyline.
Do you want to turn it into one? <Y>: *(Press* [ENTER] *to continue)*
Close/Join/Width/Edit vertex/Fit/Spline/ Decurve/Ltype gen/Undo/eXit <X>:**Join**
Select objects: **0,0**
Other corner: **12,9**
Select objects:*(Press* [ENTER] *to continue)*
21 segments added to polyline
Open/Join/Width/Edit vertex/Fit/Spline/ Decurve/Ltype gen/Undo/eXit <X>: *(Press* [ENTER] *to exit this command)*

Rather than converting individual objects into one polyline using the PEDIT command and Join option, a more efficient means of creating a polyline is through the use of the BOUNDARY command shown in the dialog box in Figure 6–85. Picking an internal point automatically traces a polyline around a closed shape in the color of the current. It must be emphasized that the shape must be completely closed for the BOUNDARY command to function correctly.

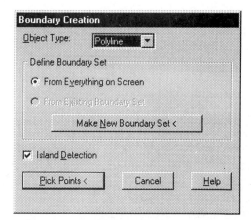

Figure 6–85

In Figure 6–86 of the Extrusion, issuing the
BOUNDARY command and clicking on the Pick
Points < button of the dialog box prompts you to
pick an internal point. Selecting a point inside of
the Extrusion at "A" traces the polyline in the current layer. Turning off the layer containing the individual objects leaves the polyline to perform various calculations.

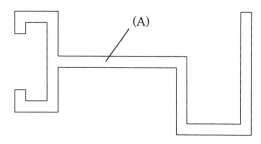

Figure 6–86

Step #3

With the entire extrusion converted into a polyline,
use the FILLET command, set a radius of 0.050,
and use the polyline option of the FILLET command to fillet all corners of the extrusion at once.
See Figure 6–87.

 Command: **FILLET**

(TRIM mode) Current fillet radius = 0.500
Polyline/Radius/Trim/<Select first object>: **R**
Enter fillet radius <0.500>: **0.050**

 Command: **FILLET**

(TRIM mode) Current fillet radius = 0.050
Polyline/Radius/Trim/<Select first object>: **P**
Select 2D polyline: *(Select the polyline in
   Figure 6–87)*
22 lines were filleted

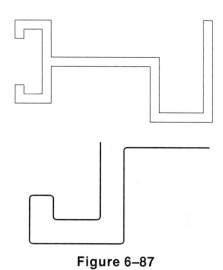

Figure 6–87

Checking the Accuracy of Extrude.Dwg

Once the extrusion has been constructed, answer the question to determine the accuracy of the drawing in figure 6–88.

Question #1

The total surface area of the extrusion is closest to

    (A) 7.020
    (B) 7.070
    (C) 7.120
    (D) 7.170
    (E) 7.220

Use the AREA command to calculate the surface area of the extrusion. This is easily accomplished since the extrusion has already been converted into a polyline.

 Command: **AREA**

<First point>/Object/Add/Subtract: **Object**
Select circle or polyline: *(Select any part of the extrusion illustrated below.)*
Area = 7.170,   Length = 38.528

Total surface area of the extrusion is "D", <u>7.170</u>.

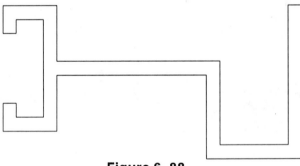

Figure 6–88

Tutorial Exercise
C-Lever.Dwg

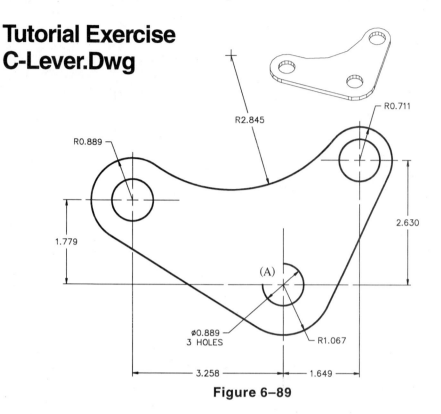

Figure 6–89

Purpose

This tutorial is designed to show you various methods to construct the c-lever object in Figure 6–89. Numerous questions will be asked about the object requiring the use of a majority of Inquiry commands.

System Settings

Keep the default drawing limits at 0.0000,0.0000 for the lower left corner and 12.0000,9.0000 for the upper right corner. Use the UNITS command and change the number of decimal places past the zero from four units to three units.

Layers

No special layers need to be created for this drawing although it is always good practice to create and draw on a separate layer for all object lines.

Suggested Commands

Begin drawing the c-lever with point "A" illustrated in Figure 6–89 at absolute coordinate 7.000,3.375. Begin laying out all circles. Then draw tangent lines and arcs. Use the TRIM command to clean up unnecessary objects. To prepare to answer the AREA command question, convert the profile of the c-lever into a polyline using the PEDIT command. Other questions pertaining to distances, angles, and point identifications follow. Do not dimension this drawing.

Whenever possible, substitute the appropriate command alias in place of the full AutoCAD command in each tutorial step; for example, use "Co" for the COPY command, "L" for the LINE command, and so on. The complete listing of all command aliases is located on pages 1–9 and 1–10.

Step #1

Construct one circle of 0.889 diameter with the center of the circle at absolute coordinate 7.000,3.375 (see Figure 6–90). Construct the remaining circles of the same diameter by using the COPY command with the Multiple option. Use of the @ symbol for the base point in the COPY command identifies the last known point, which in this case is the center of the first circle drawn at coordinate 7.000,3.375.

 Command: **CIRCLE**

3P/2P/TTR/<Center point>:**7.000,3.375**
Diameter/<Radius>: **D** *(For Diameter)*
Diameter:**0.889**

 Command: **COPY**

Select objects: **L** *(For Last)*
Select objects:*(Press* ENTER *to continue)*
<Base point or displacement>/Multiple: **M** *(For Multiple)*
Base point: **@**
Second point of displacement: **@1.649,2.630**
Second point of displacement:
@-3.258,1.779
Second point of displacement:*(Press* ENTER *to exit this command)*

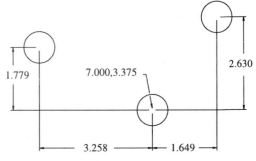

Figure 6–90

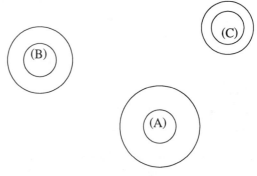

Figure 6–91

Step #2

Construct three more circles (see Figure 6–91). Even though these objects actually represent arcs, circles will be drawn now and trimmed later to form the arcs.

 Command: **CIRCLE**

3P/2P/TTR/<Center point>: **Cen**
of *(Select the edge of circle "A.")*
Diameter/<Radius>:**1.067**

 Command: **CIRCLE**

3P/2P/TTR/<Center point>: **Cen**
of *(Select the edge of circle "B.")*
Diameter/<Radius>:**0.889**

 Command: **CIRCLE**

3P/2P/TTR/<Center point>: **Cen**
of *(Select the edge of circle "C.")*
Diameter/<Radius>:**0.711**

Step #3

Construct lines tangent to the three outer circles as shown in Figure 6–92.

 Command: **LINE**

From point: **Tan**
to *(Select the outer circle near "A.")*
To point: **Tan**
to *(Select the outer circle near "B.")*
To point: *(Press* ENTER *to exit this command)*

 Command: **LINE**

From point: **Tan**
to *(Select the outer circle near "C.")*
To point: **Tan**
to *(Select the outer circle near "D.")*
To point: *(Press* ENTER *to exit this command)*

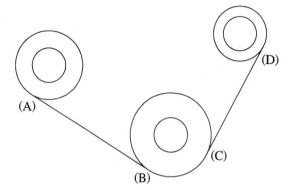

Figure 6–92

Step #4

Construct a circle tangent to the two circles in Figure 6–93 using the CIRCLE command with the Tangent-Tangent-Radius option (TTR).

 Command: **CIRCLE**

3P/2P/TTR/<Center point>: **TTR**
Enter Tangent spec: *(Select the outer circle near "A.")*
Enter second Tangent spec: *(Select the outer circle near "B.")*
Radius: **2.845**

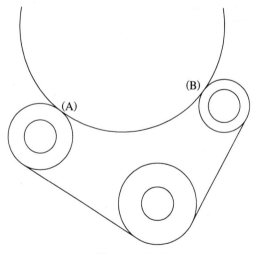

Figure 6–93

Step #5

Use the TRIM command to clean up and form the
finished drawing. Select all of the objects represented by dashed lines as cutting edges as shown
in Figure 6–94. Follow the prompts below for selecting the objects to trim.

 Command: **TRIM**

Select cutting edges: (Projmode = UCS,
  Edgemode = No extend)
Select objects: *(Select all dashed objects
  illustrated at the right.)*
Select objects: *(Press* ENTER *to continue)*
<Select object to trim>/Project/Edge/Undo:
  (Select the circle at "A.")
<Select object to trim>/Project/Edge/Undo:
  (Select the circle at "B.")
<Select object to trim>/Project/Edge/Undo:
  (Select the circle at "C.")
<Select object to trim>/Project/Edge/Undo:
  (Select the circle at "D.")
<Select object to trim>/Project/Edge/Undo:
  (Press ENTER *to exit this command)*

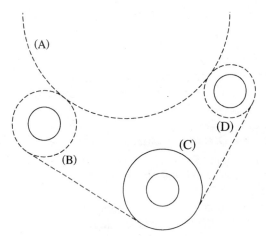

Figure 6–94

Checking the Accuracy of C-Lever.Dwg

Once the c-lever has been constructed, answer the following questions to determine the accuracy of this drawing. Use Figure 6–95 to assist in answering the questions.

1. The total area of the c-lever with all three holes removed is
   (A) 13.744
   (B) 13.749
   (C) 13.754
   (D) 13.759
   (E) 13.764

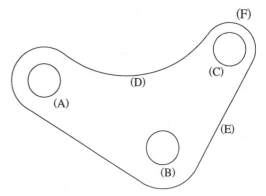

Figure 6–95

2. The total distance from the center of circle "A" to the center of circle "B" is
   (A) 3.692
   (B) 3.697
   (C) 3.702
   (D) 3.707
   (E) 3.712

3. The angle formed in the X-Y plane from the center of circle "C" to the center of circle "B" is
   (A) 223 degrees
   (B) 228 degrees
   (C) 233 degrees
   (D) 238 degrees
   (E) 243 degrees

4. The delta X-Y distances from the center of circle "C" to the center of circle "A" is
   (A) -4.907,-0.851
   (B) -4.907,-0.856
   (C) -4.907,-0.861
   (D) -4.907,-0.866
   (E) -4.907,-0.871

5. The absolute coordinate value of the center of arc "D" is
   (A) 5.869,8.218
   (B) 5.869,8.223
   (C) 5.869,8.228
   (D) 5.869,8.233
   (E) 5.869,8.238

6. The total length of line "E" is
   (A) 3.074
   (B) 3.079
   (C) 3.084
   (D) 3.089
   (E) 3.094

7. The total length of arc "F" is
   (A) 2.051
   (B) 2.056
   (C) 2.061
   (D) 2.066
   (E) 2.071

A solution for each question follows, complete with the method used to arrive at the answer. Apply these methods to any type of drawing that requires the use of Inquiry commands.

Solutions to the Questions on C-Lever

Question #1

The total area of the c-lever with all three holes removed is _____ .

The AREA command will be used to first calculate the total area of the shape and then subtract all three holes. However, before using the AREA command, all objects representing the outline of the shape must be converted into a polyline. The PEDIT command with the Join option is used to accomplish this. Use the illustration of the c-lever in Figure 6–96 to guide you in the use of the PEDIT and AREA commands.

 Command: **PEDIT**

Select polyline: *(Select the object labeled "X.")*
Object selected is not a polyline.
Do you want to turn it into one? <Y>: **Yes**
Close/Join/Width/Edit vertex/Fit/Spline/
　　Decurve/Ltype gen/Undo/eXit <X>: **Join**
Select objects: **0,0**
Other corner: **12,9**
Select objects: *(Press* ENTER *to continue)*
5 segments added to polyline
Open/Join/Width/Edit vertex/Fit/Spline/
　　Decurve/Ltype gen/Undo/eXit <X>: *(Press* ENTER *to exit this command)*

All outer segments now consist as a single polyline object. Since the circles were independent of the outer objects, they were not included in the creation of the polyline. The polyline object could have been created by selecting each individual object while using the PEDIT command; this, however, is not considered an efficient method. An alternate method of creating the polyline out of individual objects would be to use the BOUNDARY command.

Now the AREA command is used to calculate the area of the shape with the holes removed.

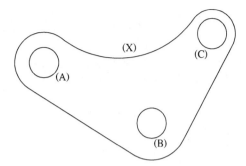

Figure 6–96

 Command: **Area**

<First point>/Object/Add/Subtract: **Add**
<First point>/Object/Subtract: **Object**
(ADD mode) Select objects: *(Select the edge of the shape near "X.")*
Area = 15.611,　Length = 17.771
Total area = 15.611
(ADD mode) Select circle or polyline: *(Press* ENTER *to continue.)*
<First point>/Object/Subtract: **Subtract**
<First point>/Object/Add: **Object**
(SUBTRACT mode) Select objects: *(Select circle"A.")*
Area = 0.621,　　Circumference = 2.793
Total area = 14.991
(SUBTRACT mode) Select objects: *(Select circle"B.")*
Area = 0.621,　　Circumference = 2.793
Total area = 14.370
(SUBTRACT mode) Select objects: *(Select circle"C.")*
Area = 0.621,　　Circumference = 2.793
Total area = 13.749
(SUBTRACT mode) Select objects: *(Press* ENTER *to continue.)*
<First point>/Object/Add: *(Press* ENTER *to exit this command.)*

The total area of the c-lever with all three holes removed is (B), 13.749 .

Question #2

The total distance from the center of circle "A" to the center of circle "B" is _____ .

Use the DIST (Distance) command to calculate the distance from the center of circle "A" to the center of circle "B" in Figure 6–97. Be sure to use the OSNAP-Center mode for locating the centers of all circles.

 Command: **DIST**

First point: **Cen**
of *(Select the edge of circle "A.")*
Second point: **Cen**
of *(Select the edge of circle "B.")*
Distance = 3.712

The total distance from the center of circle "A" to the center of circle "B" is (E), <u>3.712.</u>

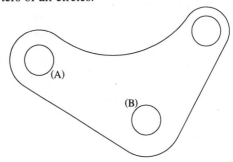

Figure 6–97

Question #3

The angle formed in the X-Y plane from the center of circle "C" to the center of circle "B" is

_____ .

Use the DIST (Distance) command to calculate the angle from the center of circle "C" to the center of circle "B" in Figure 6–98. Be sure to use the OSNAP-Center mode for locating the centers of all circles.

 Command: **DIST**

First point: **Cen**
of *(Select the edge of circle "C.")*
Second point: **Cen**
of *(Select the edge of circle "B.")*
Angle in X-Y Plane = 238

The angle formed in the X-Y plane from the center of circle "C" to the center of circle "B" is (D), <u>238 degrees</u>.

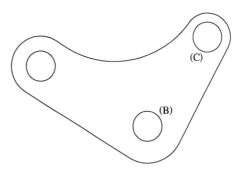

Figure 6–98

Question #4

The delta X-Y distance from the center of circle "C" to the center of circle "A" is _____

↳.

Use the DIST (Distance) command to calculate the delta X-Y distance from the center of circle "C" to the center of circle "A" in Figure 6–99. Be sure to use the OSNAP-Center mode. Notice that additional information is given when using the DIST command. For the purpose of this question, we will only be looking for the delta X-Y distance. The DIST command will display the relative X, Y, Z distances. Since this is a 2D problem, only the X and Y values will be used.

 Command: **DIST**

First point: **Cen**
of *(Select the edge of circle "C.")*
Second point: **Cen**
of *(Select the edge of circle "A.")*
Delta X = -4.907, Delta Y = -0.851, Delta Z = 0.000

The delta X-Y distance from the center of circle "C" to the center of circle "A" is (A), -4.907,-0.851.

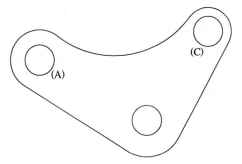

Figure 6–99

Question #5

The absolute coordinate value of the center of arc "D" is _____ .

The ID command is used to get the current absolute coordinate information on a desired point (see Figure 6–100). This command will display the X, Y, Z coordinate values. Since this is a 2D problem, only the X and Y values will be used.

 Command: **ID**

Point: **Cen**
of *(Select the edge of arc "D.")*
X = 5.869, Y = 8.223, Z = 0.000

The absolute coordinate value of the center of arc "D" is (B), 5.869,8.223.

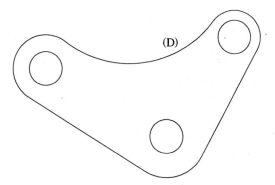

Figure 6–100

Question #6

The total length of line "E" is _____.

Use the DIST (Distance) command to find the total length of line "E" in Figure 6–101. Be sure to use the OSNAP-Endpoint mode. Notice that additional information is given when using the DIST command. For the purpose of this question, we will only be looking for the distance.

 Command: **DIST**

First point: **Endp**
of *(Select the endpoint of the line at "X.")*
Second point: **Endp**
of *(Select the endpoint of the line at "Y.")*
Distance = 3.084

The total length of line "E" is (C), 3.084.

Question #7

The total length of arc "F" is _____.

The LIST or DDMODIFY commands are used to calculate the lengths of arcs, (see Figure 6–102). However, a little preparation is needed before performing this operation. If arc "F" is selected as in Figure 6–103, notice that the entire outline is selected since it is a polyline. Use the EXPLODE command to break the outline back into individual objects. Use the DDMODIFY command to get a listing of the arc length. See also Figure 6–104.

 Command: **EXPLODE**

Select objects: *(Select the dashed polyline anywhere.)*
Select objects: *(Press [ENTER] to execute this command.)*

 Command: **DDMODIFY**

Select objects: *(Select the arc at "F")*

The total length of arc "F" is (E), 2.071.

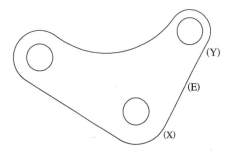

Figure 6–101

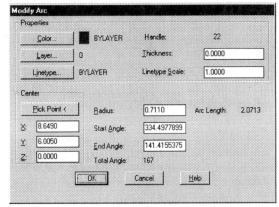

Figure 6–102

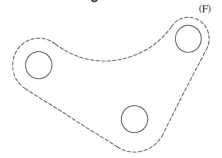

Figure 6–103

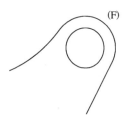

Figure 6–104

Tutorial Exercise
Fitting.Dwg

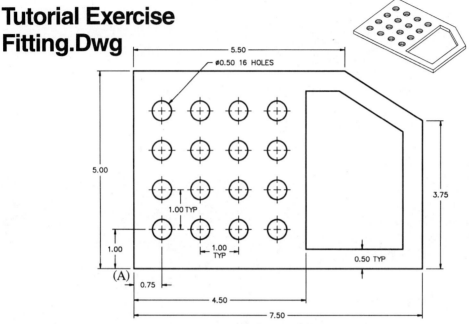

Figure 6–105

Purpose

This tutorial is designed to show you various methods to construct the fitting in Figure 6–105. Numerous questions will be asked about the object requiring the use of a majority of Inquiry commands.

System Settings

Use the UNITS command and change the number of decimal places past the zero from four units to two units. Keep the default drawing limits at 0.00,0.00 for the lower left corner and 12.00,9.00 for the upper right corner.

Layers

No special layers need to be created for this drawing although it is always good practice to create and draw on a separate layer for all object lines.

Suggested Commands

Begin drawing the fitting with point "A" in Figure 6–105 at absolute coordinate 2.24,1.91. Begin by laying out the profile of the fitting. Locate one circle and use the ARRAY command to produce four rows and columns of the circle. Use a series of OFFSET, TRIM, and FILLET commands to construct the five-sided figure on the inside of the fitting profile. To prepare for the Area question, convert the outer and inner profiles into a polyline using the PEDIT command. Other questions pertaining to distances, angles, and point identifications follow. Do not dimension this drawing.

Whenever possible, substitute the appropriate command alias in place of the full AutoCAD command in each tutorial step; for example, use "Co" for the COPY command, "L" for the LINE command, and so on. The complete listing of all command aliases is located on pages 1–9 and 1–10.

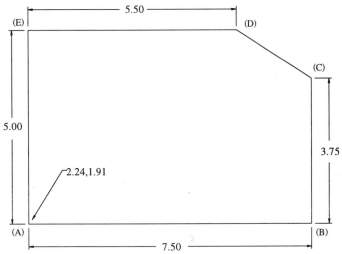

Figure 6–106

Step #1

Use the LINE command to construct the outline of the fitting using a combination of absolute, polar, and relative coordinates (see Figure 6–106). Begin the lower left corner of the fitting at absolute coordinate 2.24,1.91. Do not use the Close option of the LINE command for constructing the last side of the fitting. The Direct Distance mode could also be used to construct this shape.

 Command: **LINE**

From point: **2.24,1.91** *(Starting at "A")*
To point: **@7.50<0** *(To "B")*
To point: **@3.75<90** *(To "C")*
To point: **@-2.00,1.25** *(To "D")*
To point: **@5.50<180** *(To "E")*
To point: **@5.00<270** *(Back to "A")*
To point: *(Press* ENTER *to exit this command)*

Step #2

Construct a circle of 0.50 diameter using the CIRCLE command (see Figure 6–107). Since the last known point is at "A" from use of the previous LINE command, this point is referenced using the @ symbol followed by a coordinate value for the center point. This identifies the center of the circle from the last known point.

 Command: **CIRCLE**

3P/2P/TTR/<Center point>: **@0.75,1.00**
Diameter/<Radius>: **D** *(For Diameter)*
Diameter: **0.50**

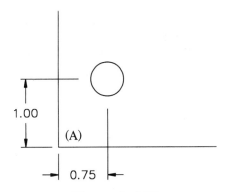

Figure 6–107

Step #3

Create multiple copies in a rectangular pattern of the last circle by using the ARRAY command as shown in Figure 6–108. There are four rows and columns each with a spacing of 1.00 unit from the center of one circle to the center of the other. Since the array is to the right and up from the existing circle, all 1.00 spacing units are positive.

 Command: **ARRAY**

Select objects: **Last** *(This should select the circle.)*
Select objects: *(Press* ENTER *to continue)*
Rectangular or Polar array (R/P): **R** *(for rectangular)*
Number of rows (---) <1>: **4**
Number of columns (|||) <1>: **4**
Unit cell or distance between rows (---): **1.00**
Distance between columns (|||): **1.00**

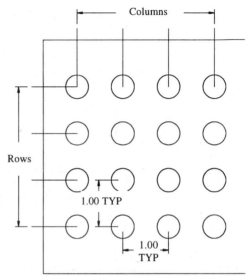

Figure 6–108

Step #4

Use the OFFSET command to copy the line at "A" parallel at a distance of 4.50 in the direction of "B", as shown in Figure 6–109.

 Command: **OFFSET**

Offset distance or Through <1.00>: **4.50**
Select object to offset: *(Select line "A.")*
Side to offset? *(Pick a point anywhere near "B.")*
Select object to offset: *(Press* ENTER *to exit this command)*

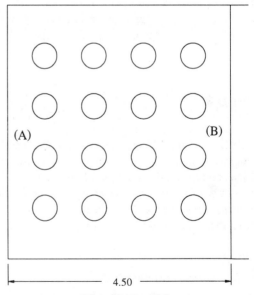

Figure 6–109

Step #5

Use the OFFSET command copy the lines at "A", "B", "C", and "D" parallel at a distance of 0.50 in the direction of "E". See Figure 6–110.

 Command:**OFFSET**

Offset distance or Through <4.50>:**0.50**
Select object to offset: *(Select line "A.")*
Side to offset? *(Pick near "E.")*
Select object to offset: *(Select line "B.")*
Side to offset? *(Pick near "E.")*
Select object to offset: *(Select line "C.")*
Side to offset? *(Pick near "E.")*
Select object to offset: *(Select line "D.")*
Side to offset? *(Pick near "E.")*
Select object to offset: *(Press* ENTER *to exit this command)*

Step #6

Use the TRIM command to partially delete the horizontal and vertical segments labeled "A", "B", "C", and "D" in Figure 6–111. Select the three dashed objects in Figure 6–111 as cutting edges.

 Command:**TRIM**

Select cutting edges: (Projmode = UCS, Edgemode = No extend)
Select objects: *(Select the three dashed objects in Figure 6–111)*
Select objects:*(Press* ENTER *to continue)*
<Select object to trim>/Project/Edge/Undo: *(Select at "A.")*
<Select object to trim>/Project/Edge/Undo: *(Select at "B.")*
<Select object to trim>/Project/Edge/Undo: *(Select at "C.")*
<Select object to trim>/Project/Edge/Undo: *(Select at "D.")*
<Select object to trim>/Project/Edge/Undo: *(Press* ENTER *to exit this command)*

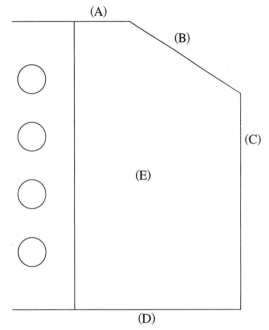

Figure 6–110

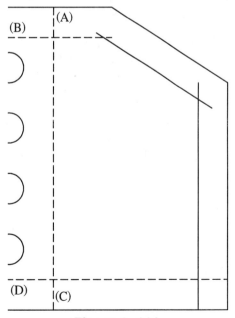

Figure 6–111

Step #7

Use the FILLET command to create corners at the intersections of the four line segments illustrated in Figure 6–112. Set the fillet radius to a value of zero to accomplish this. The TRIM command could also be used to accomplish this step.

 Command: **FILLET**

(TRIM mode) Current fillet radius = 0.50
Polyline/Radius/Trim/<Select first object>: **R**
Enter fillet radius <0.50>: **0**

 Command: **FILLET**

(TRIM mode) Current fillet radius = 0.00
Polyline/Radius/Trim/<Select first object>:
  (Select line "A")
Select second object: *(Select line "B")*

 Command: **FILLET**

(TRIM mode) Current fillet radius = 0.00
Polyline/Radius/Trim/<Select first object>:
  (Select line "B")
Select second object: *(Select line "C")*

 Command: **FILLET**

(TRIM mode) Current fillet radius = 0.00
Polyline/Radius/Trim/<Select first object>:
  (Select line "C")
Select second object: *(Select line "D")*

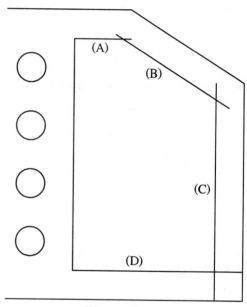

Figure 6–112

Checking the Accuracy of Fitting.Dwg

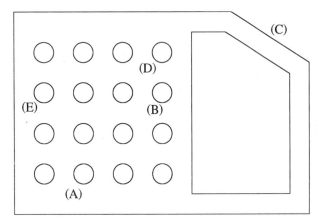

Figure 6–113

Once the fitting has been constructed, answer the following questions to determine the accuracy of the drawing. Use Figure 6–113 to assist in answering the questions.

1. The total area of the fitting with the inner slot and all holes removed is
   (A) 23.87
   (B) 23.91
   (C) 23.95
   (D) 23.99
   (E) 24.03

2. The total distance from the center of circle "A" to the center of circle "B" is
   (A) 2.79
   (B) 2.83
   (C) 2.87
   (D) 2.91
   (E) 2.95

3. The total length of line "C" is
   (A) 2.20
   (B) 2.24
   (C) 2.28
   (D) 2.32
   (E) 2.36

4. The angle formed in the X-Y plane from the center of circle "D" to the center of circle "E" is
   (A) 198 degrees
   (B) 194 degrees
   (C) 190 degrees
   (D) 186 degrees
   (E) 182 degrees

5. The absolute coordinate value of the center of circle "D" is
   (A) 5.87,5.79
   (B) 5.91,5.83
   (C) 5.95,5.87
   (D) 5.99,5.91
   (E) 6.03,5.95

A solution for each question follows complete with the method used to arrive at the answer. Apply these methods to different shapes that require the use of Inquiry commands.

Question #1

The total area of the fitting with the inner slot and all holes removed is _____.

The AREA command will be used to first calculate the total area of the object and then subtract the slot and all holes. However, before using the AREA command, all objects representing the outline of the fitting and slot must be converted into a polyline. Use the PEDIT command with the Join option to best accomplish this. Also use the illustration of the fitting in Figure 6–114 to guide you in the use of the PEDIT and AREA commands.

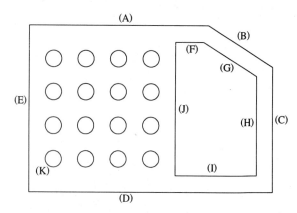

Figure 6–114

 Command: **PEDIT**

Select polyline: (*Select the object labeled "A."*)
Object selected is not a polyline.
Do you want to turn it into one? <Y>: **Y**
Close/Join/Width/Edit vertex/Fit/Spline/
   Decurve/Ltype gen/Undo/eXit <X>: **Join**
Select objects: (*Select "B," "C," "D," and "E."*)
Select objects: (*Press* ENTER *to continue*)
4 segments added to polyline
Open/Join/Width/Edit vertex/Fit/Spline/
   Decurve/Ltype gen/Undo/eXit <X>: (*Press* ENTER *to exit this command*)

All outer objects now consist of one polyline object. Repeat the above Pedit procedure for converting objects "F," "G," "H," "I," and "J" into one polyline.

 Command: **AREA**

<First point>/Object/Add/Subtract: **Add**
<First point>/Object/Subtract: **Object**
(ADD mode) Select objects: (*Select the edge of the fitting near "A."*)
Area=36.25,   Length=24.11
Total area=36.25

(ADD mode) Select objects: (*Press* ENTER *to continue*)
<First point>/Object/Subtract: **Subtract**
<First point>/Object/Add: **Object**
(SUBTRACT mode) Select objects: (*Select the slot near "F."*)
Area=9.16,   Length=12.27
Total area = 27.09
(SUBTRACT mode) Select objects: (*Select circle"K."*)
Area = 0.20,   Circumference = 1.57
Total area = 26.90
(SUBTRACT mode) Select objects: (*Carefully select the remaining 15 circles.*)
Total area = 23.95
(SUBTRACT mode) Select objects: (*Press* ENTER *to continue*)
<First point>/Object/Add: (*Press* ENTER *to exit this command*)

The total area of the fitting with the slot and all holes removed is (C), 23.95.

Question #2

The total distance from the center of circle "A" to the center of circle "B" is _____.

Use the DIST (Distance) command to calculate the distance from the center of circle "A" to the center of circle "B" as shown in Figure 6–115. Be sure to use the OSNAP-Center mode to locate the centers of both circles.

 Command: **DIST**

First point: **Cen**
of *(Select the edge of circle "A.")*
Second point: **Cen**
of *(Select the edge of circle "B.")*
Distance = 2.83

The total distance from the center of circle "A" to the center of circle "B" is (B), 2.83.

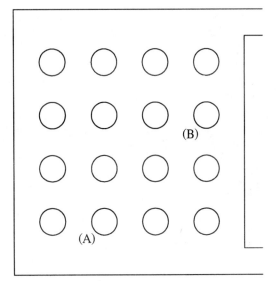

Figure 6–115

Question #3

The total length of line "C" is _____.

Use the DIST (Distance) command to find the total length of line "C" as shown in Figure 6–116. Be sure to use the OSNAP-Endpoint mode to locate the exact endpoints of line "C."

 Command: **DIST**

First point: **Endp**
of *(Select the endpoint of the line at "X.")*
Second point: **Endp**
of *(Select the endpoint of the line at "Y.")*
Distance = 2.36

The total length of line "C" is (E), 2.36.

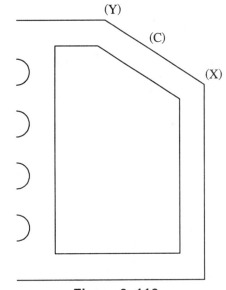

Figure 6–116

Question #4

The angle formed in the X-Y plane from the center of circle "D" to the center of circle "E" is _____.

Use the DIST (Distance) command to calculate the angle from the center of circle "D" to the center of circle "E" as shown in Figure 6–117. Be sure to use the OSNAP-Center mode to locate the exact centers of both circles.

 Command: **DIST**

First point: **Cen**
of *(Select the edge of circle "D.")*
Second point: **Cen**
of *(Select the edge of circle "E.")*
Angle in X-Y Plane = 198

The angle formed in the X-Y plane from the center of circle "D" to the center of circle "E" is (A), 198 degrees.

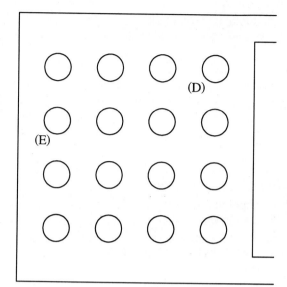

Figure 6–117

Question #5

The absolute coordinate value of the center of circle "D" is _____.

The ID command is used to get the current absolute coordinate information on a desired point. This command will display the *X,Y,Z* coordinate values. Since this is a 2D problem, only the X and Y values will be used. See Figure 6–118.

 Command: **ID**

Point: **Cen**
of *(Select the edge of circle "D.")*
X = 5.99, Y = 5.91, Z = 0.000

The absolute coordinate value of the center of circle "D" is (D), 5.99,5.91.

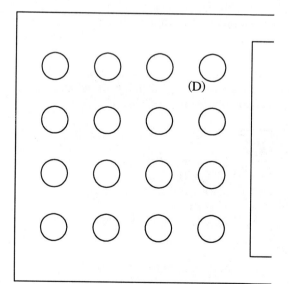

Figure 6–118

Additional Questions for Fitting.Dwg

Rotate the fitting at a -10 degree angle using "X" as the center of rotation as shown in Figure 6–119. Use the following command sequence to perform this operation.

Command: **ROTATE**
Select objects: **All**
Select objects: *(Press [ENTER] to continue)*
Base point: **Endp**
of *(Select the inclined line near "X.")*
<Rotation angle>/Reference: **-10**

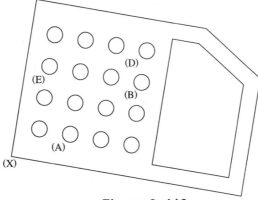

Figure 6–119

Once the fitting has been rotated, answer the following questions to determine the accuracy of the drawing. Use Figure 6–119 to assist in answering the questions.

1. The absolute coordinate value of the center of circle "D" is
   (A) 6.47,5.04
   (B) 6.51,5.08
   (C) 6.55,5.12
   (D) 6.59,5.16
   (E) 6.63,5.20

2. The absolute coordinate value of the center of circle "A" is
   (A) 4.14,2.59
   (B) 4.18,2.63
   (C) 4.22,2.67
   (D) 4.26,2.71
   (E) 4.30,2.75

3. The absolute coordinate value of the center of circle "B" is
   (A) 6.41,4.17
   (B) 6.45,4.21
   (C) 6.49,4.25
   (D) 6.53,4.29
   (E) 6.57,4.33

4. The angle formed in the X-Y plane from the center of circle "D" to the center of circle "E" is
   (A) 176 degrees
   (B) 180 degrees
   (C) 184 degrees
   (D) 188 degrees
   (E) 192 degrees

A solution for each question follows, complete with the method used to arrive at the answer. Apply these methods to any type of drawing that requires the use of Inquiry commands.

Question #1

The absolute coordinate value of the center of
circle "D" is _____.

The ID command is used to get the current absolute coordinate information on a desired point (Figure 6–120). This command will display the *X,Y,Z*
coordinate values. Since this is a 2D problem, only
the X and Y values will be used.

 Command: **ID**

Point: **Cen**
of *(Select the edge of circle "D.")*
X = 6.63, Y = 5.20, Z = 0.000

**The absolute coordinate value of the center
of circle "D" is (E), 6.63,5.20.**

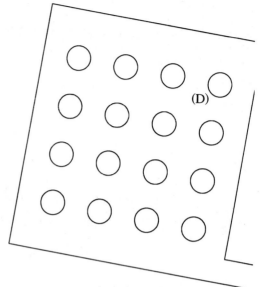

Figure 6–120

Question #2

The absolute coordinate value of the center of
circle "A" is _____.

The ID command is used to get the current absolute
coordinate information on point "A" as shown in
Figure 6–121. This command will display the *X,Y,Z*
coordinate values. Since this is a 2 dimensional problem, only the X and Y values will be used.

 Command: **ID**

Point: **Cen**
of *(Select the edge of circle "A.")*
X = 4.14, Y = 2.59, Z = 0.000

**The absolute coordinate value of the center
of circle "A" is (A), 4.14,2.59.**

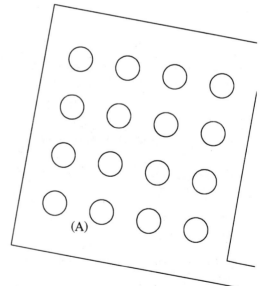

Figure 6–121

Question #3

The absolute coordinate value of the center of
circle "B" is _____.

The ID command is used to get the current absolute coordinate information on point "B" as shown
in Figure 6–122. This command will display the
X,Y,Z coordinate values. Since this is a 2D problem, only the X and Y values will be used.

 Command: **ID**

Point: **Cen**
of *(Select the edge of circle "B.")*
X = 6.45, Y = 4.21, Z = 0.000

**The absolute coordinate value of the center
of circle "B" is (B), 6.45,4.21.**

Question #4

The angle formed in the X-Y plane from the center of circle "D" to the center of circle "E" is

_____.

Use the DIST (Distance) command to calculate the
angle from the center of circle "D" to the center of
circle "E." Be sure to use the OSNAP-Center mode
to locate the center of both circles.

 Command: **DIST**

First point: **Cen**
of *(Select the edge of circle "D.")*
Second point: Cen
of *(Select the edge of circle "E.")*
Angle in X-Y Plane = 188

**The angle formed in the X-Y plane from the
center of circle "D" to the center of circle
"E" is (D), 188 degrees.**

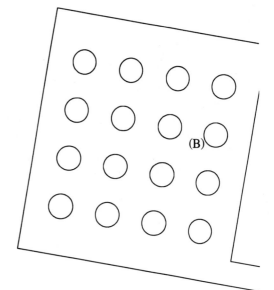

Figure 6–122

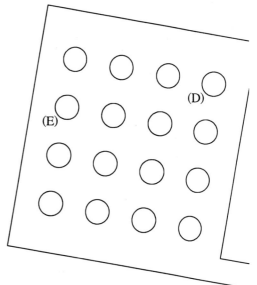

Figure 6–123

Problems for Unit 6

Problem6-1
Angleblk.Dwg

Directions for Angleblk.Dwg

Use the Units Control dialog box to set the units to decimal. Set the number of digits to the right of the decimal point from four to two. Be sure the system of angle measure is set to decimal degrees and the number of decimal places for the display of angles is zero. Keep the remaining default unit values. Keep the default settings for the drawing limits.

Begin the drawing in Figure 6–124 by locating the lower left corner of Angleblk identified by "X" at coordinate (2.35,3.17).

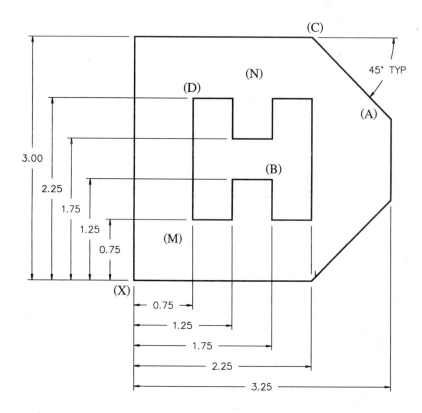

Figure 6–124

Refer to the drawing of the Angleblk in Figure 6–124 to answer the following questions.

1. The total surface area of the Angleblk with the inner "H" shape removed is
   - (A) 6.66
   - (B) 6.77
   - (C) 6.89
   - (D) 7.00
   - (E) 7.11

2. The total area of the inner "H" shape is
   - (A) 1.75
   - (B) 1.81
   - (C) 1.87
   - (D) 1.93
   - (E) 1.99

3. The total length of line "A" is
   - (A) 1.29
   - (B) 1.35
   - (C) 1.41
   - (D) 1.47
   - (E) 1.53

4. The absolute coordinate value of the end-point of the line at "B" is
   - (A) 4.04,4.42
   - (B) 4.10,4.42
   - (C) 4.16,4.42
   - (D) 4.22,4.42
   - (E) 4.28,4.42

5. The absolute coordinate value of the end-point of the line at "C" is
   - (A) 4.60,6.11
   - (B) 4.60,6.17
   - (C) 4.60,6.23
   - (D) 4.60,6.29
   - (E) 4.60,6.35

6. Use the STRETCH command and extend the inner "H" shape a distance of 0.37 units in the 180 direction. Use "N" as the first corner of the crossing window and "M" as the other corner. Use the endpoint of "D" as the base point of the stretching operation. The new surface area of Angleblk with the inner "H" shape removed is closest to
   - (A) 6.63
   - (B) 6.69
   - (C) 6.75
   - (D) 6.81
   - (E) 6.87

7. Use the SCALE command with the endpoint of the line at "D" as the base point. Reduce the size of just the inner "H" using a scale factor of 0.77. The new surface area of Angleblk with the inner "H" removed is
   - (A) 7.48
   - (B) 7.54
   - (C) 7.60
   - (D) 7.66
   - (E) 7.72

Problem 6-2
Lever1.Dwg

Directions for Lever1.Dwg
   Use the Units Control dialog box to set the units to decimal. Set the number of digits to the right of the decimal point from four to three. Be sure the system of angle measure is set to decimal degrees and the number of decimal places for the display of angles is zero. Keep the remaining default unit values.

   Begin the drawing in Figure 6–125 by locating the center of the 2.000 diameter circle at coordinate (4.500,3.250).

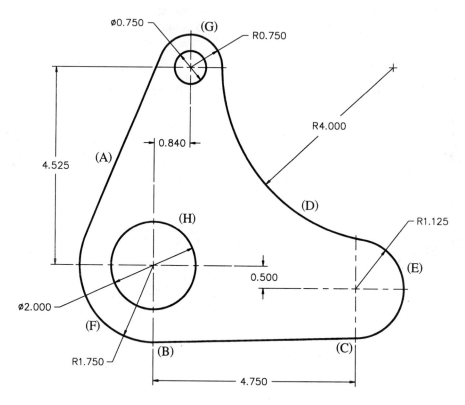

Figure 6–125

Refer to the drawing of Lever1 in Figure 6–125 to answer the following questions.

1. The total length of line segment "A" is
   (A) 4.483
   (B) 4.492
   (C) 4.499
   (D) 4.504
   (E) 4.515

2. The absolute coordinate value of the center of the 4.00 radius arc "D" is
   (A) 10.090,7.773
   (B) 10.090,7.782
   (C) 10.090,7.791
   (D) 10.090,7.800
   (E) 10.090,7.806

3. The length of the 1.125 radius arc segment "E" is
   (A) 3.299
   (B) 3.308
   (C) 3.319
   (D) 3.329
   (E) 3.337

4. The distance from the center of the 1.750 radius arc "F" to the center of the 0.750 radius arc "G" is
   (A) 4.572
   (B) 4.584
   (C) 4.596
   (D) 4.602
   (E) 4.610

5. The total area of Lever1 with the two holes removed is
   (A) 26.271
   (B) 26.282
   (C) 26.290
   (D) 26.298
   (E) 26.307

6. The circumference of the 2.000 diameter circle "H" is
   (A) 6.269
   (B) 6.276
   (C) 6.283
   (D) 6.289
   (E) 6.294

7. Use the SCALE command to reduce Lever1 in size by a scale factor of 0.333. Use the center of the 2.000 diameter hole as the base point. The absolute coordinate value of the center of the 0.750 arc "G" is
   (A) 4.780,4.757
   (B) 4.780,4.767
   (C) 4.780,4.777
   (D) 4.785,4.757
   (E) 4.793,4.777

Problem 6-3
Plate1.Dwg

Directions for Plate1.Dwg

Use the Units Control dialog box and set to decimal units. Set the number of digits to the right of the decimal point from four to three. Be sure the system of angle measure is set to decimal degrees and the number of decimal places for the display of angles is zero. Accept the defaults for the remaining unit prompts. Use the LIMITS command and set the upper right corner of the screen area to a value of 36.000,24.000.

Begin the drawing in Figure 6–126 by placing the center of the 4.000 diameter arc with keyway at coordinate (16.000,13.000).

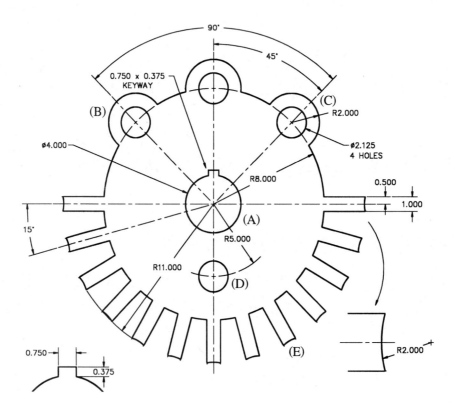

Figure 6–126

Refer to the drawing of Plate1 in Figure 6–126 to answer the following questions.

1. The distance from the center of the 2.000 radius arc "B" to the center of the 2.000 radius arc "C" is
   (A) 10.286
   (B) 10.293
   (C) 11.300
   (D) 11.307
   (E) 11.314

2. The absolute coordinate value of the center of arc "C" is
   (A) 21.657,18.657
   (B) 21.657,18.664
   (C) 21.657,18.671
   (D) 21.657,18.678
   (E) 21.657,18.685

3. The angle formed in the X-Y plane from the center of the 2.000 radius arc "C" to the center of the 2.125 diameter hole "D" is
   (A) 242 degrees
   (B) 244 degrees
   (C) 246 degrees
   (D) 248 degrees
   (E) 250 degrees

4. The total length of arc "E" is
   (A) 0.999
   (B) 1.005
   (C) 1.011
   (D) 1.017
   (E) 1.023

5. The total length of arc "C" is
   (A) 6.766
   (B) 6.772
   (C) 6.778
   (D) 6.784
   (E) 6.790

6. The total area of Plate1 with all holes including keyway removed is
   (A) 232.259
   (B) 232.265
   (C) 232.271
   (D) 232.277
   (E) 232.283

7. The distance from the center of the 2.000 radius arc "B" to the center of the 2.000 radius arc "E" is
   (A) 20.732
   (B) 20.738
   (C) 20.744
   (D) 20.750
   (E) 20.756

Problem 6-4
Hanger.Dwg

Directions for Hanger.Dwg

Use the Units Control dialog box set to decimal. Set the number of digits to the right of the decimal point from four to three. Be sure the system of angle measure is set to decimal degrees and the number of decimal places for the display of angles is zero. Accept the defaults for the remaining unit prompts. Use the LIMITS command and set the upper right corner of the screen area to a value of 250.000,150.000.

Begin drawing the hanger by locating the center of the 40.000 radius arc at coordinate (55.000,85.000). See Figure 6–127.

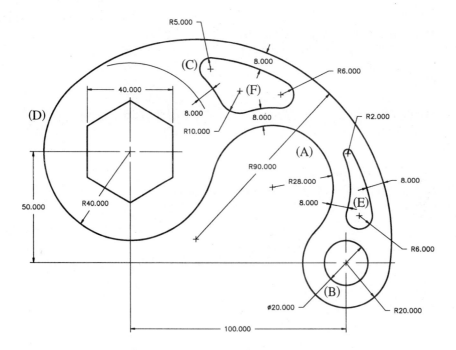

Figure 6–127

Refer to the drawing of the Hanger in Figure 6–127 to answer the following questions.

1. The area of the outer profile of the Hanger is closest to
   (A) 9970.567
   (B) 9965.567
   (C) 9975.567
   (D) 9975.005
   (E) 9980.347

2. The area of the Hanger with the polygon, circle, and irregular shapes removed is closest to
   (A) 7304.089
   (B) 7305.000
   (C) 7303.890
   (D) 7304.000
   (E) 7306.098

3. The absolute coordinate value of the center of the 28.000 radius arc "A" is closest to
   (A) 120.000,69.000
   (B) 121.082,68.964
   (C) 121.520,68.237
   (D) 121.082,69.000
   (E) 121.082,66.937

4. The absolute coordinate value of the center of the 20.000 diameter circle "B" is closest to
   (A) 154.000,35.000
   (B) 156.000,36.000
   (C) 156.147,35.256
   (D) 155.000,35.000
   (E) 156.000,37.000

5. The angle formed in the X-Y plane from the center of the 5.000 radius arc "C" to the center of the 40.000 radius arc "D" is
   (A) 219 degrees
   (B) 221 degrees
   (C) 223 degrees
   (D) 225 degrees
   (E) 227 degrees

6. The total area of irregular shape "E" is
   (A) 271.613
   (B) 271.723
   (C) 271.784
   (D) 271.801
   (E) 271.822

7. The total area of irregular shape "F" is
   (A) 698.511
   (B) 698.621
   (C) 699.817
   (D) 699.856
   (E) 699.891

Problem 6-5
Gasket1.Dwg

Directions for Gasket1.Dwg

Use the Units Control dialog box to set the units to decimal. Set the number of digits to the right of the decimal point from four to three. Be sure the system of angle measure is set to decimal degrees and the number of decimal places for the display of angles is zero. Keep the remaining default unit values.

Begin the drawing in Figure 6–128 by locating the center of the 1.500 radius circle at coordinate (5.750,4.750).

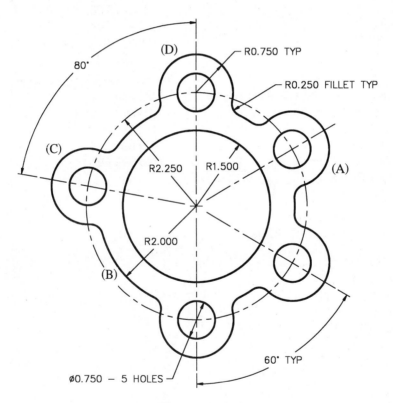

Figure 6–128

Refer to the drawing of Gasket1 in Figure 6–128 to answer the following questions.

1. The total area of Gasket1 with the all holes removed is closest to
   - (A) 9.918
   - (B) 9.921
   - (C) 9.924
   - (D) 9.927
   - (E) 9.930

2. The absolute coordinate value of the center of the 0.750 radius arc "A" is closest to
   - (A) 7.669,5.875
   - (B) 7.669,5.870
   - (C) 7.666,5.875
   - (D) 7.699,5.875
   - (E) 7.699,5.975

3. The length of arc segment "B" is closest to
   - (A) 1.698
   - (B) 1.704
   - (C) 1.710
   - (D) 1.716
   - (E) 1.722

4. The angle formed in the X-Y plane from the center of the arc "C" to the center of arc "D" is closest to
   - (A) 30 degrees
   - (B) 35 degrees
   - (C) 40 degrees
   - (D) 45 degrees
   - (E) 50 degrees

5. The total length of the 0.750 radius arc "C" is
   - (A) 2.674
   - (B) 2.680
   - (C) 2.686
   - (D) 2.692
   - (E) 2.698

Use the MOVE command to reposition Gasket1 at a distance of 1.832 in the -45-degree direction. Use the center of the 1.500 radius circle as the base point of the move. After performing this operation, answer the following questions:

6. The absolute coordinate value of the center of the 1.500 radius circle is
   - (A) 7.045,3.437
   - (B) 7.045,3.443
   - (C) 7.045,3.449
   - (D) 7.045,3.455
   - (E) 7.045,3.461

7. The absolute coordinate value of the center of the 0.750 radius arc "C" is
   - (A) 4.806,3.845
   - (B) 4.812,3.845
   - (C) 4.818,3.845
   - (D) 4.824,3.845
   - (E) 4.830,3.845

Problem 6-6
Gasket2.Dwg

Directions for Gasket2.Dwg

Use the Units Control dialog box to set the units to decimal. Set the number of digits to the right of the decimal point from four to two. Be sure the system of angle measure is set to decimal degrees and the number of decimal places for the display of angles is zero. Keep the remaining default unit values.

Begin the drawing in Figure 6–129 by locating the center of the 6.00 x 3.00 rectangle at coordinate (6.00,4.75).

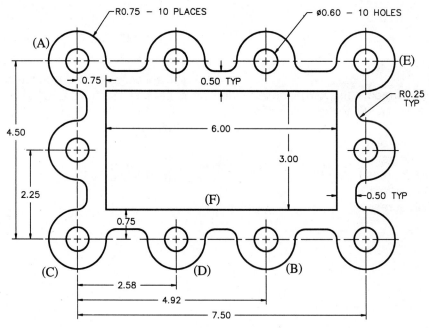

Figure 6–129

Refer to the drawing of Gasket2 in Figure 6–129 to answer the following questions.

1. The total surface area of Gasket2 with the rectangle and all ten holes removed is closest to
   (A) 21.46
   (B) 21.48
   (C) 21.50
   (D) 21.52
   (E) 21.54

2. The distance from the center of arc "A" to the center of arc "B" is closest to
   (A) 6.63
   (B) 6.67
   (C) 6.71
   (D) 6.75
   (E) 6.79

3. The length of arc segment "C" is closest to
   (A) 3.44
   (B) 3.47
   (C) 3.50
   (D) 3.53
   (E) 3.56

4. The absolute coordinate value of the center of the 0.75 radius arc "D" is closest to
   (A) 4.83,2.50
   (B) 4.83,2.47
   (C) 4.83,2.53
   (D) 4.80,2.50
   (E) 4.83,2.56

5. The angle formed in the X-Y plane from the center of the 0.75 radius arc "D" to the center of the 0.75 radius arc "A" is
   (A) 116 degrees
   (B) 118 degrees
   (C) 120 degrees
   (D) 122 degrees
   (E) 124 degrees

6. The delta X,Y distance from the intersection at "E" to the midpoint of the line at "F" is
   (A) -4.50,3.65
   (B) -4.50,-3.65
   (C) -4.50,-3.70
   (D) -4.50,3.75
   (E) -4.50,-3.75

7. Use the SCALE command to reduce the size of the inner rectangle. Use the midpoint of the line at "F" as the base point. Use a scale factor of 0.83 units. The new total surface area with the rectangle and all ten holes removed is
   (A) 26.99
   (B) 27.04
   (C) 27.09
   (D) 27.14
   (E) 27.19

Problem 6-7
Lever2.Dwg

Directions for Lever2.Dwg

Use the Units Control dialog box to set the units to decimal. Keep the number of digits to the right of the decimal point at four places. Be sure the system of angle measure is set to decimal degrees and the number of decimal places for the display of angles is zero. Keep the remaining default unit values.

Begin the drawing in Figure 6–130 by locating the center of the 1.0000 diameter circle at coordinate (2.2500,4.0000).

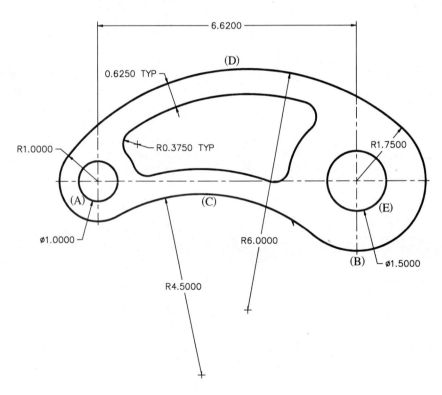

Figure 6–130

Refer to the drawing of Lever2 in Figure 6–130 to answer the following questions.

1. The total area of Lever2 with the inner irregular shape and both holes removed is
   (A) 17.6813
   (B) 17.6819
   (C) 17.6825
   (D) 17.6831
   (E) 17.6837

2. The absolute coordinate value of the center of the 4.5000 radius arc "C" is
   (A) 4.8944,-0.0822
   (B) 4.8944,-0.8226
   (C) 4.8950,-0.8232
   (D) 4.8956,-0.8238
   (E) 4.8962,-0.8244

3. The absolute coordinate value of the center of the 6.0000 radius arc "D" is
   (A) 6.0828,0.7893
   (B) 6.0834,0.7899
   (C) 6.0840,0.7905
   (D) 6.0846,0.7911
   (E) 6.0852,0.7917

4. The total length of arc "C" is
   (A) 5.3583
   (B) 5.3589
   (C) 5.3595
   (D) 5.3601
   (E) 5.3607

5. The distance from the center of the 1.0000 diameter circle "A" to the intersection of the circle and centerline at "B" is closest to
   (A) 6.8456
   (B) 6.8462
   (C) 6.8474
   (D) 6.8480
   (E) 6.8486

6. The angle formed in the X-Y plane from the upper quadrant of arc "D" to the center of the 1.5000 circle (E) is
   (A) 313 degrees
   (B) 315 degrees
   (C) 317 degrees
   (D) 319 degrees
   (E) 321 degrees

7. The delta X,Y distance from the upper quadrant of arc "C" to the center of the 1.0000 hole "A" is
   (A) -2.6444,0.3220
   (B) -2.6444,0.3226
   (C) -2.6444,0.3232
   (D) -2.6444,0.3238
   (E) -2.6444,0.3244

Problem 6-8
Flange1.Dwg

Directions for Flange1.Dwg

  Use the Units Control dialog box to set the units to decimal. Set the number of digits to the right of the decimal point from four to two. Be sure the system of angle measure is set to decimal degrees and the number of decimal places for the display of angles is zero. Keep the remaining default unit values.

  Begin the drawing in Figure 6–131 by locating the center of the 2.00 diameter circle at coordinate (6.00,5.50).

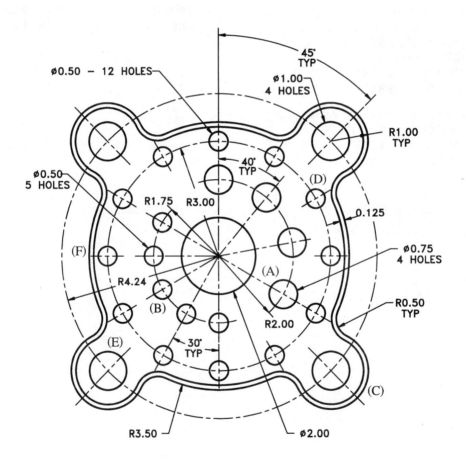

Figure 6–131

Refer to the drawing of Flange1 in Figure 6–131 to answer the following questions.

1. The total area of the 0.125 strip around the perimeter of Flange1 is
   (A) 4.10
   (B) 4.12
   (C) 4.14
   (D) 4.16
   (E) 4.18

2. The absolute coordinate value of the center of the 0.75 diameter circle "A" is
   (A) 7.73,4.45
   (B) 7.73,4.50
   (C) 7.73,4.55
   (D) 7.75,4.55
   (E) 7.75,4.60

3. The absolute coordinate value of the center of the 0.50 diameter circle "B" is closest to
   (A) 4.45,4.55
   (B) 4.48,4.55
   (C) 4.48,4.62
   (D) 4.48,4.69
   (E) 4.50,4.56

4. The length of the 1.00 radius arc "C" is
   (A) 3.82
   (B) 3.85
   (C) 3.88
   (D) 3.91
   (E) 3.97

5. The total surface area of the inner part of Flange1 with all holes removed is closest to
   (A) 35.12
   (B) 35.18
   (C) 35.24
   (D) 35.30
   (E) 35.36

6. The total length of outer arc "F" is
   (A) 2.91
   (B) 2.97
   (C) 3.03
   (D) 3.09
   (E) 3.15

7. The angle formed in the X-Y plane from the center of the 0.50 hole "D" to the center of the 0.50 hole "B" is
   (A) 204 degrees
   (B) 206 degrees
   (C) 208 degrees
   (D) 210 degrees
   (E) 212 degrees

Problem 6-9
Wedge.Dwg

Directions for Wedge.Dwg

Use the Units Control dialog box to set the units to decimal. Set the number of digits to the right of the decimal point from four to two. Be sure the system of angle measure is set to decimal degrees and the number of decimal places for the display of angles is zero. Keep all remaining default unit values.

Begin constructing the wedge with vertex "A" located at coordinate (30,30) as shown in Figure 6–132.

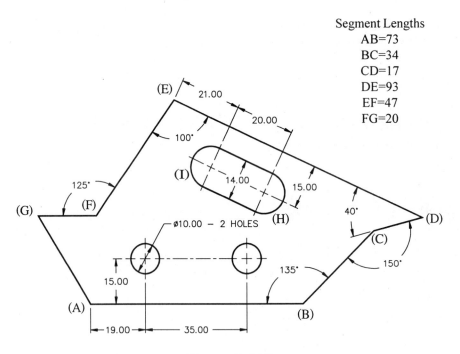

Segment Lengths
AB=73
BC=34
CD=17
DE=93
EF=47
FG=20

Figure 6–132

Refer to the drawing of Wedge in Figure 6–132 to answer the following questions.

1. The total area of the Wedge with the two holes and slot removed is
   - (A) 4367.97
   - (B) 4368.54
   - (C) 4370.12
   - (D) 4371.83
   - (E) 4374.91

2. The distance from the intersection of vertex "E" to the intersection of vertex "G" is closest to
   - (A) 60.72
   - (B) 60.74
   - (C) 60.80
   - (D) 60.85
   - (E) 60.87

3. The distance from the intersection of vertex "D" to the intersection of vertex "G" is closest to
   - (A) 131.00
   - (B) 131.12
   - (C) 131.24
   - (D) 131.36
   - (E) 131.48

4. The length of arc "H" is closest to
   - (A) 21.00
   - (B) 21.50
   - (C) 21.99
   - (D) 22.50
   - (E) 22.99

5. The overall height of the Wedge from the base of line "AB" to the peak at "E" is
   - (A) 60.72
   - (B) 65.87
   - (C) 67.75
   - (D) 69.08
   - (E) 71.98

6. The distance from the intersection of vertex "A" to the center of arc "I" is
   - (A) 61.09
   - (B) 61.67
   - (C) 61.98
   - (D) 62.93
   - (E) 63.02

7. The length of line "AG" is closest to
   - (A) 31.92
   - (B) 32.47
   - (C) 33.62
   - (D) 34.22
   - (E) 35.33

Problem 6-10
Pattern1.Dwg

Directions for Pattern1.Dwg

Use the Units Control dialog box to set the units to decimal. Set the number of digits to the right of the decimal point from four to three. Be sure the system of angle measure is set to decimal degrees and the number of decimal places for the display of angles is zero. Keep all remaining default unit values. Use the LIMITS command and set the upper right corner of the limits to 16.000,12.000.

Begin the drawing in Figure 6–133 by locating the center of the 2.500 radius arc at coordinate (4.250,5.750).

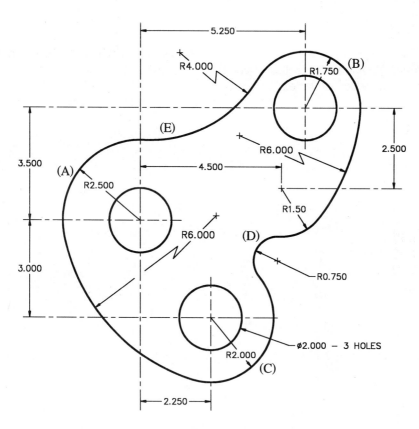

Figure 6–133

Refer to the drawing of Pattern1 in Figure 6–133 to answer the following questions.

1. The total surface area of Pattern1 with the 3 holes removed is
   - (A) 47.340
   - (B) 47.346
   - (C) 47.386
   - (D) 47.486
   - (E) 47.586

2. The distance from the center of the 2.500 radius arc "A" to the center of the 1.750 radius arc "B" is
   - (A) 6.310
   - (B) 6.315
   - (C) 6.210
   - (D) 6.321
   - (E) 6.305

3. The absolute coordinate value of the center of the 4.000 radius arc at "E" is
   - (A) 4.580,12.241
   - (B) 4.589,12.249
   - (C) 4.589,12.237
   - (D) 4.480,12.237
   - (E) 4.589,12.241

4. The perimeter of the outline of Pattern1 is
   - (A) 31.741
   - (B) 31.747
   - (C) 31.753
   - (D) 31.759
   - (E) 31.765

5. The total length of arc "A" is
   - (A) 4.633
   - (B) 4.639
   - (C) 4.645
   - (D) 4.651
   - (E) 4.657

6. The angle formed in the X-Y plane from the center of the 2.500 radius arc "A" to the center of the 2.000 radius arc "C" is
   - (A) 301 degrees
   - (B) 303 degrees
   - (C) 305 degrees
   - (D) 307 degrees
   - (E) 309 degrees

7. Use the MIRROR command to flip but not duplicate Pattern1. Use the center of the 2.000 radius arc "C" as the first point of the mirror line. Use polar coordinate @1.000<90 as the second point. The new absolute coordinate value of the center of the 0.750 radius arc "D" is
   - (A) 4.378,4.504
   - (B) 4.382,4.504
   - (C) 4.386,4.504
   - (D) 4.390,4.504
   - (E) 4.394,4.504

Problem 6-11
Bracket1.Dwg

Directions for Bracket1.Dwg

Use the Units Control dialog box to set the units to decimal. Set the number of digits to the right of the decimal point from four to three. Be sure the system of angle measure is set to decimal degrees and the number of decimal places for the display of angles is zero. Keep all remaining default unit values. Keep the default values for the limits.

Begin the drawing in Figure 6–134 by locating the center of the 1.500 radius arc "A" at coordinate (4.000,3.500).

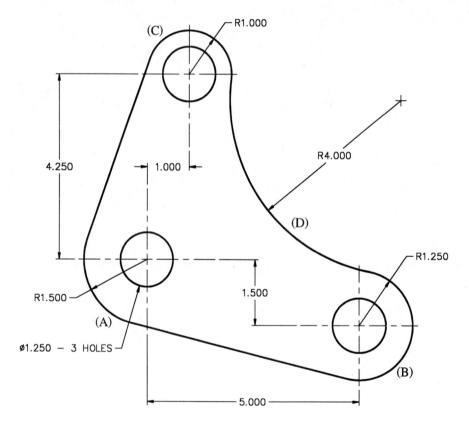

Figure 6–134

Refer to the drawing of Bracket1 in Figure 6–134 to answer the following questions.

1. The distance from the center of the 1.500 radius arc "A" to the center of the 1.250 radius arc "B" is
   - (A) 5.205
   - (B) 5.210
   - (C) 5.215
   - (D) 5.220
   - (E) 5.228

2. The distance from the center of the 1.500 radius arc "A" to the center of the 1.000 radius arc "C" is
   - (A) 4.366
   - (B) 4.370
   - (C) 4.374
   - (D) 4.378
   - (E) 4.382

3. The distance from the center of the 1.250 radius arc "B" to the center of the 1.000 radius arc "C" is
   - (A) 6.990
   - (B) 6.995
   - (C) 6.998
   - (D) 7.000
   - (E) 7.004

4. The length of arc "B" is
   - (A) 3.994
   - (B) 4.000
   - (C) 4.006
   - (D) 4.012
   - (E) 4.018

5. The absolute coordinate value of the center of the 4.000 radius arc "D" is
   - (A) 9.965,7.112
   - (B) 9.965,7.250
   - (C) 9.960,7.161
   - (D) 9.965,7.161
   - (E) 9.995,1.161

6. The total area of Bracket1 with all three 1.250 diameter holes removed is
   - (A) 27.179
   - (B) 27.187
   - (C) 27.193
   - (D) 27.198
   - (E) 28.003

7. The angle formed in the X-Y plane from the center of the 1.250 radius arc "B" to the center of the 1.000 radius arc "C" is
   - (A) 121 degrees
   - (B) 123 degrees
   - (C) 125 degrees
   - (D) 127 degrees
   - (E) 129 degrees

Problem 6-12
Lever3.Dwg

Directions for Lever3.Dwg

Begin the construction of Lever3 in Figure 6–135 by keeping the default units set to decimal but changing the number of decimal places past the zero from four to two. Be sure the system of angle measure is set to decimal degrees and the number of decimal places for the display of angles is zero. Keep the remaining default unit values. Use the LIMITS command to change the drawing limits to 0.00,0.00 for the lower left corner and 15.00,12.00 for the upper right corner.

Begin the drawing in Figure 6–135 by placing the center of the regular hexagon and 2.25 radius arc at coordinate (6.25,6.50).

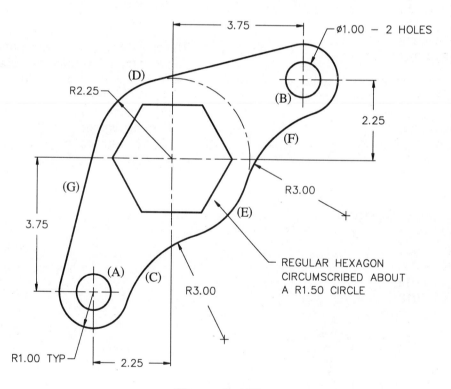

Figure 6–135

Refer to the drawing of Lever3 in Figure 6–135 to answer the following questions.

1. The distance from the center of the 1.00 diameter hole "A" to the center of the other 1.00 diameter hole "B" is
   (A) 8.39
   (B) 8.42
   (C) 8.46
   (D) 8.49
   (E) 8.52

2. The absolute coordinate value of the center of the 3.00 radius arc "C" is
   (A) 7.79,1.48
   (B) 7.79,1.51
   (C) 7.79,1.54
   (D) 7.76,1.51
   (E) 7.76,1.48

3. The total length of line "G" is
   (A) 4.14
   (B) 4.19
   (C) 4.24
   (D) 4.29
   (E) 4.34

4. The length of the 2.25 radius arc segment "E" is
   (A) 2.10
   (B) 2.13
   (C) 2.16
   (D) 2.19
   (E) 2.22

5. The angle formed in the X-Y plane from the center of the 1.00 diameter circle "B" to the center of the 2.25 radius arc "D" is closest to
   (A) 203 degrees
   (B) 205 degrees
   (C) 207 degrees
   (D) 209 degrees
   (E) 211 degrees

6. The total surface area of Lever3 with the hexagon and both 1.00 diameter holes removed is
   (A) 20.97
   (B) 21.00
   (C) 21.03
   (D) 21.06
   (E) 21.09

7. The total length of arc "D" is
   (A) 2.31
   (B) 2.36
   (C) 2.41
   (D) 2.46
   (E) 2.51

Problem 6-13
Housing1.Dwg

Directions for Housing1.Dwg

Begin constructing Housing1 in Figure 6–136 by keeping the default units set to decimal but change the number of decimal places past the zero from four to three. Be sure the system of angle measure is set to decimal degrees and the number of decimal places for the display of angles is zero. Keep all remaining default unit values. Keep the default drawing limits set to 0.000,0.000 for the lower left corner and 12.000,9.000 for the upper right corner.

Place the center of the 1.500 radius circular centerline at coordinate (6.500,5.250). Before constructing the outer ellipse, set the Pellipse variable to a value of 1. This will draw the ellipse as a polyline object.

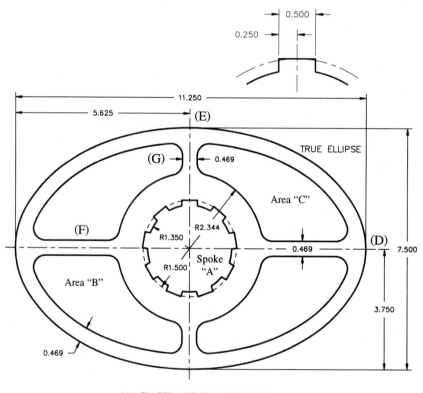

Figure 6–136

Refer to the drawing of Housing1 in Figure 6–136 to answer the following questions.

1. The perimeter of Spoke "A" is closest to
   (A) 11.564
   (B) 11.570
   (C) 11.576
   (D) 11.582
   (E) 11.588

2. The perimeter of Area "B" is closest to
   (A) 12.513
   (B) 12.519
   (C) 12.525
   (D) 12.531
   (E) 12.537

3. The total area of Area "C" is closest to
   (A) 7.901
   (B) 7.907
   (C) 7.913
   (D) 7.919
   (E) 7.927

4. The absolute coordinate value of the intersection of the ellipse and centerline at "D" is
   (A) 12.125,5.244
   (B) 12.125,5.250
   (C) 12.125,5.256
   (D) 12.125,5.262
   (E) 12.125,5.268

5. The total surface area of Housing1 with the spoke and all slots removed is
   (A) 27.095
   (B) 28.101
   (C) 28.107
   (D) 28.113
   (E) 28.119

6. The distance from the midpoint of the horizontal line segment at "F" to the midpoint of the vertical line segment at "G" is
   (A) 4.235
   (B) 4.241
   (C) 4.247
   (D) 4.253
   (E) 4.259

7. Increase Spoke "A" in size using the SCALE command. Use the center of the 1.500 radius arc as the base point. Use a scale factor of 1.115 units. The new total area of Housing1 with the spoke and all slots removed is closest to
   (A) 26.536
   (B) 26.542
   (C) 26.548
   (D) 26.554
   (E) 26.560

Problem 6-14
Cam1.Dwg

Directions for Cam1.Dwg

Start a new drawing called Cam1 (see Figure 6–137). Keep the default settings of decimal units but change the number of decimal places past the zero from four to two. Be sure the system of angle measure is set to decimal degrees and the number of decimal places for the display of angles is zero. Keep all remaining default unit values. Keep the default limit settings at 0.00,0.00 by 12.00,9.00.

Begin the drawing in Figure 6–137 by constructing the center of the 2.00 unit diameter circle at coordinate (3.00,4.00).

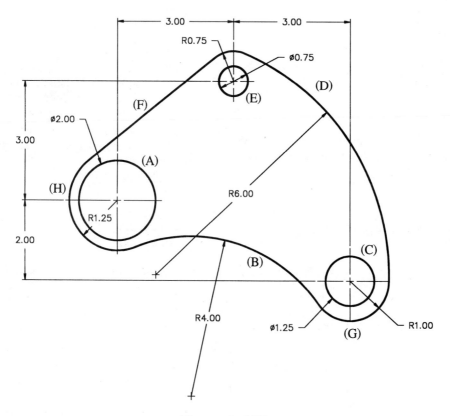

Figure 6–137

Refer to the drawing of Cam1 in Figure 6–137 to answer the following questions.

1. The absolute coordinate of the center of the 4.00 radius arc "B" is
   (A) 4.92,0.89
   (B) 4.92,-0.89
   (C) 4.95,-0.91
   (D) 4.92,-0.85
   (E) -4.95,-0.85

2. The angle formed in the X-Y plane from the center of the 2.00 diameter circle "A" to the center of the 1.25 diameter circle "C" is
   (A) 334 degrees
   (B) 336 degrees
   (C) 338 degrees
   (D) 340 degrees
   (E) 342 degrees

3. The length of arc "D" is
   (A) 7.20
   (B) 7.23
   (C) 7.26
   (D) 7.29
   (E) 7.32

4. The total area of Cam1 with all three holes removed is
   (A) 26.78
   (B) 26.81
   (C) 26.84
   (D) 26.87
   (E) 26.90

5. The total length of line "F" is
   (A) 4.05
   (B) 4.09
   (C) 4.13
   (D) 4.17
   (E) 4.21

6. The delta X distance from the quadrant of the 1.00 radius arc at "G" to the quadrant of the 1.25 radius arc at "H" is
   (A) -7.13
   (B) -7.16
   (C) -7.19
   (D) -7.22
   (E) -7.25

7. Use the ROTATE command to realign Cam1. Use the center of the 2.00 diameter circle "A" as the base point of the rotation. Rotate Cam1 from this point at a -10-degree angle. The absolute coordinate value of the center of the 0.75 diameter hole "E" is
   (A) 6.48,6.37
   (B) 6.48,6.40
   (C) 6.48,6.43
   (D) 6.51,6.46
   (E) 6.54,6.49

Problem 6-15
Pattern4.Dwg

Directions for Pattern4.Dwg

Start a new drawing called Pattern4 (see Figure 6–138). Even though this is a metric drawing, no special limits need be set. Keep the default setting of decimal units but change the number of decimal places past the zero from 4 to (zero). Be sure the system of angle measure is set to decimal degrees and the number of decimal places for the display of angles is zero. Keep all remaining default unit values.

Begin the drawing in Figure 6–138 by constructing Pattern4 with vertex "A" at absolute coordinate (50,30).

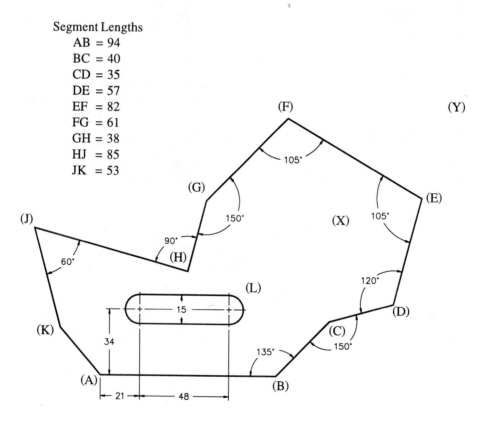

Segment Lengths
AB = 94
BC = 40
CD = 35
DE = 57
EF = 82
FG = 61
GH = 38
HJ = 85
JK = 53

Figure 6–138

Refer to the drawing of Pattern4 in Figure 6–138 to answer the following questions.

1. The total distance from the intersection of vertex "K" to the intersection of vertex "A" is
   - (A) 33
   - (B) 34
   - (C) 35
   - (D) 36
   - (E) 37

2. The total area of Pattern4 with the slot removed is
   - (A) 14493
   - (B) 14500
   - (C) 14529
   - (D) 14539
   - (E) 14620

3. The perimeter of the outline of Pattern4 is
   - (A) 570
   - (B) 578
   - (C) 586
   - (D) 594
   - (E) 602

4. The distance from the intersection of vertex "A" to the intersection of vertex "E" is
   - (A) 186
   - (B) 190
   - (C) 194
   - (D) 198
   - (E) 202

5. The absolute coordinate value of the intersection at vertex "G" is
   - (A) 104,117
   - (B) 105,118
   - (C) 106,119
   - (D) 107,120
   - (E) 108,121

6. The total length of arc "L" is
   - (A) 20
   - (B) 21
   - (C) 22
   - (D) 23
   - (E) 24

7. Stretch the portion of Pattern4 around the vicinity of angle "E." Use "Y" as the first corner of the crossing box. Use "X" as the other corner. Use the endpoint of "E" as the base point of the stretching operation. For the new point, enter a polar coordinate value of 26 units in the 40-degree direction. The new degree value of the angle formed at vertex "E" is
   - (A) 78 degrees
   - (B) 79 degrees
   - (C) 80 degrees
   - (D) 81 degrees
   - (E) 82 degrees

Problem 6-16
Rotor.Dwg

Directions for Rotor.Dwg

Start a new drawing called Rotor as shown in Figure 6–139. Keep the default setting of decimal units but change the number of decimal places past the zero from four to three. Be sure the system of angle measure is set to decimal degrees and the number of decimal places for the display of angles is zero. Keep all remaining default unit values. Keep the default limit settings at 0.000,0.000 for the lower left corner and 12.000,9.000 for the upper right corner.

Begin the drawing in Figure 6–139 by constructing the center of the 6.250 unit diameter circle at "A" at coordinate (5.500,5.000).

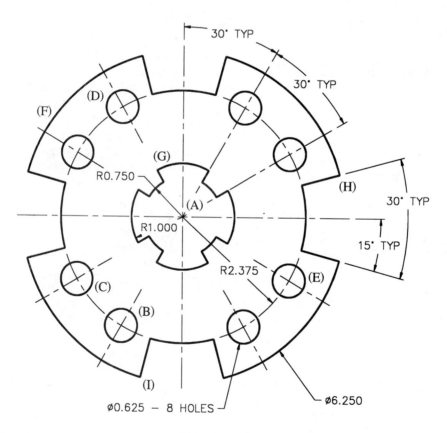

Figure 6–139

Refer to the drawing of the Rotor in Figure 6–139 to answer the following questions.

1. The absolute coordinate value of the center of the 0.625 diameter circle "B" is closest to
   (A) 4.294,2.943
   (B) 4.300,2.943
   (C) 4.306,2.943
   (D) 4.312,2.943
   (E) 4.318,2.943

2. The total area of the Rotor with all eight holes and the center slot removed is
   (A) 21.206
   (B) 21.210
   (C) 21.214
   (D) 21.218
   (E) 21.222

3. The total length of arc "F" is
   (A) 3.260
   (B) 3.264
   (C) 3.268
   (D) 3.272
   (E) 3.276

4. The distance from the center of the 0.625 circle "C" to the center of the 0.625 circle "D" is
   (A) 3.355
   (B) 3.359
   (C) 3.363
   (D) 3.367
   (E) 3.371

5. The angle formed in the X-Y plane from the center of the 0.625 circle "B" to the center of the 0.625 circle "E" is
   (A) 11 degrees
   (B) 13 degrees
   (C) 15 degrees
   (D) 17 degrees
   (E) 19 degrees

6. The delta X,Y distance from the intersection at "H" to the intersection at "I" is
   (A) -3.827,-3.827
   (B) -3.827,3.827
   (C) -3.834,3.820
   (D) -3.841,3.813
   (E) -3.848,-3.806

7. Use the SCALE command to increase the size of just the center slot "G." Use the center of arc "F" as the base point. Use a scale factor of 1.500 units. The new surface area of the Rotor with the center slot and all eight holes removed is
   (A) 17.861
   (B) 17.868
   (C) 17.875
   (D) 17.882
   (E) 17.889

Problem 6-17
Template.Dwg

Directions for Template.Dwg

Use the Units Control dialog box to set the units to decimal. Set the number of digits to the right of the decimal point from four to three. Change the system of degrees from "Decimal" to "Degrees/Minutes/Seconds" along with the following angular precision: 0D00'00". Keep the default values for the limits of the drawing.

Begin the drawing in Figure 6–140 by locating Point "A" at coordinate (4.500,2.750).

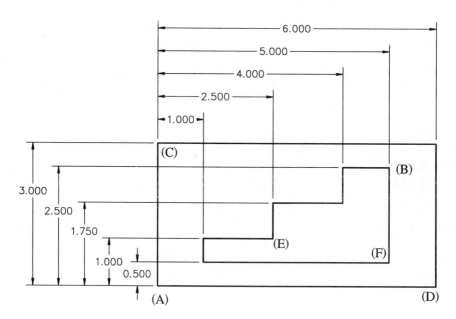

Figure 6–140

Refer to the drawing of the Template in Figure 6–140 to answer the following questions.

1. The angle formed in the X-Y plane from the intersection at "A" to the intersection at "B" is closest to
   (A) 26d25'21"
   (B) 26d28'3"
   (C) 26d30'10"
   (D) 26d32'37"
   (E) 26d33'54"

2. The distance from the intersection at "C" to the intersection at "D" is
   (A) 6.704
   (B) 6.708
   (C) 6.712
   (D) 6.716
   (E) 6.720

3. The angle formed in the X-Y plane from the intersection at "E" to the intersection at "F" is closest to
   (A) 347d31'10"
   (B) 347d47'32"
   (C) 348d1'53"
   (D) 348d20'20"
   (E) 348d41'24"

4. The total area of the Template with the step pattern removed is
   (A) 13.367
   (B) 13.371
   (C) 13.375
   (D) 13.379
   (E) 13.383

Using the SCALE command and select all objects to reduce the size of the Template. Use Point "A" as the base point and 0.950 as the scale factor and answer the following questions.

5. The perimeter of the inside step pattern is
   (A) 11.384
   (B) 11.388
   (C) 11.392
   (D) 11.396
   (E) 11.400

6. The new total area of the Template with the step pattern removed is
   (A) 12.059
   (B) 12.063
   (C) 12.067
   (D) 12.071
   (E) 12.075

7. The absolute coordinate value of the endpoint of the line at "B" is
   (A) 9.200,5.100
   (B) 9.250,5.125
   (C) 9.250,5.130
   (D) 9.260,5.135
   (E) 9.260,5.140

Problem 6-18
S-Cam.Dwg

Directions for S-Cam.Dwg

Begin the construction of S-Cam in Figure 6–141 by keeping the default units set to decimal but chang-ing the number of decimal places past the zero from four to three. Be sure the system of angle measure is set to decimal degrees and the number of decimal places for the display of angles is zero. Keep all remaining default unit values. Keep the default drawing limits set to 0.000,0.000 for the lower left corner and 12.000,9.000 for the upper right corner.

Begin the drawing in Figure 6–141 by placing the center of the 1.750 diameter circle at coordinate (2.500,3.750).

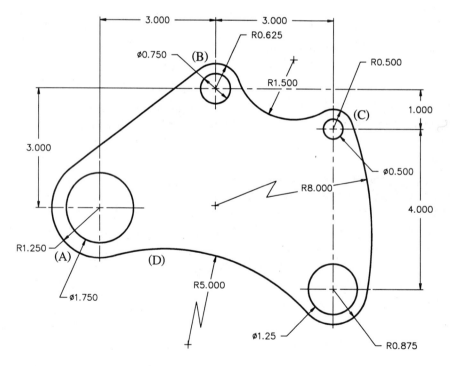

Figure 6–141

Refer to the drawing of the S-Cam in Figure 6–141 to answer the following questions.

1. The total surface area of the S-Cam with all four holes removed is
    (A) 27.654
    (B) 27.660
    (C) 27.666
    (D) 27.672
    (E) 28.678

2. The distance from the center of the 1.250 radius arc "A" to the center of the 0.625 radius arc "B" is
    (A) 4.237
    (B) 4.243
    (C) 4.249
    (D) 4.255
    (E) 4.261

3. The absolute coordinate value of the center of arc "D" is
    (A) 4.208,-2.258
    (B) 4.208,-2.262
    (C) 4.208,-2.266
    (D) 4.208,-2.270
    (E) 4.208,-2.274

4. The total length of arc "D" is
    (A) 5.480
    (B) 5.486
    (C) 5.492
    (D) 5.498
    (E) 6.004

5. The angle formed in the X-Y plane from the center of the 0.625 arc "B" to the center of the 0.500 arc "C" is closest to
    (A) 334 degrees
    (B) 336 degrees
    (C) 338 degrees
    (D) 340 degrees
    (E) 342 degrees

6. The delta X,Y distance from the center of the 0.500 arc "C" to the quadrant of the 5.000 arc "D" is
    (A) -4.284,-3.012
    (B) -4.288,-3.012
    (C) -4.292,-3.012
    (D) -4.296-3.012
    (E) -4.300,-3.012

7. Use the SCALE command to reduce S-Cam in size. Use the center of the 1.250 radius arc "A" as the base point. Use a scale factor of 0.822 units. The new surface area of S-Cam with all four holes removed is
    (A) 18.694
    (B) 18.698
    (C) 18.702
    (D) 18.706
    (E) 18.710

Problem 6-19
Housing2.Dwg

Directions for Housing2.Dwg

Begin the construction of Housing2 in Figure 6–142 by keeping the default units set to decimal, but change the number of decimal places past the zero from four to three. Be sure the system of angle measure is set to decimal degrees and the number of decimal places for the display of angles is zero. Keep all remaining default unit values. Keep the default drawing limits set to 0.000,0.000 for the lower left corner and 12.000,9.000 for the upper right corner. Before constructing the outer ellipse, set the PELLIPSE variable to a value of 1. This will draw the ellipse as a polyline object.

Begin the drawing in Figure 6–142 by placing the center of the ellipse at coordinate (5.500,4.500).

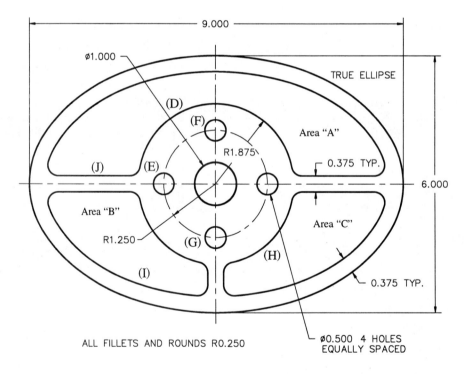

Figure 6–142

Refer to the drawing of Housing2 in Figure 6–142 to answer the following questions.

1. The total area of Housing2 with all five holes and Areas "A," "B," and "C" removed is
   (A) 20.082
   (B) 20.088
   (C) 20.094
   (D) 20.100
   (E) 20.106

2. The perimeter of Area "A" is
   (A) 19.837
   (B) 19.843
   (C) 19.849
   (D) 19.855
   (E) 19.861

3. The length of arc "D" is
   (A) 5.107
   (B) 5.113
   (C) 5.119
   (D) 5.125
   (E) 5.131

4. The distance from the center of the 0.500 diameter hole "E" to the center of the 0.500 diameter hole "F" is
   (A) 1.762
   (B) 1.768
   (C) 1.774
   (D) 1.780
   (E) 1.786

5. The total length of polyarc "I" is
   (A) 4.374
   (B) 4.380
   (C) 4.386
   (D) 4.392
   (E) 4.398

6. The delta X,Y distance from the center of circle "G" to the midpoint of the horizontal line at "J" is
   (A) -2.943,1.438
   (B) 2.943,1.438
   (C) 2.937,-1.438
   (D) -2.931,1.431
   (E) -2.931,-1.425

7. Use the MOVE command to relocate Housing2 at a distance of 0.375 units in a 45-degree angle direction. The new absolute coordinate value of the center of the 0.500 diameter hole "G" is
   (A) 5.759,3.515
   (B) 5.765,3.503
   (C) 5.759,3.509
   (D) 5.765,3.509
   (E) 5.765,3.515

Problem 6-20
Pattern5.Dwg

Directions for Pattern5.Dwg

Begin the construction of Pattern5 in Figure 6–143 by keeping the default units set to decimal, but change the number of decimal places past the zero from four to zero. Be sure the system of angle measure is set to decimal degrees and the number of decimal places for the display of angles is zero. Keep all remaining default unit values. Use the LIMITS command to set the drawing limits to 0,0 for the lower left corner and 250,200 for the upper right corner.

Begin the Figure 6–143 drawing by placing Vertex "A" at absolute coordinate (190,30).

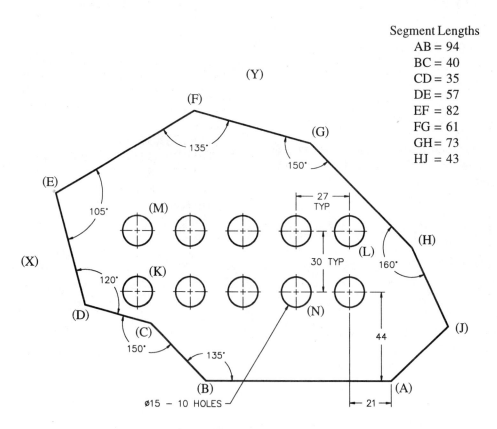

Segment Lengths
AB = 94
BC = 40
CD = 35
DE = 57
EF = 82
FG = 61
GH = 73
HJ = 43

Figure 6–143

Refer to the drawing of Pattern5 in Figure 6–143 to answer the following questions.

1. The distance from the intersection of vertex "J" to the intersection of vertex "A" is
   (A) 38
   (B) 39
   (C) 40
   (D) 41
   (E) 42

2. The perimeter of Pattern5 is
   (A) 523
   (B) 524
   (C) 525
   (D) 526
   (E) 527

3. The total area of Pattern5 with all 10 holes removed is
   (A) 16369
   (B) 16370
   (C) 16371
   (D) 16372
   (E) 16373

4. The distance from the center of the 15 diameter hole "K" to the center of the 15 diameter hole "L" is
   (A) 109
   (B) 110
   (C) 111
   (D) 112
   (E) 113

5. The angle formed in the X-Y plane from the center of the 15 diameter hole "M" to the center of the 15 diameter hole "N" is closest to
   (A) 340 degrees
   (B) 342 degrees
   (C) 344 degrees
   (D) 346 degrees
   (E) 348 degrees

6. The absolute coordinate value of the intersection at "F" is
   (A) 90,163
   (B) 92,163
   (C) 94,165
   (D) 96,167
   (E) 98,169

7. Use the STRETCH command to lengthen Pattern5. Use "Y" as the first point of the stretch crossing box. Use "X" as the other corner. Pick the intersection at "F" as the base point and stretch Pattern5 a total of 23 units in the 180 direction. The new total area of Pattern5 with all ten holes removed is
   (A) 17746
   (B) 17753
   (C) 17760
   (D) 17767
   (E) 17774

Problem 6-21
Bracket5.Dwg

Directions for Bracket5.Dwg

   Begin the construction of Bracket5 in Figure 6–144 by keeping the default units set to decimal but change the number of decimal places past the zero from four to two. Be sure the system of angle measure is set to decimal degrees and the number of decimal places for the display of angles is zero. Keep all remaining default unit values. Use the LIMITS command to set the drawing limits to 0.00,0.00 for the lower left corner and 15.00,12.00 for the upper right corner.

   Begin the drawing in Figure 6–144 by placing the center of the 1.50 diameter hole at coordinate (7.50,5.75).

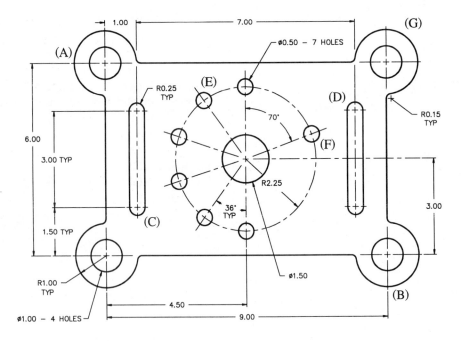

Figure 6–144

Refer to the drawing of Bracket5 in Figure 6–144 to answer the following questions.

1. The distance from the quadrant of the 1.00 radius arc at "A" to the quadrant of the 1.00 radius arc at "B" is
   - (A) 12.09
   - (B) 12.12
   - (C) 12.15
   - (D) 12.18
   - (E) 12.21

2. The distance from the center of the 0.25 radius arc at "C" to the center of the 0.25 radius arc at "D" is
   - (A) 7.62
   - (B) 7.65
   - (C) 7.68
   - (D) 7.71
   - (E) 7.74

3. The distance from the center of the 0.50 diameter circle at "E" to the center of the 0.50 diameter circle at "F" is
   - (A) 3.50
   - (B) 3.53
   - (C) 3.56
   - (D) 3.59
   - (E) 3.62

4. The length of arc "G" is
   - (A) 4.42
   - (B) 4.45
   - (C) 4.48
   - (D) 4.51
   - (E) 4.54

5. The total area of Bracket5 with all holes and slots removed is
   - (A) 53.72
   - (B) 53.75
   - (C) 53.78
   - (D) 53.81
   - (E) 53.84

6. The delta X,Y distance from the center of the 0.25 radius arc "C" to the center of the 0.50 circle "E" is
   - (A) 2.12,3.38
   - (B) 2.18,3.32
   - (C) 2.24,3.26
   - (D) 2.30,3.20
   - (E) 2.36,3.14

7. The angle formed in the X-Y plane from the center of the 0.25 radius arc "D" to the center of the1.00 radius arc "G" is
   - (A) 50 degrees
   - (B) 52 degrees
   - (C) 54 degrees
   - (D) 56 degrees
   - (E) 58 degrees

Problem 6-22
Plate2A.Dwg

Directions for Plate2A.Dwg

Begin Plate2A by keeping the default units set to decimal but changing the number of decimal places past the zero from four to two. Keep the system of angle measure set to decimal degrees, but change the fractional places for display of angles to two. Keep the remaining default unit values. Keep the default drawing limits set to 0.00,0.00 for the lower left corner and 12.00,9.00 for the upper right corner.

Begin constructing Plate2A by starting the lower left corner "X" at absolute coordinate (2.25,2.25). See Figure 6–145.

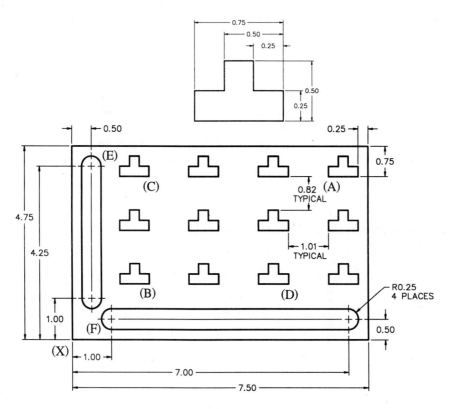

Figure 6–145

Refer to the drawing of Plate2A in Figure 6–145 to answer the following questions.

1. The total perimeter of the horizontal slot is
   (A) 13.42
   (B) 13.47
   (C) 13.52
   (D) 13.57
   (E) 13.62

2. The total perimeter of the vertical slot is
   (A) 7.98
   (B) 8.03
   (C) 8.07
   (D) 8.14
   (E) 8.20

3. The total area of the horizontal slot is
   (A) 3.02
   (B) 3.08
   (C) 3.14
   (D) 3.20
   (E) 3.26

4. The total area of the vertical slot is
   (A) 1.82
   (B) 1.88
   (C) 1.94
   (D) 2.00
   (E) 2.06

5. The distance from the endpoint of the line at "C" to the endpoint of the line at "D" is
   (A) 4.25
   (B) 4.30
   (C) 4.35
   (D) 4.40
   (E) 4.45

6. The angle formed in the X-Y plane formed from the center of arc "E" to the endpoint of the line at "B" is
   (A) 296.90 degrees
   (B) 296.96 degrees
   (C) 297.02 degrees
   (D) 297.08 degrees
   (E) 297.14 degrees

7. The angle formed in the X-Y plane from the endpoint of the line at "A" to the center of arc "F" is
   (A) 212.47 degrees
   (B) 212.53 degrees
   (C) 212.59 degrees
   (D) 212.65 degrees
   (E) 212.71 degrees

Problem 6-23
Plate2B.Dwg

Directions for Plate2B.Dwg

Change the number of decimal places past the zero from two to three. Change from decimal degrees to degrees/minutes/seconds. Change the number of places for display of angles to four.

Rotate Plate2B at a 32 degree, 0', 0" angle using absolute coordinate (2.250,2.250) at "X" below as the base point of rotation. Perform a ZOOM-Extents. Also see Figure 6–146.

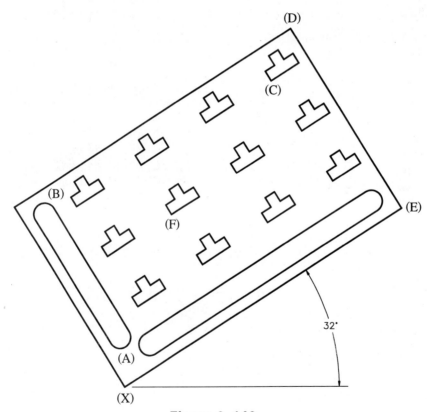

Figure 6–146

Refer to the drawing of Plate2B in Figure 6–146 to answer the following questions.

1. The angle in the X-Y plane from the center of arc "A" to the center of arc "B" is
   (A) 122d0'0"
   (B) 122d20'0"
   (C) 122d40'0"
   (D) 122d40'30'
   (E) 123d0'0"

2. The absolute coordinate value of the end-point of the line at "C" is
   (A) 5.627,9.071
   (B) 5.631,9.075
   (C) 5.635,9.079
   (D) 5.639,9.083
   (E) 5.643,9.087

3. The absolute coordinate value of the end-point of the line at "D" is
   (A) 6.077,10.237
   (B) 6.081,10.241
   (C) 6.085,10.245
   (D) 6.089,10.249
   (E) 6.093,10.253

4. The absolute coordinate value of the end-point of the line at "E" is
   (A) 8.606,6.220
   (B) 8.610,6.224
   (C) 8.614,6.228
   (D) 8.618,6.232
   (E) 8.622,6.236

5. The angle formed in the X-Y plane from the endpoint of the line at "F" to the center of arc "B" is
   (A) 178d1'2"
   (B) 178d27'37"
   (C) 179d15'12"
   (D) 179d39'49"
   (E) 180d34'29"

6. The delta X,Y distance from the intersection at "D" to the center of the 0.250 radius arc "A" is
   (A) -3.937,-6.878
   (B) -3.941,-6.882
   (C) -3.945,-6.886
   (D) -3.949,-6.890
   (E) -3.953,-6.894

7. Use the SCALE command to reduce Plate2B in size. Use absolute coordinate 2.250,2.250 as the base point. Use 0.777 as the scale factor. The area of Plate2B with both slots and the 12 T-shaped objects removed is
   (A) 16.659
   (B) 16.663
   (C) 16.667
   (D) 16.671
   (E) 16.675

Problem 6-24
Ratchet.Dwg

Directions for Ratchet.Dwg

Use the Units Control dialog box to change the number of decimal places past the zero from four to two. Be sure the system of angle measure is set to decimal degrees and the number of decimal places for the display of angles is zero. Keep the remaining default unit values. Use the LIMITS command to set the upper right corner of the display screen to 13.00,10.00.

Begin by drawing the center of the 1.00 radius arc of the Ratchet at absolute coordinate (6.00,4.50), as shown in Figure 6–147.

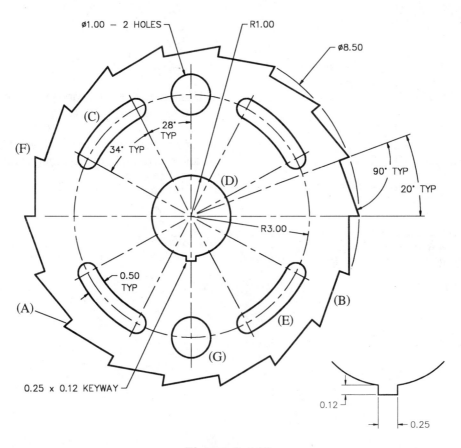

Figure 6–147

Refer to the drawing of the Ratchet in Figure 6–147 to answer the following questions.

1. The total length of the short line segment "A" is
    (A) 0.22
    (B) 0.24
    (C) 0.26
    (D) 0.28
    (E) 0.30

2. The total length of line "B" is
    (A) 1.06
    (B) 1.19
    (C) 1.33
    (D) 1.45
    (E) 1.57

3. The total length of arc "C" is
    (A) 1.81
    (B) 1.93
    (C) 2.08
    (D) 2.19
    (E) 2.31

4. The perimeter of the 1.00 radius arc "D" with the 0.25 x 0.12 keyway is
    (A) 6.52
    (B) 6.63
    (C) 6.77
    (D) 6.89
    (E) 7.02

5. The total surface area of the Ratchet with all 4 slots, the two 1.00 diameter holes, and the 1.00 radius arc with the keyway removed is
    (A) 42.98
    (B) 43.04
    (C) 43.10
    (D) 43.16
    (E) 43.22

6. The absolute coordinate value of the end-point at "F" is
    (A) 2.01,5.10
    (B) 2.01,5.63
    (C) 2.01,5.95
    (D) 2.20,5.95
    (E) 2.37,5.95

7. The angle formed in the X-Y plane from the endpoint of the line at "F" to the center of the 1.00 diameter hole "G" is
    (A) 304 degrees
    (B) 306 degrees
    (C) 308 degrees
    (D) 310 degrees
    (E) 312 degrees

Problem 6-25
Slide.Dwg

Directions for Slide.Dwg

Use the current unit settings and limits settings for this drawing (Figure 6–148). The number of decimal places past the zero should already be set to four. Be sure the system of angle measure is set to decimal degrees and the number of decimal places for the display of angles is zero.

Begin by drawing the center of the 0.5000 diameter circle of the Slide at absolute coordinate (2.0000,2.2500).

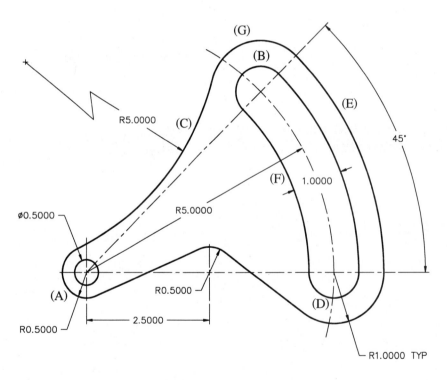

Figure 6–148

Refer to the drawing of the Slide in Figure 6–148 to answer the following questions.

1. The absolute coordinate value of the center of the 5.0000 radius arc "C" is
    (A) -0.2800,6.5603
    (B) 0.2887,-5.4393
    (C) -0.3953,7.2503
    (D) -0.2819,7.2543
    (E) -0.3953,6.2543

2. The total length of arc "E" is
    (A) 4.7264
    (B) 4.7124
    (C) 5.4302
    (D) 4.8710
    (E) 4.6711

3. The distance from the center of the 0.5000 diameter circle "A" to the center of the 5.0000 radius arc "C" is
    (A) 5.5000
    (B) 5.0043
    (C) 4.8768
    (D) 4.6691
    (E) 4.0001

4. The total area of the of the 1.0000 diameter slot is
    (A) 4.7124
    (B) 4.7863
    (C) 6.8370
    (D) 4.7625
    (E) 5.9102

5. The angle formed in the X-Y plane from the upper quadrant of the 1.0000 diameter slot at "B" to the lower quadrant of the slot at "D" is
    (A) 282 degrees
    (B) 284 degrees
    (C) 286 degrees
    (D) 288 degrees
    (E) 290 degrees

6. The total area of the Slide with the 0.5000 diameter hole and slot "F" removed is
    (A) 8.9246
    (B) 13.6370
    (C) 13.8750
    (D) 13.8333
    (E) 14.0297

7. The absolute coordinate value of the center of the 1.0000 radius arc "G" is
    (A) 5.5120,5.5621
    (B) 5.5237,5.5551
    (C) 5.5355,5.5590
    (D) 5.5355,5.6123
    (E) 5.5355,5.7855

Problem 6-26
Geneva.Dwg

Directions for Geneva.Dwg

   Start a new drawing called Geneva (Figure 6–149). Keep the default settings of decimal units, but change the number of decimal places past the zero from four to two. Be sure the system of angle measure is set to decimal degrees and the number of decimal places for the display of angles is zero. Keep the remaining default unit values.

   Begin the drawing by constructing the 1.50 diameter arc at absolute coordinate (7.50,5.50).

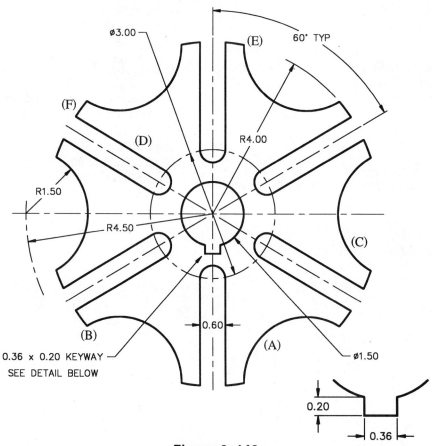

Figure 6–149

Refer to the drawing of the Geneva in Figure 6–149 to answer the following questions.

1. The total length of arc "A" is closest to
   (A) 3.00
   (B) 3.10
   (C) 3.20
   (D) 3.30
   (E) 3.40

2. The angle formed in the X-Y plane from the intersection at "B" to the center of arc "C" is
   (A) 11 degrees
   (B) 13 degrees
   (C) 15 degrees
   (D) 17 degrees
   (E) 19 degrees

3. The absolute coordinate value of the mid-point of line "D" is
   (A) 5.27,7.13
   (B) 5.27,7.17
   (C) 5.23,7.13
   (D) 5.31,7.13
   (E) 5.31,7.09

4. The total area of the Geneva with the 1.50 diameter hole and keyway removed is closest to
   (A) 27.20
   (B) 27.30
   (C) 27.40
   (D) 27.50
   (E) 27.60

5. The total distance from the midpoint of arc "F" to the center of arc "A" is
   (A) 8.24
   (B) 8.29
   (C) 8.34
   (D) 8.39
   (E) 8.44

6. The delta X,Y distance from the intersection at "E" to the center of arc "C" is
   (A) 3.93,-3.75
   (B) 3.93,3.75
   (C) 3.75,3.80
   (D) 3.75,3.93
   (E) 3.75,-3.93

7. Use the SCALE command to reduce the Geneva in size. Use 7.50,5.50 as the base point; use a scale factor of 0.83 units. The absolute coordinate value of the intersection at "E" is
   (A) 8.12,8.71
   (B) 8.12,8.76
   (C) 8.12,8.81
   (D) 8.12,8.86
   (E) 8.12,8.91

Problem 6-27
Rotor2.Dwg

Directions for Rotor2.Dwg

    Start a new drawing called Rotor2 (Figure 6–150). Keep the default settings of decimal units, but change the number of decimal places past the zero from four to three. Be sure the system of angle measure is set to decimal degrees and the number of decimal places for the display of angles is zero. Keep all remaining default unit values.

    Begin by constructing the 2.550 diameter circle at absolute coordinate (11.125,9.225).

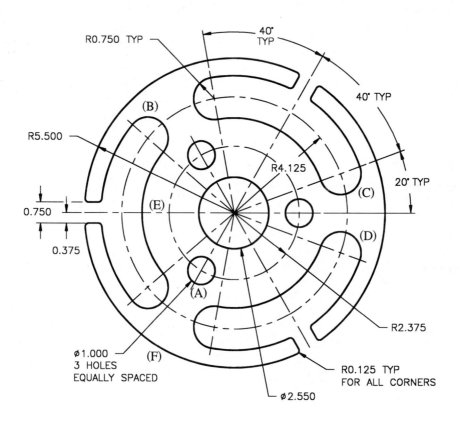

Figure 6–150

Refer to the drawing of Rotor2 in Figure 6–150 to answer the following questions.

1. The absolute coordinate value of the center of hole "A" is closest to
   (A) 9.937,7.158
   (B) 9.937,7.163
   (C) 9.937,7.168
   (D) 9.942,7.168
   (E) 9.947,7.173

2. The perimeter of Rotor2 is
   (A) 81.850
   (B) 81.855
   (C) 81.860
   (D) 81.865
   (E) 81.870

3. The distance from the center of arc "B" to the center of arc "C" is
   (A) 7.125
   (B) 7.130
   (C) 7.135
   (D) 7.140
   (E) 7.145

4. The absolute coordinate value of the center of arc "D" is closest to
   (A) 15.001,7.814
   (B) 15.001,7.819
   (C) 15.001,7.824
   (D) 15.006,7.829
   (E) 15.011,7.834

5. The total length of arc "F" is
   (A) 10.474
   (B) 10.479
   (C) 10.484
   (D) 10.489
   (E) 10.494

6. The total area of Rotor2 with all four holes removed is
   (A) 54.902
   (B) 54.907
   (C) 54.912
   (D) 54.917
   (E) 54.922

7. Change the diameter of all three 1.000 diameter holes to 0.700 diameter. Change the diameter of the 2.550 hole to a new diameter of 1.625. The new area of Rotor2 with all four holes removed is
   (A) 59.122
   (B) 59.127
   (C) 59.132
   (D) 59.137
   (E) 59.142

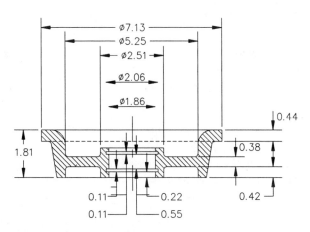

Region Modeling Techniques

As an alternate method of performing geometric constructions, a region model can be constructed. A region model represents a closed two-dimensional shape that is treated as a single object similar to a block. Properties such as Area and Perimeter are associated with regions. In some cases, the use of a region modeler may make geometric constructions easier and less time consuming. The method of constructing regions is through the use of "Boolean operations" which makes it completely different than using conventional construction methods. These operations of Union, Subtraction, and Intersection allow objects to be joined together into a single object or subtracted from each other leaving a difference in both objects. Once a region is constructed, a separate mass property utility is available to perform calculations such as area, perimeter, centroid, and even moments of inertia.

Locating the REGION Command

The REGION command can either be entered in from the keyboard using the word "REGION"; or it can be selected from the Draw pull-down menu shown in Figure 7–1.

The Boolean operations of Union, Subtraction, and Intersection may be selected from the Modify pull-down menu under the heading of "Boolean". These three commands do have custom command buttons designed for accessing the command from the Modify II toolbar, as shown in Figure 7–2.

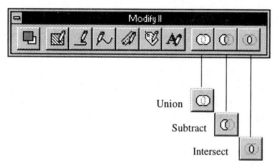

Union

Subtract

Intersect

Figure 7–2

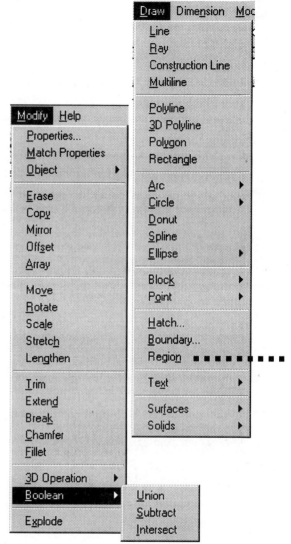

Figure 7–1

Creating a Region

Use the REGION command to create region objects. The object created by the REGION command must be a closed shape such as the circle in Figure 7–3. Regions cannot be made from closed objects consisting of individual line segments; they must be made from lightweight polyline objects or from objects converted into a lightweight polyline. The rectangle in Figure 7–3 was not constructed as individual lines; it consists of a lightweight polyline made with the RECTANG command where all objects are considered one object. See the following prompts for the use of the REGION command.

Command:**REGION**
Select objects: *(Select the circle at the right)*
Select objects: *(Select the rectangular polyline at the right)*
Select objects: *(Press* ENTER *to convert both objects into regions)*
2 loops extracted.
2 regions created.

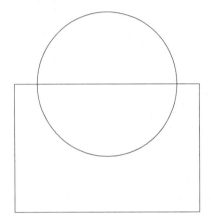

Figure 7–3

Using Boolean Operations

One of the most fundamental methods of creating regions is through the use of the Boolean operations of union, subtraction, and intersection. All three operations are explained in the following pages and utilize the circle and rectangle in Figure 7–4. To aid in the construction process, the four lines representing the rectangle have been converted into a single polyline. The RECTANG command automatically draws the rectangle as a polyline. These Boolean operations only operate on regions; therefore, convert the circle and rectangle shown in Figure 7–4 into separate regions using the REGION command.

Command:**REGION**
Select objects: *(Select the circle at the right)*
Select objects: *(Select the rectangular polyline at the right)*
Select objects: *(Press* ENTER *to convert both objects into regions)*
2 loops extracted.
2 regions created.

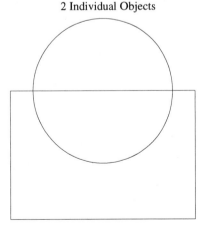

2 Individual Objects

Figure 7–4

Union

Use this command to join a group of objects into one complete region. Selecting the circle and rectangular polyline creates the region in Figure 7–5.

 Command: **UNION**

Select objects: *(Select the circle and rectangle)*
Select objects: *(Press [ENTER] to create a union of both objects)*

Subtraction

Use this command to subtract an object or series of objects from a source object. To create the cove in Figure 7–6, select the rectangle as the source object; the circle is subtracted from the rectangle to form the cove shape.

 Command: **SUBTRACT**

Select solids and regions to subtract from…
Select objects: *(Select the rectangle in Figure 7–4 as the source object)*
Select objects: *(Press [ENTER] to continue)*
Select solids and regions to subract…
Select objects: *(Select the circle in Figure 7–4)*
Select objects: *(Press [ENTER] to subtract the circle from the rectangle)*

Intersection

Use this command to produce the region from the intersection of two or more objects. The region shape in Figure 7–7 is common to both the circle and rectangle.

 Command: **INTERSECT**

Select objects: *(Select the circle and rectangle)*
Select objects: *(Press [ENTER] to create the intersection)*

Creating a Region by Union

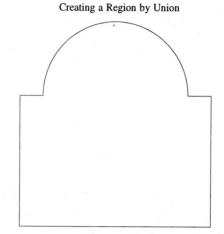

Figure 7–5

Creating a Region by Subtraction

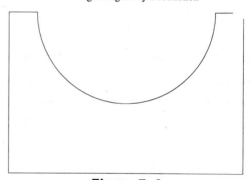

Figure 7–6

Creating a Region by Intersection

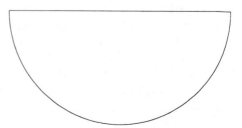

Figure 7–7

Creating a Region Boundary

So far, the Boolean operations of union, subtraction, and intersection operate only on objects converted into regions. You can use another method to create a series of regions automatically. This is through the use of the BOUNDARY command which displays the Boundary Creation dialog box shown in Figure 7–8. This command is located in the Draw pull-down menu area. Clicking in the "Object Type:" edit box displays two types of boundaries; namely, polyline and region boundaries. The polyline boundary will be discussed later in this unit.

After updating the "Object Type:" to reflect the Region option, you may now pick a point inside of an object. The result will be a region traced around the entire shape. It is also very important to create a new layer before performing this operation. The region object will be created in the current layer.

The object shown in Figure 7–9A was constructed using a series of lines and circles. The conventional editing commands of TRIM and OFFSET were used to create the shape the way it is shown in Figure 7–9A. Next, activate the Boundary command and be sure the current "Object Type:" is set to Region. Click on the "Pick Points<" button and identify a point at "A".

Notice all outlines and objects highlighted. All of these shapes will be made into three separate regions based on the current layer. In other words, the original objects in addition to the regions are present in the drawing. See Figure 7–9B.

Once the three regions are created, the circle and inner rectangle may now be subracted from the outer shape using the Boolean SUBTRACT command. The result is the creation of a single object called a region. When selecting the region in Figure 7–9C, the whole region highlights. A mass property calculation may then be run on the single region. If the need exists to return the object back to its original objects, use the EXPLODE command to accomplish this.

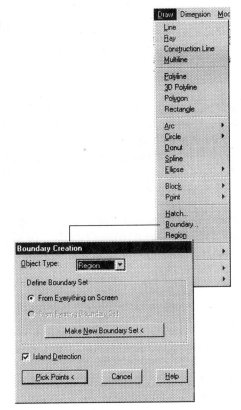

Figure 7–8

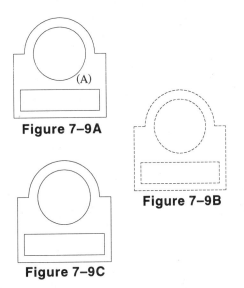

Figure 7–9A

Figure 7–9B

Figure 7–9C

Creating a Polyline Boundary

As with Figure 7–8 of creating a region when using the BOUNDARY command, a polyline boundary can also be created from the same Boundary Creation dialog box shown in Figure 7–10. The operation is identical to that of creating a region boundary; first the "Object Type:" is selected; then the "Pick Points <" button is selected which returns to the drawing editor; an internal point on the object to be traced is picked and the boundary is created.

The same set of rules for creating the region boundary exist for the polyline boundary; that is, the boundary is always created on the current layer. It is therefore good practice to create a layer to hold all boundary information and make it current before creating the boundary.

As with Figure 7–9A, the object shown in Figure 7–11A was constructed using a series of lines and circles. The conventional editing commands of TRIM and OFFSET were used to create the shape the way it is shown in Figure 7–11A. Next, activate the BOUNDARY command and be sure the current "Object Type:" is set to Polyline (this should be the default value). Click on the "Pick Points <" button and identify a point at "A".

Notice all outlines and objects highlighted. All of these shapes will be made into three separate polyline objects based on the current layer. In other words, the original objects, in addition to the polylines, are present in the drawing. See Figure 7–11B.

Once the three polylines are created, the circle and inner rectangle may now be subracted from the outer shape using the Subtract option of the AREA command. If the need exists to return the object back to its original objects, use the EXPLODE command to accomplish this. See Figure 7–11C.

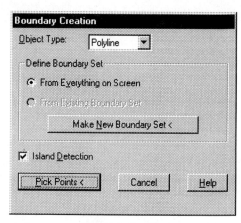

Figure 7–10

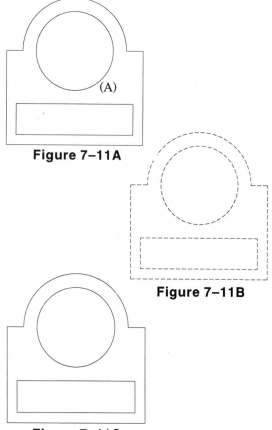

Figure 7–11A

Figure 7–11B

Figure 7–11C

Analyzing a Region Model using the MASSPROP Command

The following properties of the region selected are calculated and listed below:

- Area
- Perimeter
- Bounding Box
- Centroid
- Moments of Inertia
- Product of Inertia
- Radii of Gyration
- Principal Moments and X-Y directions about centroid.

The Area option displays the enclosed area of the region.

The Perimeter represents the total length of the inside or outside of the region.

The Bounding Box is represented by a rectangular box that totally encloses the region. The box is identified by two diagonal points as in "A" and "B" in the illustration at the right. See Figure 7–12.

The Centroid represents the center of the regions area.

If the "File..." button is selected from the "Mass Properties" dialog box, the following prompt appears:

Write to a file? <N>:
Answering "Yes" displays the next prompt:
File name <Default drawing name>:

After entering a file name, all mass property calculations are written out to a text file similar to the illustration in Figure 7–13. The text file that is created has the extension .MPR.

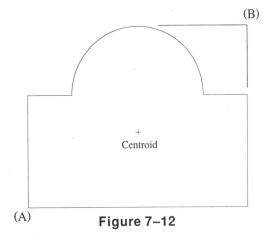

(B)

(A)   **Figure 7–12**

——————— REGIONS ———————	
Area:	18.7838
Perimeter:	49.3000
Bounding box:	X: 14.2915 — 20.6972
	Y: -19.6862 — -12.9640
Centroid:	X: 17.4944
	Y: -16.7342
Moments of inertia:	X: 5323.3672
	Y: 5828.7436
Product of inertia:	XY: -5499.0384
Radii of gyration:	X: 16.8345
	Y: 17.6155
Principal moments and X-Y directions about centroid:	
	I: 63.2719 along
	[1.0000 0.0000]
	J: 79.9080 along
	[0.0000 1.0000]

Figure 7–13

Tutorial Exercise
Gasket.Dwg

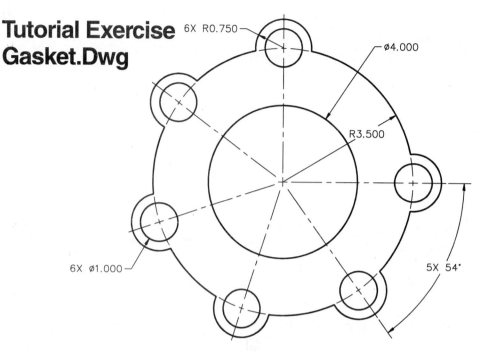

Figure 7–14

Purpose

This tutorial is designed to construct the gasket in Figure 7–14 using region modeling techniques. Once the object is constructed, it will be analyzed for accuracy.

System Settings

Begin a new drawing called "Region". Use the Units Control dialog box to change the number of decimal places past the zero from four to two. Keep the remaining default unit values. Keep the current limits set to 0,0 for the lower left corner and 12,9 for the upper right corner. The GRID or SNAP commands do not need to be set to any certain values.

Layers

Create the following layers with the format:

Name	Color	Linetype
Object	Green	Continuous
Boundary	Magenta	Continuous

Suggested Commands

Begin by locating the center of the large 4.00 diameter hole at absolute coordinate 6.00,4.50. Construct a circle of 0.75 radius and 1.00 diameter; then convert all circles in the drawing into regions using the REGION command. Use the ARRAY command to create the circular pattern of six holes spaced 54 degrees away from each other. Begin joining the outer perimeter of the gasket using the UNION command. To create the holes as cut-outs, use the SUBTRACT command. Use the MASSPROP command to analyze the region.

Whenever possible, substitute the appropriate command alias in place of the full AutoCAD command in each tutorial step. For example, use "Co" for the COPY command, "L" for the LINE command, and so on. The complete listing of all command aliases is located on pages 1–9 and 1–10.

Step #1

Make the "object" layer the new current layer. Construct two circles, one of radius 3.50 and the other of diameter 4.00 (see Figure 7–15). Use the center point of 6.00,4.50 for both circles. The "@" symbol is used for the center of the second circle to use the last known coordinate as the center.

 Command:**CIRCLE**

3P/2P/TTR/<Center point>:**6.00,4.50**
Diameter/<Radius>:**3.50**

 Command:**CIRCLE**

3P/2P/TTR/<Center point>: **@**
Diameter/<Radius>: **D** *(For Diameter)*
Diameter:**4.00**

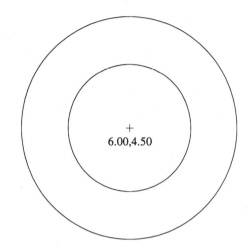

Figure 7–15

Step #2

Begin forming the lugs and holes of the region by constructing a circle of radius 0.75 and circle of diameter 1.00 from the upper quadrant of the large circle at "A" in Figure 7–16.

 Command:**CIRCLE**

3P/2P/TTR/<Center point>:**Qua**
of *(Select the quadrant of the large circle at "A")*
Diameter/<Radius>:**0.75**

 Command:**CIRCLE**

3P/2P/TTR/<Center point>: **@**
Diameter/<Radius>: **D** *(For Diameter)*
Diameter:**1.00**

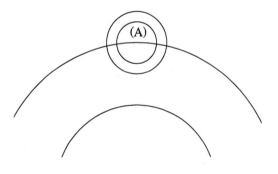

Figure 7–16

Step #3

Before copying the last two circles drawn in a circular pattern using the ARRAY command, convert all circles into regions using the REGION command.

Command: **REGION**
Select objects: *(Select all four circles shown in Figure 7–17)*
Select objects: *(Press [ENTER] key to execute this command)*
4 loops extracted.
4 regions created.

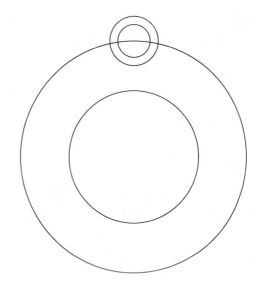

Figure 7–17

Step #4

Use the ARRAY command to duplicate the two regions six times using an included angle of 270 degrees. Since this is a positive angle, the direction of the array will be in the counterclockwise direction. See Figure 7–18.

 Command: **ARRAY**
Select objects: *(Select circles "A" and "B")*
Select objects: *(Press [ENTER] to continue)*
Rectangular or Polar array (R/P) <R>: **Polar**
Base/<Specify center point of array>: **Cen**
of *(Select the center of the circle at "C")*
Number of items: **6**
Angle between items (+=CCW, -=CW): **270**
Rotate object as they are copied? <Y>: *(Press [ENTER] to accept this default value)*

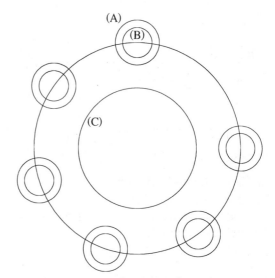

Figure 7–18

Step #5

Begin forming the region by using the UNION command to join the large circle with the six smaller 0.750-radius circles. See Figure 7–19.

 Command: **UNION**

Select objects: *(Select the large dashed circle at the right)*

Select objects: *(Select the six dashed circles at the right)*

Select objects: *(Press* ENTER *to create a union of both objects)*

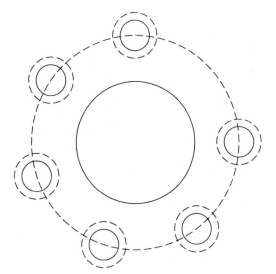

Figure 7–19

Step #6

Use the SUBTRACT command to subtract the seven holes in Figure 7–20 from the outer region.

 Command: **SUBTRACT**

Select solids and regions to subract from...

Select objects: *(Select the region at "A" as the source object)*

Select objects: *(Press* ENTER *to continue)*

Select solids and regions to subtract from...

Select objects: *(Select all seven circles "B" through "H" shown in Figure 7–20. This will subtract them from the original region)*

Select objects: *(Press* ENTER *to perform the subtraction operation)*

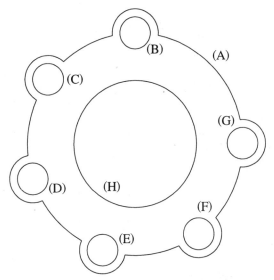

Figure 7–20

Step #7

An alternate step would be to construct the large
and inner circles as in Step #1. Begin forming the
lugs and holes of the region by constructing a circle
of radius 0.75 and circle of diameter 1.00 from the
upper quadrant of the large circle as in Step #2. Instead of creating regions of the two small circles,
use conventional AutoCAD commands to array the
two circles to form the six lugs and holes shown in
Figure 7–21A.

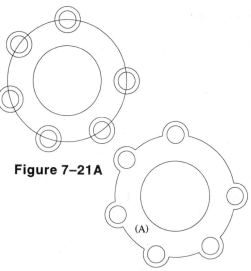

Figure 7–21A

Then, use the TRIM command to trim the excess
lines to form the object shown in Figure 7–21B.
The outline of the object will now consist of a series of arcs. The REGION command requires a
polyline shape before being converted into a region.
The BOUNDARY command will be used to trace a
region around all objects. Before performing this
command, make "Boundary" the new current layer.

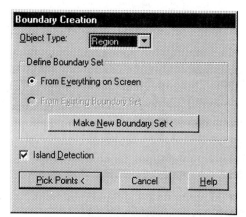

Figure 7–21B

Instead of using the REGION command, issue the
BOUNDARY command, change the "Object Type:"
to Region, and click on the "Pick Points <" button
shown in Figure 7–22. When prompted to pick an
internal point, pick a point located at "A" shown in
Figure 7–21B.

When picking an internal point inside of the object,
the entire object highlights as in Figure 7–23. Using the BOUNDARY command automatically performs an island detection based on the internal point
selected. This command automates the use of the
REGION command since each individual object
must be selected before a region is created of the
object. Also, all regions are now placed on the
"Boundary" layer.

Figure 7–22

With all highlighted objects in Figure 7–23 now
converted to regions, use the SUBTRACT command
to delete the seven circular regions from the outer
profile; one region is now created which is similar
to the object created in Step #6.

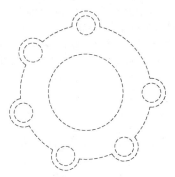

Figure 7–23

Step #8

To perform a calculation on a region, the MASSPROP command is used. This command is very useful to extract the area, perimeter, bounding box, and centroid information of a region. Other information such as Moments of inertia, Product of inertia, Radii of gyration, and Principal moments are also calculated.

 Command:**MASSPROP**

Select objects: *(Select the region model shown in Figure 7–24)*

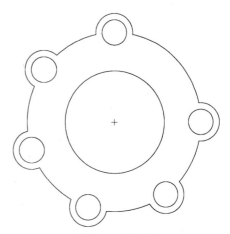

Figure 7–24

—————— REGIONS ——————

Area:	26.75
Perimeter:	59.49
Bounding box:	X: 1.92—0.25
	Y: 0.42—8.75
Centroid:	X: 5.99
	Y: 4.49
Moments of inertia:	X: 655.10
	Y: 1075.29
Product of inertia:	XY: 717.02
Radii of gyration:	X: 4.95
	Y: 6.34

Principal moments and X-Y directions about centroid:

I: 115.42 along [0.71 -0.71]
J: 118.08 along [0.71 0.71]

Step #9

The complete region after editing is shown in Figure 7–25. Dimensions may be added to the model at a later date.

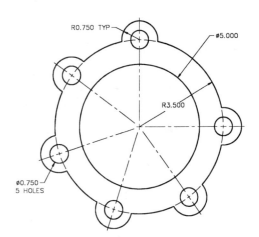

Figure 7–25

Problems for Unit 7

Problem 7–1

Construct Problem 3–1 on page 3–44 as a region model. Begin by constructing the center of the 1.50 radius circle at absolute coordinate 6.50,4.25. When completed, provide your answers to the following questions on the lines below. Round off all answers to two decimal places.

The area of the region is closest to: _____
The perimeter of the region is closest to: _____
The centroid of the region is located at: _____

Problem 7–2

Construct Problem 3–2 on page 3–44 as a region model. Begin by constructing the center of the 6.00 x 3.00 rectangle at absolute coordinate 6.54,4.21. When completed, provide your answers to the following questions on the lines below. Round off all answers to two decimal places.

The area of the region is closest to: _____
The perimeter of the region is closest to: _____
The centroid of the region is located at: _____

Problem 7–3

Construct Problem 3–3 on page 3–45 as a region model. Begin by constructing the center of the 1.875 radius arc at absolute coordinate 6.36,4.61. When completed, provide your answers to the following questions on the lines below. Round off all answers to two decimal places.

The area of the region is closest to: _____
The perimeter of the region is closest to: _____
The centroid of the region is located at: _____

Problem 7–4

Construct Problem 3–4 on page 3–45 as a region model. Begin by constructing the center of the 1.00 radius arc at absolute coordinate 7.12,5.38. When completed, provide your answers to the following questions on the lines below. Round off all answers to two decimal places.

The area of the region is closest to: _____
The perimeter of the region is closest to: _____
The centroid of the region is located at: _____

Problem 7–5

Construct Problem 3–5 on page 3–46 as a region model. Begin by constructing the center of the 2.25 radius arc at absolute coordinate 8.40,5.90. When completed, provide your answers to the following questions on the lines below. Round off all answers to two decimal places.

The area of the region is closest to: _____
The perimeter of the region is closest to: _____
The centroid of the region is located at: _____

Problem 7–6

Construct Problem 3–6 on page 3–46 as a region model. Begin by constructing the center of the 0.38 diameter circle at absolute coordinate 6.67,4.19. When completed, provide your answers to the following questions on the lines below. Round off all answers to two decimal places.

    The area of the region is closest to: _____
    The perimeter of the region is closest to: _____
    The centroid of the region is located at: _____

Problem 7–7

Construct Problem 3–7 on page 3–47 as a region model. Begin by constructing the center of the 1.0 radius arc at absolute coordinate 6.87,8.42. When completed, provide your answers to the following questions on the lines below. Round off all answers to two decimal places.

    The area of the region is closest to: _____
    The perimeter of the region is closest to: _____
    The centroid of the region is located at: _____

Problem 7–8

Construct Problem 3–8 on page 3–47 as a region model. Begin by constructing the center of the 25 radius arc at absolute coordinate 92,45. When completed, provide your answers to the following questions on the lines provided below. Round off all answers to two decimal places.

    The area of the region is closest to: _____
    The perimeter of the region is closest to: _____
    The centroid of the region is located at: _____

Problem 7–9

Construct Problem 3–9 on page 3–48 as a region model. Begin by constructing the center of the 0.53 radius arc at absolute coordinate 1.156,3.594. When completed, provide your answers to the following questions on the lines below. Round off all answers to two decimal places.

    The area of the region is closest to: _____
    The perimeter of the region is closest to: _____
    The centroid of the region is located at: _____

Problem 7–10

Construct Problem 3–10 on page 3–48 as a region model. Begin by constructing the center of the 2.50 diameter arc at absolute coordinate 12.70,9.29. When completed, provide your answers to the following questions on the lines below. Round off all answers to two decimal places.

    The area of the region is closest to: _____
    The perimeter of the region is closest to: _____
    The centroid of the region is located at: _____

Problem 7–11

Construct Problem 3–11 on page 3–49 as a region model. Begin by constructing the center of the 3.0 diameter arc at absolute coordinate 4.74,4.19. When completed, provide your answers to the following questions on the lines below. Round off all answers to two decimal places.

    The area of the region is closest to: _____
    The perimeter of the region is closest to: _____
    The centroid of the region is located at: _____

Problem 7–12

Construct Problem 3–12 on page 3–49 as a region model. Begin by constructing the center of the 50 diameter arc at absolute coordinate 100,90. When completed, provide your answers to the following questions on the lines below. Round off all answers to two decimal places.

The area of the region is closest to: _____

The perimeter of the region is closest to: _____

The centroid of the region is located at: _____

Problem 7–13

Construct Problem 3–13 on page 3–50 as a region model. Begin by constructing the center of the 3.12 radius arc at absolute coordinate 7.69,2.55. When completed, provide your answers to the following questions on the lines below. Round off all answers to two decimal places.

The area of the region is closest to: _____

The perimeter of the region is closest to: _____

The centroid of the region is located at: _____

Problem 7–14

Construct Problem 3–14 on page 3–50 as a region model. Begin by constructing the center of the 38 diameter circle at absolute coordinate 51,64. When completed, provide your answers to the following questions on the lines provided below. Round off all answers to two decimal places.

The area of the region is closest to: _____

The perimeter of the region is closest to: _____

The centroid of the region is located at: _____

Problem 7–15

Construct Problem 3–15 on page 3–51 as a region model. Begin by constructing the center of the 2.0 radius center arc at absolute coordinate 4.22,5.95. When completed, provide your answers to the following questions on the lines below. Round off all answers to two decimal places.

The area of the region is closest to: _____

The perimeter of the region is closest to: _____

The centroid of the region is located at: _____

Problem 7–16

Construct Problem 3–16 on page 3–51 as a region model. Begin by constructing the center of the leftmost 1.38 major diameter ellipse at absolute coordinate 4.49,3.28. When completed, provide your answers to the following questions on the lines below. Round off all answers to two decimal places.

The area of the region is closest to: _____

The perimeter of the region is closest to: _____

The centroid of the region is located at: _____

Problem 7–17

Construct Problem 3–17 on page 3–52 as a region model. Begin by constructing the center of the 14 unit hexagon at absolute coordinate 141,77. When completed, provide your answers to the following questions on the lines below. Round off all answers to two decimal places.

The area of the region is closest to: _____

The perimeter of the region is closest to: _____

The centroid of the region is located at: _____

Problem 7–18

Construct Problem 3–18 on page 3–52 as a region model. Begin by constructing the center of the 3.00 diameter circle at absolute coordinate 12.13,8.87. When completed, provide your answers to the following questions on the lines below. Round off all answers to two decimal places.

The area of the region is closest to: _____

The perimeter of the region is closest to: _____

The centroid of the region is located at: _____

Problem 7–19

Construct Problem 3–19 on page 3–53 as a region model. Begin by constructing the center of the 4.18 diameter arc at absolute coordinate 6.34,6.48. When completed, provide your answers to the following questions on the lines below. Round off all answers to two decimal places.

The area of the region is closest to: _____

The perimeter of the region is closest to: _____

The centroid of the region is located at: _____

Problem 7–20

Construct Problem 3–20 on page 3–53 as a region model. Begin by constructing the center of the 40 radius arc at absolute coordinate 150,69. When completed, provide your answers to the following questions on the lines provided below. Round off all answers to two decimal places.

The area of the region is closest to: _____

The perimeter of the region is closest to: _____

The centroid of the region is located at: _____

Problem 7–21

Construct Problem 3–21 on page 3–54 as a region model. Begin by constructing the center of the 1.50 radius arc at 6.40,10.73. When completed, provide your answers to the following questions on the lines provided below. Round off all answers to two decimal places.

The area of the region is closest to: _____

The perimeter of the region is closest to: _____

The centroid of the region is located at: _____

UNIT **8**

The figure at the top of the page shows a construction section drawing with the following labels:

35°

MEZZ. FL.
ELEV. 110'

2"

4 5/8"

4" bolts

1'-1 1/8"

8"

1-1/2" lam. oak

3/4" oak

8"

2X12

2X12

2X12C

1 3/4"

5/8" GYP BD.

10 7/8"

Section Views

Principles of orthographic projections remain the key method for the production of engineering drawings, whether using manual methods or CAD. As these drawings get more complicated in nature, the job of the operator or designer becomes more difficult in the interpretation of views, especially where hidden features are involved. The concept of slicing a view to expose these interior details is the purpose of performing a section. Section views then follow the same rules as orthographic or multiview drawings except that the creation of a section makes the drawing easier to read since hidden features are converted to visible features. In this unit you will learn how sections are formed in addition to the many types of sections available to the designer. Three tutorial exercises at the end of the unit are designed to give you experience using different methods of cross-hatching when using AutoCAD as a drafting tool.

Section View Basics

Figure 8–1 is a pictorial representation of a typical flange consisting of eight bolt holes and counterbore hole in the center.

The drawings in Figure 8–2 show a typical solution to a multiview problem complete with front and side views. The front view displaying the eight bolt holes is obvious to interpret; however, the numerous hidden lines in the side view make the drawing difficult to understand, and this is considered a relatively simple drawing. To relieve the confusion

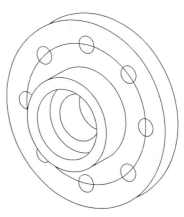

Figure 8–1

associated with a drawing too difficult to understand because of numerous hidden lines, a section
is made of the part. Orthographic methods are followed up to the creation of the side view.

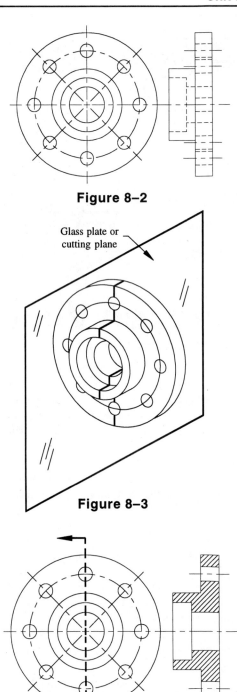

Figure 8–2

To understand section views more, see Figure 8–3.
Creating a section view slices an object in such a
way so as to expose what used to be hidden features and convert them into visible features. This
slicing or cutting operation can be compared to that
of using a glass plate or cutting plane to perform
the section. In the object in Figure 8–3, the glass
plate cuts the object in half. It is the responsibility
of the designer or CAD operator to convert one half
of the object into a section and to discard the other
half. Surfaces that come in contact with the glass
plane are crosshatched to show where the actual
cutting took place.

Glass plate or
cutting plane

Figure 8–3

A completed section view drawing is shown in Figure 8–4. Two new types of lines are also shown, a
cutting plane line and section lines. The cutting plane
line performs the cutting operation on the front view.
In the side view, section lines show the surfaces
that were cut. Notice that holes are not section lined
since the cutting plane passes across the center of
the hole. Notice also that hidden lines are not displayed in the side view. It is poor practice to merge
hidden lines into a section view although there are
always exceptions. The arrows of the cutting plane
line tell the designer to view the section in the direction of the arrows and discard the other half.

Figure 8–4

The cutting plane line consists of a very thick line at a series of dashes approximately 0.25" in length (see Figure 8–5A). A polyline is used to create this line of 0.05 thickness. The arrows point in the direction of sight used to create the section with the other half generally discarded. Assign this line one of the dashed linetypes; the hidden linetype is reserved for detailing invisible features in views. The section line, by contrast with the cutting plane line, is a very thin line (see Figure 8–5B). This line identifies the surfaces being cut by the cutting plane line. The section line is usually drawn at an angle and at a specified spacing.

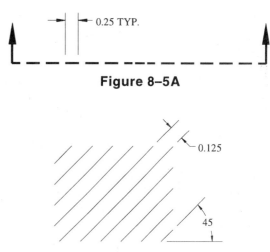

Figure 8–5A

Figure 8–5B

A wide variety of hatch patterns are already supplied with the software. One of these patterns, ANSI31, is displayed in Figure 8–6. This is one of the more popular patterns with lines spaced 0.125 units apart from each other and at a 45-degree angle.

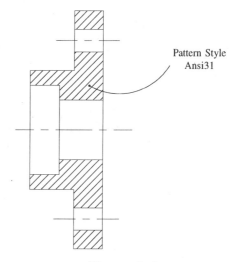

Pattern Style
Ansi31

Figure 8–6

The object in Figure 8–7 illustrates proper section lining techniques. Much of the pain of spacing the section lines apart from each other and at angles has been eased considerably by using the computer as a tool. However, the designer must still practice proper section lining techniques at all times for clarity of the section.

The next four examples illustrate common errors involved with section lines. In Figure 8–8A, section lines run the correct directions and at the same angle; however, the hidden lines have not been converted into object lines. This will confuse the more experienced designer since the presence of hidden lines in the section means more complicated invisible features. Figure 8–8B is yet another error encountered when creating sections. Again, the section lines are properly placed; however, all surfaces representing holes have been removed, which displays the object as a series of sectioned blocks unconnected, implying four separate parts.

Figure 8–8C appears to be a properly sectioned object; however, upon closer inspection, we see that the angle of the crosshatch lines in the upper half differ from the same lines in the lower left half. This suggests two different parts, when in actuality, it is the same part. In Figure 8–8D, all section lines run the correct direction. The problem is the lines run through areas that were not sliced by the cutting plane line. These areas in Figure 8–8D represent drill and counterbore holes and are left unsectioned.

These have been identified as the most commonly made errors when crosshatching an object. Remember just a few rules to follow: section lines are present only on surfaces that are cut by the cutting plane line; section lines are drawn in one direction when crosshatching the same part; hidden lines are usually omitted when creating a section view; areas such as holes are not sectioned since the cutting line only passes across this feature.

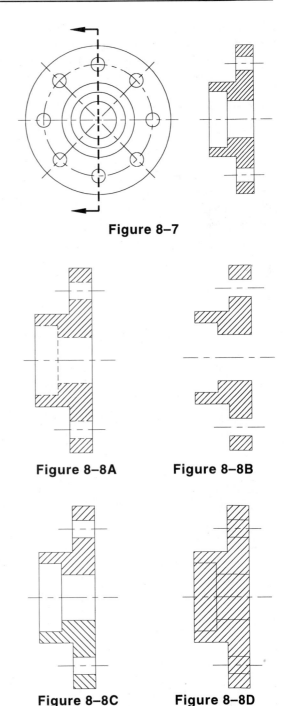

Figure 8–7

Figure 8–8A

Figure 8–8B

Figure 8–8C

Figure 8–8D

Full Sections

When the cutting plane line passes through the entire object, a full section is formed. In Figure 8–9, a full section would be the same as taking an object and cutting it completely in half. Depending on the needs of the designer, one half is kept, the other half is discarded. The half that is kept is section lined.

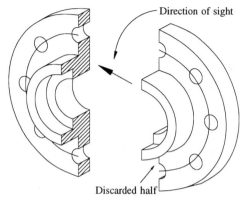

Figure 8–9

The multiview solution to the problem in Figure 8–9 is shown in Figure 8–10. The front view is drawn with lines projected across to form the side view. To show that the side view is in section, a cutting plane line is added to the front view. This line performs the physical cut. You have the option of keeping either half of the object. This is the purpose of adding arrowheads to the cutting plane line. The arrowheads define the direction of sight you view the object to form the section. You must then interpret what surfaces are being cut by the cutting plane line in order to properly add crosshatching lines to the section which is located in the right side view. Hidden lines are not necessary once a section has been made.

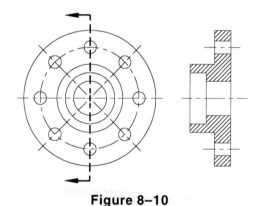

Figure 8–10

Numerous examples illustrate a cutting plane line with the direction of sight off to the left. This does not mean that a cutting plane line cannot have a direction of sight going to the right as in Figure 8–11. In this example, the section is formed from the left side view if the circular features are located in the front view.

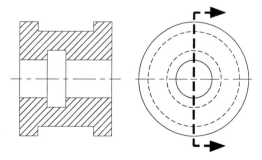

Figure 8–11

Half Sections

When symmetrical-shaped objects are involved, sometimes it is unnecessary to form a full section by cutting straight through the object. Instead, the cutting plane line passes only halfway through the object which makes the drawing in Figure 8–12 a half section. The rules for half sections are the same as for full sections; namely, a direction of sight is established, part of the object is kept, and part is discarded.

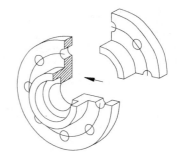

Figure 8–12

The views are laid out in Figure 8–13 in the usual multiview format. To prepare the object as a half section, the cutting plane line passes halfway through the front view before being drawn off to the right. The right side view is converted into a half section by crosshatching the upper half of the side view while leaving hidden lines in the lower half.

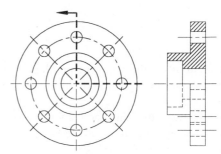

Figure 8–13

Depending on office practices, some designers prefer to omit hidden lines entirely from the side view similar to the drawing in Figure 8–14. In this way, the lower half is drawn of only what is visible.

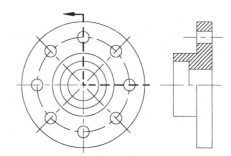

Figure 8–14

Figure 8–15 shows another way of drawing the cutting plane line to conform to the right side view drawn in section. Hidden lines have been removed from the lower half; only those lines visible are displayed.

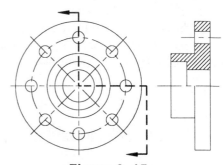

Figure 8–15

Assembly Sections

It would be unfair to give designers the impression that section views are only used for displaying internal features of individual parts. Yet another advantage of using section views is that it permits the designer to create numerous objects, assemble them, and then slice the assembly to expose internal details of all parts. This type of section is an assembly section similar to Figure 8–16. For all individual parts, notice the section lines running the same directions. This follows one of the basic rules of section views: keep section lines at the same angle for each individual part.

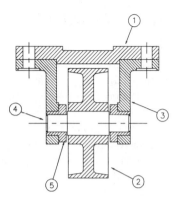

Figure 8–16

Figure 8–17 shows the difference of assembly sections and individual parts that have been sectioned. For parts in an assembly that contact each other, it is good practice to alternate the directions of the section lines and make the assembly much more clear and to distinguish the parts from each other. This can be accomplished by changing the angle of the hatch pattern or even the scale of the pattern.

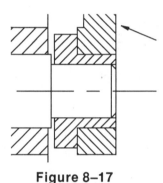

Figure 8–17

To identify parts in an assembly, an identifying part number along with a circle and arrowhead line are used as shown in Figure 8–18. The line is very similar to a leader line used to call out notes for specific parts on a drawing. The addition of the circle highlights the part number. Sometimes this type of call out is referred to as a "bubble."

Figure 8–18

In the enlarged assembly shown in Figure 8–19, the large area in the middle is actually a shaft used to support a pulley system. With the cutting plane passing through the assembly including the shaft, refrain from crosshatching features such as shafts, screws, pins, or other types of fasteners. The overall appearance of the assembly is actually enhanced by not crosshatching these items.

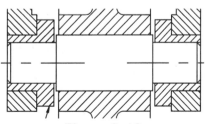

Figure 8–19

Aligned Sections

Aligned sections take into consideration the angular position of details or features of a drawing. Instead of drawing the cutting plane line vertically through the object in Figure 8–20, the cutting plane is angled or aligned with the same angle the elements are at. Aligned sections are also made to produce better clarity of a drawing. In Figure 8–20, with the cutting plane forming a full section of the object, it is difficult to obtain the true size of the angled elements. In the side view, they appear foreshortened or not to scale. Hidden lines were added as an attempt to better clarify the view.

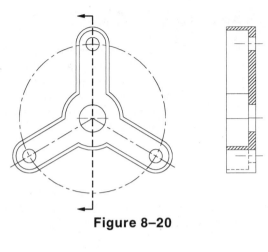

Figure 8–20

Instead of drawing the cutting plane line all the way through the object, the line is bent at the center of the object before being drawn through one of the angled legs (see Figure 8–21). The direction of sight arrows on the cutting plane line not only determines which direction the view will be sectioned, but also shows another direction for rotating the angled elements so they line up with the upper elements. This rotation is usually never more than 90 degrees. As lines are projected across to form the side view, the section appears as if it were a full section. This is only because the features were rotated and projected in section for greater clarity of the drawing.

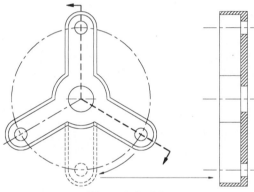

Figure 8–21

Offset Sections

Offset sections take their name from offsetting the cutting plane line to pass through certain details in a view (see Figure 8–22). If the cutting plane line passes straight through any part of the object, details would be exposed while others would remain hidden. By offsetting the cutting plane line, the designer controls its direction and which features of a part it passes through. The view to section follows the basic section rules.

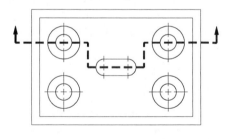

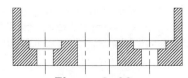

Figure 8–22

Sectioning Ribs

Contrary to section view principles, parts made out of cast iron with webs or ribs used for reinforcement do not follow basic rules of sections. In Figure 8–23, the front view has the cutting plane line passing through the entire view; the side view at "A" is crosshatched according to section view basics. However, it is difficult to read the thickness of the base since the crosshatching includes the base along with the web. A more efficient method is to ignore crosshatching webs as in "B." Therefore, not crosshatching the web exposes other important details such as thicknesses of bases and walls.

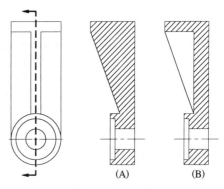

Figure 8–23

The object in Figure 8–24 is another example of performing a full section on an area consisting of webbed or ribbed features. By not crosshatching the webbed areas, more information is available such as the thickness of the base and wall areas around the cylindrical hole. This may not be considered true projection; however, it is considered good practice.

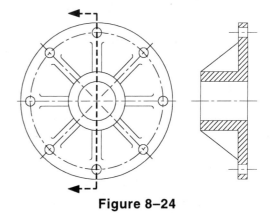

Figure 8–24

Broken Sections

At times only a partial section of an area needs to be created. For this reason, a broken section might be used. The object in Figure 8–25A shows a combination of sectioned areas and conventional areas outlined by the hidden lines. When converting an area to a broken section, you create a break line, crosshatch one area, and leave the other area in a conventional drawing. Break lines may take the form of short freehanded line segments shown in Figure 8–25B or a series of long lines separated by break symbols as in Figure 8–25C. The LINE command can be used with Ortho-Off to produce the desired effect as in Figures 8–25A and 8–25B.

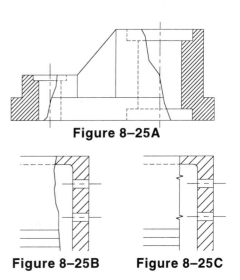

Figure 8–25A

Figure 8–25B   **Figure 8–25C**

Revolved Sections

Section views may be constructed as part of a view by using revolved sections. In Figure 8–26, the elliptical shape is constructed while it is revolved into position and then crosshatched.

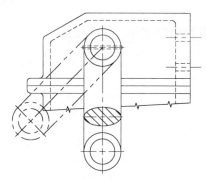

Figure 8–26

Figure 8–27 is another example of a revolved section where a cross-section of the C-clamp was cut away and revolved to display its shape.

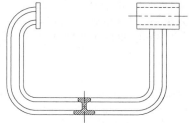

Figure 8–27

Removed Sections

Removed sections are very similar to revolved sections with the exception that instead of drawing the section somewhere inside of the object, as is the case of a revolved section, the section is placed elsewhere or removed to a new location in the drawing. The cutting plane line is present with the arrows showing the direction of sight. Identifying letters are placed on the cutting plane and underneath the section to keep track of the removed sections especially when there are a number of them on the same drawing sheet. See Figure 8–28.

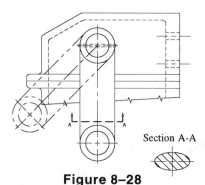

Section A-A

Figure 8–28

Another way of displaying removed sections is to use centerlines as a substitute for the cutting plane line. In Figure 8–29, the centerlines determine the three shapes of the chisel and display the basic shapes from circle to octagon to rectangle. Identification numbers are not required in this particular example.

Figure 8–29

Isometric Sections

Section views may be incorporated into pictorial drawings that appear in Figure 8–30A and 8–30B. Figure 8–30A is an example of a full isometric section with the cutting plane passing through the entire object. In keeping with basic section rules, only those surfaces sliced by the cutting plane line are crosshatched. Isometric sections make it easy to view cut edges compared to holes or slots. Figure 8–30B is an example of an isometric drawing converted into a half section.

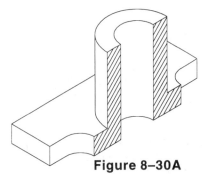

Figure 8–30A

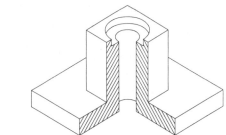

Figure 8–30B

Architectural Sections

Mechanical representations of machine parts are not the only type of drawings where section views are used. Architectural drawings rely on sections to show the type of building materials that go into the construction of foundation plans, roof details, or wall sections as shown in Figure 8–31. Here numerous types of crosshatching symbols are used to call out the different types of building materials such as brick veneer at "A," insulation at "B," finished flooring at "C," floor joists at "D," concrete block at "E," earth at "F," and poured concrete at "G." Section symbols provided in AutoCAD were used to crosshatch most of the building components with the exception of the floor joists and insulation.

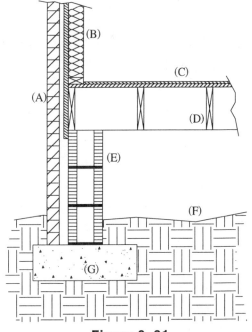

Figure 8–31

Using the BHATCH Command

Picking "Hatch…" from the "Draw" pull-down menu activates the main Boundary Hatch dialog box shown in Figure 8–32. Use this dialog box to pick a point identifying the area to be crosshatched, select objects using one or more of the popular selection set modes, select hatch patterns supported in the software, and change the scale and angle of the hatch pattern. The "Advanced…" button activates a second dialog box which controls the various hatching styles in addition to controlling the automatic island detection feature. All areas grayed out in Figure 8–32 are currently unavailable. Only when additional parameters have been satisfied will these buttons activate for use.

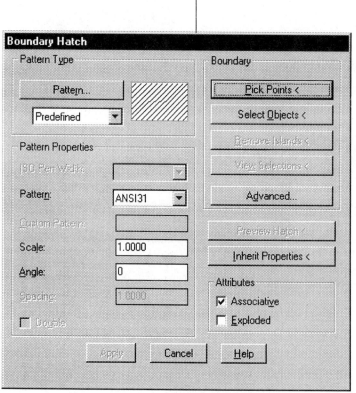

Figure 8–32

The object shown in Figure 8–33 will be used to demonstrate the boundary crosshatching method. The object needs to have areas "A," "B," and "C" crosshatched.

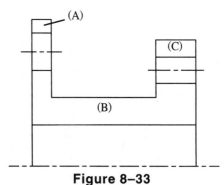

Figure 8–33

In the past, the BREAK command was one technique used to isolate the three areas shown in Figure 8–34. Another technique to use in crosshatching is to trace a closed polyline on top of the three areas, performing the hatch operation, and then erase each polyline boundary. The BHATCH command automatically highlights the areas to be crosshatched.

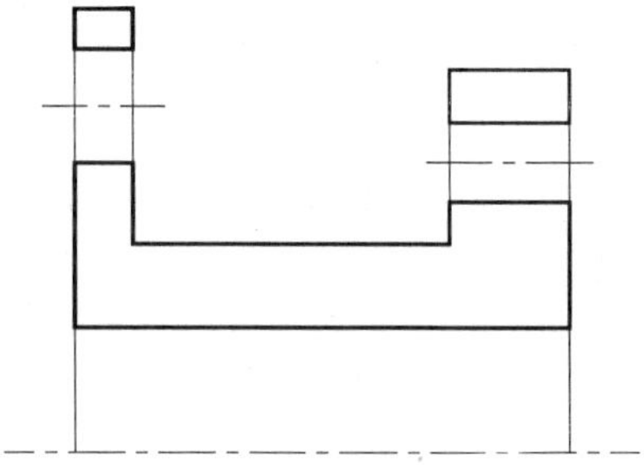

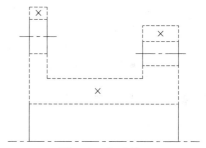

Figure 8–34

The BHATCH command outlines the boundaries to be crosshatched using a method of ray tracing. Selecting an internal point in the area to be crosshatched sends out tracers in all directions. As a tracer indentifies the edge of an object, it begins highlighting the closed shape. In Figure 8–35 are three areas with internal points identified by the "Xs." The dashed areas identify the highlighted areas.

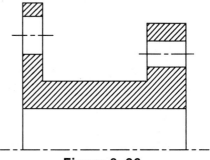

Figure 8–35

In the final step of the BHATCH command, as the crosshatch pattern is applied to the areas, the highlighted outlines deselect leaving the crosshatch patterns. See Figure 8–36.

Figure 8–36

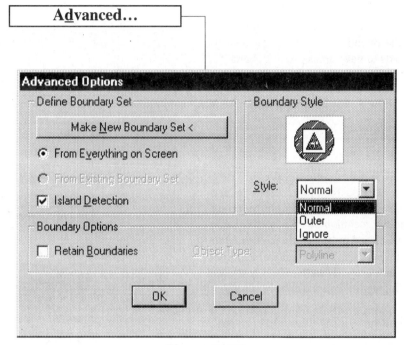

Figure 8–37

Selecting the "Advanced ..." button from the main Boundary Hatch dialog box activates the dialog box shown in Figure 8–37. This dialog box holds a number of crosshatching options such as Making a New Boundary Set, Boundary Style, Boundary Options, and an Island Detection check box.

Figure 8–38 shows three boundary styles and how they affect levels of crosshatching. The Normal boundary style is the default hatching method where upon selecting all objects within a window, the hatching begins with the outermost boundary, skipping the next inside boundary, hatching the next innermost boundary, and so on. Notice the hatching pattern still exposing the text objects for easy reading. The outermost boundary style hatches only the outermost boundary of the object. The ignore hatch style ignores the default hatching methods of alternating crosshatching and hatches the entire object.

You have the option of retaining the hatch boundary by clicking inside of the check box called Retain Boundaries.

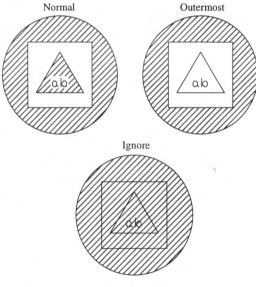

Figure 8–38

By default, Island Detection is turned on. In this way, selecting an internal point identifies all boundaries including islands.

The Boundary Hatch dialog box holds numerous hatch patterns in a drop down edit box in Figure 8–39. Clicking on a pattern in this listing makes the pattern current.

Selecting the "Pattern ..." button from the Boundary Hatch dialog box displays a series of crosshatching patterns already created and ready for use in Figure 8–40. Select a particular pattern by picking on the pattern itself. If the wrong pattern was selected, simply pick the correct one. Selecting the "Next" button takes you to the next hatch pattern palette screen. Once on the second hatch pattern collection, the "Previous" button activates, allowing you to either go back to the previous hatch pattern palette or to the next hatch pattern palette by selecting the "Next" button.

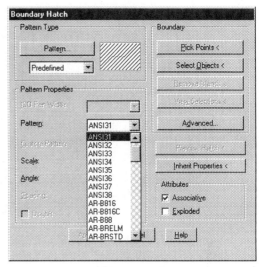

Figure 8–39

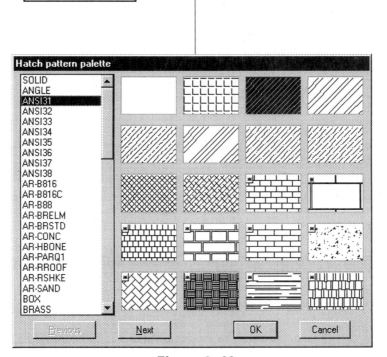

Figure 8–40

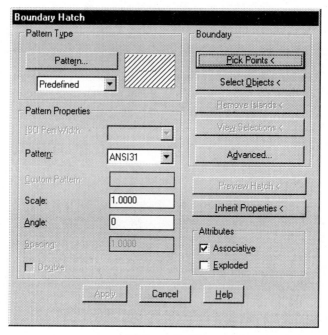

Figure 8–41

Once a hatch pattern is selected, the name is displayed in the Boundary Hatch dialog box in Figure 8–41. If the scaling and angle settings look favorable, select the "Pick Points <" button. The command prompt requires you to pick an internal point or points. In Figure 8–42, click in the three areas marked by the "X." Notice these areas highlight. If these areas are correct, press the ENTER key to return to the Boundary Hatch dialog box.

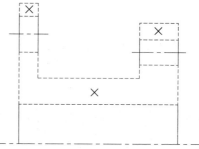

Figure 8–42

Once back in the Boundary Hatch dialog box, you now have the option of first previewing the hatch to see if all settings and the appearance of the pattern are desirable. Pick the "Preview Hatch <" button to accomplish this. The results appear in Figure 8–43. Technically, the pattern is still not placed on the object. Striking the ENTER key to exit preview mode returns you back to the main Boundary Hatch dialog box. If the hatch pattern is correct in appearance, pick the "Apply" button to place the pattern with the drawing.

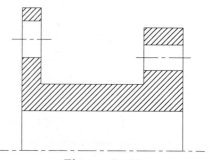

Figure 8–43

Solid Fill Hatch Patterns

Past versions of AutoCAD allowed individuals to fill in a confined space using the SOLID command (see Figure 8–44A). This worked especially well for vertice corners. Four points had to be picked to define a rectangle before the solid was placed. An additional four points defined the next rectangle and so on.

A new solid fill pattern is now available through the BHATCH command on the Hatch Pattern Palette. By picking an internal point, the entire closed area will be filled with a solid pattern and placed on the current layer. The color of the pattern will take on the same color assigned to the layer.

The solid hatch pattern in Figure 8–44B was placed with one internal point pick.

Creating a solid fill for the outline of the object presented numerous difficulties. As the SOLID command was effective where pronounced vertices were involved, such as a triangle, rectangle, or multisided polygon, it was very ineffective on curved outlines as in the graphic in Figure 8–45A. To simulate a solid fill, the BHATCH command was used along with a hatch pattern such as ANSI31 where the scale of the pattern was reduced very small. The end result was a solid fill; however, upon closer inspection, the crosshatch lines were placed at very small increments from each other. This tended to increase the size of the drawing file for even the simplest of fill operations.

The solid hatch pattern also works for curved outlines as well as outlines involving know vertice corners. Figure 8–45B displays the solid pattern completely filling in the outline where all corners consist of some type of curve generated by the FILLET command.

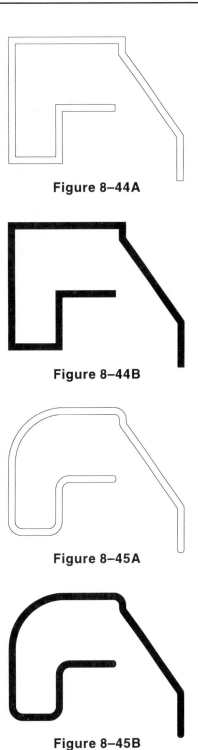

Figure 8–44A

Figure 8–44B

Figure 8–45A

Figure 8–45B

Using BHATCH to Hatch Islands

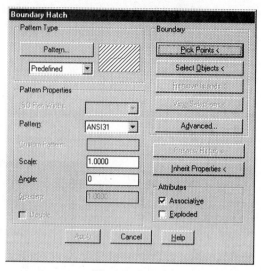

Figure 8–47

When confronted with the task of crosshatching islands in Figure 8–46, first issue the BHATCH command (see Figure 8–47), select a pattern, and pick a point at "A" which will not only define the outer perimeter of the object, but the inner shapes as well (see Figure 8–46). This result is due to "Island Detection" being activated in the "Advanced Options" dialog box of the BHATCH command.

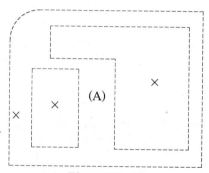

Figure 8–46

Selecting "Preview Hatch <" should display the hatch pattern shown in Figure 8–48. If changes need to be made, such as a change in the hatch scale or angle, preview allows these changes to be made. After changes, be sure to preview the pattern once again to check if the results are desirable. Selecting "Apply" places the pattern and exits back to the command prompt.

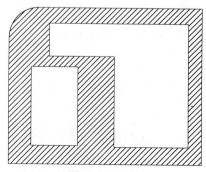

Figure 8–48

Hatch Pattern Scaling

Hatching patterns are already predefined in size and angle. When using the BHATCH command, the pattern used is assigned a scale value of 1.00 which will draw the pattern exactly the way it was orginally created. Figure 8–49 shows the effects of the BHATCH command when accepting the default value for the pattern scale.

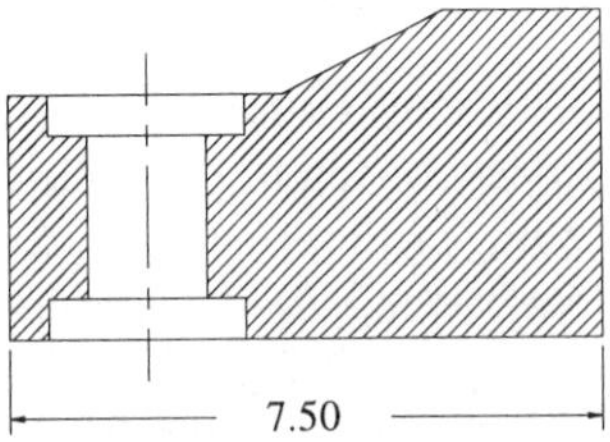

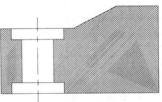

Figure 8–49

Entering a different scale value for the pattern in the Boundary Hatch dialog box will either increase or decrease the spacing in between crosshatch lines. Figure 8–50 is an example of the ANSI31 pattern with a new scale value of 0.50.

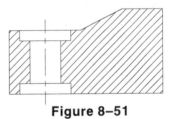

Figure 8–50

As the scale of a pattern can be decreased to hatch small areas, so also may the pattern be scaled up for large areas. Figure 8–51 has a hatch scale of 2.00 which doubles all distances in between hatch lines.

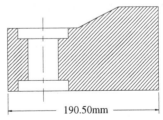

Figure 8–51

Use care when hatching large areas. In Figure 8–52, the distance measures 190.50 millimeters. If the hatch scale of 1.00 were used, the pattern would take on a filled appearance. This results in numerous lines being generated which will increase the size of the drawing file. A value of 25.4 is used to scale hatch lines for metric drawings.

Figure 8–52

Inherit Hatching Properties

The image in Figure 8–56 consists of a simple assembly drawing. At least three different hatch patterns are displayed to define the different parts of the assembly. Unfortunately, a segment of one of the parts was not crosshatched and it is unclear what pattern, scale, and angle were used to place the pattern. Whenever faced with this problem, click on the "Inherit Properties <" button in the main Boundary Hatch dialog box in Figure 8–57. Clicking on the pattern at "A" returns you to the Boundary Hatch dialog box where the pattern, scale, and angle are automatically set from the selected pattern.

To complete the assembly, click on the "Pick Points <" button of the Boundary Hatch dialog box and click an internal point in the empty area. The hatch pattern is placed in this area and matches that of the other patterns to identify the common parts as shown in Figure 8–58.

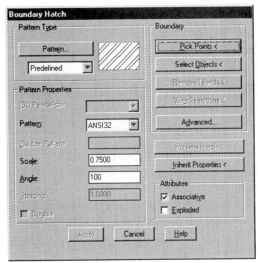

Figure 8–57

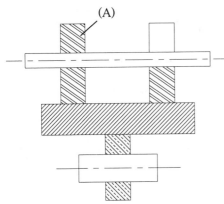

Figure 8–56

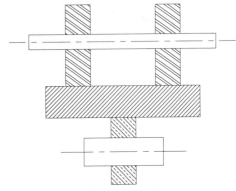

Figure 8–58

Properties of Associative Crosshatching

The following is a review of associative crosshatching. In Figure 8–59, the plate needs to be hatched in the ANSI31 pattern at a scale of two units and an angle of zero; the two slots and three holes are to be considered islands. Enter the BHATCH command, make the necessary changes inside of the Boundary Hatch dialog box, click the OK button, and mark an internal point somewhere inside of the object (such as at "A").

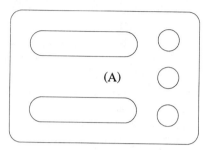

Figure 8–59

When all objects inside and including the outline highlight, press the ENTER key to return back to the Boundary Hatch dialog box. It is very important to realize that all objects highlighted are tied to the associative crosshatch object. Each shape works directly with the hatch pattern to insure that the outline of the object is being read by the hatch pattern and that the hatching is performed outside of the outline. You have the option of first previewing the hatch pattern or applying the hatch pattern. In either case, the results appear similar to the object in Figure 8–60.

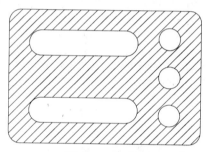

Figure 8–60

Associative crosshatch objects may be edited and the results of this editing will have an immediate impact on the future appearance of the hatch pattern. For example, the two outer holes need to be increased in size by a factor of 2; also, the middle hole needs to be repositioned to the other side of the object (see Figure 8–61). Using the MOVE command not only allows you to reposition the hole, but when the move is completed, the crosshatch pattern updates itself automatically to the moved circle. In the same manner, using the SCALE command to increase the size of the two outer circles by a value of 2 units makes the circles large and automatically updates the hatch pattern.

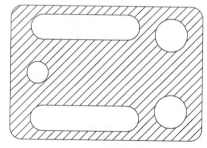

Figure 8–61

In Figure 8-62, the length of the slots need to be increased. Also, holes "B" and "C" need to be deleted. The STRETCH command is used to lengthen the slots. Use the crossing box at "A" to select the slots and stretch them one unit to the right. Use the ERASE command to delete holes "B" and "C." The result is shown in Figure 8-63 and in all cases, the associative crosshatch pattern updates to these changes.

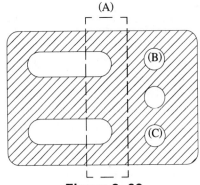

Figure 8—62

In past uses of associative crosshatching, if a member of the group of objects that helped define the boundary of the hatch pattern was erased, the hatch pattern would remain in the drawing; however, it would lose its associativity. Also, the crosshatch pattern would not mend to the erased hole resulting in a blank spot on the drawing where the hole once existed. In Figure 8–63, associative crosshatching has been enhanced for members to be erased while the hatch pattern still maintains its associativity. The two outer holes were deleted using the ERASE command. After performing the erase, the hatch pattern updated itself and hatched through the areas once occupied by the holes.

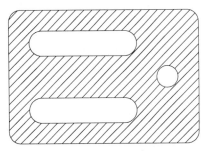

Figure 8—63

Once a hatch pattern is placed in the drawing, it will not update itself to any new additions in the form of closed shapes in the drawing. In Figure 8–64, a rectangle was added at the center of the object. Notice, however, the crosshatch pattern cuts directly through the rectangle. Since the crosshatch pattern does not have the intelligence to recognize the new boundary, the entire hatch pattern must be deleted and crosshatched again. In this way, the rectangular boundary will be recognized.

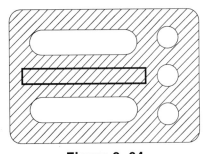

Figure 8—64

Using the HATCHEDIT Command

In several past versions of AutoCAD, whenever a hatch pattern was placed in a drawing, it remained uneditable. If the pattern needed to change or if the scale needed to be increased, the original pattern had to be erased or rehatched. In Figure 8–65, the pattern needs to be increased to a new scale factor of 3 units and the angle of the pattern needs to be rotated 90 degrees. The current scale value of the pattern is 2 units and the angle is 0 degrees. Issuing the HATCHEDIT command prompts you to select the hatch pattern to edit. Clicking on the hatch pattern anywhere inside of the object in Figure 8–65 displays the Hatchedit dialog box, in Figure 8–66.

With the dialog box displayed in Figure 8–66, click in the Scale edit box to change the scale from the current value of 2 units to the new value of 3 units. Next, click on the Angle edit box and change the angle of the hatch pattern from the current value of 0 degrees to the new value of 90 degrees. This will increase the spacing in between each hatch line and rotate the pattern 90 degrees in the counterclockwise direction. Clicking on the Apply button in the Hatchedit dialog box returns to the drawing editor and updates the hatch pattern to these changes (see Figure 8–67). Be aware of the following rules to practice for the HATCHEDIT command to function; HATCHEDIT only works on an associative hatch pattern. If the pattern loses associativity through the use of the MIRROR command or if the hatch pattern is exploded, the HATCHEDIT command will cease to function.

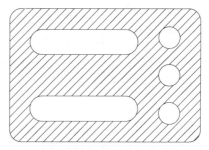

Figure 8–65

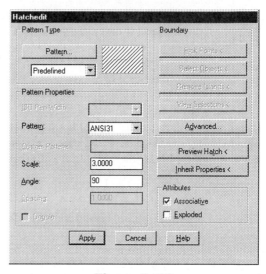

Figure 8–66

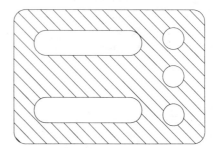

Figure 8–67

Tutorial Exercise
Dflange.Dwg

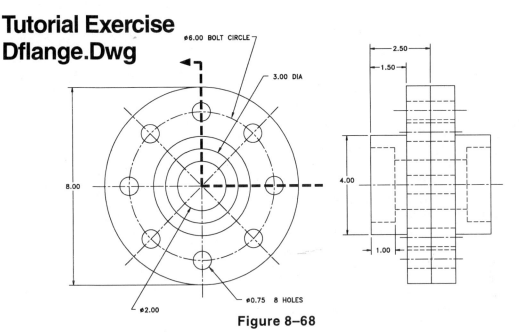

Figure 8–68

Purpose

This tutorial is designed to use the MIRROR, CHPROP, DDCHPROP, and BHATCH commands to convert the Dflange (double flange), into a half section.

System Settings

This drawing is complete except for converting the upper half into a half section. All Units, Limits, Grid, and Snap have already been set. From the drawing disk supplied with this text, call up a drawing called Dflange.

Layers

The following layers have already been created:

Name	Color	Linetype
Object	White	Continuous
Cen	Yellow	Center
Hid	Red	Hidden
Xhatch	Magenta	Continuous
Cpl	Yellow	Dashed
Pline	Green	Continuous
Dim	Yellow	Continuous

Suggested Commands

Begin this tutorial by converting one half of the object into a section by erasing unnecessary hidden lines. Next use the CHPROP or DDCHPROP commands and change the remaining hidden lines to the Object layer. Issue the BHATCH command and use the ANSI31 hatch pattern to hatch the upper half of the object on the Xhatch layer. Finally, duplicate the view just crosshatched to form a matching flange using the MIRROR command.

Whenever possible, substitute the appropriate command alias in place of the full AutoCAD command in each tutorial step. For example, use "Co" for the COPY command, "L" for the LINE command, and so on. The complete listing of all command aliases is located on pages 1–9 and 1–10.

Step #1

Zoom into the bottom half of the side view. Then, use the MIRROR command to create a duplicate copy of the side view of the Dflange. Use the vertical line from "D" to "E" as the mirror line.

 Command:**MIRROR**

Select objects: *(Pick a point at "A")*
Other corner: *(Pick a point at "B")*
Select objects: **R** *(For Remove)*
Remove objects: *(Select line "C" to remove it from the selection set)*
Remove objects: *(Press* ENTER *to continue)*
First point of mirror line: **Endp**
of *(Select the endpoint of the line at "D")*
Second point: **Endp**
of *(Select the endpoint of the line at "E")*
Delete old objects? <N> *(Press* ENTER *to exit this command)*

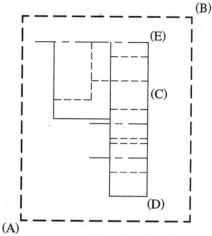

Figure 8–69

Step #2

Again use the MIRROR command to copy and duplicate one-half of the bottom of the Dflange. It is this half that will be converted into a half section.

 Command:**MIRROR**

Select objects: **P** *(For Previous)*
Select objects: *(Also select the vertical line at "C" to add this object to the current selection set)*
Select objects: **R** *(For Remove)*
Remove objects: *(Select the horizontal centerline at "D" to remove it from the selection set)*
Remove objects: *(Press* ENTER *to continue)*
First point of mirror line: **Endp**
of *(Select the endpoint of the centerline at "A")*
Second point: **Endp**
of *(Select the other endpoint of the centerline at "B")*
Delete old objects? <N> *(Press* ENTER *to exit this command)*

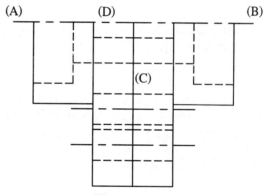

Figure 8–70

Step #3

Use the ZOOM-Previous option to display the previous view. Magnify the upper half of the side view using the ZOOM command. Begin the preparation of the section by deleting any unnecessary objects such as the hidden and centerlines located at "A," "B," "C," and "D" shown in Figure 8–71.

 Command: **ERASE**

Select objects: *(Select the 4 lines at "A," "B," "C," and "D")*
Select objects: *(Press the* [ENTER] *key to execute this command)*

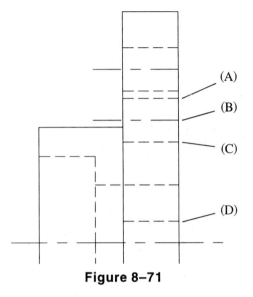

Figure 8–71

Step #4

Use the CHPROP or DDCHPROP commands to convert the five hidden lines shown in Figure 8–72 from hidden lines to object lines. Perform the change from the layer "Hid" to the layer "Object."

 Command: **CHPROP**

Select objects: *(Select the five lines labeled "A," "B," "C," "D," and "E")*
Select objects: *(Press* [ENTER] *to continue with this command)*
Change what property (Color/LAyer/LType/ltScale/Thickness) ? **LA**
New layer <HID>: **Object**
Change what property (Color/LAyer/LType/ltScale/Thickness) ? *(Press* [ENTER] *to exit this command)*

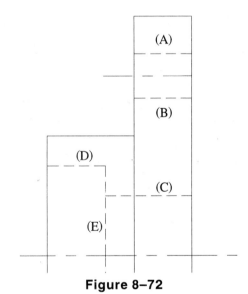

Figure 8–72

Step #5

Use the TRIM command, select the horizontal line at "A" as the cutting edge, and select the vertical line at "B" as the object to trim, as seen in Figure 8–73.

 Command:**TRIM**

Select cutting edges: (Projmode = UCS,
   Edgemode = No extend)
Select objects: *(Select the horizontal line at "A")*
Select objects: *(Press* ⌷ENTER⌷ *to continue with this
   command)*
<Select object to trim>/Project/Edge/Undo:
   (Select the vertical line at "B")
<Select object to trim>/Project/Edge/Undo:
   (Press ⌷ENTER⌷ *to exit this command)*

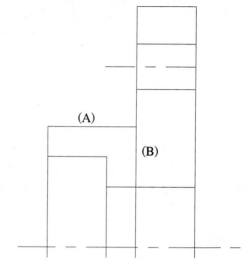

Figure 8–73

Step #6

The side view will now have the surfaces sliced by the cutting plane line crosshatched. The two areas labeled "A" and "B" in Figure 8–74 need to be prepared for crosshatching by using the BHATCH command.

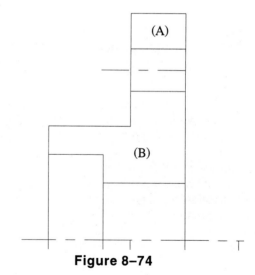

Figure 8–74

Step #7

Make the "Xhatch" layer the new current layer. This
layer will be used to hold all crosshatch information.

 Command:**-LAYER**

?/Make/Set/New/ON/OFF/Color/Ltype/Freeze/
Thaw/LOck/Unlock:**Set**
New current layer<0>:**Xhatch**
?/Make/Set/New/ON/OFF/Color/Ltype/Freeze/
Thaw/LOck/Unlock: *(Press [ENTER] to exit this
command)*

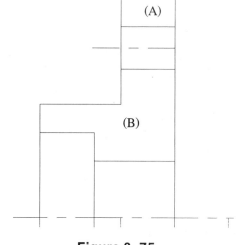

Figure 8–75

Step #8

Use the BHATCH command to display the Boundary Hatch dialog box shown in Figure 8–76. Use
the pattern "ANSI31" and keep all default settings.
Click on the "Pick Points <" button. When the drawing reappears, click inside of areas "A" and "B" as
shown in Figure 8–75. When finished selecting these
internal points, press the ENTER key to return to
the Boundary Hatch dialog box.

 Command:**BHATCH**

*(The Boundary Hatch dialog box appears.
Make changes to match Figure 8–76. When
finished, click the "Pick Points<" button.)*
Select internal point: *(Pick a point at "A)*
Selecting everything…
Selecting everything visible…
Analyzing the selected data…
Analyzing internal islands…

Select internal point: *(Pick a point at "B")*
Analyzing internal islands…

Select internal point: *(Press [ENTER] to exit this area
and return to the Boundary Hatch dialog box)*

While in the Boundary Hatch dialog box, click on
the "Preview Hatch <" button to view the results.
Click "Continue" and return to the Boundary Hatch
dialog box. If the hatch results are acceptable, click
on the "Apply" button to place the hatch pattern.

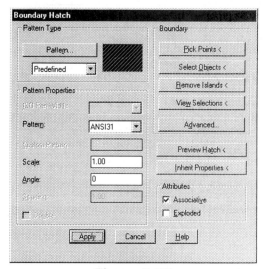

Figure 8–76

Step #9

Duplicate and flip the upper half of the side view to form the opposite half of the flange assembly using the MIRROR command. Use the crossing option to select the objects illustrated in Figure 8–77. Select "C" and "D" as the first and second points of the mirror lines. Notice that after performing this operation, a message appears after the use of the MIRROR command stating that the hatch boundary associativity was removed.

 Command: **MIRROR**

Select objects: **C** *(For Crossing)*
First corner: *(Pick a point at "A")*
Other corner: *(Pick a point at "B")*
Select objects: **R** *(For Remove)*
Remove objects: *(Select the vertical line at "E" to remove it from the selection set)*
Remove objects: *(Press [ENTER] to continue with this command)*
First point of mirror line: **Endp**
of *(Select the endpoint of the centerline at "C")*
Second point: **Int**
of *(Select the intersection of the lines at "D")*
Delete old objects? <N> *(Press [ENTER] to accept the default)*

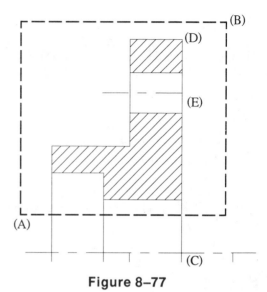

Figure 8–77

Step #10

Dimensions may be added at a later time similar to Figure 8–78. Notice the bottom halves of the side views show conventional hidden lines. This is one method of representing half sections.

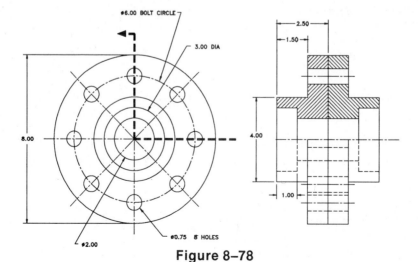

Figure 8–78

Tutorial Exercise
Coupler.Dwg

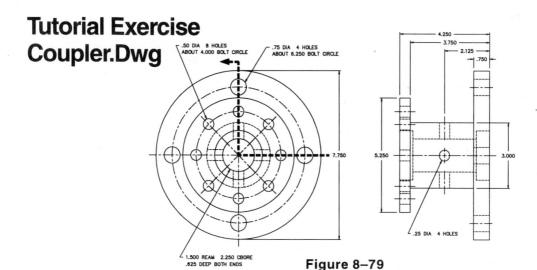

Figure 8–79

Purpose

This tutorial is designed to use the MIRROR, CHPROP, DDCHPROP, and BHATCH commands to convert the coupler into a half section. The methods will be similar to the previous tutorial, with the exception of the crosshatching segment being different.

System Settings

Use the Units Control dialog box to change the number of decimal places from four to two. Keep the remaining defaults settings. The limits of the drawing are already preset to 0,0 for the lower left corner and 21,17 for the upper right corner. The GRID command may be used to change the spacing from 1.00 to 0.50.

Layers

Create the following layers with the format:

Name	Color	Linetype
Object	White	Continuous
Cen	Yellow	Center
Hid	Red	Hidden
Xhatch	Magenta	Continuous
Cpl	Yellow	Dashed
Dim	Yellow	Continuous

Suggested Commands

This tutorial begins similar in procedure to the previous exercise, Dflange, with the exception of the crosshatching segment. Convert one-half of the object into a section by erasing unnecessary hidden lines. Use the CHPROP or DDCHPROP commands and change the remaining hidden lines to the Object layer. Issue the BHATCH command and use the ANSI31 hatch pattern to hatch the upper half of the coupler on the Xhatch layer.

Whenever possible, substitute the appropriate command alias in place of the full AutoCAD command in each tutorial step. For example, use "Co" for the COPY command, "L" for the LINE command, and so on. The complete listing of all command aliases is located on pages 1–9 and 1–10.

Step #1

Use the MIRROR command to copy and flip the upper half of the side view and form the lower half. When in object selection mode, use the Remove option to deselect the main centerline and hole. If these objects are included in the mirror operation, a duplicate copy of these objects will be created. See Figure 8–80.

 Command:**MIRROR**

Select objects: *(Pick a point at "A")*
Other corner: *(Pick a point at "B")*
Select objects: **R** *(To Remove)*
Remove objects: *(Select the centerlines "C" and "D")*
Remove objects: *(Pick a point at "E")*
Other corner: *(Pick a point at "F")*
Remove objects: *(Press* [ENTER] *to continue with this command)*
First point of mirror line: **Endp**
of *(Select the endpoint of the centerline near "C")*
Second point: **Endp**
of *(Select the endpoint of the centerline near "D")*
Delete old objects? <N> *(Press* [ENTER] *to execute this command)*

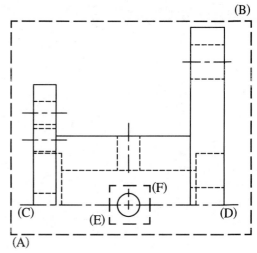

Figure 8–80

Step #2

Begin converting the upper half of the side view to a half section by using the ERASE command to remove any unnecessary hidden lines and centerlines from the view. See Figure 8–81.

 Command:**ERASE**

Select objects: *(Carefully select the hidden lines labeled "A," "B," "C," and "D")*
Select objects: *(Select the centerline labeled "E")*
Select objects: *(Press* [ENTER] *to execute the ERASE command)*

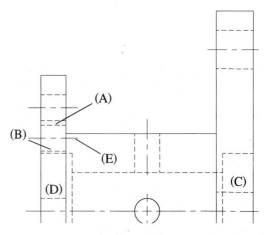

Figure 8–81

Step #3

Since the remaining hidden lines actually represent object lines when shown in section, use the CHPROP or DDCHPROP commands to convert all hidden lines labeled in Figure 8–82 from the "Hid" layer to the "Object" layer. See Figure 8–82.

 Command: **CHPROP**

Select objects: *(Select all hidden lines labeled "A" through "K")*
Select objects: *(Press* ENTER *to continue with this command)*
Change what property (Color/LAyer/LType/ ltScale/Thickness) ? **LA**
New layer <HID>: **Object**
Change what property (Color/LAyer/LType/ ltScale/Thickness) ? *(Press* ENTER *to exit this command)*

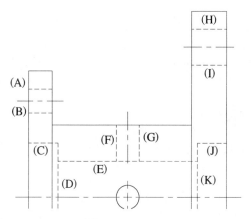

Figure 8–82

Step #4

Remove unnecessary line segments from the upper half of the converted section using the TRIM command. Use the horizontal line at "A" as the cutting edge, and select the two vertical segments at "B" and "C" as the objects to trim. See Figure 8–83.

 Command: **TRIM**

Select cutting edges: (Projmode = UCS, Edgemode = No extend)
Select objects: *(Select horizontal line "A")*
Select objects: *(Press* ENTER *to continue with this command)*
<Select object to trim>/Project/Edge/Undo: *(Select the vertical line at "B")*
<Select object to trim>/Project/Edge/Undo: *(Select the vertical line at "C")*
<Select object to trim>/Project/Edge/Undo: *(Press* ENTER *to exit this command)*

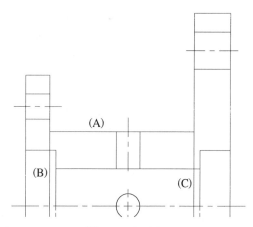

Figure 8–83

Step #5

Before crosshatching areas "A," "B," "C," and "D" in Figure 8–84, make the "Xhatch" layer the new current layer. This layer will be used to hold all crosshatch information. Then issue the BHATCH command.

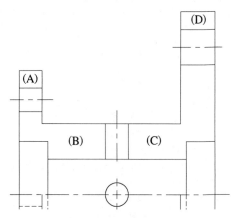

Figure 8–84

 Command:**-LAYER**

?/Make/Set/New/ON/OFF/Color/Ltype/Freeze/
   Thaw/LOck/Unlock:**Set**
New current layer<Pline>:**Xhatch**
?/Make/Set/New/ON/OFF/Color/Ltype/Freeze/
   Thaw/LOck/Unlock: *(Press* [ENTER] *to exit this
   command)*

Step #6

Use the BHATCH command to display the Boundary Hatch dialog box shown in Figure 8–85. Use the pattern "ANSI31" and keep all default settings. Click on the "Pick Points <" button. When the drawing reappears, click inside of areas "A," "B," "C," and "D" in Figure 8–84. When finished selecting these internal points, press the ENTER key to return to the Boundary Hatch dialog box.

 Command:**BHATCH**

*(The Boundary Hatch dialog box appears.
   Make changes to match Figure 8–85. When
   finished, click the "Pick Points <" button.)*
Select internal point: *(Pick a point at "A")*
Selecting everything…
Selecting everything visible…
Analyzing the selected data…
Analyzing internal islands…

Select internal point: *(Pick a point at "B")*
Analyzing internal islands…

Select internal point: *(Pick a point at "C")*
Analyzing internal islands…

Select internal point: *(Pick a point at "D")*
Analyzing internal islands…

Select internal point: *(Press* [ENTER] *to exit this area
   and return to the Boundary Hatch dialog box)*

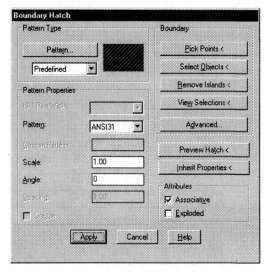

Figure 8–85

While in the Boundary Hatch dialog box, click on the "Preview Hatch <" button to view the results. Click "Continue" and return to the Boundary Hatch dialog box. If the hatch results are acceptable, click on the "Apply" button to place the hatch pattern.

Step #7

The complete crosshatched view is shown in Figure 8–86. The full drawing is also shown in Figure 8–87 along with dimensions. As with other half section examples, it is the option of the operator or designer to show all hidden lines in the lower half or delete all hidden and centerlines from the lower half and simply interpret the section in the upper half.

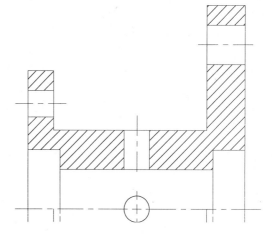

Figure 8–86

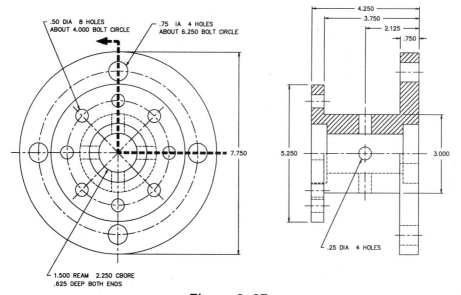

Figure 8–87

Tutorial Exercise
Assembly.Dwg

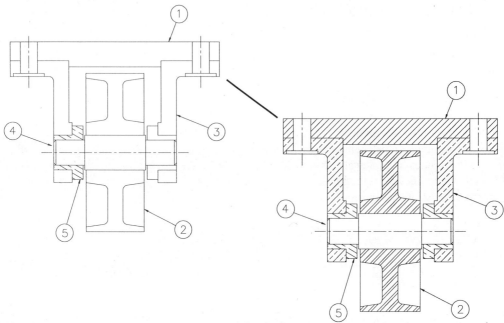

Figure 8–88

Purpose

This tutorial is designed to use the BHATCH command to crosshatch an assembly drawing.

System Settings

All settings have been made since this exercise is available on disk.

Layers

Layers have already been created for this tutorial exercise.

Name	Color	Linetype
Center	Yellow	Center
Leader	Cyan	Continuous
Object	White	Continuous
Section	Magenta	Continuous

Suggested Commands

The BHATCH command will be used exclusively during this tutorial exercise.

Whenever possible, substitute the appropriate command alias in place of the full AutoCAD command in each tutorial step. For example, use "Co" for the COPY command, "L" for the LINE command, and so on. The complete listing of all command aliases is located on pages 1–9 and 1–10.

Hatch Pattern Angle Manipulation

As with the scale of the hatch pattern, the angle for the hatch pattern can be controlled by the designer depending on the effect the pattern has with the area being hatched. By default, the BHATCH command displays a "0 degree" angle for all patterns. In Figure 8–53, the angle for "ANSI31" is drawn at 45 degrees—the angle the pattern was originally created in.

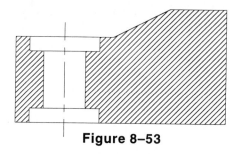

Figure 8–53

Entering any angle different from the default value of "0" will rotate the hatch pattern by that value. This means if a pattern was originally designed at a 45-degree angle like "ANSI31," entering a new angle for the pattern would begin rotating the pattern starting at the 45-degree position. In Figure 8–54, a new angle of 45 degrees is entered in the Boundary Hatch dialog box. Since the original angle was already 45 degrees, this new angle value is added to the original to obtain a vertical crosshatch pattern; positive angles rotate the hatch pattern in the counterclockwise direction.

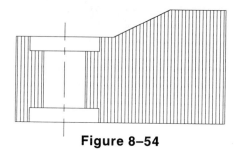

Figure 8–54

Again, entering an angle other than the default rotates the pattern from the original angle to a new angle. In Figure 8–55, an angle of 90 degrees has been applied to the "ANSI31" pattern in the Boundary Hatch dialog box.

Providing different angles for patterns is useful when creating section assemblies where different parts are in contact with each other and patterns are placed at different angles making the parts easy to see.

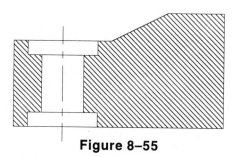

Figure 8–55

Step #1

Before beginning the crosshatching operations, turn off the layers called "Leader" and "Center." The Leader layer holds all bubbles identifying numbers and leaders pointing to the specific parts. The Center layer will turn off all centerlines in the drawing. See Figure 8–89.

 Command:**-LAYER**

?/Make/Set/New/ON/OFF/Color/Ltype/Freeze/
   Thaw/LOck/Unlock: **Off**
Layer name(s) to turn Off <>: **Leader,Center**
?/Make/Set/New/ON/OFF/Color/Ltype/Freeze/
   Thaw/LOck/Unlock: *(Press* ENTER *to exit this
   command)*

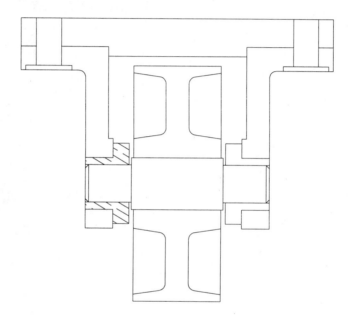

Figure 8–89

Step #2

Issue the BHATCH command and begin crosshatching the assembly; first crosshatch Part 1, the plate (see Figure 8–90A). Use the default hatch pattern of ANSI31 in addition to a scale factor of 1.0000. Change the angle to 90 degrees. Pick three internal points "A," "B," and "C" to identify the areas to crosshatch (see Figure 8–90B). When the Boundary Hatch dialog box reappears, click on the Preview Hatch button to see if the object correctly crosshatched. If the results are acceptable, click on the Apply button to place the hatch pattern.

 Command: **BHATCH**

(The Boundary Hatch dialog box appears. Make changes to match Fig. 8-90A. When finished, click the "Pick Points <" button.)
Select internal point: *(Pick a point at "A")*
Selecting everything...
Selecting everything visible...
Analyzing the selected data...
Analyzing internal islands...

Select internal point: *(Pick a point at "B")*
Analyzing internal islands...
Select internal point: *(Pick a point at "C")*
Analyzing internal islands...

Select internal point: *(Press* [ENTER] *to exit this area and return to the Boundary Hatch dialog box)*

While in the Boundary Hatch dialog box, click on the "Preview Hatch <" button to view the results. Click "Continue" and return to the Boundary Hatch dialog box. If the hatch results are acceptable, click on the "Apply" button to place the hatch pattern.

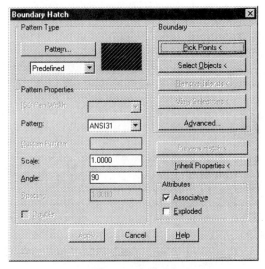

Figure 8–90A

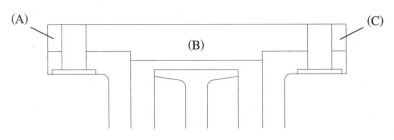

Figure 8–90B

Step #3

Next, crosshatch the pulley in the assembly drawing. Execute the BHATCH command; when the Boundary Hatch dialog box appears, choose the pattern "ANSI32" (see Figure 8–91A); click on two internal points inside of the pulley (see Figure 8–91B). If the proper boundaries highlight, press the ENTER key to return back to the Boundary Hatch dialog box. Click on the "Preview" button to preview the hatch pattern; if the results are desirable, click on the "Apply" button to place the hatch pattern (see Figure 8–91A).

 Command: **BHATCH**

(The Boundary Hatch dialog box appears. Make changes to match Fig. 8-91A. When finished, click on the "Pick Points <" button.)
Select internal point: *(Pick a point at "A")*
Selecting everything...
Selecting everything visible...
Analyzing the selected data...
Analyzing internal islands...

Select internal point: *(Pick a point at "B")*
Analyzing internal islands...

Select internal point: *(Press* ENTER *to exit this area and return to the Boundary Hatch dialog box)*

While in the Boundary Hatch dialog box, click on the "Preview Hatch <" button to view the results. Click "Continue" and return to the Boundary Hatch dialog box. If the hatch results are acceptable, click on the "Apply" button to place the hatch pattern.

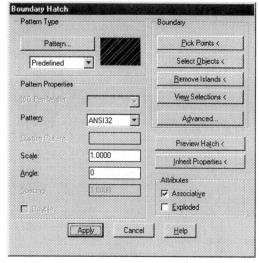

Figure 8–91A

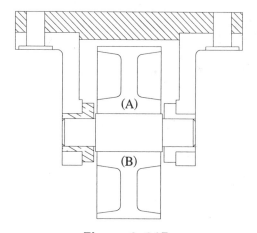

Figure 8–91B

Step #4

Next, crosshatch the two brackets that support the pulley axle and are connected to the plate. Execute the BHATCH command and select the "ANSI33" hatch pattern; keep all other defaults (see Figure 8–92A). When prompted to select internal points, click inside of the areas identified by "A" through "F" (see Figure 8–92B). When returning to the Boundary Hatch dialog box, click on the "Preview Hatch" button to visually inspect the hatch pattern. If the preview appears correct, click on the "Apply" button to place the hatch pattern (see Figure 8–92A).

 Command: **BHATCH**

(The Boundary Hatch dialog box appears. Make changes to match Fig. 8-92A. When finished, click on the "Pick Points <" button.)
Select internal point: *(Pick a point at "A")*
Selecting everything…
Selecting everything visible…
Analyzing the selected data…
Analyzing internal islands…

Select internal point: *(Pick a point at "B")*
Analyzing internal islands…

Select internal point: *(Pick a point at "C")*
Analyzing internal islands…

Select internal point: *(Pick a point at "D")*
Analyzing internal islands…

Select internal point: *(Pick a point at "E")*
Analyzing internal islands…

Select internal point: *(Pick a point at "F")*
Analyzing internal islands…

Select internal point: *(Press [ENTER] to exit this area and return to the Boundary Hatch dialog box)*

While in the Boundary Hatch dialog box, click on the "Preview Hatch <" button to view the results. Click "Continue" and return to the Boundary Hatch dialog box. If the hatch results are acceptable, click on the "Apply" button to place the hatch pattern.

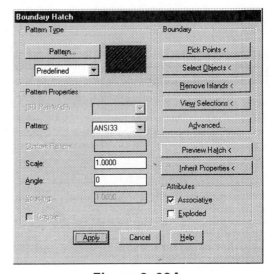

Figure 8–92A

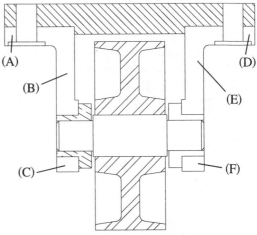

Figure 8–92B

Step #5

Issue the BHATCH command and click on the "Inherit Properties" button as shown in Figure 8–93A. This will allow you to select an existing hatch pattern; the pattern, scale and angle are now current in the Boundary Hatch dialog box in Figure 8–93A.

Notice that although one of the bushings has already been crosshatched, it is difficult to determine the hatch pattern used along with scale and angle. Pick internal points in the drawing and the object will be crosshatched to these new current parameters.

  Command: **BHATCH**

(When the Boundary Hatch dialog box appears, click on the "Inherit Properties <" button.)

Select hatch object: *(Select the hatch pattern already visible in the bushing at "A" in Figure 8–93B)*

(When the Boundary Hatch dialog box reappears, click on the "Pick Points <" button.)

Select internal point: *(Pick a point at "B")*
Selecting everything…
Selecting everything visible…
Analyzing the selected data…
Analyzing internal islands…

Select internal point: *(Pick a point at "C")*
Analyzing internal islands…

Select internal point: *(Press ⌜ENTER⌝ to exit this area and return to the Boundary Hatch dialog box)*

While in the Boundary Hatch dialog box, click on the "Preview Hatch <" button to view results. Click "Continue" and return to the Boundary Hatch dialog box. If the hatch results are acceptable, click on the "Apply" button to place the hatch pattern.

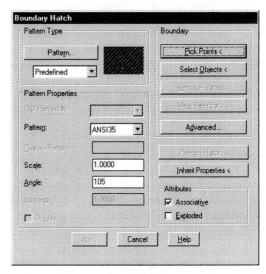

Figure 8–93A

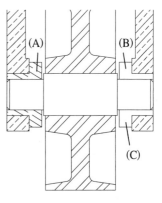

Figure 8–93B

Step #6

The hatch pattern currently displayed in the pulley is too large for the area it occupies; the spacing needs to be scaled in half (see Figure 8–94A). Issue the HATCHEDIT command and select the hatch pattern anywhere in the pulley. When the Hatchedit dialog box appears, change the scale from a value of 1.0000 to 0.5000 units (see Figure 8–94B). Click on the "Apply" button to update the hatch pattern to the new changes.

 Command:**HATCHEDIT**

Select hatch object: *(Select the hatch pattern inside of the pulley)*

When the Hatchedit dialog box appears, change the scale to 0.5000 units in Fig. 8-94B. Then click on the "Preview Hatch <" button to view the results. Click "Continue" and return to the Hatchedit dialog box. If the hatch results are acceptable, click on the "Apply" button to place the hatch pattern.

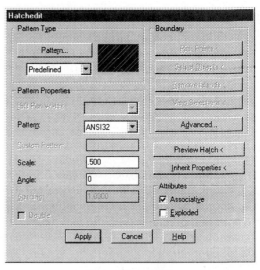

Figure 8–94B

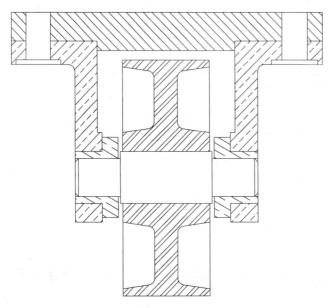

Figure 8–94A

Step #7

Stretch the base to make it 0.25 units longer on both sides. With Ortho mode turned on, use the direct distance mode to accomplish this task. When selecting the objects to stretch, take care not to accidentally select the edge of the counterbore hole. Use the ZOOM command to magnify this part of the screen.

 Command:**STRETCH**

Select objects to stretch by crossing-window or crossing-polygon...
Select objects: *(Pick a point at "B")*
Other corner: *(Pick a point at "A")*
Select objects: *(Press* ENTER *to continue with this command)*
Base point or displacement: *(Pick a blank part of the display screen at "C")*

Second point of displacement: *(With Ortho on, move the cursor directly to the left and enter a value of 0.25)*

 Command:**STRETCH**

Select objects to stretch by crossing-window or crossing-polygon...
Select objects: *(Pick a point at "D")*
Other corner: *(Pick a point at "E")*
Select objects: *(Press* ENTER *to continue with this command)*
Base point or displacement: *(Pick a blank part of the display screen at "F")*
Second point of displacement: *(With Ortho on, move the cursor directly to the right and enter a value of 0.25)*

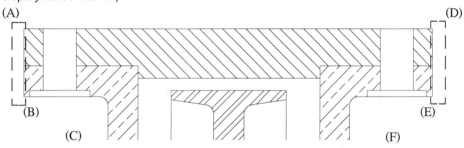

(A)   (D)

(B)   (E)

(C)   (F)

Figure 8–95

Step #8

Turn all layers back on; this crosshatching exercise is completed (see Figure 8–96).

 Command:**-LAYER**

?/Make/Set/New/ON/OFF/Color/Ltype/Freeze/
Thaw/LOck/Unlock: **On**
Layer name(s) to turn On <>: **Leader,Center**
?/Make/Set/New/ON/OFF/Color/Ltype/Freeze/
Thaw/LOck/Unlock: *(Press* ENTER *to exit this command)*

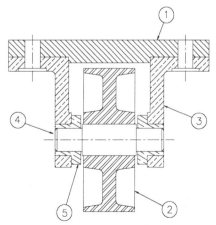

Figure 8–96

Problems for Unit 8

Directions for Problem 8-1:

Based on the following isometric and orthographic drawings, identify which of the following best depicts a half-section of the object. Place your answer in the box provided.

Answer

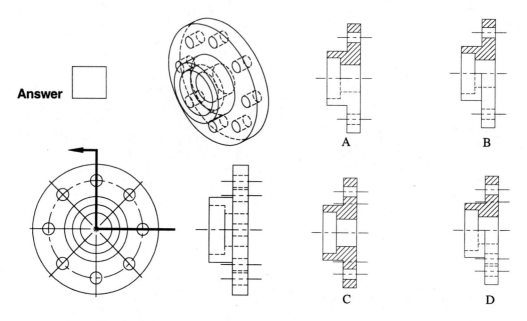

Directions for Problem 8-2:

Based on the following isometric drawing and orthographic drawing, identify which of the following best depicts a full section of the object. Place your answer in the box provided.

Answer

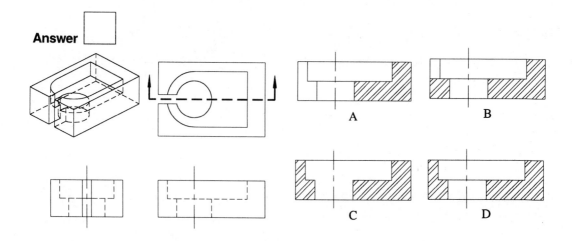

Directions for Problem 8-3

    Based on the following isometric drawing and orthographic front view, identify which of the following best depicts a full section of the object. Place your answer in the box provided.

Answer

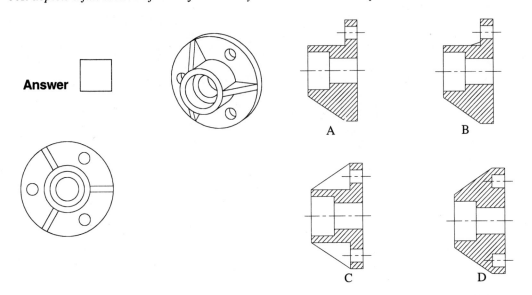

Directions for Problem 8-4:

    Based on the following isometric and orthographic drawings, identify which of the following best depicts an offset section of the object. Place your answer in the box provided.

Answer

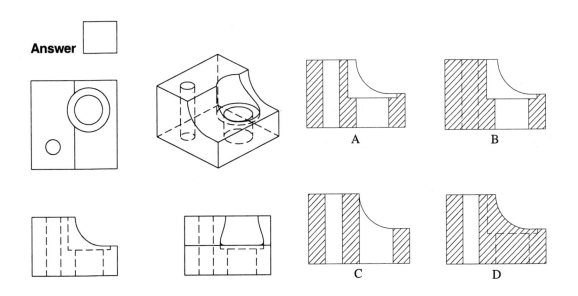

Problem 8–5

Center a three-view drawing and make the front view a full section.

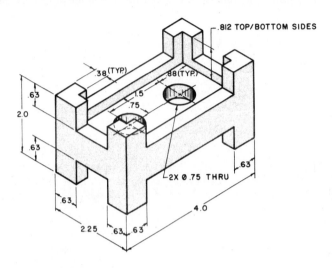

Problem 8–6

Center two views within the work area, and make one view a full section. Use correct drafting practices for the ribs.

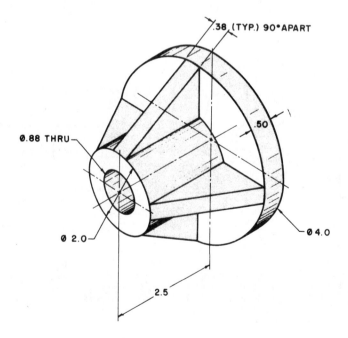

Problem 8–7

Center two views within the work area, and make one view a full section.

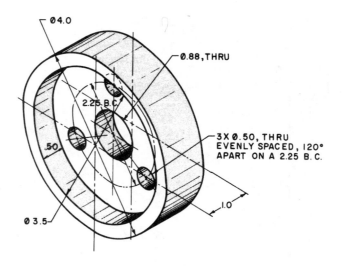

Problem 8–8

Center the front view and top view within the work area. Make one view a full section.

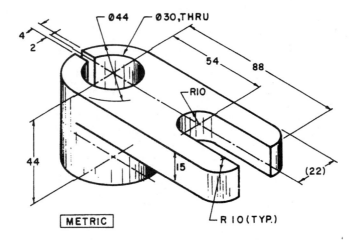

Problem 8–9

Center the required views within the work area, and add removed section A-A.

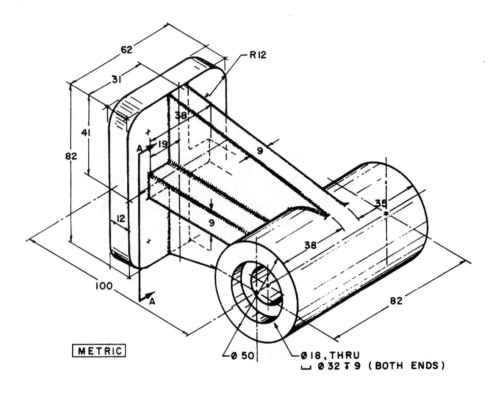

Problem 8–10

Center three views within the work area, and make one view an offset section.

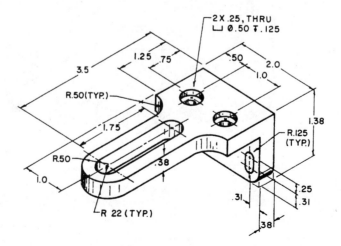

Problem 8–11

Center three views within the work area, and make one view an offset section.

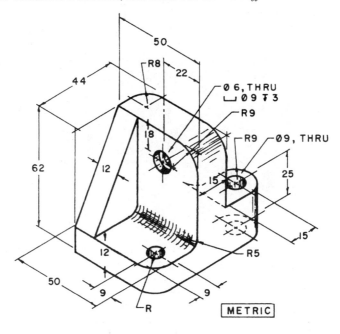

Problem 8–12

Center three views within the work area, and make one view an offset section.

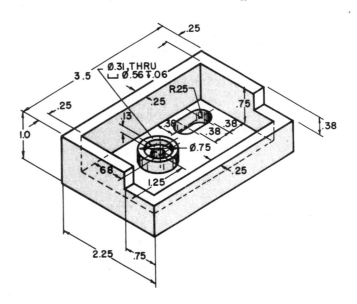

Problem 8–13

Center three views within the work area, and make one view an offset section.

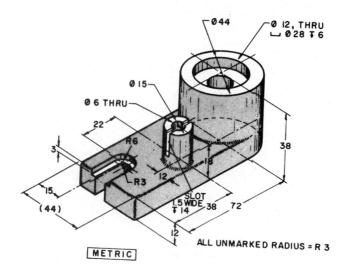

Problem 8–14

Center two views within the work area, and make one view an offset section.

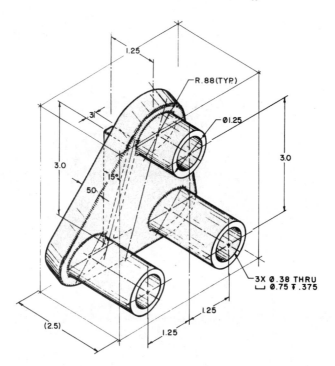

Problem 8–15

Center the front view and top view within the work area. Make one view a half section.

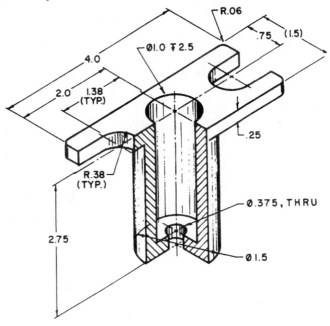

Problem 8–16

Center two views within the work area, and make one view a half section.

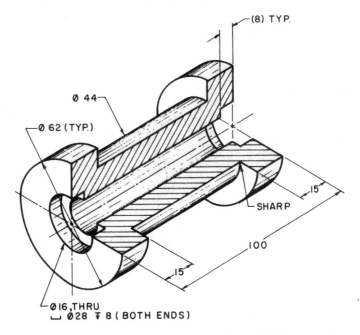

METRIC

Problem 8–17

Center the two views within the work area, and make one view a half section.

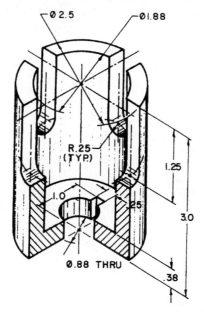

Problem 8–18

Center two views within the work area, and make one view a half section.

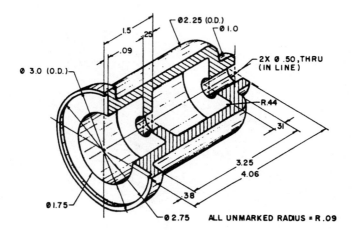

Problem 8–19

Center two views within the work area, and make one view a half section.

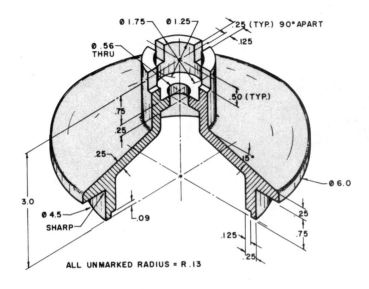

Problem 8–20

Center the required views within the work area, and make one view a broken-out section to illustrate the complicated interior area.

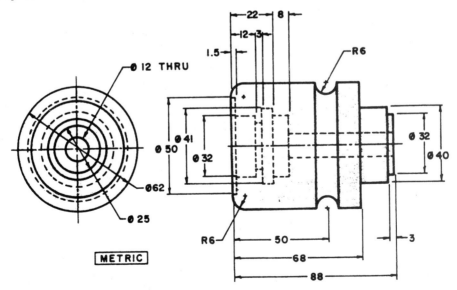

Problem 8–21

Center the required views within the work area, and add removed section A-A.

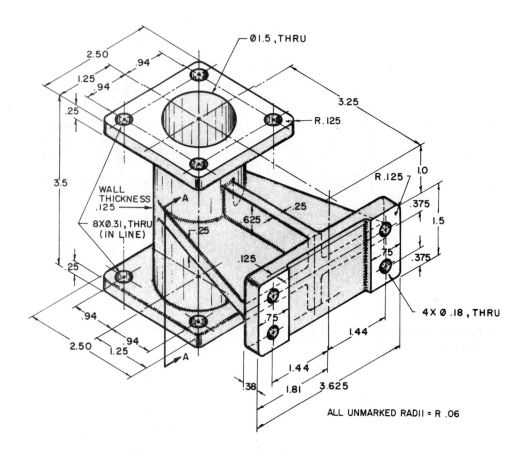

ALL UNMARKED RADII = R .06

Problem 8–22

*Center the required views within the work area,
and add removed section A-A.*

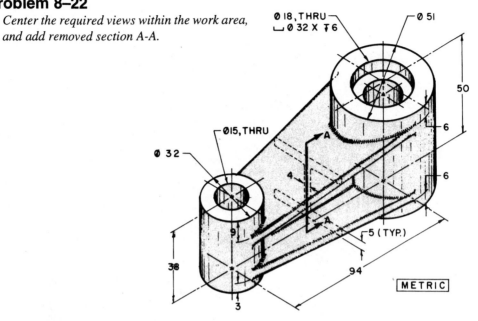

Directions for Problems 8-23 through 8-25:

*Center required views within the work area.
Leave a 1-inch or 25-mm space between views.
Make one view into a section view to fully illustrate the object. Use a full half, offset, broken-
out, revolved, or removed section. Consult your
instructor if dimensions are to be added.*

Problem 8–23

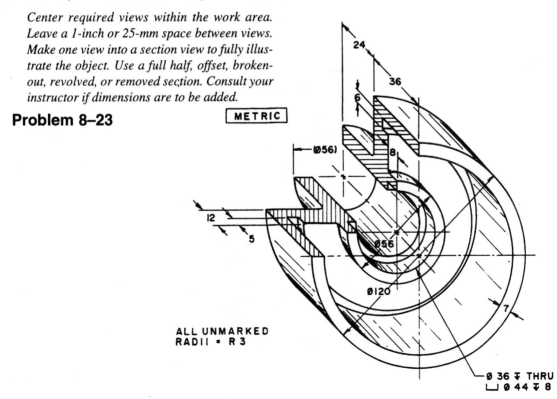

Problem 8–24

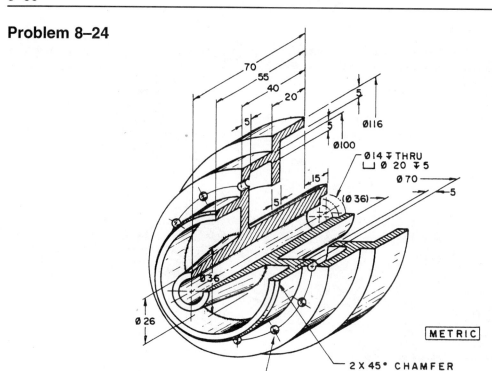

70
55
40
20
5
5
5
Ø116
Ø100
Ø14 ⊤ THRU
⊔ Ø 20 ⊤ 5
15
Ø 70
5
(Ø 36)
5
Ø 36
Ø 26

METRIC

ALL FILLETS = R2

2 X 45° CHAMFER

12 X Ø 5 ⊤ THRU
EVENLY SPACED ON A Ø 80 B.C.

Problem 8–25

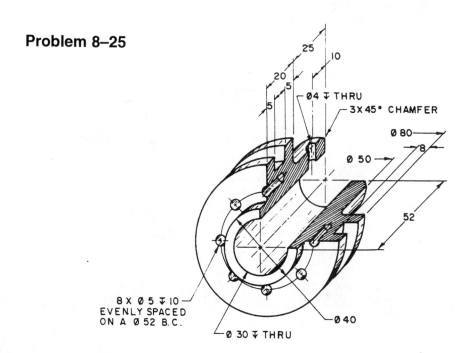

25
10
20
5
Ø4 ⊤ THRU
3 X 45° CHAMFER
5
Ø 80
8
Ø 50
52

8 X Ø 5 ⊤ 10
EVENLY SPACED
ON A Ø 52 B.C.

Ø 40

Ø 30 ⊤ THRU

Directions for Problem 8-26:

Using the background grid as a guide, reproduce each problem on a CAD system. Add all dimensions. Use a grid spacing of 0.25 units.

Problem 8–26

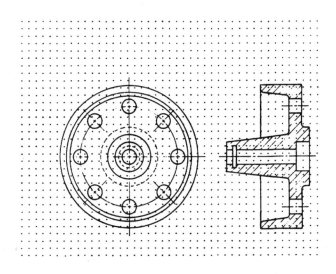

Auxiliary Views

During the discussion of multiview drawings, we determined that you need to draw enough views of an object to accurately describe it. This requires a front, top, and right side view in most cases. Sometimes additional views are required such as left side, bottom, and back views to show features not visible in the three primary views. Other special views like sections are taken to expose interior details for better clarity. Sometimes all of these views are still not enough to describe the object, especially when features are located on an inclined surface. To produce a view perpendicular to this inclined surface, an auxiliary view is drawn. This unit will describe where auxiliary views are used and how they are projected from one view to another. A tutorial exercise is presented to go through the steps in the construction of an auxiliary view. Additional problems are provided at the end of this unit for further study of auxiliary views.

Auxiliary View Basics

Figure 9–1 presents interesting results if constructed as a multiview drawing or orthographic projection. Let us see how this object differs from others previously discussed in Unit 4.

Figure 9–2 should be quite familiar; it represents the standard glass box with object located in the center. Again, the purpose of this box is to prove how orthographic views are organized and laid out. Figure 9–2 is no different. First the primary views, front, top, and right side views are projected from

Figure 9–1

the object to intersect perpendicular with the glass plane. Under normal circumstances, this procedure would satisfy most multiview drawing cases. Remember, only those views necessary to describe the object are drawn. However, under closer inspection, we notice that the object in the front view consists of an angle forming an inclined surface.

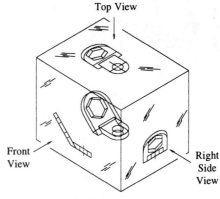

Figure 9–2

When laying out the front, top, and right side views, a problem occurs. The front view shows the basic shape of the object, the angle of the inclined surface (see Figure 9–3). The top view shows true size and shape of the surface formed by the circle and arc. The right side view shows the true thickness of the hole from information found in the top view. However, there does not exist a true size and shape of the features found in the inclined surface at "A." We see the hexagonal hole going through the object in the top and right side views. These views, however, show the detail not to scale, or foreshortened. For that matter, the entire inclined surface is foreshortened in all views. This is one case where the front, top, and right side views are not enough to describe the object. An additional view, or auxiliary view, is used to display the true shape of surfaces along an incline.

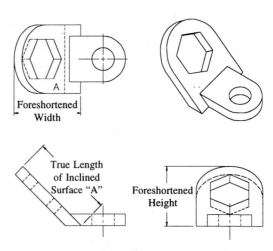

Figure 9–3

To prove the formation of an auxiliary view, let's create another glass box; this time an inclined plane is formed. This plane is always parallel to the inclined surface of the object. Instead of just projecting the front, top, and right side views, the geometry describing the features along the inclined surface is projected to the auxiliary plane similar to Figure 9–4A.

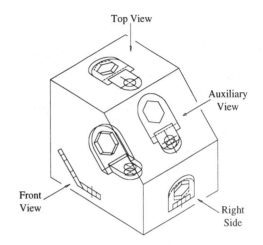

Figure 9–4A

As in multiview projection, the edges of the glass box are unfolded out using the edges of the front view plane as the pivot. See Figure 9–4B.

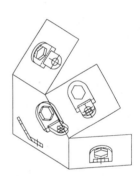

Figure 9–4B

All planes are extended perpendicular to the front view where the rotation stops. The result is the organization of the multiview drawing complete with an auxiliary viewing plane, as shown in Figure 9–4C.

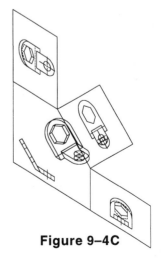

Figure 9–4C

Figure 9–5 is the final layout complete with auxiliary view. This figure shows the auxiliary being formed as a result of the inclined surface being in the front view. An auxiliary view may be made in relation to any inclined surface located in any view. Also, the figure displays circles and arcs in the top view which appear as ellipses in the auxiliary view.

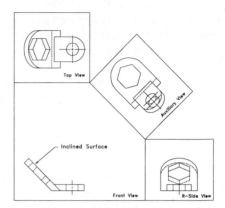

Figure 9–5

It is usually not required to draw elliptical shapes in one view where the feature is shown true size and shape in another. The resulting view minus these elliptical features is called a partial view which is used extensively in auxiliary views. An example of the top view converted into a partial view is shown in Figure 9–6.

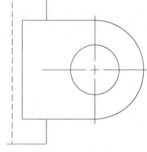

Figure 9–6

A few rules to follow when constructing auxiliary views are displayed pictorially in Figure 9–7. First, the auxiliary plane is always constructed parallel to the inclined surface. Once this is established, visible as well as hidden features are projected from the incline to the auxiliary view. These projection lines are always drawn perpendicular to the inclined surface and the auxiliary view.

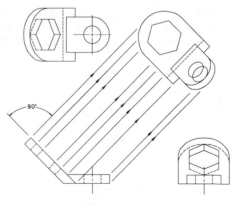

Figure 9–7

Constructing an Auxiliary View

Figure 9–8 is a multiview drawing consisting of front, top, and right side views. The inclined surface in the front view is displayed in the top and right side views; however, the surface appears foreshortened in both adjacent views. An auxiliary view of the incline needs to be made to show its true size and shape. Currently the display screen has the Grid On in addition to the position of the typical AutoCAD cursor. Follow Figures 9–9 through 9–18 for one suggested method for projecting to find auxiliary views.

To assist with the projection process, it would help if the current grid display could be rotated parallel and perpendicular to the inclined surface (see Figure 9–9). This is accomplished using the SNAP command and the Rotate option.

Command: **SNAP**
Snap spacing or ON/OFF/Aspect/Rotate/
    Style<0.50>: **R** *(For Rotate)*
Base point <0,0,0>: **Endp**
of *(Select the endpoint of the line at "A" as the new base point)*
Rotation angle <0>: **Endp**
of *(Select the endpoint of the line at "B" to define the rotation angle by clicking on key points along an incline)*

Figure 9–10 shows the results of rotating the grid through the use of the SNAP command. This operation has no effect on the already existing views; however, the grid is now placed rotated in relation to the incline located in the front view. Notice the appearance of the standard AutoCAD cursor has also changed to conform to the new grid orientation. In addition to snapping to these new grid dots, lines are easily drawn perpendicular to the incline using Ortho On, which will draw lines in relation to the current cursor.

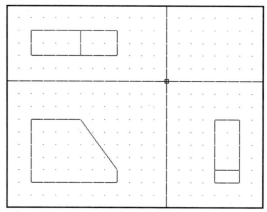

Figure 9–8

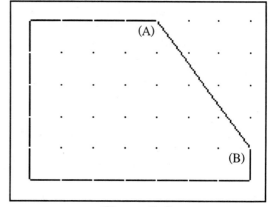

Figure 9–9

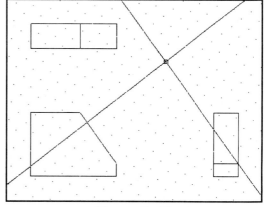

Figure 9–10

Use the OFFSET command to construct a reference line at a specified distance away from the incline in the front view (see Figure 9–11). This reference line becomes the start for the auxiliary view.

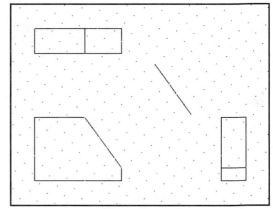

Figure 9–11

Use the LENGTHEN command to extend the two endpoints of the previous line (see Figure 9–12). The exact distances are not critical; however, the line should be long enough to accept projector lines from the front view.

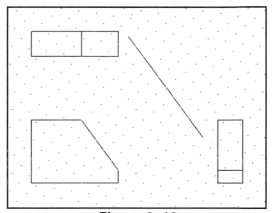

Figure 9–12

Use the OFFSET command to copy the auxiliary reference line the thickness of the object (see Figure 9–13). This distance may be retrieved from the depth of the top or right side views since they both contain the depth measurement of the object.

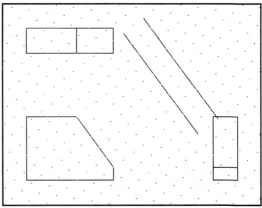

Figure 9–13

Use the LINE command and connect each intersection on the front view perpendicular to the outer
line on the auxiliary (see Figure 9–14). Draw the
lines starting with the OSNAP-Intersec option and
ending with the OSNAP-Perpend option.

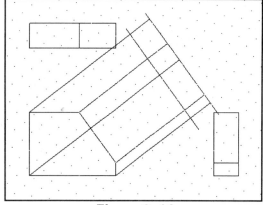

Figure 9–14

Before editing the auxiliary view, analyze the drawing to see if any corners in the front view are to be
represented as hidden lines in the auxiliary view
(see Figure 9–15). It turns out that the lower left
corner of the front view is hidden in the auxiliary
view. Use the CHPROP or DDCHPROP commands
to convert this projection line from a linetype of
continuous to hidden. This is best accomplished by
assigning the hidden linetype to a layer and then
using CHPROP or DDCHPROP to convert the line
to the different layer.

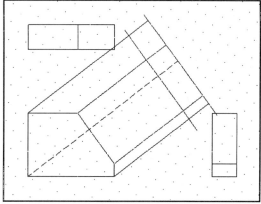

Figure 9–15

Use the TRIM command to partially delete all projection lines and corners of the auxiliary view. Another method would be to use the FILLET command set to a radius of 0. Selecting two lines in a
corner autmatically trims the excess lines away (see
Figure 9–16).

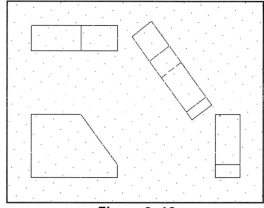

Figure 9–16

The result is a multiview drawing complete with auxiliary view displaying the true size and shape of the inclined surface. See Figure 9–17.

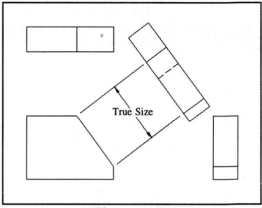

Figure 9–17

For dimensioning purposes, use the SNAP command and set the grid and cursor appearance back to normal. See figure 9–18.

Command:**SNAP**
Snap spacing or ON/OFF/Aspect/Rotate/Style
   <0.50>:**R** *(For Rotate)*
Base point <Current default>:**0,0,0**
Rotation angle <Current default>:**0**

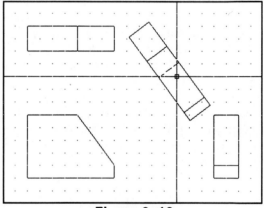

Figure 9–18

Tutorial Exercise
Bracket.Dwg

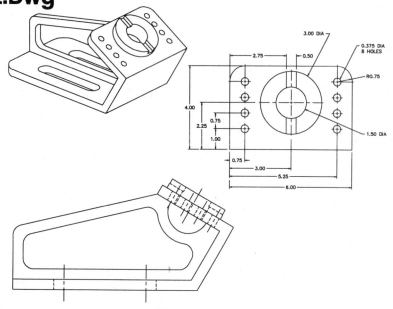

Figure 9–19

Purpose

This tutorial is designed to allow you to construct an auxiliary view of the inclined surface in the bracket shown in Figure 9–19.

System Settings

Since this drawing is provided on CD, edit an existing drawing called "Bracket." Follow the steps in this tutorial for the creation of an auxiliary view. All Units, Limits, Grid, and Snap values have been previously set.

Layers

The following layers have already been created with the following format:

Name	Color	Linetype
Cen	Yellow	Center
Dim	Yellow	Continuous
Hid	Red	Hidden
Obj	Cyan	Continuous

Suggested Commands

Begin this tutorial by using the OFFSET command to construct a series of lines parallel to the inclined surface containing the auxiliary view. Next construct lines perpendicular to the inclined surface. Use the CIRCLE command to begin laying out features that lie in the auxiliary view. Use ARRAY to copy the circle in a rectangular pattern. Add centerlines using the DIMCENTER command. Insert a predefined view called "Top." A three-view drawing consisting of front, top, and auxiliary views is completed.

Whenever possible, substitute the appropriate command alias in place of the full AutoCAD command in each tutorial step. For example, use "Co" for the COPY command, "L" for the LINE command, and so on. The complete listing of all command aliases is located on pages 1–9 and 1–10.

Step #1

Before you begin, understand that an auxiliary view will be taken from a point of view illustrated in Figure 9–20. This direction of sight is always perpendicular to the inclined surface. This perpendicular direction ensures the auxiliary view of the inclined surface will be of true size and shape. Begin this tutorial by turning Ortho off for the next few steps using the command from the keyboard or by double-clicking on the Ortho button located in the status bar of the display screen. Restore a previously saved view called "Front" using the VIEW or DDVIEW commands.

Command: **ORTHO**
ON/OFF <On>: **Off**

 Command: **VIEW**
?/Delete/Restore/Save/Window: **R**
View name to restore: **Front**

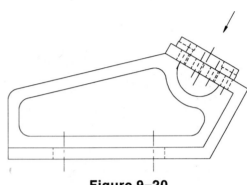

Figure 9–20

Step #2

Use the SNAP command to rotate the grid perpendicular to the inclined surface. For the base point, identify the endpoint of the line at "A" in Figure 9–21. For the rotation angle, use the "rubberband"

cursor and mark a point at the endpoint of the line at "B." The grid should change along with the standard AutoCAD cursor. Restore the view called "Overall" using the VIEW or DDVIEW commands.

Command: **SNAP**
Snap spacing or ON/OFF/Aspect/Rotate/Style
    <0.50>: **Rotate**
Base point <0,0>: **Endp**
of (Select the endpoint of the line at "A")
Rotation angle <0>: **Endp**
of (Select the endpoint of the line at "B")

 Command: **VIEW**
?/Delete/Restore/Save/Window: **R**
View name to restore: **Overall**

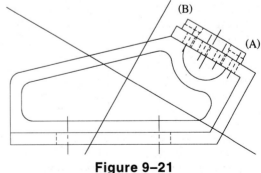

Figure 9–21

Step #3

Turn Snap off by striking the F9 function key. Begin the construction of the auxiliary view by using the OFFSET command to copy a line parallel to the inclined line (see Figure 9–22). Use an offset distance of 8.50 as the distance between the front and auxiliary view.

Step #4

Refer to the working drawing Figure 9–19 for the necessary dimensions required to construct the auxiliary view. Perform a ZOOM-All if the image on the screen is not similar to 9–22. Use the OFFSET command again to add the depth of the auxiliary. Remember, the depth of the auxiliary view is the same dimension found in the top and right side views. Set the offset distance to 6.00. Set the new current layer to "Obj" using the -LAYER command from the keyboard or the Layer and Linetype Properties dialog box. See also Figure 9–23.

 Command:**OFFSET**

Offset distance or Through <8.50>:**6.00**
Select object to offset: *(Select the inclined line at "A")*
Side to offset? *(Select a point anywhere near "B")*
Select object to offset: *(Press* ⌈ENTER⌉ *to exit this command)*

 Command:**-LAYER**

?/Make/Set/New/ON/OFF/Color/Ltype/Freeze/
    Thaw/LOck/Unlock:**Set**
New current layer <0>:**Obj**
?/Make/Set/New/ON/OFF/Color/Ltype/Freeze/
    Thaw/LOck/Unlock: *(Press* ⌈ENTER⌉ *to exit this command)*

 Command:**OFFSET**

Offset distance or Through <Through>:**8.50**
Select object to offset: *(Select the inclined line at "A")*
Side to offset? *(Select a point anywhere near "B")*
Select object to offset: *(Press* ⌈ENTER⌉ *to exit this command)*

(B)

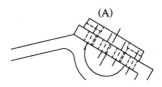
(A)

Figure 9–22

(B)

(A)

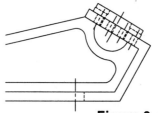
Figure 9–23

Step #5

Project two lines from the endpoints of the front view at "A" and "B" in Figure 9–24. These lines should extend past the outer line of the auxiliary view. Turn the Snap off and Ortho on. This should aid in this operation.

Command: **ORTHO**
ON/OFF <Off>: **On**

 Command: **LINE**

From point: **Endp**
of *(Pick the endpoint of the line at "A")*

To point: *(Pick a point anywhere near "B")*
To point: *(Press ⌜ENTER⌝ to exit this command)*

 Command: **LINE**

From point: **Endp**
of *(Pick the endpoint of the line at "C")*
To point: *(Pick a point anywhere near "D")*
To point: *(Press ⌜ENTER⌝ to exit this command)*

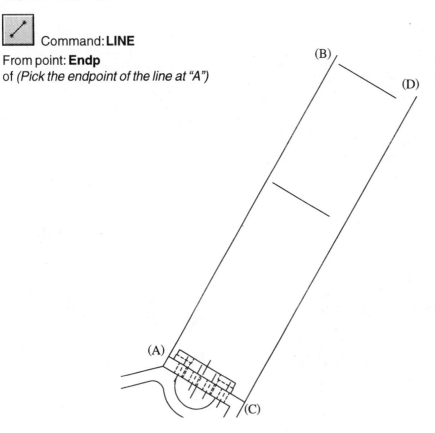

Figure 9–24

Step #6

Use the ZOOM-Window option to magnify the display of the auxiliary view similar to Figure 9–25. Then use the MULTIPLE command and enter FILLET to create four corners of the view in Figure 9–25. When completed with this operation, save the display to a new name of "Aux" using the VIEW or DDVIEW commands.

Command:**MULTIPLE**
Multiple command:**FILLET**
(TRIM mode) Current fillet radius = 0.00
Polyline/Radius/Trim/<Select first object>:
   (Select the line at "A")
Select second object: *(Select the line at "B")*
FILLET
(TRIM mode) Current fillet radius = 0.00
Polyline/Radius/Trim/<Select first object>:
   (Select the line at "B")
Select second object: *(Select the line at "C")*
FILLET
(TRIM mode) Current fillet radius = 0.00
Polyline/Radius/Trim/<Select first object>:
   (Select the line at "C")
Select second object: *(Select the line at "D")*
FILLET
(TRIM mode) Current fillet radius = 0.00
Polyline/Radius/Trim/<Select first object>:
   (Select the line at "D")
Select second object: *(Select the line at "A")*
FILLET
(TRIM mode) Current fillet radius = 0.00
Polyline/Radius/Trim/<Select first object>:
   (Press the ESC key to cancel the command)

 Command:**VIEW**
?/Delete/Restore/Save/Window:**Save**
View name to save: **Aux**

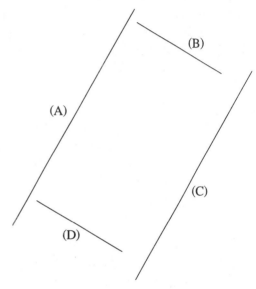

Figure 9–25

Step #7

Use the ZOOM-Previous option to demagnify the
screen back to the original display. Once here, draw
a line from the endpoint of the centerline in the
front view to a point past the auxiliary view (see
Figure 9–26). Check to see that Ortho mode is on.
This line will assist in constructing circles in the
auxiliary view.

 Command: **LINE**

From point: **Endp**
of *(Select the endpoint of the centerline at "A")*
To point: *(Select a point anywhere near "B")*
To point: *(Press* ENTER *to exit this command)*

Step #8

Use the OFFSET command and offset the line at
"A" a distance of 3.00 units (see Figure 9–27). The
intersection of this line and the previous line form
the center for placing two circles.

 Command: **OFFSET**

Offset distance or Through <6.00>: **3.00**
Select object to offset: *(Select the line at "A")*
Side to offset? *(Select a point anywhere near "B")*
Select object to offset: *(Press* ENTER *to exit this
   command)*

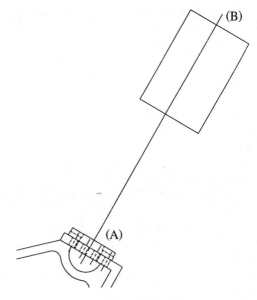

Figure 9–26

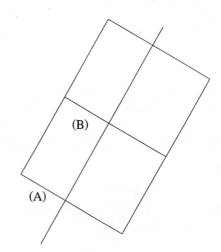

Figure 9–27

Step #9

Use the VIEW or DDVIEW commands and restore the view "Aux." Then, draw two circles of diameters 3.00 and 1.50 from the center at "A" using the CIRCLE command (see Figure 9–28). For the cen-

ter of the second circle, the @ option may be used to pick up the previous point that was the center of the 3.00 diameter circle.

 Command: **VIEW**

?/Delete/Restore/Save/Window: **Restore**
View name: **Aux**

 Command: **CIRCLE**

3P/2P/TTR/<Center point>: **Int**
of *(Select the intersection of the two lines at "A")*
Diameter/<Radius>: **D**
Diameter: **3.00**

 Command: **CIRCLE**

3P/2P/TTR/<Center point>: **@** *(To reference the last point)*
Diameter/<Radius>: **D**
Diameter: **1.50**

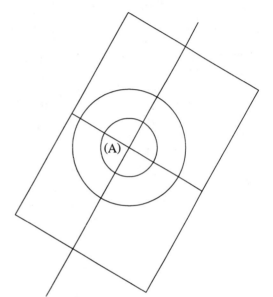

Figure 9–28

Step #10

Use the OFFSET command to offset the centerline the distance of 0.25 units (see Figure 9–29). Perform this operation on both sides of the centerline. Both offset lines form the width of the 0.50 slot.

 Command: **OFFSET**

Offset distance or Through <3.00>: **0.25**
Select object to offset: *(Select the middle line at "A")*
Side to offset? *(Select a point anywhere near "B")*
Select object to offset: *(Select the middle line at "A" again)*
Side to offset? *(Select a point anywhere near "C")*
Select object to offset: *(Press* ENTER *to exit this command)*

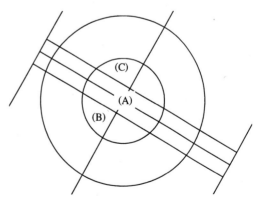

Figure 9–29

Step #11

Use the TRIM command to trim away portions of
the lines as shown in Figure 9–30.

 Command:**TRIM**

Select cutting edges: (Projmode = UCS,
   Edgemode = No extend)
Select objects: *(Select both circles as cutting
   edges)*
Select objects: *(Press [ENTER] to continue)*
<Select object to trim>/Project/Edge/Undo:
   (Select the line at "A")
<Select object to trim>/Project/Edge/Undo:
   (Select the line at "B")
<Select object to trim>/Project/Edge/Undo:
   (Select the line at "C")
<Select object to trim>/Project/Edge/Undo:
   (Select the line at "D")
<Select object to trim>/Project/Edge/Undo:
   (Select the line at "E")
<Select object to trim>/Project/Edge/Undo:
   (Select the line at "F")
<Select object to trim>/Project/Edge/Undo:
   (Press [ENTER] to exit this command)

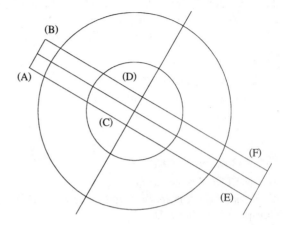

Figure 9–30

Step #12

Use the ERASE command to delete the two lines at
"A" and "B" in Figure 9–31. Standard centerlines will
be placed here later marking the center of both circles.

 Command:**ERASE**

Select objects: *(Select the lines at "A" and "B")*
Select objects: *(Press [ENTER] to execute this
   command)*

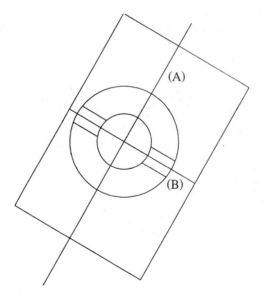

Figure 9–31

Step #13

To identify the center of the small 0.375 diameter circle, use the OFFSET command to copy parallel the line at "A" a distance of 0.75 units and the line at "C" the distance of 1.00 units. See Figure 9–32.

 Command:**OFFSET**

Offset distance or Through <0.25>: **0.75**
Select object to offset: *(Select the line at "A")*
Side to offset? *(Select a point anywhere near "B")*
Select object to offset: *(Press* ⌨ENTER *to exit this command)*

 Command:**OFFSET**

Offset distance or Through <0.75>: **1.00**
Select object to offset: *(Select the line at "C")*
Side to offset? *(Select a point anywhere near "D")*
Select object to offset: *(Press* ⌨ENTER *to exit this command)*

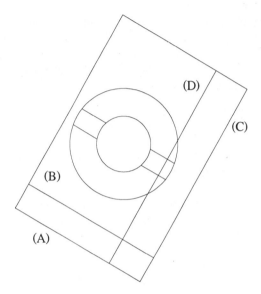

Figure 9–32

Step #14

Draw a circle of 0.375 diameter from the intersection of the two lines created in the last OFFSET command as shown in Figure 9–33. Use the ERASE command to delete the two lines at "B" and "C." A standard center marker will be placed at the center of this circle.

 Command:**CIRCLE**

3P/2P/TTR/<Center point>: **Int**
of *(Select the intersection of the two lines at "A")*
Diameter/<Radius>: **D**
Diameter: **0.375**

Command: **ERASE**
Select objects: *(Select the two lines at "B" and "C")*
Select objects: *(Press* ⌨ENTER *to execute this command)*

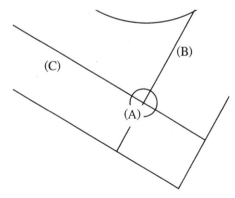

Figure 9–33

Step #15

Set the new current layer to "Cen." Prepare the following parameters before placing a center marker at the center of the 0.375 diameter circle (see Figure 9–34). Set the dimension variable DIMCEN to a value of -0.07 units. The negative value will construct lines that are drawn outside of the circle. Use the DIMCENTER command to place the center marker.

 Command:**-LAYER**

?/Make/Set/New/ON/OFF/Color/Ltype/Freeze/
   ThawLOck/Unlock: **Set**
New current layer <OBJ>: **Cen**
?/Make/Set/New/ON/OFF/Color/Ltype/Freeze/
   ThawLOck/Unlock: *(Press* [ENTER] *to exit this*
   command)

Command:**DIMCEN**
New value for DIMCEN <0.09> New value: **-0.07**

 Command:**DIMCENTER**

Select arc or circle: *(Select the small circle at "A")*

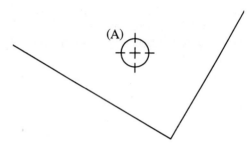

Figure 9–34

Step #16

Use the ROTATE command to rotate the center marker parallel to the edges of the auxiliary view (see Figure 9–35). Select the center marker and circle as the objects to rotate. Check to see that Ortho is on.

 Command:**ROTATE**

Select objects: *(Select the small circle and all objects that make up the center marker; the Window option is recommended here)*
Select objects: *(Press* [ENTER] *to continue)*
Base point: **Cen**
of *(Select the edge of the small circle at "A")*
<Rotation angle>/Reference: *(Pick a point anywhere at "B")*

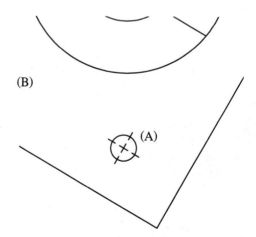

Figure 9–35

Step #17

Since the remaining seven holes are along a set pattern, use the ARRAY command and perform a rectangular array, as shown in Figure 9–36. The number of rows are two and number of columns four.

 Command: **ARRAY**

Select objects: *(Select the small circle and center marker)*
Select objects: *(Press [ENTER] to continue)*
Rectangular or Polar array (R/P) <R>: **R**
Number of rows(---) <1>: **2**
Number of columns(III) <1>: **4**
Unit cell or distance between rows (---): **4.50**
Distance between columns (III): **-0.75**

Step #18

Use the FILLET command set to a radius of 0.75 to place a radius along the two corners of the auxiliary following the prompts and Figure 9–37.

 Command: **FILLET**

(TRIM mode) Current fillet radius = 0.00
Polyline/Radius/Trim/<Select first object>: **R**
Enter fillet radius <0.00>: **0.75**

 Command: **FILLET**

(TRIM mode) Current fillet radius = 0.75
Polyline/Radius/Trim/<Select first object>:
  (Select line "A")
Select second object: *(Select line "B")*

 Command: **FILLET**

(TRIM mode) Current fillet radius = 0.75
Polyline/Radius/Trim/<Select first object>:
  (Select line "B")
Select second object: *(Select line "C")*

Distance between rows is 4.50 units and between columns is -0.75 units; this will force the circles to be patterned to the left, which is where we want them to go.

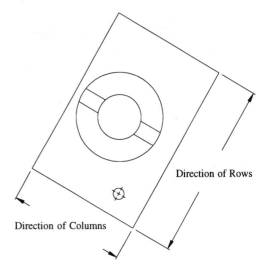

Direction of Rows

Direction of Columns

Figure 9–36

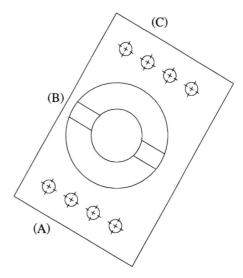

Figure 9–37

Step #19

Place a center marker in the center of the two large circles using the existing value of the dimension variable DIMCEN (see Figure 9–38). Since the center marker is placed in relation to the World coordinate system, use the ROTATE command to rotate it parallel to the auxiliary view. Ortho should be on.

 Command:**DIMCENTER**

Select arc or circle: *(Select the large circle at "A")*

 Command:**ROTATE**

Select objects: *(Select all lines that make up the large center marker)*
Select objects: *(Press* ⌷ENTER⌷ *to continue)*
Base point: **Cen**
of *(Select the edge of the large circle at "A")*
<Rotation angle>/Reference: *(Pick a point anywhere at "B")*

Step #20

Restore the view named "Overall" using the VIEW or DDVIEW commands. Complete the multiview drawing of the bracket by inserting an existing block called "Top" into this drawing (see Figure 9–39). This block represents the complete top view of the drawing. Use an insertion point of 0,0 for placing this view in the drawing.

 Command:**VIEW**

?/Delete/Restore/Save/Window:**R**
View name to restore: **Overall**

 Command:**INSERT**

Block name (or ?): **Top**
Insertion point: **0,0**
X scale factor <1>/Corner/XYZ: *(Press* ⌷ENTER⌷ *to accept default)*
Y scale factor (default=X): *(Press* ⌷ENTER⌷ *to accept default)*
Rotation angle <0>: *(Press* ⌷ENTER⌷ *to accept this default)*

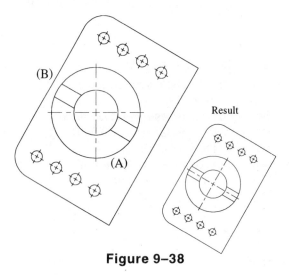

Figure 9–38

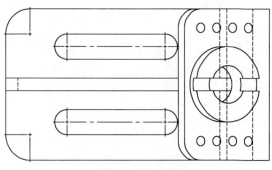

Figure 9–39

Step #21

Return the grid back to its original orthographic form using the Rotate option of the SNAP command. Use a base point of 0,0 and a rotation angle of 0 degrees (see Figure 9–40). This is especially helpful when adding dimensions to the drawing.

Command: **SNAP**
Snap spacing or ON/OFF/Aspect/Rotate/Style
   <0.50>: **R** *(For Rotate)*
Base point <0,0>: **0,0**
Rotation angle <330>: **0**

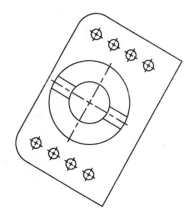

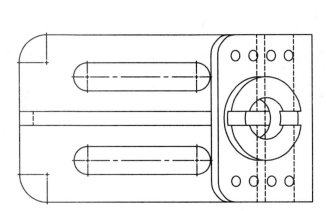

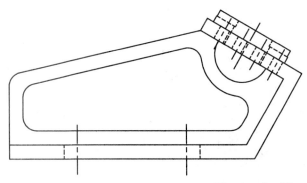

Figure 9–40

Problems for Unit 9

Directions for Problems 9-1 through 9-9

Draw the required views to fully illustrate each object. Be sure to include an auxiliary view.

Problem 9–1   **Problem 9–2**

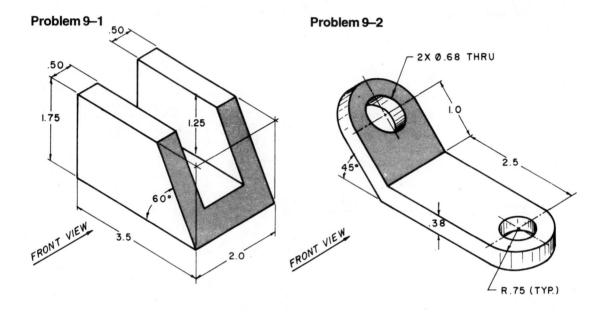

Problem 9–3

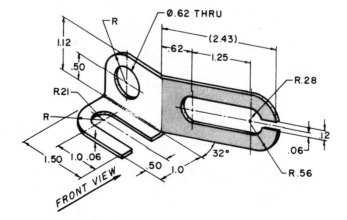

Problem 9–4

Problem 9–5

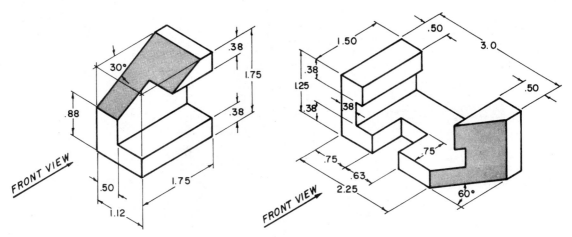

Problem 9–6

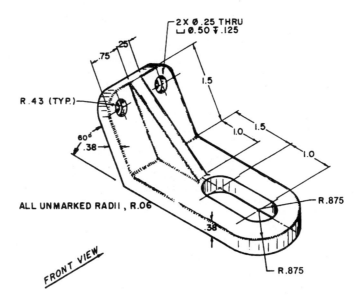

Problem 9–7

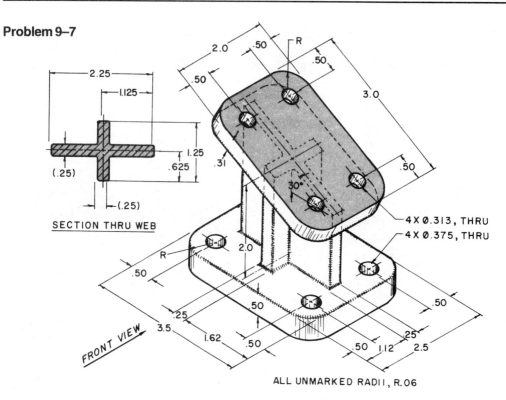

2.25
1.125
1.25
.625
(.25)
(.25)

SECTION THRU WEB

2.0   .50   R
.50   .50
3.0
.50
.31
30°
.50
4X Ø.313, THRU
4X Ø.375, THRU

R
.50
.25
3.5   1.62   .50
2.0
.50
.50
.25
.50   1.12   2.5

FRONT VIEW

ALL UNMARKED RADII, R.06

Problem 9–8

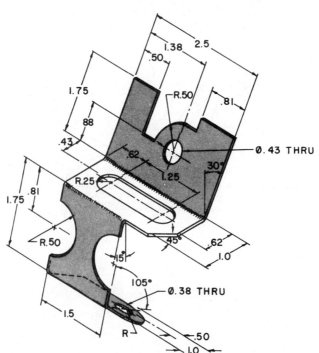

2.5
1.38
.50
1.75
.88
.43
R.50
.81
62
Ø.43 THRU
1.25
30°
R.25
1.75   .81
R.50
45°   .62
15°   1.0
105°   Ø.38 THRU
1.5
R
.50
1.0

Problem 9–9

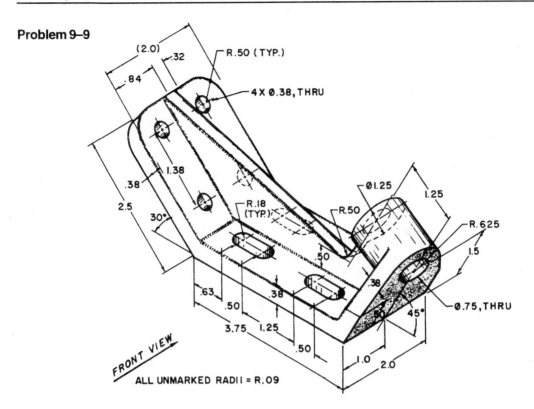

FRONT VIEW

ALL UNMARKED RADII = R.09

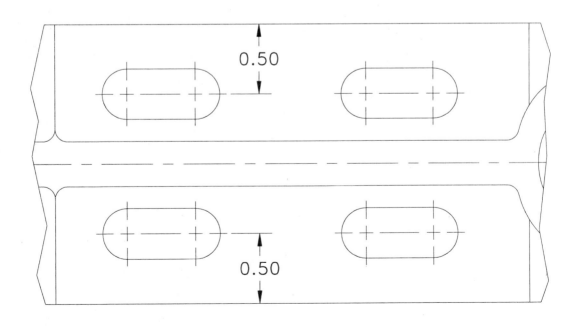

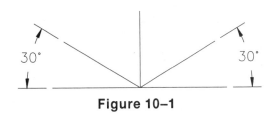

UNIT 10

Isometric Drawings

Multiview or orthographic projections are necessary to produce parts that go into the construction of all kinds of objects. Skill is involved in laying out the primary views, projecting visible objects into other views, and adding dimensions to describe the size of the object being made. Yet another skill involves reading or interpreting these engineering drawings, which for some individuals is extremely difficult and complex. If only there existed some type of picture of the object, then the engineering drawing might make sense. Isometric drawings be-

come a means of drawing an object in picture form for better clarification of what the object looks like. These types of drawings resemble a picture of an object that is drawn in two dimensions. As a result, existing AutoCAD commands such as LINE and COPY are used for producing isometric drawings. This unit will explain isometric basics including how regular, angular, and circular objects are drawn in isometric. Numerous isometric aids such as Snap and isometric axes will be explained to assist in the construction of isometric drawings.

Isometric Basics

Isometric drawings consist of two-dimensional drawings that are tilted at some angle to expose other views and give the viewer the illusion that what he or she is viewing is a three-dimensional drawing. The tilting occurs with two 30-degree angles that are struck from the intersection of a horizontal baseline and a vertical line (see Figure 10–1). The directions formed by the 30-degree angles represent actual dimensions of the object; this may be

30°   30°

Figure 10–1

either the width or depth. The vertical line in most cases represents the height dimension.

Figure 10–2 is a very simple example of how an object is aligned to the isometric axis. Once the horizontal baseline and vertical line are drawn, the 30-degree angles are projected from this common point, which becomes the reference point of the isometric view. In this example, once the 30-degree lines are drawn, the baseline is no longer needed and is usually discarded through erasing. Depending on how the object is to be viewed, width and depth measurements are made along the 30 degree lines. Height is measured along the vertical line. Figure 10–2 has the width dimension measured off to the left 30-degree line while the depth dimension measures to the right along the right 30-degree line. Once the object is blocked with overall width, depth, and height, details are added, and lines are erased and trimmed, leaving the finished object. Holes no longer appear as full circles but rather as ellipses. Techniques of drawing circles in isometric will be discussed later in this unit.

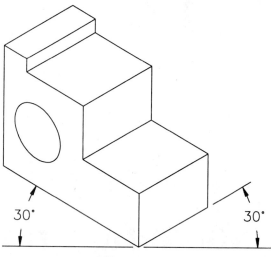

Figure 10–2

Notice the objects in Figure 10–3 both resemble the object in Figure 10–2 except they appear from a different vantage point. The problem with isometric drawings is that if an isometric of an object is drawn from one viewing point and you want an isometric from another viewing point, an entirely different isometric drawing must be generated from scratch. Complex isometric drawings from different views can be very tedious to draw. Another interesting observation concerning the objects in Figure 10–3 is that one has hidden lines while the other does not. Usually only the visible surfaces of an object are drawn in isometric leaving out hidden lines. Since this is considered the preferred practice, there are always times that hidden lines are needed on very complex isometric drawings.

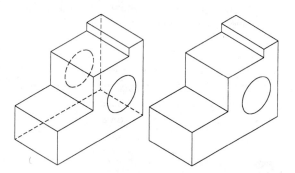

Figure 10–3

Creating an Isometric Grid

Figure 10–4 shows the current AutoCAD screen complete with cursor and grid on. In manual drawing and sketching days, an isometric grid was used to lay out all lines before transferring the lines to paper or Mylar for pen and ink drawings. An isometric grid may be defined in an AutoCAD drawing using the SNAP command. This would be the same grid found on isometric grid paper.

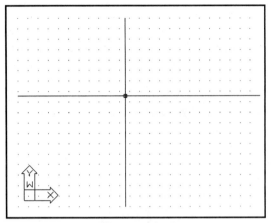

Figure 10–4

The display screen in Figure 10–5 reflects the use of the SNAP command and how this command affects the current grid display:

Command: **SNAP**
Snap spacing or ON/OFF/Aspect/Rotate/Style
    <0.2500>: **S** *(for Style)*
Standard/Isometric <S>: **I** *(for Isometric)*
Vertical spacing <0.5000>: *(Press* ENTER *to accept default value)*

Choosing an isometric style of Snap changes the grid display from orthographic to isometric, shown in Figure 10–5. The grid distance conforms to a vertical spacing height that you specify. As the grid changes, notice the display of the typical AutoCAD cursor; it conforms to an isometric axis plane and is used as an aid in constructing isometric drawings.

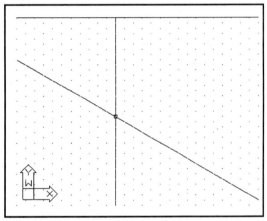

Figure 10–5

To see how this vertical spacing distance affects the grid changing it to isometric, see Figure 10–6. The grid dot at "A" becomes the reference point where the horizontal baseline is placed followed by the vertical line represented by the dot at "B." At dots "A" and "B," 30-degree lines are drawn; points "C" and "D" are formed where they intersect. This is how an isometric screen display is formed.

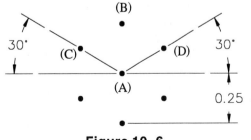

Figure 10–6

Resetting Grid and Snap Back to Their Default Values

Once an isometric drawing is completed, it may be necessary to change the Grid, Snap, and cursor back to normal. This might result from the need to place text on the drawing, and the isometric axis now confuses instead of assists the drawing process. Use the following prompts to reset the Snap back to the Standard spacing. See Figure 10-10.

Command: **SNAP**
Snap spacing or ON/OFF/Rotate/Style
    <0.2500>: **S** *(for Style)*
Standard/Isometric <I>: **S** *(for Standard)*
Spacing/Aspect <0.5000>: *(Press* ENTER *to accept the default value)*

Notice when changing the Snap style to Standard, the AutoCAD cursor changes back to its original display.

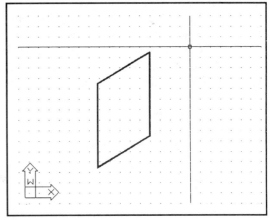

Figure 10–10

Isoplane Aids

We established that there are three isometric axis modes to draw on: Left, Top, and Right. It was not established how to make a mode current to draw on. The Tools area of the pull-down menu exposes the "Drawing Aids…" area, which displays the dialog box in Figure 10–11. Here in addition to Snap

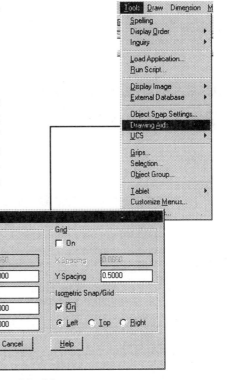

Figure 10–11

and Grid, the three isometric modes may be made current by simply placing a check in the appropriate box. Only one isometric mode may be current at a time. In addition to setting these modes, an isometric area exists to automatically set up an isometric grid by placing a check in the box and put it back to normal by removing the check. This has the same effect as using the SNAP-Style-Isometric option. This dialog box may be brought up through the keyboard by entering "DDRMODES" at the command prompt or by entering "'DDRMODES" while inside of a command.

There is a quicker method to move from one isometric axis mode to another. By default, after setting up an isometric grid, the Left isometric axis is active (see Figure 10–12). By typing from the keyboard CTRL-E or ^E, the Left axis changes to the Top axis. Typing another ^E changes from the Top axis to the Right axis. Typing a third ^E changes from the Right axis back to the Left axis and the pattern repeats from here. Using this keyboard entry, it is possible to switch or toggle from one mode to another.

The F5 function key also allows you to scroll through the different Isoplane modes.

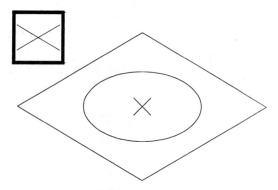

$$Default =$$

$$First\ \wedge E =$$

$$Second\ \wedge E =$$

Figure 10–12

Creating Isometric Circles

Circles appear as ellipses when drawn in any of the three isometric axes. The ELLIPSE command has a special Isocircle option to assist in drawing isometric circles; the Isocircle option will appear in the ELLIPSE command only if the current SNAP-Style is Isometric (see Figure 10–13). The prompt sequence for this command is:

 Command:**ELLIPSE**

Arc/Center/Isocircle/<Axis endpoint 1>:**I** *(for Isocircle)*
Center of circle: *(Select a center point)*
<Circle radius>/Diameter: *(Enter a value for the radius or type "D" for diameter and enter a value)*

Figure 10–13

When drawing isometric circles using the ELLIPSE command, it is important to match the isometric axis with the isometric plane the circle is to be drawn in. Figure 10–14 shows a cube displaying all three axes.

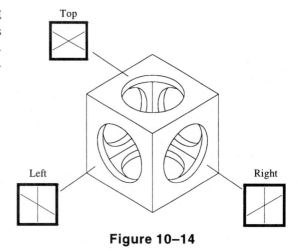

Figure 10–14

Figure 10–15 shows the result of drawing an isometric circle using the wrong isometric axis. The isometric box is drawn in the top isometric plane while the current isometric axis is Left. An isometric circle can be drawn to the correct size, but notice it does not match the box it was designed for. If you notice halfway through the ELLIPSE command that you are in the wrong isometric axis, type ^E until the correct axis appears.

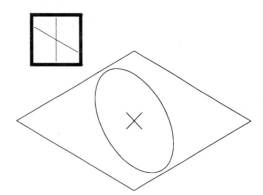

Figure 10–15

Basic Isometric Construction

Any isometric drawing, no matter how simple or complex, has an overall width, height, and depth dimension. Start laying out the drawing with these three dimensions to create an isometric box illustrated in the example in Figure 10–16A. Some techniques rely on piecing the isometric drawing together by views; unfortunately, it is very easy to get lost in all of the lines using this method. Once a box is created from overall dimensions, somewhere inside the box is the object.

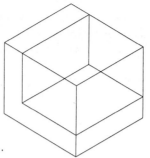

Figure 10–16A

With the box as a guide, begin laying out all visible features in the primary planes. Use the Left, Top, or Right isometric axis modes to assist you in this construction process. See Figure 10–16B.

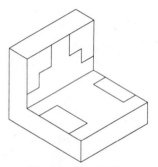

Figure 10–16B

You may use existing AutoCAD editing commands, especially COPY, to duplicate geometry to show depth of features, as shown in figure 10–16C. The OFFSET command should not be attempted for performing isometric drawings. Next, use the TRIM command to partially delete geometry where objects are not visible. Remember, most isometric objects do not require hidden lines.

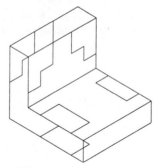

Figure 10–16C

Use the LINE command to connect intersections of surface corners. The resulting isometric drawing is illustrated in Figure 10–16D.

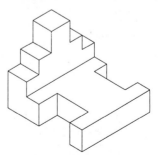

Figure 10–16D

Creating Angles in Isometric—Method #1

Drawing angles in isometric is a little tricky but not impossible. The two-view drawing in Figure 10–17 has an angle of unknown size; however, one endpoint of the angle measures 2.06 units from the top horizontal line of the front view at "B" and the other endpoint measures 1.00 unit from the vertical line of the front view at "A." This is more than enough information to lay out the endpoints of the angle in isometric. The MEASURE command can be used to easily lay out these distances. The LINE command is then used to connect the points to form the angle.

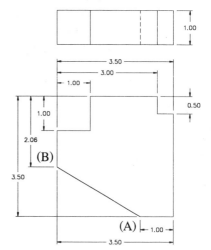

Figure 10–17

Before using the MEASURE command, set the PDMODE variable to a new value of 3. Points will appear as an X instead of a dot, as shown in Figure 10–18A. Points may also be set using the Point Style dialog box found under the Format area of the pull-down menu. Now use the MEASURE command to set off the two distances.

Command:**PDMODE**
New value for Pdmode <0>:**3**

Command:**MEASURE**
Select object to measure: *(Select the inclined line at "A")*
<Segment length>/Block:**1.00**

Command:**MEASURE**
Select object to measure: *(Select the vertical line at "B")*
<Segment length>/Block:**2.06**

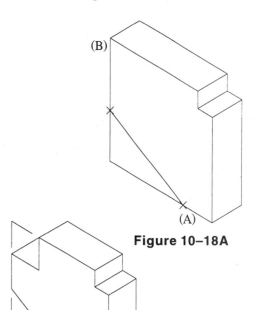

Figure 10–18A

The interesting part of the MEASURE command is that measuring will occur at the nearest endpoint of the line where the line was selected from. Therefore, it is important which endpoint of the line is selected. Once the points have been placed, the LINE command is used to draw a line from one point to the other using the OSNAP-Node option (see Figure 10–18B).

Figure 10–18B

Creating Angles in Isometric—Method #2

The exact same two-view drawing is illustrated in Figure 10–19. This time, one distance is specified along with an angle of 30 degrees. Even with the angle given, the position of the isometric axes makes any angle construction by degrees inaccurate. The distance XY is still needed to construct the angle in isometric. Use the MEASURE command to find distance XY, place a point, and connect the first distance with the second to form the 30-degree angle in isometric. It is always best to set the PDMODE system variable to a new value in order to visibly view the point. Points may also be set using the Point Style dialog box found under the Format area of the pull-down menu. A new value of 3 will assign the point as an X.

See Figures 10–20A and the following prompt sequences to change the point mode and measure the appropriate distances in the previous step.

Command: **PDMODE**
New value for Pdmode <0>: **3**

Command: **MEASURE**
Select object to measure: *(Select the inclined line at "A")*
<Segment length>/Block: **1.00**

Command: **MEASURE**
Select object to measure: *(Select the vertical line at "B")*
<Segment length>/Block: **Endp**
of *(Select the endpoint of the line at X in the 2-view drawing in Figure 10–19)*
Second point: **Int**
of *(Select the intersection at Y in the 2-view drawing in Figure 10–19)*

Line "B" is selected as the object to measure (see Figure 10–19). Since this distance is unknown, the MEASURE command may be used to set off the distance XY by identifying an endpoint and intersection from the front view in Figure 10–19. This means the view must be constructed only enough to lay out the angle and project the results to the isometric using the MEASURE command and the preceeding prompts.

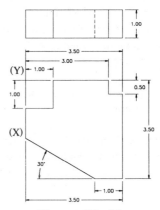

Figure 10–19

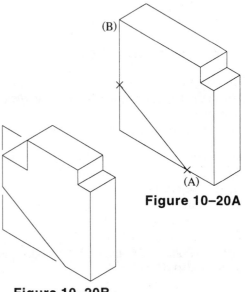

Figure 10–20A

Figure 10–20B

Isometric Construction Using Ellipses

Constructing circles as part of isometric drawing is possible using one of the three isometric axes positions. It is up to you to decide which axis to use. Before this, however, an isometric box consisting of overall distances is first constructed, as shown in Figure 10–21A. Use the ELLIPSE command to place the isometric circle at the base. To select the correct axis type CTRL-E or F5 until the proper axis appears in the form of the cursor. Place the ellipses. Lay out any other distances.

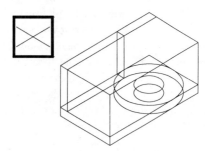

Figure 10–21A

For ellipses at different positions, type CTRL-E or F5 to select another isometric axis (see Figure 10–21B). Remember these axis positions may also be selected from the Drawing Aids dialog box from the Tools area of the menu bar.

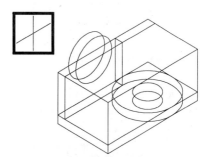

Figure 10–21B

Use the TRIM command to trim away any excess objects that are considered unnecessary, as shown in Figure 10–21C.

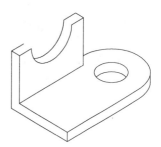

Figure 10–21C

Use the LINE command to connect endpoints of edges that form surfaces, as shown in Figure 10–21D.

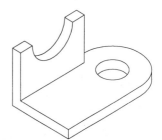

Figure 10–21D

Creating Isometric Sections

In some cases, it is necessary to cut an isometric drawing to expose internal features. The result is an isometric section similar to a section view formed from an orthographic drawing. The difference with the isometric section, however, is the cutting plane is usually along one of the three isometric axes. Figure 10–22A displays an orthographic section in addition to the isometric drawing.

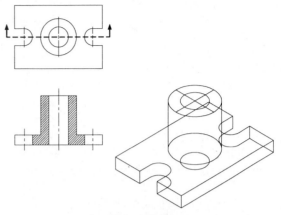

Figure 10–22A

The isometric in Figure 10–22B has additional lines representing surfaces cut by the cutting plane line. The lines to define these surfaces were formed using the LINE command in addition to a combination of top and right isometric axes modes. During this process, Ortho mode was toggled on and off numerous times depending on the axis direction. Use the TRIM command to remove objects from the half that will eventually be discarded.

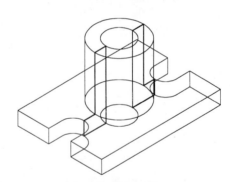

Figure 10–22B

Once ellipses were trimmed the remaining objects representing the front half of the isometric were removed exposing the back half. The front half is then discarded. This has the same effect as conventional section views where the direction of sight dictates which half to keep. For a full section, the BHATCH command is used to crosshatch the surfaces being cut by the cutting plane line, as in Figure 10–22C. Surfaces designated holes or slots not cut are not crosshatched. This same procedure is followed for converting an object into a half section.

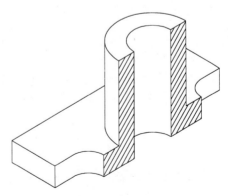

Figure 10–22C

Exploded Isometric Views

Isometric drawings are sometimes grouped together to form an exploded drawing of how a potential or existing product is assembled (see Figure 10–23). This involves aligning parts that fit with line segments, usually in the form of centerlines. Bubbles identifying the part number are attached to the drawing. Exploded isometric drawings come in handy for creating bill of material information and, for this purpose, have an important application to manufacturing. Once the part information is identified in the drawing and title block area, this information is extracted and brought into a third-party business package where important data collection information is able to actually track the status of parts in production in addition to the shipping date for all finished products.

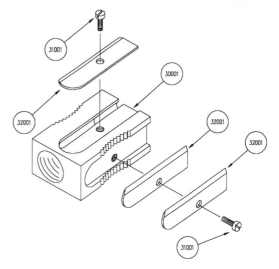

Figure 10–23

Isometric Assemblies

Assembly drawings show the completed part as if it were to be assembled (see Figure 10–24). Sometimes this drawing has an identifying number placed with a bubble for bill of material needs. Assembly drawings commonly are placed on the same display screen as the working drawing. With the assembly along the side of the working drawing, you have a pictorial representation of what the final product will look like and this can help you the understand the orthographic views.

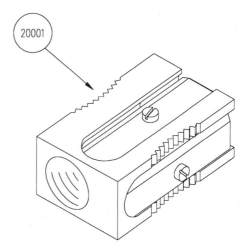

Figure 10–24

Tutorial Exercise
Plate.Dwg

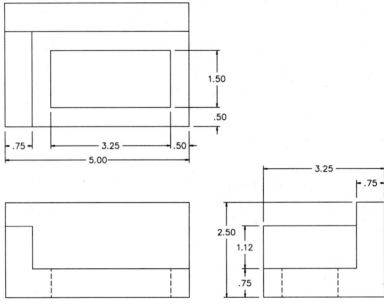

Figure 10–25

Purpose

The purpose of this tutorial exercise is to use a series of coordinates along with AutoCAD editing commands to construct an isometric drawing of the plate.

System Settings

Begin a new drawing called "Plate." Use the Units Control dialog box to change the number of decimal places past the zero from four to two. Keep the remaining default unit values. Using the LIMITS command, keep 0,0 for the lower left corner and change the upper right corner from 12,9 to 10.50,8.00. Use the GRID command and change the grid spacing from 1.00 to 0.25 units. Do not turn the Snap or Ortho on.

Layers

Special layers do not have to be created for this tutorial exercise.

Suggested Commands

Begin this exercise by changing the grid from the standard display to an isometric display using the SNAP-Style option. Remember both the Grid and Snap can be manipulated by the SNAP command only if the current Grid value is 0. Use Absolute and Polar coordinates to lay out the base of the plate. Then begin using the COPY command followed by TRIM to duplicate objects and clean up or trim unnecessary geometry.

Whenever possible, substitute the appropriate command alias in place of the full AutoCAD command in each tutorial step. For example, use "Co" for the COPY command, "L" for the LINE command, and so on. The complete listing of all command aliases is located on pages 1–9 and 1–10.

Step #1

Use the LINE command to draw the object in Figure 10–26.

 Command: **LINE**

From point: **5.629,0.750**
To point: **@3.25<30**
To point: **@5.00<150**
To point: **@3.25<210**
To point: **C**

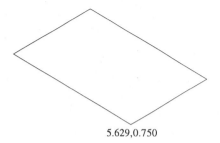

5.629,0.750

Figure 10–26

Step #2

Copy the four lines drawn in Step #1 up at a distance of 2.50 units in the 90-degree direction. See Figure 10–27.

 Command: **COPY**

Select objects: *(Select lines "A," "B," "C," and "D")*
Select objects: *(Press ⏎ to continue)*
<Base point or displacement>/Multiple: **Endp**
of *(Select the line at "A")*
Second point of displacement: **@2.50<90**

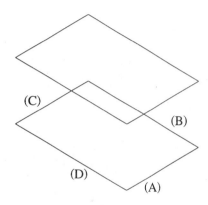

Figure 10–27

Step #3

Connect the top and bottom isometric boxes with line segments, as shown in Figure 10–28. Draw one segment using the LINE command. Use the COPY-Multiple command to duplicate and form the remaining segments at "C" and "D". Erase the two dashed lines since they are not visible in an isometric drawing.

 Command: **LINE**

From point: **Endp**
of *(Select the endpoint of the line at "A")*
To point: **Endp**
of *(Select the endpoint of the line at "B")*
To point: *(Press ⏎ to exit this command)*

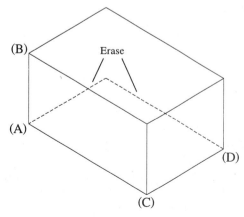

Figure 10–28

Step #4

Copy the two dashed lines in Figure 10–29 a distance of 0.75 in the 210-degree direction.

 Command: **COPY**

Select objects: *(Select the two dashed lines in Figure 10–29)*
Select objects: *(Press* ENTER *to continue)*
<Base point or displacement>/Multiple: **Endp**
of *(Select the endpoint at "A")*
Second point of displacement: **@0.75<210**

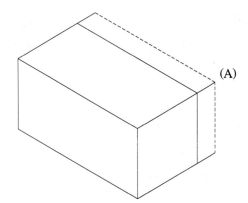

Figure 10–29

Step #5

Copy the two dashed lines in Figure 10–30 a distance of 0.75 in the 90-degree direction. This forms the base of the plate.

 Command: **COPY**

Select objects: *(Select the two dashed lines in Figure 10–30)*
Select objects: *(Press* ENTER *to continue)*
<Base point or displacement>/Multiple: **Endp**
of *(Select the endpoint at "A")*
Second point of displacement: **@0.75<90**

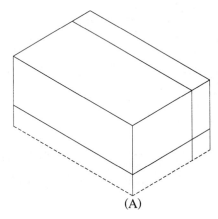

Figure 10–30

Step #6

Copy the dashed line in Figure 10–31 a distance of 0.75 in the -30-degree direction.

 Command: **COPY**

Select objects: *(Select the dashed line in Figure 10–31)*
Select objects: *(Press* ENTER *to continue)*
<Base point or displacement>/Multiple: **Endp**
of *(Select the endpoint at "A")*
Second point of displacement: **@0.75<-30**

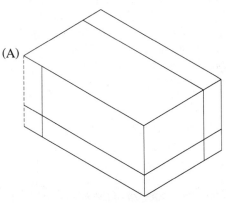

Figure 10–31

Step #7

Use the FILLET command to place a corner between the two dashed lines at "A" and "B" and at "C" and "D" in Figure 10–32. The current fillet radius should already be set to a value of 0.

 Command: **FILLET**

(TRIM mode) Current fillet radius = 0.5000
Polyline/Radius/Trim/<Select first object>: **R**
Enter fillet radius <0.5000>: **0**

 Command: **FILLET**

(TRIM mode) Current fillet radius = 0.0000
Polyline/Radius/Trim/<Select first object>:
  (Select "A")
Select second object: *(Select "B")*

 Command: **FILLET**

(TRIM mode) Current fillet radius = 0.0000
Polyline/Radius/Trim/<Select first object>:
  (Select "C")
Select second object: *(Select "D")*

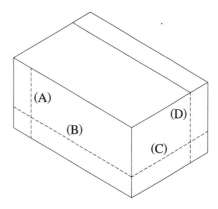

Figure 10–32

Step #8

Copy the dashed line in Figure 10–33 to begin forming the top of the base.

 Command: **COPY**

Select objects: *(Select the dashed line in Figure 10–33)*
Select objects: *(Press* ENTER *to continue)*
<Base point or displacement>/Multiple: **Endp**
of *(Select the endpoint at "A")*
Second point of displacement: **Endp**
of *(Select the endpoint at "B")*

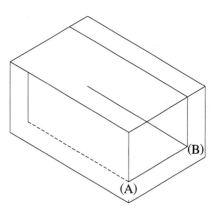

Figure 10–33

Step #9

Copy the dashed line in Figure 10–34. This forms the base of the plate.

 Command: **COPY**

Select objects: *(Select the dashed line in Figure 10–34)*
Select objects: *(Press ENTER to continue)*
<Base point or displacement>/Multiple: **Endp**
of *(Select the endpoint at "A")*
Second point of displacement: **Endp**
of *(Select the endpoint at "B")*

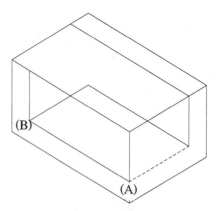

Figure 10–34

Step #10

Use the TRIM command to clean up the excess lines in Figure 10–35.

 Command: **TRIM**

Select cutting edges: (Projmode = UCS, Edgemode = No extend)
Select objects: *(Select the three dashed lines shown in Figure 10–35)*
Select objects: *(Press ENTER to continue)*
<Select object to trim>/Project/Edge/Undo:
 (Select the inclined line at "A")
<Select object to trim>/Project/Edge/Undo:
 (Select the inclined line at "B")
<Select object to trim>/Project/Edge/Undo:
 (Select the vertical line at "C")
<Select object to trim>/Project/Edge/Undo:
 (Press ENTER to exit this command)

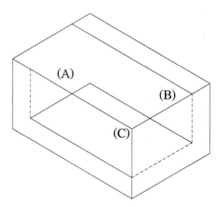

Figure 10–35

Step #11

Your display should be similar to Figure 10–36.

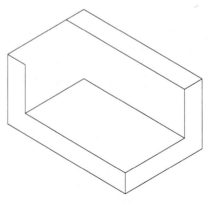

Figure 10–36

Step #12

Copy the dashed line in Figure 10–37 a distance of
1.12 units in the 90-degree direction.

 Command: **COPY**

Select objects: *(Select the dashed line in*
  Figure 10–37)
Select objects: *(Press* ⎡ENTER⎤ *to continue)*
 <Base point or displacement>/Multiple: **Endp**
of *(Select the endpoint at "A")*
Second point of displacement: **@1.12<90**

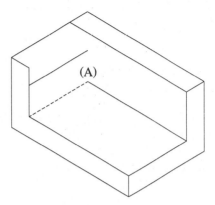

Figure 10–37

Step #13

As shown in Figure 10–38, copy the dashed line at
"A" to new positions at "B" and "C." Then delete
the line at "A" using the ERASE command.

 Command: **COPY**

Select objects: *(Select the dashed line in*
  Figure 10–38)
Select objects: *(Press* ⎡ENTER⎤ *to continue)*
 <Base point or displacement>/Multiple: **M**
Base point: **Endp**
of *(Select the endpoint at "A")*
Second point of displacement: **Endp**
of *(Select the endpoint at "B")*
Second point of displacement: **Endp**
of *(Select the endpoint at "C")*
Second point of displacement: *(Press* ⎡ENTER⎤ *to*
  exit this command)

 Command: **ERASE**

Select objects: *(Select the dashed line at "A")*
Select objects: *(Strike* ⎡ENTER⎤ *to execute this*
  command)

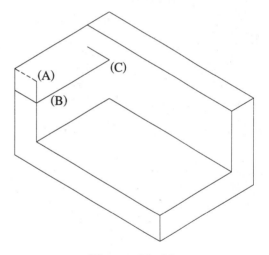

Figure 10–38

Step #14

Use the TRIM command to clean up the excess lines
in Figure 10–39.

 Command: **TRIM**

Select cutting edges: (Projmode = UCS,
    Edgemode = No extend)
Select objects: *(Select the two dashed lines in*
    Figure 10–39)
Select objects: *(Press* [ENTER] *to continue)*
 <Select object to trim>/Project/Edge/Undo:
    (Select the vertical line at "A")
 <Select object to trim>/Project/Edge/Undo:
    (Select the vertical line at "B")
 <Select object to trim>/Project/Edge/Undo:
    (Select the inclined line at "C")
 <Select object to trim>/Project/Edge/Undo:
    (Press [ENTER] *to exit this command)*

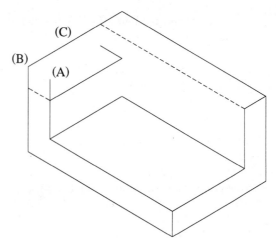

Figure 10–39

Step #15

Use the COPY command to duplicate the dashed
line in Figure 10–40 from the endpoint of "A" to
the endpoint at "B."

 Command: **COPY**

Select objects: *(Select the dashed line in*
    Figure 10–40)
Select objects: *(Press* [ENTER] *to continue)*
 <Base point or displacement>/Multiple: **Endp**
of *(Select the endpoint at "A")*
Second point of displacement: **Endp**
of *(Select the endpoint at "B")*

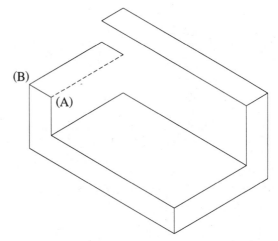

Figure 10–40

Isoplane Modes

The AutoCAD cursor has always been the vehicle for drawing objects or constructing windows for object selection mode. Once in isometric Snap mode, AutoCAD supports three axes to assist in the construction of isometric drawings. The first axis is the Left axis and may control that part of an object falling into the left projection plane. The left axis cursor displays a vertical line intersected by a 30 degree angle line, which is drawn to the left. This axis is displayed in the illustration in Figure 10–7 and in the following drawing:

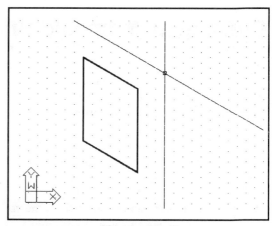

Figure 10–7

The next isometric axis is the Top mode. Objects falling into the top projection plane may be drawn using this isometric axis. This cursor consists of two 30-degree angle lines intersecting each other forming the center of the cursor. This mode is displayed in Figure 10–8 and in the following drawing:

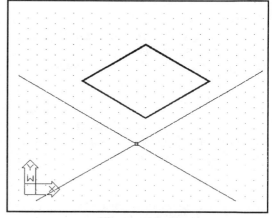

Figure 10–8

The final isometric axis is called the Right mode and is formed by the intersection of a vertical line and a 30-degree angle drawn off to the right. As with the previous two modes, objects that fall along the right projection plane of an isometric drawing may be drawn using this cursor. It is displayed in Figure 10–9 and in the following drawing. The current Ortho mode affects all three modes. If Ortho is on, and the current isometric axis is right, lines and other operations requiring direction will be forced to be drawn vertical or at a 30-degree angle to the right as shown in the following drawing:

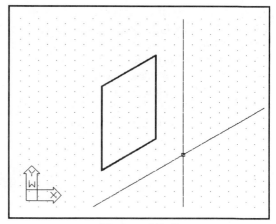

Figure 10–9

Step #16

Use the COPY command to duplicate the dashed line in Figure 10–41 from the endpoint of "A" to the endpoint at "B."

 Command:**COPY**

Select objects: *(Select the dashed line in Figure 10–41)*
Select objects: *(Press* ENTER *to continue)*
 <Base point or displacement>/Multiple: **Endp**
of *(Select the endpoint at "A")*
Second point of displacement: **Endp**
of *(Select the endpoint at "B")*

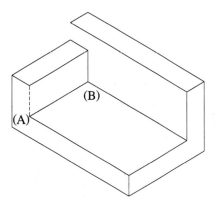

Figure 10–41

Step #17

Use the LINE command to connect the endpoints of the segments at "A" and "B" in Figure 10–42.

 Command:**LINE**

From point: **Endp**
of *(Select the endpoint at "A")*
To point: **Endp**
of *(Select the endpoint at "B")*
To point: *(Press* ENTER *to exit this command)*

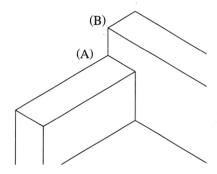

Figure 10–42

Step #18

Use the COPY command to duplicate the dashed line in Figure 10–43 from the endpoint of "A" to the distance of 0.50 units specified by a polar coordinate. This value begins the outline of the rectangular hole through the object.

 Command:**COPY**

Select objects: *(Select the dashed line in Figure 10–43)*
Select objects: *(Press* ENTER *to continue)*
 <Base point or displacement>/Multiple: **Endp**
of *(Select the endpoint at "A")*
Second point of displacement: **@0.50<210**

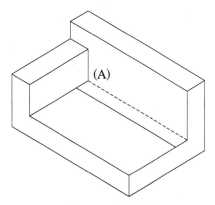

Figure 10–43

Step #19

Use the COPY command to duplicate the dashed line in Figure 10–44 from the endpoint of "A" to the distance of 0.50 units using a polar coordinate.

 Command: **COPY**

Select objects: *(Select the dashed line in Figure 10–44)*
Select objects: *(Press* ENTER *to continue)*
 <Base point or displacement>/Multiple: **Endp**
of *(Select the endpoint at "A")*
Second point of displacement: **@0.50<-30**

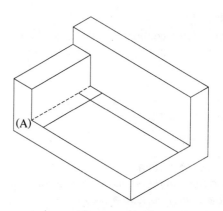

Figure 10–44

Step #20

Use the COPY command to duplicate the dashed line in Figure 10–45 from the endpoint of "A" to the distance of 0.50 units using a polar coordinate.

 Command: **COPY**

Select objects: *(Select the dashed line in Figure 10–45)*
Select objects: *(Press* ENTER *to continue)*
 <Base point or displacement>/Multiple: **Endp**
of *(Select the endpoint at "A")*
Second point of displacement: **@0.50<30**

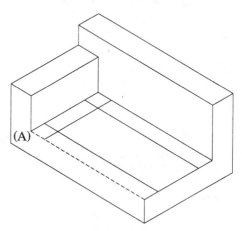

Figure 10–45

Step #21

Use the COPY command to duplicate the dashed line in Figure 10–46 from the endpoint of "A" to the distance of 0.50 units using a polar coordinate.

 Command: **COPY**

Select objects: *(Select the dashed line in Figure 10–46)*
Select objects: *(Press* ENTER *to continue)*
 <Base point or displacement>/Multiple: **Endp**
of *(Select the endpoint at "A")*
Second point of displacement: **@0.50<150**

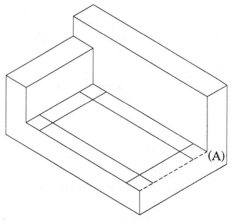

Figure 10–46

Step #22

Use the FILLET command with a radius of 0 to corner the four dashed lines in Figure 10–47. Use the MULTIPLE command to remain in the FILLET command. To exit the FILLET command prompts, use the ESC key to cancel the command when finished.

Command:**MULTIPLE**
Multiple command:**FILLET**
(TRIM mode) Current fillet radius = 0.0000
Polyline/Radius/Trim/<Select first object>:
   (Select line "A")
Select second object: *(Select line "B")*

FILLET
(TRIM mode) Current fillet radius = 0.0000
Polyline/Radius/Trim/<Select first object>:
   (Select line "B")
Select second object: *(Select line "C")*

FILLET
(TRIM mode) Current fillet radius = 0.0000
Polyline/Radius/Trim/<Select first object>:
   (Select line "C")
Select second object: *(Select line "D")*

FILLET
(TRIM mode) Current fillet radius = 0.0000
Polyline/Radius/Trim/<Select first object>:
   (Select line "D")
Select second object: *(Select line "A")*

FILLET
(TRIM mode) Current fillet radius = 0.0000
Polyline/Radius/Trim/<Select first object>:
   (Press the ESC key to cancel Multiple mode)

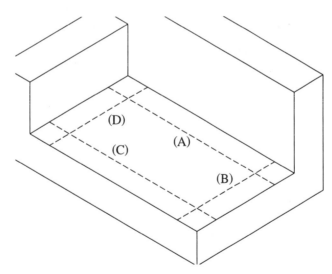

Figure 10–47

Step #23

Use the COPY command to duplicate the dashed line in Figure 10–48. This will begin forming the thickness of the base inside of the rectangular hole.

 Command: **COPY**

Select objects: *(Select the dashed line in Figure 10–48)*
Select objects: *(Press* ⌈ENTER⌋ *to continue)*
<Base point or displacement>/Multiple: **Endp**
of *(Select the endpoint at "A")*
Second point of displacement: **Endp**
of *(Select the endpoint at "B")*

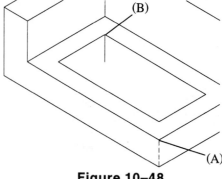

Figure 10–48

Step #24

Use the COPY command to duplicate the dashed lines in Figure 10–49. These lines form the inside surfaces to the rectangular hole.

 Command: **COPY**

Select objects: *(Select the dashed lines in Figure 10–49)*
Select objects: *(Press* ⌈ENTER⌋ *to continue)*
<Base point or displacement>/Multiple: **Endp**
of *(Select the endpoint at "A")*
Second point of displacement: **Endp**
of *(Select the endpoint at "B")*

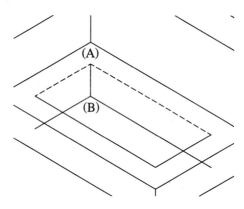

Figure 10–49

Step #25

Use the TRIM command to clean up excess lines in
Figure 10–50.

Command:**TRIM**
Select cutting edges: (Projmode = UCS,
   Edgemode = No extend)
Select objects: *(Select the two dashed lines in*
   Figure 10–50)
Select objects: *(Press* ENTER *to continue)*
  <Select object to trim>/Project/Edge/Undo:
   (Select the line at "A")
  <Select object to trim>/Project/Edge/Undo:
   (Select the line at "B")
  <Select object to trim>/Project/Edge/Undo:
   (Press ENTER *to exit this command)*

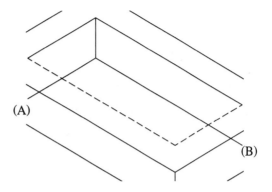

(A)

(B)

Figure 10–50

Step #26

The completed isometric is illustrated in Figure
10–51. This drawing may be dimensioned using
the Oblique option of the DIMEDIT command. Consult your instructor if this next step is necessary.

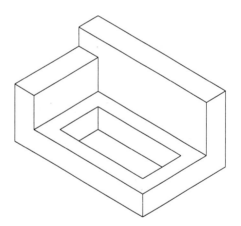

Figure 10–51

Tutorial Exercise
Hanger.Dwg

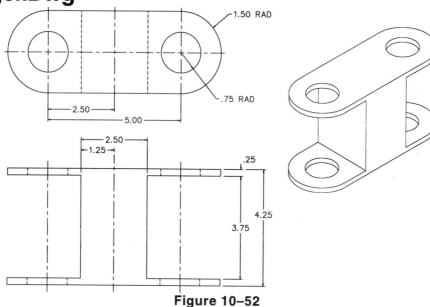

Figure 10–52

Purpose

The purpose of this tutorial exercise is to use a series of coordinates along with AutoCAD editing commands to construct an isometric drawing of the hanger in Figure 10–52.

System Settings

Begin a new drawing called "Hanger." Use the Units Control dialog box to change the number of decimal places past the zero from four to two. Keep the remaining default unit values. Using the LIMITS command, keep 0,0 for the lower left corner and change the upper right corner from 12,9 to 15.50,9.50. Use the GRID command and change the Grid spacing from 1.00 to 0.25 units. Do not turn the Snap or Ortho on.

Layers

Special layers do not have to be created for this tutorial exercise.

Suggested Commands

Begin this exercise by changing the grid from the standard display to an isometric display using the SNAP-Style option. Remember both the Grid and Snap can be manipulated by the SNAP command only if the current grid value is 0. Use absolute and polar coordinates to lay out the base of the Hanger. Then begin using the COPY command followed by TRIM to duplicate objects and clean up or trim unnecessary geometry.

Whenever possible, substitute the appropriate command alias in place of the full AutoCAD command in each tutorial step. For example, use "Co" for the COPY command, "L" for the LINE command, and so on. The complete listing of all command aliases is located on pages 1–9 and 1–10.

Step #1

Set the SNAP-Style option to Isometric with a vertical spacing of 0.25 units. Type CTRL-E or F5 to switch to the Top Isoplane mode. Use the LINE command to draw the rectangular isometric box representing the total depth of the object along with the center-to-center distance of the holes and arcs that will be placed in the next step. See Figure 10–53.

Command:**SNAP**
Snap spacing or ON/OFF/Aspect/Rotate/Style
    <0.50>:**S**
Standard/Isometric <S>:**I** *(For Isometric)*
Vertical spacing <0.50>:**0.25**

Command: ^**E** *(To switch to the Top Isoplane mode)*

 Command:**LINE**
From point:**6.28,0.63**
To point: **@5.00<30**
To point: **@ 3.00<150**
To point: **@5.00<210**
To point: **C**

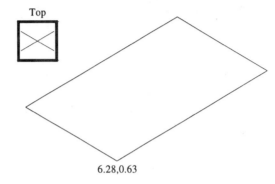

Figure 10–53

Step #2

While in the Top Isoplane mode, use the ELLIPSE command to draw two isometric ellipses of 0.75 and 1.50 radii each, as shown in Figure 10–54. Identify the midpoint of the inclined line at "A" as the center of the first ellipse. To identify the center of the second ellipse, use the @ option which stands for "last point" and will identify the center of the small circle as the same center as the large circle.

 Command:**ELLIPSE**
Arc/Center/Isocircle/<Axis endpoint 1>:**I** (for Isocircle)
Center of circle: **Mid**
of *(Select the inclined line at "A")*
 <Circle radius>/Diameter:**0.75**

 Command:**ELLIPSE**
Arc/Center/Isocircle/<Axis endpoint 1>:**I** (for Isocircle)
Center of circle: **@**
 <Circle radius>/Diameter:**1.50**

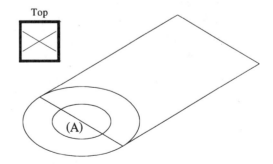

Figure 10–54

Step #3

Copy both ellipses from the midpoint of the inclined line at "A" to the midpoint of the inclined line at "B" in Figure 10–55.

 Command: **COPY**

Select objects: *(Select both ellipses in Figure 10–55)*
Select objects: *(Press ENTER to continue)*
<Base point or displacement>/Multiple: **Mid**
of *(Select the midpoint of the inclined line at "A")*
Second point of displacement: **Mid**
of *(Select the midpoint of the inclined line at "B")*

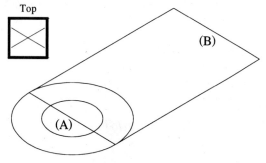

Figure 10–55

Step #4

Turn the Snap off by striking the F9 function key. Use the TRIM command to delete segments of the ellipses. See Figure 10–56.

 Command: **TRIM**

Select cutting edges: (Projmode = UCS,
    Edgemode = No extend)
Select objects: *(Select dashed lines "A" and "B")*
Select objects: *(Press ENTER to continue)*
<Select object to trim>/Project/Edge/Undo:
    (Select the ellipse at "C")
<Select object to trim>/Project/Edge/Undo:
    (Select the ellipse at "D")
<Select object to trim>/Project/Edge/Undo:
    (Press ENTER to exit this command)

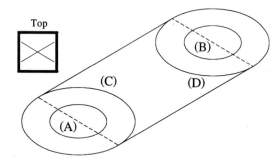

Figure 10–56

Step #5

Copy all objects in Figure 10–57 up the distance of
0.25 units to form the bottom base of the hanger.
The Right Isoplane mode can be activated by typing CTRL-E or F5.

Command: **^E**
Right Isoplane

 Command: **COPY**

Select objects: *(Select all objects in Figure
10–57)*
Select objects: *(Press* ENTER *to continue)*
<Base point or displacement>/Multiple: **Mid**
of *(Select the midpoint of the inclined line at "A")*
Second point of displacement: **@0.25<90**

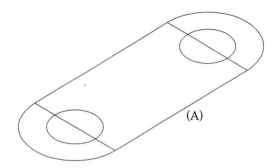

(A)

Figure 10–57

Step #6

Use the ERASE command to delete the three dashed
lines in Figure 10–58. These lines are not visible at
this point of view and should be erased.

Command: **ERASE**
Select objects: *(Select the three dashed lines
in Figure 10–58)*
Select objects: *(Press* ENTER *to execute this
command)*

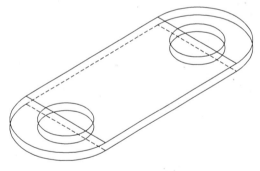

Figure 10–58

Step #7

Your display should appear similar to the illustration in Figure 10–59. Begin partially deleting other objects to show only visible features of the isometric drawing. Check that the snap has been turned off to better assist in the next series of operations. The next few steps that follow refer to the area outlined in Figure 10–59. Use the ZOOM-Window option to magnify this area.

Command: **SNAP**
Snap spacing or ON/OFF/Aspect/Rotate/Style
    <0.25>: **Off**

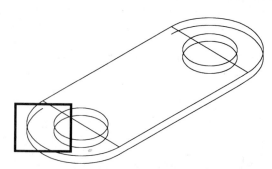

Figure 10–59

Step #8

Carefully draw a line tangent to both ellipses. Use the OSNAP-Quadrant option to assist in constructing the line. See Figure 5–60.

 Command: **LINE**

From point: **Qua**
of *(Select the quadrant at "A")*
To point: **Qua**
of *(Select the quadrant at "B")*
To point: *(Press ⟨ENTER⟩ to exit this command)*

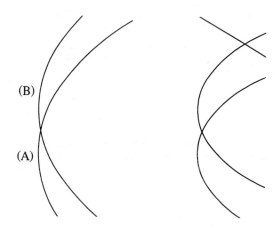

Figure 10–60

Step #9

Use the TRIM command, select the short dashed line as the cutting edge, and select the arc segment in Figure 10–61 to trim.

 Command: **TRIM**

Select cutting edges: (Projmode = UCS,
    Edgemode = No extend)
Select objects: *(Select dashed line at "A")*
Select objects: *(Press ⟨ENTER⟩ to continue)*
<Select object to trim>/Project/Edge/Undo:
    (Select the ellipse at "B")
<Select object to trim>/Project/Edge/Undo:
    (Press ⟨ENTER⟩ to exit this command)

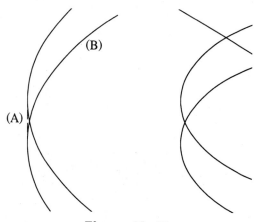

Figure 10–61

Step #10

The completed operation is in Figure 10–62. Use ZOOM-Previous to return to the previous display.

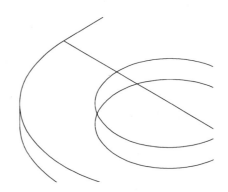

Figure 10–62

Step #11

Use the ZOOM-Window option to magnify the right half of the base in Figure 10–63. Prepare to construct the tangent edge to the object using the previous steps.

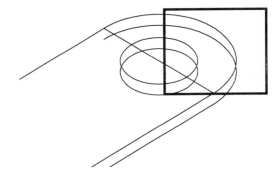

Figure 10–63

Step #12

Follow the same procedure as in Steps 8 and 9 to construct a line from the quadrant point on the top ellipse to the quadrant point on the bottom ellipse (Figure 10–64). Then use the TRIM command to clean up any excess objects. Use the ERASE command to delete any elliptical arc segments that may have been left untrimmed.

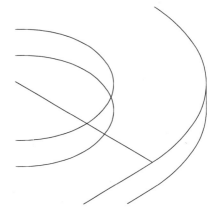

Figure 10–64

Step #13

Use the TRIM command to partially delete the ellipses in Figure 10–65 to expose the thickness of the base.

 Command: **TRIM**

Select cutting edges: (Projmode = UCS,
  Edgemode = No extend)
Select objects: *(Select dashed ellipses "A" and
  "B")*
Select objects: *(Press [ENTER] to continue)*
 <Select object to trim>/Project/Edge/Undo:
  (Select the lower ellipse at "C")
 <Select object to trim>/Project/Edge/Undo:
  (Select the lower ellipse at "D")
 <Select object to trim>/Project/Edge/Undo:
  (Press [ENTER] to exit this command)

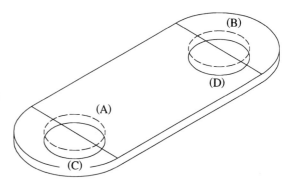

Figure 10–65

Step #14

Use the COPY command to duplicate the bottom base and form the upper plate of the hanger as shown in Figure 10–66. Copy the base a distance of 4 units straight up.

 Command: **COPY**

Select objects: *(Select all dashed objects in
  Figure 10–66)*
Select objects: *(Press [ENTER] to continue)*
 <Base point or displacement>/Multiple: **Nea**
of *(Select the nearest point along the bottom
  arc at "A")*
Second point of displacement: **@4.00<90**

When completed with this step, perform a ZOOM-All to display the entire isometric.

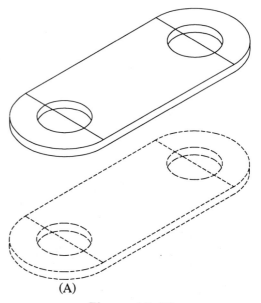

Figure 10–66

Step #15

Use the COPY command to duplicate the inclined line at "A" the distance of 2.50 units to form the line represented by a series of dashes in Figure 10–67. This line happens to be located at the center of the object.

 Command:**COPY**

Select objects: *(Select the line at "A")*
Select objects: *(Press* [ENTER] *to continue)*
<Base point or displacement>/Multiple: **Mid**
of *(Select the midpoint of the inclined line at "A")*
Second point of displacement: **@2.50<30**

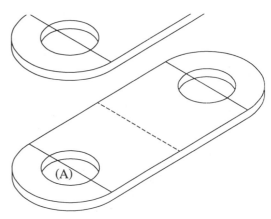

Figure 10–67

Step #16

Duplicate the line represented by dashes in Figure 10–68 to form the two inclined lines at "B" and "C." These lines will begin the construction of the sides of the hanger. Use the Copy-Multiple option to accomplish this.

 Command:**COPY**

Select objects: *(Select the dashed line in Figure 10–68)*
Select objects: *(Press* [ENTER] *to continue)*
<Base point or displacement>/Multiple: **M**
Base point: **Mid**
of *(Select the midpoint of the dashed line at "A")*
Second point of displacement: **@1.25<30**
Second point of displacement: **@1.25<210**
Second point of displacement: *(Press* [ENTER] *to exit this command)*

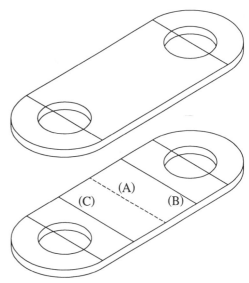

Figure 10–68

Step #17

Use the COPY command to duplicate the two dashed lines in Figure 10–69 straight up at a distance of 3.75 units. The polar coordinate mode is used to accomplish this.

 Command: **COPY**

Select objects: *(Select both dashed lines in Figure 10–69)*
Select objects: *(Press [ENTER] to continue)*
<Base point or displacement>/Multiple: **Mid**
of *(Select the midpoint of the inclined line at "A")*
Second point of displacement: **@3.75<90**

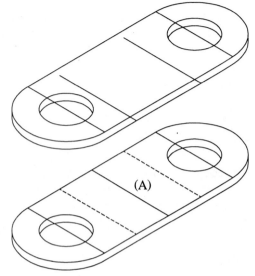

Figure 10–69

Step #18

Use the LINE command along with the OSNAP-Endpoint option to draw a line from endpoint "A" to endpoint "B" in Figure 10–70.

 Command: **LINE**

From point: **Endp**
of *(Select the endpoint of the inclined line at "A")*
To point: **Endp**
of *(Select the endpoint of the inclined line at "B")*
To point: *(Press [ENTER] to exit this command)*

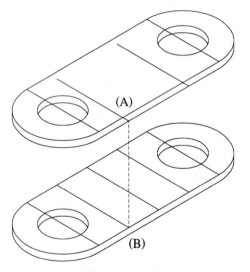

Figure 10–70

Step #19

Use the COPY command and the Multiple option to copy the dashed line at "A" to "B," "C" and "D" in Figure 10–71.

 Command: **COPY**

Select objects: *(Select the dashed line in Figure 10–71)*
Select objects: *(Press* ENTER *to continue)*
<Base point or displacement>/Multiple: **M**
Base point: **Endp**
of *(Select the endpoint of the vertical line at "A")*
Second point of displacement: **Endp**
of *(Select the endpoint of the line at "B")*
Second point of displacement: **Endp**
of *(Select the endpoint of the line at "C")*
Second point of displacement: **Endp**
of *(Select the endpoint of the line at "D")*
Second point of displacement: *(Press* ENTER *to exit the command)*

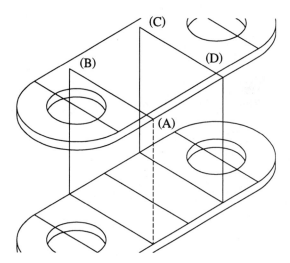

Figure 10–71

Step #20

Use the ERASE command to delete all lines represented as dashed lines in Figure 10–72.

 Command: **ERASE**

Select objects: *(Select all dashed objects in Figure 10–72)*
Select objects: *(Press* ENTER *to execute this command)*

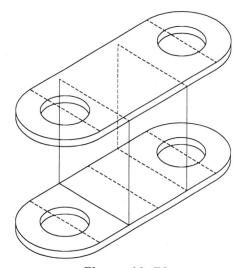

Figure 10–72

Step #21

Use the TRIM command to partially delete the vertical line in Figure 10–73. The segment to be deleted is hidden and not shown in an isometric drawing.

 Command:**TRIM**

Select cutting edges: (Projmode = UCS,
  Edgemode = No extend)
Select objects: *(Select dashed objects "A" and
  "B")*
Select objects: *(Press [ENTER] to continue)*
<Select object to trim>/Project/Edge/Undo:
  (Select the vertical line at "C")
<Select object to trim>/Project/Edge/Undo:
  (Press [ENTER] to exit this command)

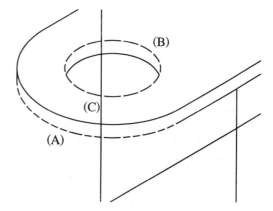

Figure 10–73

Step #22

Use the TRIM command to partially delete the vertical line in Figure 10–74. The segment to be deleted is hidden and not shown in an isometric drawing.

 Command:**TRIM**

Select cutting edges: (Projmode = UCS,
  Edgemode = No extend)
Select objects: *(Select dashed elliptical arc at "A")*
Select objects: *(Press [ENTER] to continue)*
<Select object to trim>/Project/Edge/Undo:
  (Select the vertical line at "B")
<Select object to trim>/Project/Edge/Undo:
  (Press [ENTER] to exit this command)

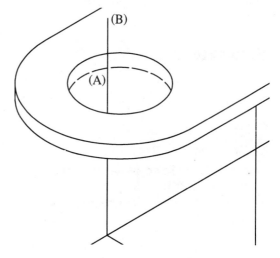

Figure 10–74

Step #23

Use the TRIM command to partially delete the inclined line in Figure 10–75.

 Command: **TRIM**

Select cutting edges: (Projmode = UCS, Edgemode = No extend)

Select objects: *(Select dashed objects "A" and "B")*

Select objects: *(Press* ENTER *to continue)*

<Select object to trim>/Project/Edge/Undo: *(Select the line at "C")*

<Select object to trim>/Project/Edge/Undo: *(Select the line at "D")*

<Select object to trim>/Project/Edge/Undo: *(Press* ENTER *to exit this command)*

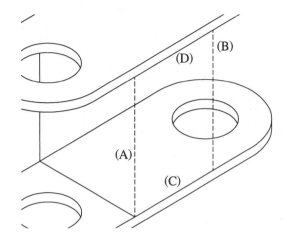

Figure 10–75

Step #24

Use the TRIM command to partially delete the objects in Figure 10–76. The segments to be deleted are hidden and not shown in an isometric drawing. Use ERASE to delete any leftover elliptical arc segments.

 Command: **TRIM**

Select cutting edges: (Projmode = UCS, Edgemode = No extend)

Select objects: *(Select dashed objects "A" and "B")*

Select objects: *(Press* ENTER *to continue)*

<Select object to trim>/Project/Edge/Undo: *(Select the line at "C")*

<Select object to trim>/Project/Edge/Undo: *(Select the elliptical arc at "D")*

<Select object to trim>/Project/Edge/Undo: *(Select the ellipse at "E")*

<Select object to trim>/Project/Edge/Undo: *(Select the elliptical arc at "F")*

<Select object to trim>/Project/Edge/Undo:

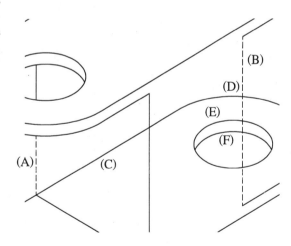

Figure 10–76

Step #25

The completed isometric appears in Figure 10–77. This drawing may be dimensioned using the Oblique option of the DIMEDIT command. Consult your instructor if this next step is necessary.

Step #26

As an extra step, convert the completed hanger into an object with a rectangular hole through it (see Figure 10–78). Begin by copying lines "A," "B," "C," and "D" at a distance of 0.25 units to form the inside rectangle using polar coordinates. Since the lines will overlap at the corners, use the FILLET command set to a radius of 0 to create corners of the hole. Use the LINE command to draw the inclined line "E" at any distance with a 150-degree angle. Use either TRIM or EXTEND to complete the new version of the hanger.

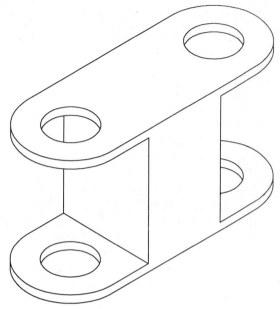

Figure 10–77

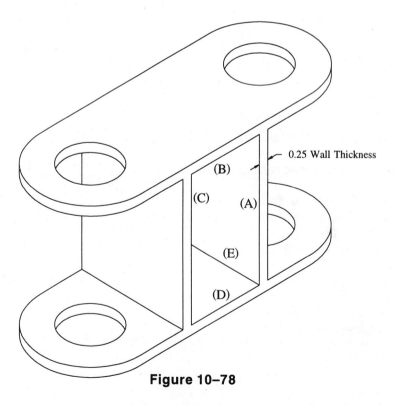

Figure 10–78

Problems for Unit 10

Directions for Problem 10–1:

Construct an isometric drawing of the object.

Problem 10–1

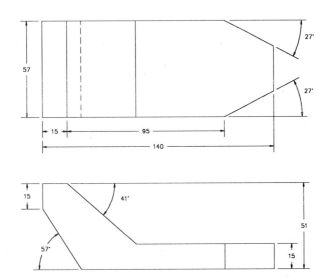

Directions for Problem 10–2:

Construct an isometric drawing of the object. Begin the corner of the isometric at "A."

Problem 10–2

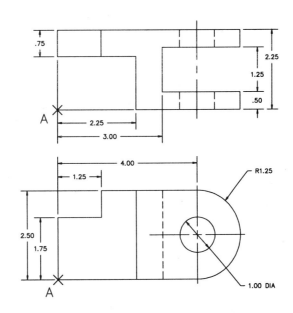

Directions for Problems 10–3 through 10–17:
    Construct an isometric drawing of the object.

Problem 10–3

Problem 10–4

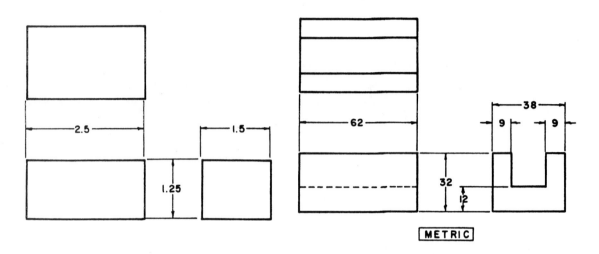

METRIC

Problem 10–5

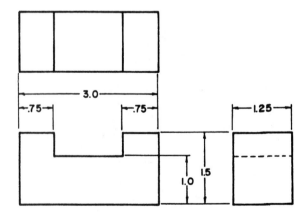

Problem 10–6

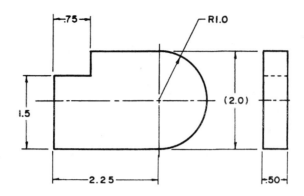

Problem 10–7

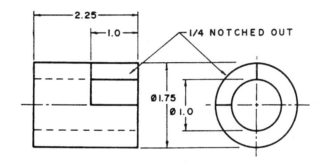

Problem 10–8

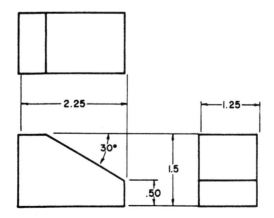

Problem 10–9

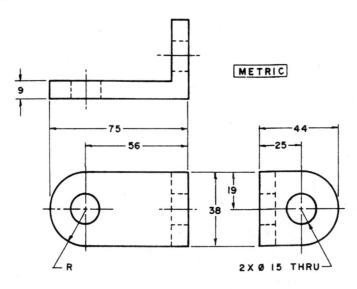

Problem 10–10

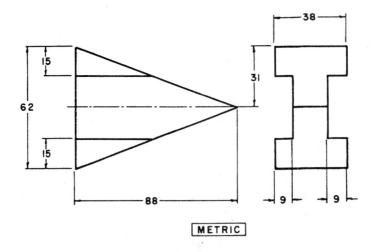

Problem 10–11

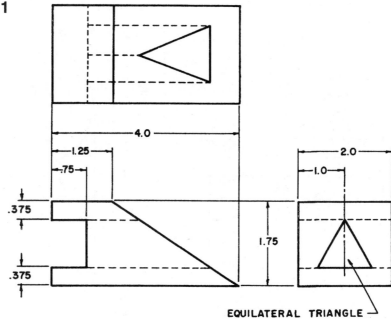

EQUILATERAL TRIANGLE

Problem 10–12

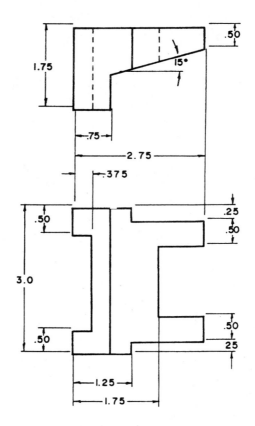

Problem 10–13

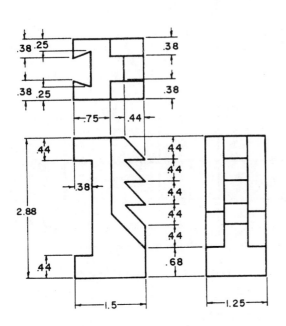

Problem 10–14

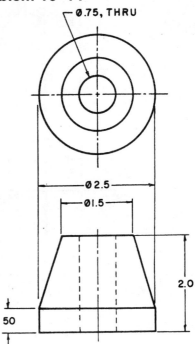

Problem 10–15

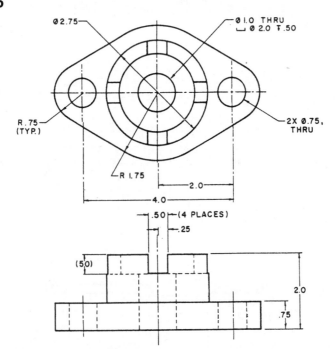

Problem 10–16

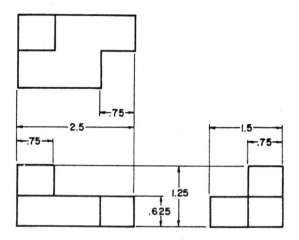

Problem 10–17

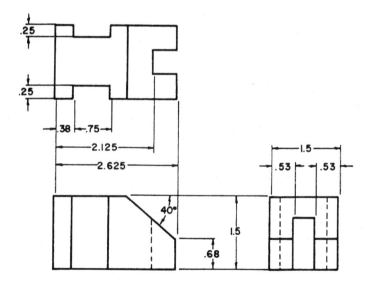

11

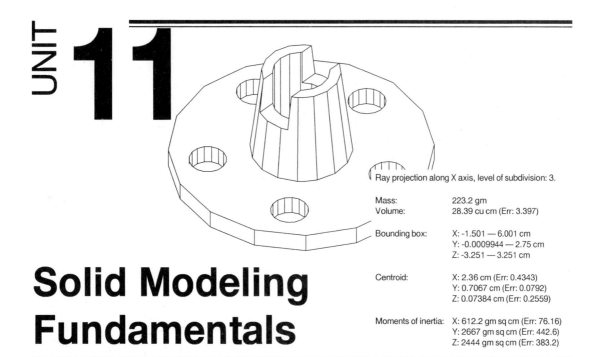

Ray projection along X axis, level of subdivision: 3.

Mass:	223.2 gm
Volume:	28.39 cu cm (Err: 3.397)
Bounding box:	X: -1.501 — 6.001 cm
	Y: -0.0009944 — 2.75 cm
	Z: -3.251 — 3.251 cm
Centroid:	X: 2.36 cm (Err: 0.4343)
	Y: 0.7067 cm (Err: 0.0792)
	Z: 0.07384 cm (Err: 0.2559)
Moments of inertia:	X: 612.2 gm sq cm (Err: 76.16)
	Y: 2667 gm sq cm (Err: 442.6)
	Z: 2444 gm sq cm (Err: 383.2)

Solid Modeling Fundamentals

It is said that humans see, hear, and exist in a 3D world. Why not draw in three dimensions, and visualize the object before placing objects on the computer screen? Part of learning the art of visualization is the study and construction of models in 3D in order to obtain as accurate as possible an image of an object undergoing design. Solid models are more informationally correct than wireframe models since a solid object may be analyzed by calculating such items as mass properties, center of gravity, surface area, moments of inertia, and much more. The solid model starts the true design process by defining objects as a series of primitives.

Boxes, cubes, cylinders, spheres, and wedges are all examples of primitives. These building blocks are then joined together or subtracted from each other using certain modifying commands. Fillets and chamfers may be created to give the solid model a more realistic appearance and functionality. Two-dimensional views may be extracted from the solid model along with a cross section of the model. Study the pages that explain basic solid modeling concepts before completing the tutorial exercises at the end of this unit. First, a brief description on methods of representing drawings will be discussed.

Orthographic Projection

It is no secret that the heart of any engineering draw-ing is the ability to break up the design into three main views of front, top, and right side views rep-resenting what is called orthographic or multiview projection (see Figure 11–1). The engineer or de-signer is then required to interpret the views and their dimensions, to paint a mental picture of a graphic version of the object if it were already made or constructed.

As most engineers have the skill to convert the multiview drawing into a picture in their mind, a vast majority of us would be confused by the nu-merous hidden and centerlines of a drawing and their meaning to the overall design. We need some type of picture to help us interpret the multiview drawing and get a feel for what the part looks like, including the functionality of the part. This may be the major advantage of constructing an object in 3D.

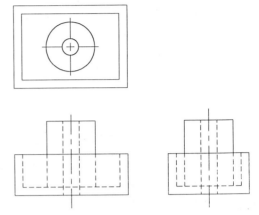

Figure 11–1

Isometric Drawings

The isometric drawing is the easiest 3D representa-tion to produce. It is based on tilting two axes at 30-degree angles along with a vertical line to rep-resent the height of the object. As easy as the iso-metric is to produce, it is also one of the most inac-curate methods of producing a 3D image.

Figure 11–2 shows a view aligned to see the top of the object, front view, and right side view. If we wished to see a different view of the object in iso-metric, a new drawing would have to be produced from scratch. This is the major disadvantage of us-ing isometric drawings. Still, because of ease in drawing, isometrics remain popular in many school and technical drafting rooms.

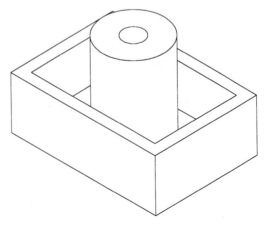

Figure 11–2

3D Extruded Models

Previous AutoCAD versions afforded the user the capability of drawing objects in the Z direction and then viewing the model at any angle. This method was called 3D visualization. Objects are assigned an elevation for starting the surface and a thickness which extrudes the object and produces opaque sides. The thickness can be entered in either a positive direction (Up), or negative direction (Down).

To view the model, the VPOINT command is used to identify a position on a 2D globe which serves as a means of identifying a 3D position. Advantages of this type of construction are the use of the VPOINT command and how the model is able to be viewed from any angle. See Figure 11–3.

A disadvantage of this method is the inability to place top and bottom surfaces on the model. As a result, the model looks hollow when viewed from above or underneath.

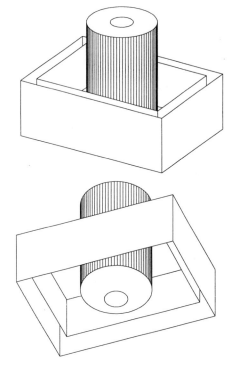

Figure 11–3

3D Wireframe Models

As shown in Figure 11–4, wireframes are fundamental types of models. However, as lines intersect from the front and back of the object, it is sometimes difficult to interpret the true design of the wireframe.

Using a user-definable coordinate system, previously defined as the UCS, a new coordinate system can be defined along any plane on which to place objects. Depth can be controlled by a number of aids, especially XYZ point filters. In this method, values are temporarily saved for later use inside a command. The values can be retrieved and a coordinate value entered to complete the command.

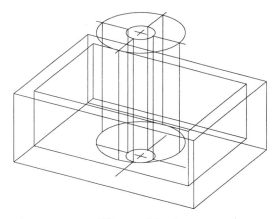

Figure 11–4

3D Surfaced Models

Surfacing picks up where the wireframe model leaves off. Because of the complexity of the wireframe and the number of intersecting lines, surfaces are applied to the wireframe. The surfaces are in the form of opaque objects, called 3D faces. As always, 3D faces are placed using Osnap options to assist in point selection. A hidden line removal is performed to view the model without the interference of other objects. Again, the VPOINT command is used to view the model in different positions to make sure all sides of the model have been surfaced. The format of the 3DFACE command is outlined using the following prompts and Figure 11–5.

Command: **OSNAP** *(or use the Osnap dialog box)*
Object snap modes: **Endp**

Command: **3DFACE**
First point: *(Select the point at "A")*
Second point: *(Select the point at "B")*
Third point: *(Select the point at "C")*
Fourth point: *(Select the point at "D")*
Third point: *(Press [ENTER] to exit this command)*

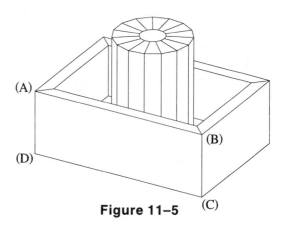

Figure 11–5

Solid Models

The creation of a solid model remains the most exact way to represent a model in 3D. It is also the most versatile representation of an object. Solid models may be viewed as identical to wireframe and surfaced construction models. Wireframe models and surfaced models may be analyzed by taking distance readings and identifying key points of the model. Key orthographic views such as front, top, and right side views may be extracted from wireframe models (see Figure 11–6A). Surfaced models may be imported into shading packages for increased visualization (see Figure 11–6B). Solid models do all of these operations and more. This is because the solid model, as it is called, is a solid representation of the actual object. From cylinders to slabs, wedges to boxes, all objects that go in the creation of a solid model have volume. This allows a model to be constructed of what are referred to as primitives. These primitives are then merged into one using addition and subtraction operations. What remains is the most versatile of 3D drawings. This method of creating models will be the main focus of this unit.

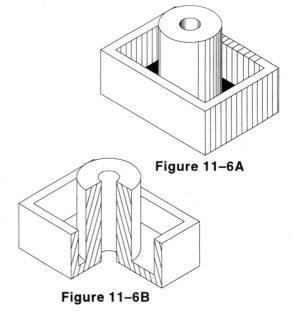

Figure 11–6A

Figure 11–6B

Solid Modeling Basics

All objects, no matter how simple or complex, are composed of simple geometric shapes or primitives. The shapes range from boxes to cylinders to cones and so on. Solid modeling allows for the creation of these primitives. Once created the shapes are either merged or subtracted to form the final object. Follow the next series of steps to form the object in Figure 11–7.

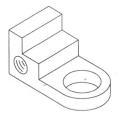

Figure 11–7

Begin the process of solid modeling by constructing a solid slab that will represent the base of the object. You can do this by using the BOX command. Supply the length, width, and height of the box, and the result is a solid slab, as shown in Figure 11–8A.

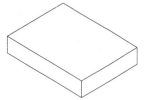

Figure 11–8A

Next, construct a cylinder using the CYLINDER command. This cylinder will eventually form the curved end of the object. One of the advantages of constructing a solid model is the ability to merge primitives together to form composite solids. Using the UNION command, combine the cylinder and box to form the complete base of the object in Figure 11–8B.

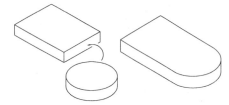

Figure 11–8B

As the object progresses, create another solid box and move it into position on top of the base. There, use the UNION command to join this new block with the base. As you add new shapes during this process, they all become part of the same solid (see Figure 11–8C).

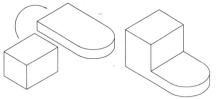

Figure 11–8C

Yet another box is created and moved into position. However, instead of combining the block, this new box is removed from the solid. This process is called subtraction and when complete, creates the step shown in Figure 11–8D. The AutoCAD command used during this process is SUBTRACT.

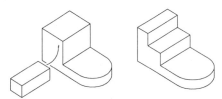

Figure 11–8D

You can form holes in a similar fashion. First, use the CYLINDER command to create a cylinder the diameter and depth of the desired hole. Next, move the cylinder into the solid and then subtract it using the SUBTRACT command. Again, the object in Figure 11–8E represents a solid object.

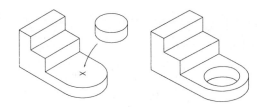

Figure 11–8E

Using existing AutoCAD tools such as user coordinate systems, create another cylinder with the CYLINDER command. It too is moved into position where it is subtracted using the SUBTRACT command. The complete solid model of the object is shown in Figure 11–8F (right).

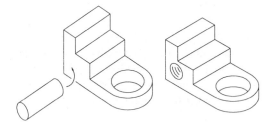

Figure 11–8F

Rewards from constructing a solid model out of an object come in many forms. Profiles of different surfaces of the solid model may be taken. Section views of solids may be automatically formed and crosshatched as in Figure 11–9. A very important analysis tool is the MASSPROP command, which is short for mass property. Information such as the mass and volume of the solid object may be calculated along with centroids and moments of inertia, components that are used in computer aided engineering (CAE) and in finite element analysis (FEA) of the model. See Figure 11–10.

Before beginning to create solid models, a thorough understanding of the user coordinate system must first be discussed.

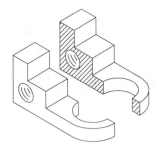

Figure 11–9

Mass:	223.2 gm
Volume:	28.39 cu cm (Err: 3.397)
Bounding box:	X:-1.501—6.001 cm
	Y:-0.0009944—2.75 cm
	Z:-3.251—3.251 cm
Centroid:	X: 2.36 cm
	Y: 0.7067 cm
	Z: 0.07384 cm
Moments of inertia:	X: 612.2 gm sq cm (Err: 76.16)
	Y: 2667 gm sq cm (Err: 442.6)
	Z: 2444 gm sq cm (Err: 383.2)

Figure 11–10

The UCS Command

Two-dimensional computer-aided design still remains the most popular form of representing drawings for most applications. However in applications such as manufacturing, 3D models are becoming increasingly popular for creating rapid prototype models or to create tool paths from the 3D model. To assist in this creation process, a series of user coordinate systems are used to create construction planes where features such as holes and slots are located. In Figure 11–11, a model of a box is displayed along with the user coordinate icon. The presence of the "W" signifies the World Coordinate System or home position of drawing. Figure 11–12 identifies the directions of the three user coordinate system axes. Notice that as "X" and "Y" are identified in the user coordinate system icon, the positive "Z" direction is straight up; a negative "Z" direction is down. The UCS command is used to create different user defined coordinate systems. The command sequence follows along with all options; the UCS command and options can also be selected from the UCS toolbar illustrated in Figure 11–13.

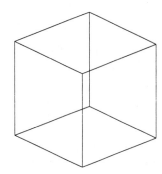

Figure 11–11

 Command: **UCS**

Origin/ZAxis/3point/OBject/View/X/Y/Z/Prev/
Restore/Save/Del/?/<World>:

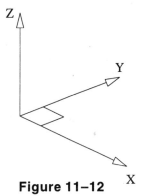

Figure 11–12

The following options are available to you when you use the UCS command:

Origin—Identifies a new user coordinate system at origin 0,0,0.

ZAxis—Identifies a user coordinate system from two points defining the Z axis.

3point—Identifies a user coordinate system by 3 points.

Object—Identifies a user coordinate system in relation to an object selected.

View—Identifies a user coordinate system by the current display.

X/Y/Z—Identifies a user coordinate system by rotation along the X, Y, or Z axis.

Prev—Sets the user coordinate system icon to the previously defined user coordinate system.

Restore—Restores a previously saved user coordinate system.

Save—Saves the position of a user coordinate system under a unique name given by the CAD operator.

Del—Deletes a user coordinate system from the database of the current drawing.

?—Lists all previously saved user coordinate systems.

<World>—Switches to the World Coordinate System from any previously defined user coordinate system.

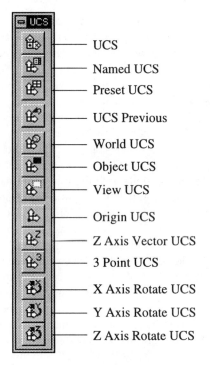

UCS
Named UCS
Preset UCS
UCS Previous
World UCS
Object UCS
View UCS
Origin UCS
Z Axis Vector UCS
3 Point UCS
X Axis Rotate UCS
Y Axis Rotate UCS
Z Axis Rotate UCS

Figure 11–13

The UCS—Origin Option

The Origin option of the UCS command defines a new user coordinate system by moving the current UCS to a new 0,0,0 position while leaving the direction of the X, Y, and Z axes unchanged. The command sequence for using the Origin option is as follows:

 Command: **UCS**

Origin/ZAxis/3point/OBject/View/X/Y/Z/Prev/
   Restore/Save/Del/?/<World>: **O** *(For Origin)*
Origin point <0,0,0>: *(Identify a point for the
  new origin)*

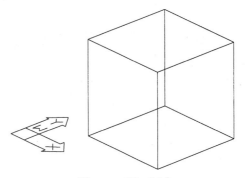

Figure 11–14A

Figure 11–14A is a sample model with the current coordinate system being the World Coordinate System. Follow Figures 11–14B and 11–14C to define a new user coordinate system using the Origin option.

Using the Origin option of the UCS command, identify a new origin point for 0,0,0 at "A" as shown in Figure 11–14B. This should move the user coordinate system icon to the point that you specify. Once the icon has moved, the "W" is removed from the icon and a "plus" appears at the intersection of the X and Y axis on the icon. If the icon remains in its previous location and does not move, use the UCSICON command with the Origin option to snap the icon to its new origin point.

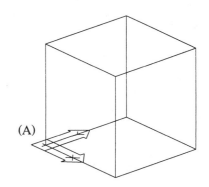

(A)

Figure 11–14B

 Command: **UCS**

Origin/ZAxis/3point/OBject/View/X/Y/Z/Prev/
   Restore/Save/Del/?/<World>: **O** *(For Origin)*
Origin point <0,0,0>: **Endp**
of *(Select the endpoint of the line at "A")*

Objects may now be drawn parallel to this new coordinate system. Remember that 0,0,0 is identified by the "plus" on the user coordinate system icon. See Figure 11–14C.

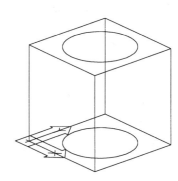

Figure 11–14C

The UCS—3 Point Option

Use the 3 Point option of the UCS command to
specify a new user coordinate system by identifying an origin and new directions of its positive X
and Y axes. Figure 11–15A shows a 3D cube in the
world coordinate system. To construct objects on
the front panel, a new user coordinate system must
first be defined parallel to the front. Use the following command sequence to accomplish this task.

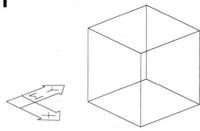

Figure 11–15A

 Command: **UCS**

Origin/ZAxis/3point/OBject/View/X/Y/Z/Prev/
    Restore/Save/Del/?/<World>: **3** *(For 3 point)*
Origin point <0,0,0>: **Endp**
of *(Select the endpoint of the model at "A" as
    shown in Figure 11–15B)*
Point on positive portion of the X axis <>: **Endp**
of *(Select the endpoint of the model at "B")*
Point on positive-Y portion of the UCS X-Y
    plane <>: **Endp**
of *(Select the endpoint of the model at "C")*

With the user coordinate system standing straight
up, or aligned with the front of the cube, any type
of object may be constructed along this plane as in
Figure 11–15C.

The 3 Point method of defining a new user coordinate system is quite useful in the example shown in
Figure 11–16 where a UCS needs to be aligned with
the inclined plane. Use the intersection at "A" as the
center of the new UCS, the intersection at "B" as the
direction of the positive X axis, and the intersection
at "C" as the direction of the positive Y axis.

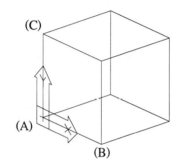

Figure 11–15B

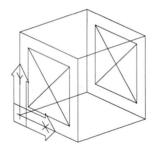

Figure 11–15C

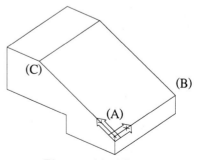

Figure 11–16

The UCS—X/Y/Z Rotation Options

Using the X/Y/Z rotation options will rotate the current user coordinate around the specific axis. Once a letter is selected as the pivot, a prompt appears asking for the rotation angle about the pivot axis. The right-hand rule is used to determine the positive direction of rotation around an axis. Think of the right hand gripping the pivot axis with the thumb pointing in the positive X, Y, or Z direction. The curling of the fingers on the right hand determines the direction of rotation. All positive rotations occur in the counter-clockwise directions. Figure 11–17A, "A" shows a rotation about the X axis; "B" shows a rotation about the Y axis; and "C" shows a rotation about the Z axis.

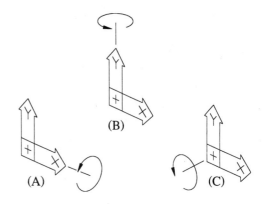

Figure 11–17A

Given the same cube shown in Figure 11–17B in the world coordinate system, the X option of the UCS command will be used to stand the icon straight up by entering a 90-degree rotation value as in the following prompt sequence.

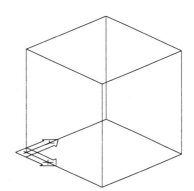

Figure 11–17B

 Command: **UCS**

Origin/ZAxis/3point/OBject/View/X/Y/Z/Prev/
   Restore/Save/Del/?/<World>:**X**
Rotation angle about the X axis: **90**

The X axis is used as the pivot of rotation; entering a value of 90 degrees rotates the icon the desired degrees in the counterclockwise direction as in Figure 11–17C.

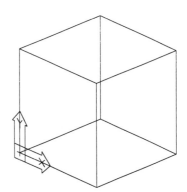

Figure 11–17C

The UCS—Object Option

Given the 3D cube in Figure 11–18A, another option of defining a new user coordinate system is by selecting an object and having the user coordinate system align to that object. Use the following command sequence and Figure 11–18B to accomplish this.

 Command:**UCS**
Origin/ZAxis/3point/OBject/View/X/Y/Z/Prev/
    Restore/Save/Del/?/<World>:**OB** *(For the
    Object option)*
Select object to align UCS: *(Select the circle in
    Figure 11–18B)*

Selecting the circle at the right conforms the user coordinate system to the object selected. Objects determine the alignment of the user coordinate system in many ways. In the case of the circle, the center of the circle becomes the origin of the user coordinate system. Where the circle was selected becomes the point through which the positive X axis aligns.

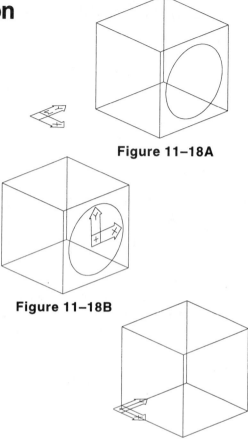

Figure 11–18A

Figure 11–18B

 # The UCS—View Option

The View option of the user coordinate system command allows you to establish a new coordinate system where the XY plane is perpendicular to the current screen viewing direction; in other words, it

is parallel to the display screen. Given the current user coordinate system in Figure 11–19A, follow the prompt below along with Figure 11–19B to align the user coordinate system with the View option.

Figure 11–19A

 Command:**UCS**
Origin/ZAxis/3point/OBject/View/X/Y/Z/Prev/
    Restore/Save/Del/?/<World>:**V** *(For View)*

The results are displayed in Figure 11–19B with the user coordinate system aligned parallel to the display screen.

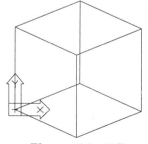

Figure 11–19B

Applications of Rotating the UCS

Follow the next series of steps to perform numerous rotations of the user coordinate system in the example of the cube shown in Figure 11–20A.

 Command:**UCS**

Origin/ZAxis/3point/OBject/View/X/Y/Z/Prev/
　　Restore/Save/Del/?/<World>:**X**
Rotation angle about the X axis:**90**

Next align the user coordinate system with one of the sides of the cube by performing the rotation using the Y axis as the pivot axis (see Figure 11–20B). The positive angle rotates the icon in the counterclockwise direction.

 Command:**UCS**

Origin/ZAxis/3point/OBject/View/X/Y/Z/Prev/
　　Restore/Save/Del/?/<World>:**Y**
Rotation angle about the Y axis:**90**

Next rotate the user coordinate system using the Z axis as the pivot axis. The degree of rotation entered at 45 degrees tilts the user coordinate system shown in Figure 11–20D.

 Command:**UCS**

Origin/ZAxis/3point/OBject/View/X/Y/Z/Prev/
　　Restore/Save/Del/?/<World>:**Z**
Rotation angle about the Z axis:**45**

To tilt the UCS icon along the Z axis pointing down at a 45-degree angle, enter a rotation angle of -90 degrees. The results are shown in Figure 11–20E.

 Command:**UCS**

Origin/ZAxis/3point/OBject/View/X/Y/Z/Prev/
　　Restore/Save/Del/?/<World>:**Z**
Rotation angle about the Z axis:**-90**

Begin rotating the user coordinate system along the X axis at a rotation angle of 90 degrees using the following prompt sequence. This will align the user coordinate system with the front of the cube shown in Figure 11–20B.

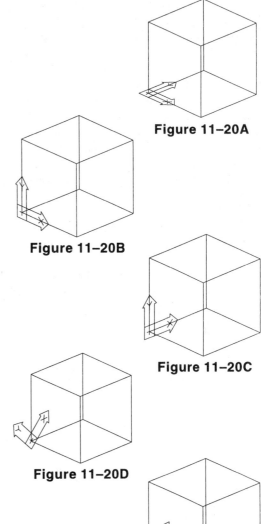

Figure 11–20A

Figure 11–20B

Figure 11–20C

Figure 11–20D

Figure 11–20E

Using the UCS Control Dialog Box

It is considered good practice to assign a name to a user coordinate system once it has been created. Once numerous user coordinate systems have been defined in a drawing, they are easily restored by using their name instead of recreating each coordinate system; this is easily accomplished using the Restore option of the UCS command. Another method is to pick "Named UCS…" from the "UCS" area of the "Tools" pull-down menu illustrated in Figure 11–21 (top). This displays the dialog box in Figure 11–21 (bottom right), that lists all user coordinate systems defined in the drawing. To make one of these coordinate systems current, highlight the desired UCS name and pick the "Current" button shown in Figure 11–21 (bottom right). This dialog box provides a quick method of restoring previously defined coordinate systems without entering them in at the keyboard.

Picking "Preset UCS…" under the "UCS" area of the "Tools" pull-down menu displays the dialog box shown in Figure 11–21 (bottom left). Use this dialog box to automatically align the user coordinate system icon to sides of an object such as front view, top view, back view, right side view, left side view, and bottom view.

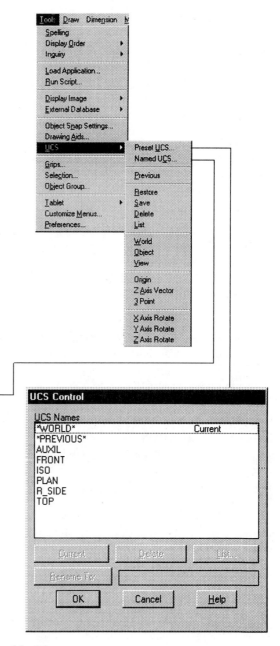

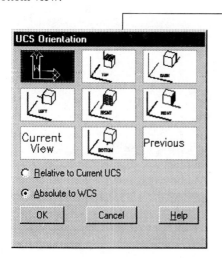

Figure 11–21

Using the VPOINT Command

The object in Figure 11–22 represents the plan view of a 3D model already created. Unfortunately, it becomes very difficult to understand what the model looks like in its present condition. Use the VPOINT command to view a wireframe, surfaced, or solid model in three dimensions. The following is a typical prompt sequence for the VPOINT command:

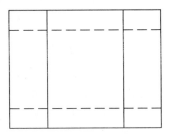

Figure 11–22

Command: **VPOINT**
Rotate/<Viewpoint> <0.0000,0.0000,1.0000>:

The VPOINT command stands for "View point." A point is identified in 3D space. This point becomes a location where the model is viewed from. The point may be entered in at the keyboard as an *X,Y,Z* coordinate or picked with the aid of a 2D globe (see Figure 11–26). The object shown in Figure 11–23 is a typical result of using the VPOINT command.

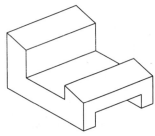

Figure 11–23

Numerous areas exist to pick the VPOINT command. Picking the View area of the pull-down menu exposes 3D Viewpoint in Figure 11–24. Pick here to activate the cascading menu containing numerous viewpoint options. Items listed as Top, Bottom, SW Isometric, and NE Isometric are view point presets used to view a 3D model from these different locations. They provide a fast way of viewing a model in a primary or isometric view.

A Viewpoint toolbox is also available as shown in Figure 11–25. It contains similar options in the 3D Viewpoint pull-down menu. The Viewpoint toolbox has the extra advantage of having a picture in the

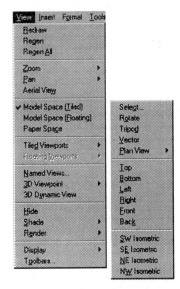

Figure 11–24

form of an icon that guides you through which view-point to pick for the desired viewing effect. If this toolbox is not present, it may be activated from the pull-down menu by picking the following sequence: View > Toolbars > Viewpoint.

Besides picking from the pull-down menu or toolbox, one method of defining a viewpoint is through the use of an *X, Y, Z* coordinate. In the fol-lowing command sequence, a new viewing point is established 1 unit in the positive X direction, 1 unit in the negative Y direction, and 1 unit in the posi-tive Z direction.

Command: **VPOINT**
Rotate/<Viewpont><0.0000,0.0000,1.0000>:
  1,-1,1

Another method of defining a viewing point is to define two angular axes by rotation. The first angle defines the view point in the X-Y axis. However this is only 2D. The second angle defines the view-point from the X-Y axis. This tilts the viewing point up for a positive angle or down for a negative angle.

Command: **VPOINT**
Rotate/<Viewpont><0.0000,0.0000,1.0000>:**R**
  (For Rotate)
Enter angle in X-Y plane from X axis < >: **45**
Enter angle from X-Y plane < >: **30**

If the first command prompt is followed by the ENTER key, a graphic image consisting of globe and tripod appear in Figure 11–26. Although the globe appears two-dimensional, it provides you with the ability to pick a viewpoint depending on how you read the globe. The intersection of the hori-zontal and vertical lines form the North Pole of the globe. The inner circle forms the equator and the outer circle forms the South Pole. The following examples illustrate numerous viewing points.

Command: **VPOINT**
Rotate/<Viewpont><0.0000,0.0000,1.0000>:
  (Press ENTER *to display the viewpoint globe shown in Figure 11–26).*

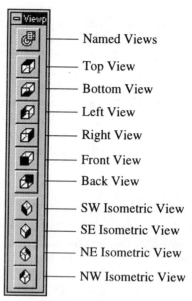

Named Views
Top View
Bottom View
Left View
Right View
Front View
Back View
SW Isometric View
SE Isometric View
NE Isometric View
NW Isometric View

Figure 11–25

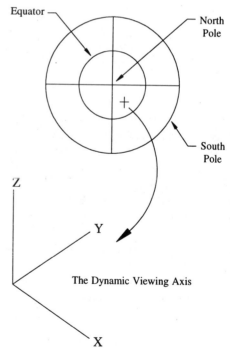

Equator
North Pole
South Pole
Z
Y
X
The Dynamic Viewing Axis

Figure 11–26

Viewing Along the Equator

To obtain the results in Figure 11–27, use the VPOINT command and mark a point at "A" to view the front view; mark a point at "B" to view the top view; and mark a point at "C" to view the right side view. Coordinates could also have been entered to achieve the same results:

Command: **VPOINT**
Rotate/<Viewpoint><0.0000,0.0000,1.0000>:
   0,-1,0 *(At "A")*

Command: **VPOINT**
Rotate/<Viewpoint><0.0000,0.0000,1.0000>:
   0,0,1 *(At "B")*

Command: **VPOINT**
Rotate/<Viewpoint><0.0000,0.0000,1.0000>:
   1,0,0 *(At "C")*

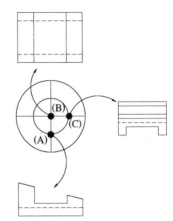

Figure 11–27

Viewing Near the North Pole

Picking the four points shown in Figure 11–28 results in the different viewing points for the object. Since all points are inside the inner circle, the results are aerial views, or views from above. Remember that the Equator is symbolized by the inner circle. Depending on which quadrant you select, you will look up at the object from the right corner, left corner, or either of the rear corners.

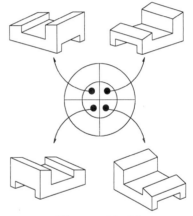

Figure 11–28

Viewing Near the South Pole

Picking the four points shown in Figure 11–29 results in underground views, or viewing the object from underneath. This is true since all points selected lie between the small and large circles. Again, remember that the small circle symbolizes the Equator, while the large circle symbolizes the South Pole.

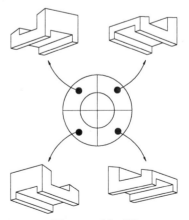

Figure 11–29

Using the Viewpoint Presets Dialog Box

Picking "Select…" from the "3D Viewpoint" area of
the View pull-down menu shown in Figure 11–30
displays the Viewpoint Presets dialog box shown
in Figure 11–31. Use this dialog box to help define a new point of viewing a 3D model. The square
image with circle in the center allows you to define a viewing point in the X-Y plane. The semi-
circular-shaped image allows you to define a viewing point from the X-Y plane. The combination of
both directions form the 3D viewpoint.

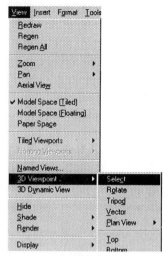

Figure 11–30

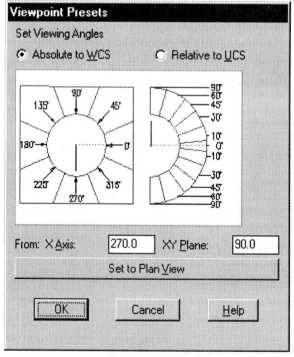

Figure 11–31

When selecting a viewing point in the X-Y axis, you have two options for selecting the desired point of view. When picking outside of the circle on 0, 90, 180, etc., as in "A" in Figure 11–32A, the viewpoint snaps from one angle to another in increments of 45 degrees. The resulting angle is displayed at the bottom of the square. If you want a more detailed viewpoint in the X-Y axis, pick a point inside and near the center of the circle at "B." The resulting angle will display all values in between the 45-degree increment from before. If you already know the angle in the X-Y axis, you may enter it in the edit box next to the prompt "X Axis."

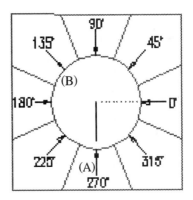

Figure 11–32A

Figure 11–32A illustrates setting a viewing angle in the X-Y plane. Figure 11–32B shows the second half of the Viewpoint presets dialog box, which defines an angle from the X-Y plane. Selecting a point in between both semicircles at "A" snaps the viewpoint in 10-, 30-, 45-, 60-, and 90-degree increments. If you want a more detailed selection, pick the viewing point inside of the smaller semicircle and near its center at "B." This will allow you to select an angle different from the five default values listed above. An alternate method of selecting an angle from the XY plane would be to place the value in the edit box next to the prompt "XY Plane."

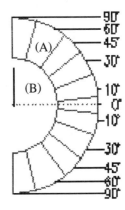

Figure 11–32B

How View Point can affect the UCS Icon

The User Coordinate System icon may take on different forms depending upon how it is viewed using the VPOINT command. Viewing the icon in
plan view or in an aerial view displays the image
in Figure 11–33A. The presence of the "+" sign
indicates the icon is located in the current viewport
at the current UCS origin. The UCS icon must be
totally displayed on the screen; if the icon is unable to be totally shown, the "+" sign is removed
and the icon locates itself in the lower left corner
of the display screen.

Figure 11–33A

When viewing a model from below, or in the negative Z direction, the User Coordinate System icon
takes the form of the image in Figure 11–33B. The
absence of the two lines near the intersection of the
X and Y axes acts as a visual cue alerting you that
you are either viewing the model from below or
from the icon's back side.

Figure 11–33B

When working with multiple viewports, defining
the UCS in one viewport displays the normal icon
in Figure 11–33A; yet in an adjacent viewport, the
icon may disappear and be replaced with the image in Figure 11–33C. This is commonly referred
to as the "broken pencil" icon; it appears when the
current UCS is viewed orthogonal or perpendicular to the to the viewing angle in a particular
viewport. This icon also warns you that drawing
lines or other objects may not be obvious in the
viewport it is present.

Figure 11–33C

In all previous examples (Figures 11–33A through
11–33C), the UCS icon has been displayed while
working in the Model Space environment. The icon
displayed in Figure 11–33D represents the Paper
Space environment. This environment is strictly two
dimensional and is used as a layout tool for arrang-
ing views and displaying title block information.
This icon will be implemented when performing
the last tutorial exercise in this unit.

Figure 11–33D

Selecting Solid Modeling Commands

Figures 11–34 and 11–35 show two of the more popular locations for accessing solid modeling commands. In Figure 11–34, solid modeling commands may be picked from the pull-down menu area under the main heading of Draw. Selecting "Solids >" displays four groupings of commands. The first grouping—Box, Sphere, Cylinder, Cone, Wedge, and Torus—are considered the building blocks of the solid model and are used to construct basic primitives. The second grouping displays the EXTRUDE and REVOLVE commands, which are additional ways to construct solid models. Polyline outlines or circles may be extruded to a thickness that you designate. You may also revolve other polyline outlines about an object representing a centerline of rotation. The third grouping of solid modeling commands enables you to slice the solid model in half, cut the solid model into what is called a section, and perform an interference check where two solids are constructed near each other but must not touch or intersect. The last grouping, Setup >, displays three commands designed to extract the orthographic views from the solid model. The three commands are SOLPROF, SOLVIEW, and SOLDRAW. Notice also in Figure 11–35 the complete grouping of solid modeling commands into a single toolbar. This is another convenient way to select solid modeling commands.

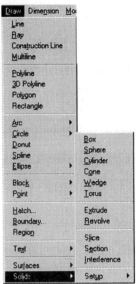

Figure 11–34

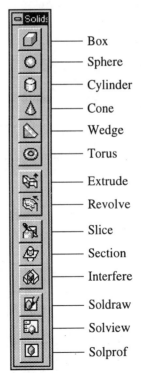

Figure 11–35

Using the BOX Command

Use this command to create a 3D box. One corner of the box is located along with a diagonal corner. A height is assigned to complete the definition of the box. If you know all three dimensions, you may construct the solid box by entering values for its length, width, and height. You may construct a cube if all three dimensions are the same value. Figure 11–36 shows examples of solid boxes.

 Command: **BOX**

Center/<Corner of box> <0,0,0>: *(Pick a point at "A")*
Cube/Length/<other corner>: *(Pick a point at "B")*
　Height: **1**

 Command: **BOX**

Center/<Corner of box> <0,0,0>: *(Pick a point at "C")*
Cube/Length/<other corner>: **L** *(To define the length of the box)*
　Length: **5**
　Width: **2**
　Height: **1**

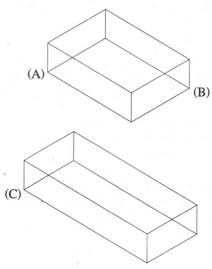

Figure 11–36

Using the CONE Command

Use the CONE command and Figure 11–37 to construct a cone of specified height with the base of the cone either circular or elliptical.

 Command: **CONE**

Elliptical/<center point> <0,0,0>: *(Pick a point designating the center of the cone at "A")*
Diameter/<Radius>: **1**
Apex/<Height>: **4**

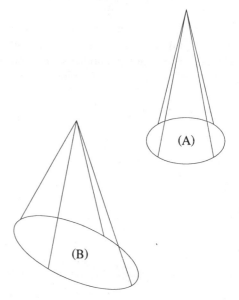

Figure 11–37

Command: **CONE**

Elliptical/<center point> <0,0,0>: **E** (For an elliptical base)
Center/<Axis endpoint>: **C** (To define the center of the ellipse)
Center of ellipse <0,0,0>: (Pick a point

designating the center of the elliptical base at "B" in Figure 11–37)
Axis endpoint: **@2<0**
Other axis distance: **@1<90**
Apex/<Height>:**4**

Using the WEDGE Command

Yet another solid primitive is the wedge that consists of a box that has been diagonally cut. Use the WEDGE command to create a wedge with the base parallel to the current user coordinate system and the slope of the wedge constructed along the X axis. Use the following prompts and Figure 11–38 as examples of how to use this command.

 Command:**WEDGE**

Center/<Corner of wedge> <0,0,0>: (Pick a point at "A")
Cube/Length/<other corner>: (Pick a point at "B")
   Height: **3**

 Command:**WEDGE**

Center/<Corner of wedge> <0,0,0>: (Pick a point at "C")
Cube/Length/<other corner>: **L** (To define the length of the wedge)
   Length: **5**
   Width: **1**
   Height: **3**

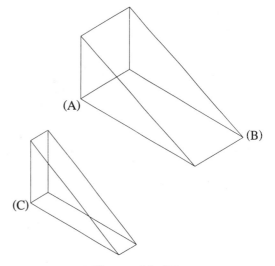

Figure 11–38

Using the CYLINDER Command

The CYLINDER command is similar to the CONE command except that the cylinder is drawn without a taper. The central axis of the cylinder is along the Z axis of the current user coordinate system. Use this command and Figure 11–39 to construct a cylinder with either a circular or elliptical base.

 Command:**CYLINDER**

Elliptical/<center point> <0,0,0>: *(Pick a point at "A" to specify the center of the cylinder)*
Diameter/<Radius>:**D**
Diameter:**2**

 Command:**CYLINDER**

Elliptical/<center point> <0,0,0>:**E** *(For an elliptical base)*
Center/<Axis endpoint>: **C** *(To define the center of the ellipse)*
Center of ellipse <0,0,0>: *(Pick a point at "B" designating the center of the elliptical base)*
Axis endpoint: **@2<0**
Other axis distance: **@1<90**
Center of other end/<Height>:**4**

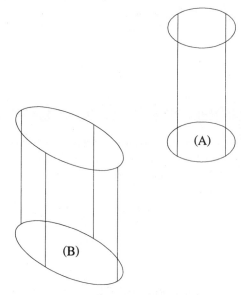

Figure 11–39

Using the SPHERE Command

Use the SPHERE command to construct a sphere by defining the center of the sphere along with a radius or diameter. As with the cylinder, the central axis of a sphere is along the Z axis of the current user coordinate system. In Figure 11–40, the sphere is constructed in wireframe mode and does not look like much of a sphere. Perform a hidden line removal using the HIDE command to get a better view of the sphere.

 Command:**SPHERE**

Center of sphere <0,0,0>:
Diameter/<Radius> of sphere: **D**
Diameter:**4**

Command:**HIDE**

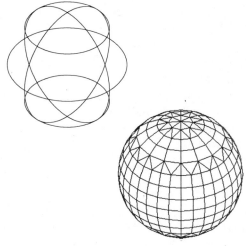

Figure 11–40

The TORUS Command

A torus is formed when a circle is revolved about a line in the same plane as the circle. In other words, a torus is similar to a 3D doughnut. The torus may be constructed using either a radius or diameter method. When using the radius method, two radius values must be used to define the torus; one for the radius of the tube and the other for the radius from the center of the torus to the center of the tube. Use two diameter values when specifying a torus by diameter. Once you construct the torus, it lies parallel to the current user coordinate system. Use the following prompts to construct a torus similar to Figure 11–41.

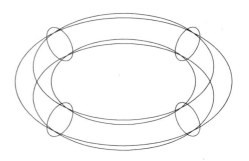

 Command:**TORUS**

Center of torus <0,0,0>: *(Identify the center of the torus through coordinate entry or by picking)*
Diameter/<Radius> of torus: **5** *(For the radius of the torus at "A")*
Diameter/<Radius> of tube: **1** *(For the radius of the tube at "B")*

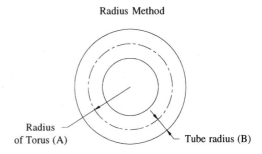

Radius Method

Radius of Torus (A)   Tube radius (B)

 Command:**TORUS**

Center of torus <0,0,0>: *(Identify the center of the torus through coordinate entry or by picking)*
Diameter/<Radius> of torus: **D** *(For diameter of the torus at "C")*
Diameter: **4**
Diameter/<Radius> of tube: **D** *(For diameter of the tube at "D")*
Diameter: **2**

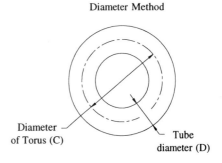

Diameter Method

Diameter of Torus (C)   Tube diameter (D)

Figure 11–41

Using Boolean Operations

To combine of one or more primitives to form a common solid, a system is available to illustrate the relationship between the individuals that make up the solid model. This system is called a Boolean operation. Boolean operations must act on at least a pair of primitives, regions, or solids. These operations in the form of commands are located in the main pull-down menu under "Modify > Boolean." Boolean operations allow you to add two or more objects together, subtract a single or group of objects from another, or find the overlapping volume; in other words, to form the solid common to both primitives. Highlighted in Figure 11–42 are the UNION, SUBTRACT, and INTERSECT commands that you use to perform the Boolean operations previously explained.

In Figure 11–43A, a cylinder has been constructed along with a square block. Depending on which Boolean operation you use, the results could be quite different. In Figure 11–43B, both the square slab and cylinder are one object and considered one solid object. This is the purpose of the UNION command: to join or unite two solid primitives into one. Figure 11–43C illustrates the intersection of the two solid primitives or the area that both solids have in common. This solid is obtained by using the INTERSECT command. Figure 11–43D shows the results of removing or subtracting the cylinder from the square slab—a hole is formed inside the square slab as a result of using the SUBTRACT command. All Boolean operation commands may work on numerous solid primitives; that is, if you subtract numerous cylinders from a slab, you may subtract all cylinders at the same time.

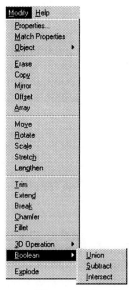

Figure 11–42

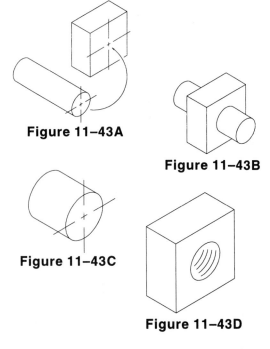

Figure 11–43A

Figure 11–43B

Figure 11–43C

Figure 11–43D

The UNION Command

This construction operation joins two or more selected solid objects together into a single solid object. See Figure 11–44.

 Command:**UNION**

Select objects: *(Pick the box and cylinder)*
Select objects: *(Press* ⌜ENTER⌝ *to perform the union operation)*

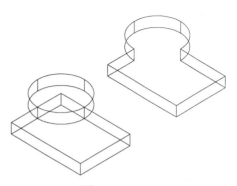

Figure 11–44

The SUBTRACT Command

Use this command to subtract one or more solid objects from a source object. See Figure 11–45.

 Command:**SUBTRACT**

Select solids and regions to subtract from...
Select objects: *(Pick the box)*
Select objects: *(Press* ⌜ENTER⌝ *to continue with this command)*
Select solids and regions to subtract...
Select objects: *(Pick the cylinder)*
Select objects: *(Press* ⌜ENTER⌝ *to perform the subtraction operation)*

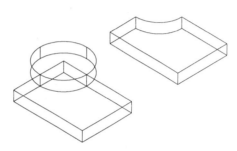

Figure 11–45

The INTERSECT Command

Use this command to find the solid common to a group of selected solid objects. See Figure 11–46.

 Command:**INTERSECT**

Select objects: *(Pick the box and cylinder)*
Select objects: *(Press* ⌜ENTER⌝ *to perform the intersection operation)*

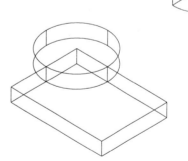

Figure 11–46

3D Applications of Unioning Solids

Figure 11–47 shows an object consisting of one horizontal solid box, two vertical solid boxes, and two extruded semi-circular shapes. All primitives have been positioned with the MOVE command. To join all solid primitives into one solid object, use the UNION command. The order of selection of these solids for this command is not important. Use the following prompts and Figure 11–47.

 Command:**UNION**

Select objects: *(Select the solid extrusion at "A")*
Select objects: *(Select the vertical solid box at "B")*
Select objects: *(Select the horizontal solid box at "C")*
Select objects: *(Select the vertical solid box at "D")*
Select objects: *(Select the solid extrusion at "E")*
Select objects: *(Press ENTER to perform the union operation)*

3D Applications of Moving Solids

Using the same problem from the previous example, let us now add a hole in the center of the base. The cylinder was already created using the CYLINDER command. It now needs to be moved to the exact center of the base. You may use the MOVE command along with the OSNAP-Tracking mode to accomplish this. Tracking mode will automatically activate the Ortho mode when it is in use.

 Command:**MOVE**

Select objects: *(Select the cylinder at "A")*
Select objects: *(Press ENTER to continue with this command)*
Base point or displacement: **Cen**
of *(Select the bottom of the cylinder at "A")*
Second point of displacement: **TK** *(To activate tracking mode)*
First tracking point: **Mid**
of *(Select the midpoint of the bottom of the base at "B")*
Next point *(Press ENTER to end tracking):* **Mid**
of *(Select the midpoint of the bottom of the base at "C")*
Next point (Press ENTER to end tracking): *(Press ENTER to end tracking and perform the move operation)*

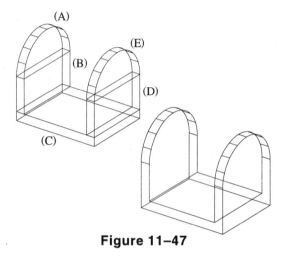

Figure 11–47

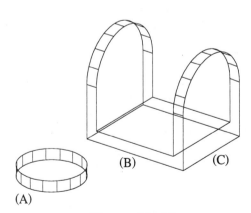

Figure 11–48

3D Applications of Subtracting Solids

Now that the solid cylinder is in position, use the SUBTRACT command to remove the cylinder from the base of the main solid and create a hole in the base. See Figure 11–49.

 Command: **SUBTRACT**

Select solids and regions to subtract from...
Select objects: *(Select the main solid as source at "A")*
Select objects: *(Press ENTER to continue with this command)*
Select solids and regions to subtract...
Select objects: *(Select the cylinder at "B")*
Select objects: *(Press ENTER to perform the subtraction operation)*

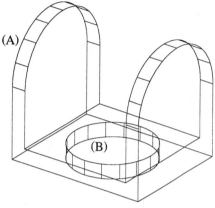

Figure 11–49

3D Applications of Aligning Solids

Two more holes need to be added to the vertical sides of the solid object. A cylinder was already constructed. However it is in the vertical position. This object needs to be rotated and moved into position. The ALIGN command would be a good com- mand to use in this situation. A series of source and destination points guide the placement of one ob- ject on another. The object rotates and moves into position. Use the following command prompt se- quence and Figure 11–50 for the ALIGN command.

Command: **ALIGN**
Select objects: *(Select the cylinder)*
Select objects: *(Press ENTER to continue with this command)*
Specify 1st source point: **Cen**
of *(Select the bottom of the cylinder at "A")*
Specify 1st destination point: **Cen**
of *(Select the outside of the main solid at "B")*
Specify 2nd source point: **Cen**
of *(Select the top of the cylinder at "C")*
Specify 2nd destination point: **Cen**
of *(Select the outside of the main solid at "D")*
Specify 3rd source point or <continue>: *(Press ENTER to continue)*
Scale objects to alignment points? [Yes/No] <No>: *(Press ENTER to accept the default value and perform the align operation)*

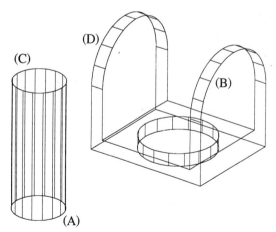

Figure 11–50

The cylinder is removed to create holes in each of the vertical sides of the solid object using the SUB- TRACT command. The completed object is shown in Figure 11–51.

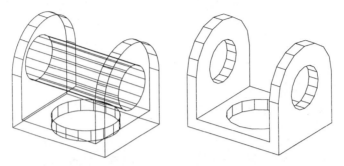

Figure 11–51

Creating Solid Extrusions

The EXTRUDE command creates a solid by extrusion. Only polylines and circles may be extruded. Once objects are polylines, use the following prompts to construct a solid extrusion of the object in Figure 11–52. For the height of the extrusion, a positive numeric value may be entered or the distance may be determined by picking two points on the display screen.

 Command:**EXTRUDE**

Select objects: *(Select the polyline object at "A"in Figure 11–52)*
Select objects: *(Press* ⏎ *to continue with this command)*
Path/<Height of extrusion>:**1.00**
Extrusion taper angle <0>: *(Press* ⏎ *to accept the default and perform the extrusion operation)*

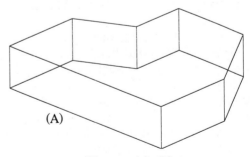

(A)

Figure 11–52

You may create an optional taper along with the extrusion by entering an angle value for the prompt, "Extrusion taper angle."

 Command:**EXTRUDE**

Select objects: *(Select the polyline object at "B" in Figure 11–53)*
Select objects: *(Press* ENTER *to continue with this command)*
Path/<Height of extrusion>: **1.00**
Extrusion taper angle <0>: **15**

You may also create a solid extrusion by selecting a path to be followed by the object being extruded. Typical paths include regular and elliptical arcs, 2D and 3D polylines, or splines. The object in Figure 11–54 was created by first contructing the polyline path. Then, a new user coordinate system was created using the UCS command along with the ZAxis option; the new UCS is positioned at the end of the polyline with the Z axis extending along the polyline. Finally, a circle was constructed with its center point at the end of the polyline before being extruded along the polyline path.

 Command:**EXTRUDE**

Select objects: *(Select the small circle as the object to extrude)*
Select objects: *(Press* ENTER *to continue)*
Path/<Height of Extrusion>: **P** *(For Path)*
Select path: *(Select the polyline object representing the path)*

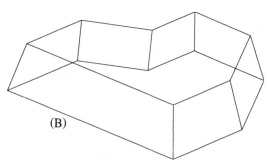

(B)

Figure 11–53

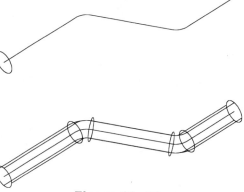

Figure 11–54

Creating Revolved Solids

The REVOLVE command creates a solid by re-volving an object about an axis of revolution. Only polylines, polygons, circles, ellipses, and 3D polylines may be revolved. If a group of objects are not in the form of polylines, group them to-gether using the PEDIT command. The polyline must form a closed shape. The resulting image in Figure 11– 55 represents a solid object.

 Command: **REVOLVE**

Select objects: *(Select profile "A" as the object to revolve)*
Select objects: *(Press* [ENTER] *to continue with this command)*
Axis of revolution—Object/X/Y/<Start point of axis>: **O** *(For Object)*
Select an object: *(Select line "B")*
Angle of revolution <full circle>: *(Press* [ENTER] *to accept the default and perform the revolving operation)*

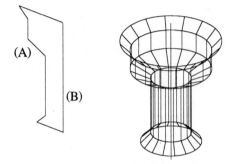

Figure 11–55

A practical application of this type of solid would be to first construct an additional solid consisting of a cylinder using the CYLINDER command. Be sure this solid is larger in diameter than the revolved solid. Existing Osnap options are fully supported in solid modeling. Use the Center option of OSNAP along with the MOVE command to position the revolved solid inside of the cylinder.

 Command: **MOVE**

Select objects: *(Select the revolved solid in Figure 11–56)*
Select objects: *(Press* [ENTER] *to continue with this command)*
Base point or displacement: **Cen**
of *(Select the center of the revolved solid at "A")*
Second point of displacement: **Cen**
of *(Select the center of the cylinder at "B")*

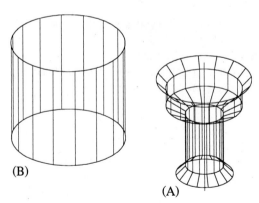

Figure 11–56

Then, once the revolved solid is positioned inside of the cylinder, use the SUBTRACT command to subtract the revolved solid from the cylinder (see Figure 11–57). Use the Hide command to perform a hidden line removal at "B" to check that the solid is correct (this would be difficult to interpret in wireframe mode).

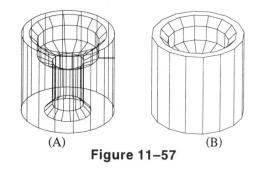

(A)   (B)

Figure 11–57

Using the ISOLINE and FACETRES Commands

Tessellation refers to the lines that are displayed on any curved surface to help visualize the surface shown in Figure 11–58. Tessellation lines are automatically formed when you construct solid primitives such as cylinders and cones. These lines are also calculated when you performing solid modeling operations such as SUBTRACT and UNION.

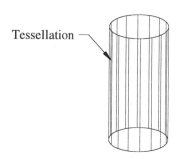

Tessellation —

Figure 11–58

The ISOLINE Command

The number of tessellation lines per curved object is controlled by the command called ISOLINES. By default, this variable is set to a value of 4. Figure 11–59 shows the results of setting this variable to other values such as "8" and "12." The more lines used to describe a curved surface, the more accurate the surface will look; however, it will take longer to process hidden line removals using the HIDE command and Boolean operation commands such as UNION and SUBTRACT.

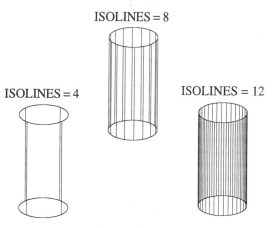

ISOLINES = 8

ISOLINES = 4

ISOLINES = 12

Figure 11–59

The FACETRES Command

As shown in Figure 11–59, tessellation lines on cylinders can affect wireframe models. When performing hidden line removals on these objects, the results are displayed in Figure 11–60. The cylinder with FACETRES set to 0.50 processes much quicker than the cylinder with FACETRES set to 1 since there are fewer surfaces to process in such operations as hidden line removals. However the image with FACETRES set to 1 shows a more defined circle. The default value for FACETRES is 0.50, which seems adequate for most applications.

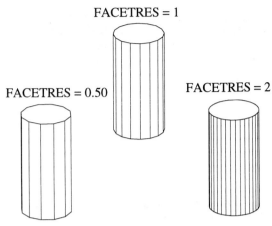

Figure 11–60

Editing Solid Models

Filleting Operations

Filleting of complex objects is easily handled with the FILLET command. This is the same command used to create a 2D fillet. Use the following prompt sequence and Figure 11–61.

 Command: **FILLET**

(TRIM mode) Current fillet radius = 0.5000
Polyline/Radius/Trim/<Select first object>:
   (Select the edge at "A" which represents the
   intersection of both cylinders)
Enter radius <0.5000>:**0.25**
Chain/Radius/<Select edge>: *(Press* ENTER *to*
   perform the fillet operation)
1 edges selected for fillet.

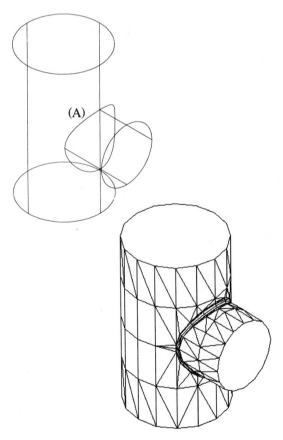

Figure 11–61

A group of objects with a series of edges may be filleted using the Chain option of the FILLET com-

 Command: **FILLET**

(TRIM mode) Current fillet radius = 0.5000
Polyline/Radius/Trim/<Select first object>: *(Select the edge at "A")*
Enter radius <0.5000>: *(Press* ENTER *to accept the default value)*
Chain/Radius/<Select edge>: **C** *(For chain mode)*
Edge/Radius/<Select edge chain>: *(Select all edges that form the top of the plate)*
Edge/Radius/<Select edge chain>: *(Press* ENTER *when finished selecting all edges to perform the fillet operation)*
7 edges selected for fillet.

Chamfering Operations

As the FILLET command uses the Chain mode to group a series of edges together, the CHAMFER

 Command: **CHAMFER**

(TRIM mode) Current chamfer Dist1 = 0.5000, Dist2 = 0.5000
Polyline/Distance/Angle/Trim/Method/<Select first line>: *(Pick the edge at "A")*
Select base surface: *(The top surface should be highlighted. If not, continue with the next series of prompts)*
Next/<OK>: **N** *(To select the next surface; the top surface should highlight)*
Next/<OK>: *(Press* ENTER *to accept the top surface)*
Enter base surface distance <0.5000>: *(Press* ENTER *to accept the default)*
Enter other surface distance <0.5000>: *(Press* ENTER *to accept the default)*
Loop/<Select edge>: **L** *(To loop all edges together into one)*
Edge/<Select edge loop>: *(Pick any edge)*
Edge/<Select edge loop>: *(Press* ENTER *to perform the chamfering operation)*

mand. Use the following prompt sequence and Figure 11–62.

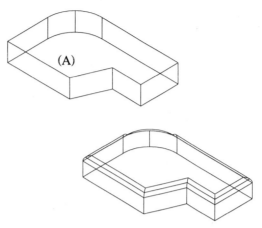

Figure 11–62

command uses the Loop option to do the same type of operation. Use the following prompt sequence and Figure 11–63.

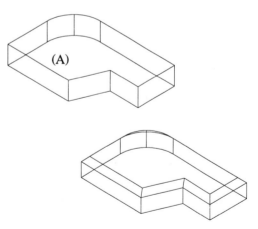

Figure 11–63

Editing Solid Boxes

Figure 11–64A shows a solid box constructed with the BOX command at a height of 1.00 units. After you construct the box, you may decide to increase the height to 1.50 units or decrease the height to 0.50 units. Unfortunately, there is not an easy, more refined way to accomplish this.

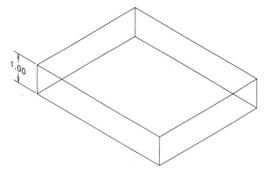

Figure 11–64A

To increase the height of the box in Figure 11–64B, construct another box 0.50 units high and move it on top of the existing box. Then use the UNION command to join both boxes into one for a total height of 1.50 units.

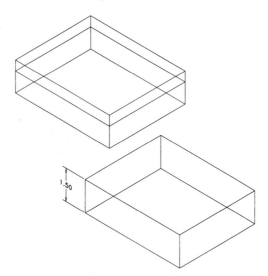

Figure 11–64B

To reduce the height of the slab to 0.50 units, construct a box of 0.50 units and move it to the top of the existing 1.00 unit box (see Figure 11–64C). Use the SUBTRACT command to remove the top box from the other and create the remaining box 0.50 units high. This represents the only way to edit a box object outside of deleting the box from the beginning and starting a new box with the proper dimensions.

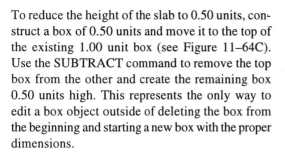

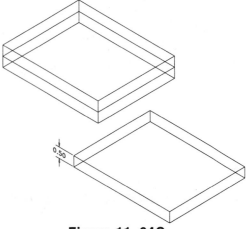

Figure 11–64C

Editing Holes

Figure 11–65A shows an outer cylinder 4.00 units in diameter and a hole 2.00 units in diameter. Unfortunately, the hole needs to be 1.00 units in diameter. To edit the hole without reconstructing the cylinder from the beginning, create another cylinder to match the size of the hole that needs to be replaced. Next, line up the bottom of the small cylinder with the bottom of the hole using the MOVE command. Then, use the UNION command to join the two cylinders together.

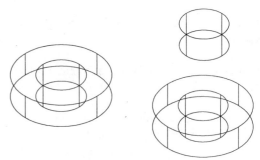

Figure 11–65A

The result is the single cylinder illustrated in Figure 11–65B; the small cylinder that matched the size of the hole was used as a plug to fill the void of the hole. Construct a cylinder 1.00 units in diameter and subtract it from the large cylinder to create a hole 1.00 units in diameter. While these do not represent the best ways to edit solid models, they remain the only way without deleting the model and starting again.

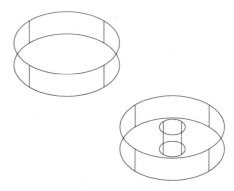

Figure 11–65B

Modeling Techniques Using Viewports

Using viewports can be very important and beneficial when laying out a solid model. Use the VPORTS command to lay out the display screen into a number of individually different display areas. This command is selected by clicking on the View area of the pull-down menu shown in Figure 11–66 and selecting Tiled Viewports. A cascading menu appears with the different types of viewports available. Clicking on the word "Layout…" displays the Viewport Layout dialog box shown in Figure 11–67.

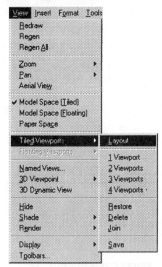

Figure 11–66

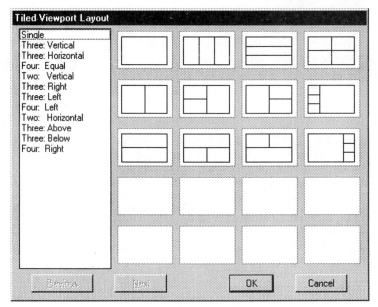

Figure 11–67

A number of viewport configurations may be selected that will convert the display screen into the VPORT configuration desired. Figure 11–68 shows an application of how viewports are used in the construction of a 3D solid model. The image in the right viewport represents the solid model viewed in three dimensions. The image on the left represents the model in 2D mode. The image on the left is viewed based on the current user coordinate system. This gives you two viewports to help visualize the construction of the solid model.

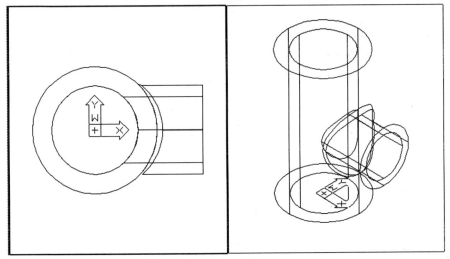

Figure 11–68

Obtaining Mass Properties

Use the MASSPROP command to calculate the mass properties of a selected solid (see Figure 11–69A). All calculations in Figure 11–69B are based on the current position of the user coordinate system.

 Command:**MASSPROP**

Select objects: *(Select the model)*

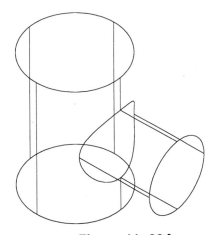

Figure 11–69A

```
_____SOLIDS_____

Mass:                 26.8978
Volume:               26.8978
Bounding box:         X: 5.8821 — 11.6617
                      Y: 3.1262 — 6.6855
                      Z: 0.0000 — 5.0000
Centroid:             X: 8.2360
                      Y: 4.9058
                      Z: 2.5000
Moments of inertia:
                      X: 895.3560
                      Y: 2108.3669
                      Z: 2569.2755
Products of inertia:
                      XY: 1086.7870
                      YZ: 329.8900
                      ZX: 553.8236
Radii of gyration:
                      X: 5.7695
                      Y: 8.8535
                      Z: 9.7734

Principal moments and X-Y-Z directions about centroid:
                      I: 79.8905 along [1.0000 0.0000 0.0000]
                      J: 115.7440 along [0.0000 1.0000 0.0000]
                      K: 97.4092 along [0.0000 0.0000 1.0000]
```

Figure 11–69B

The Interfere Command

Use this command to find any interference shared by a series of solids. As an interference is identified, a solid may be created out of the common volume similar to Figure 11–70.

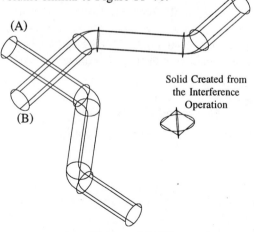

Command:**INTERFERE**
Select the first set of solids: *(Select solid "A")*
Select objects: 1 found
Select objects: *(Press ENTER to continue)*
Select the second set of solids: *(Select solid "B")*
Select objects: 1 found
Select objects: *(Press ENTER to continue)*
Comparing 1 solid against 1 solid.
Interfering solids (first set): 1
 (second set): 1
Interfering pairs: 1
Create interference solids ? <N>:**Y**

(A)

(B)

Solid Created from
the Interference
Operation

Figure 11–70

Cutting Sections from Solid Models

The SECTION Command

Use this command to create a cross section of a selected solid object similar to Figure 11–71. Note
that section lines are not used to define the surfaces
cut by the operation. You have control of how the
section will be created; in Figure 11–71, the location of the user coordinate system defines the cutting plane.

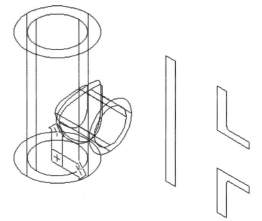

 Command:**SECTION**

Select objects: *(Select the object in Figure
   11–71)*
Select objects: *(Press* ⟦ENTER⟧ *to continue with this
   command)*
Section plane by Object/Zaxis/View/XY/YZ/ZX/
   <3points>: **XY**
Point on XY plane <0,0,0>: *(Press* ⟦ENTER⟧ *to
   perform the sectioning operation)*

Figure 11–71

Slicing Solid Models

The SLICE Command

This command is similar to the SECTION command with the exception that the solid model is
actually cut or sliced at a location that you define;
in the example in Figure 11–72, this location is
defined by the user coordinate system. Before the
slice is made, you also have the option of keeping
the desired half while discarding the other.

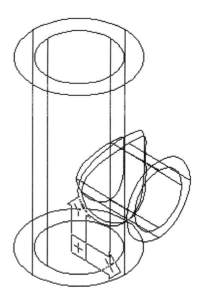

Figure 11–72

Command:**SLICE**

Select objects: *(Select the solid object at "A" in Figure 11–72)*

Select objects: *(Press* ENTER *to continue with this command)*

Slicing plane by Object/Zaxis/View/XY/YZ/ZX/ <3points>:**XY**

Point on XY plane <0,0,0>: *(Press* ENTER *to accept this default value)*

Both sides/<Point on desired side of the plane>: **B** *(To keep both sides)*

The MOVE command was used to separate both halves in Figure 11–73.

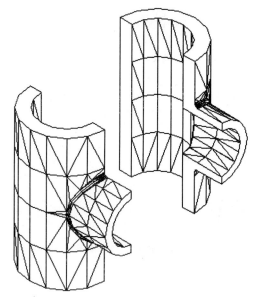

Figure 11–73

Shading Solid Models

As 3D models are created, it is often difficult to interpret their surfaces without performing a hidden line removal or a shaded rendering. Selecting "View" from the pull-down menu area displays the RENDER, SHADE, and HIDE commands as shown in Figure 11–74. The SHADE and HIDE commands will be explained further.

The quickest way to verify the model is to perform a hidden line removal using the HIDE command. No shading occurs; however, only those surfaces in view will be displayed while others will be removed from view.

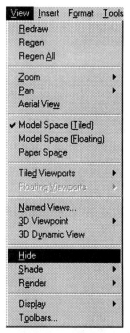

Figure 11–74

For a quick look at what the object would look like as a shaded image, use the SHADE command. The results will be similar to using the HIDE command except that the current color of the objects will take on that color in a shaded form.

This command automatically produces a hidden line removal in addition to displaying the shaded image in the current color. When you execute this command, the screen temporarily will go blank while the system calculates the surfaces to shade. The length of time to accomplish this is of course dependent on the complexity of the object you are shading. Once the shaded image is displayed on the screen, it cannot be plotted; it can, however, be made into a slide using the MSLIDE command (see Figure 11–75). This image consists of a colorization of the solid model and is not meant for different surface tones or casting of shadows.

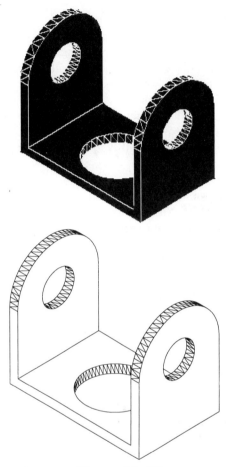

Figure 11—75

Using the SOLVIEW and SOLDRAW Comands

Once you create the solid model, you can lay out and draw its orthogonal views using the commands located in the Draw area of the pull-down menu under Solids and Setup. There are three commands that deal with the ability to draw, view, or profile orthogonal views. Only two areas will be discussed in this section, namely the ability to lay out 2D views and then the ability to construct a drawing of these views.

As shown in Figure 11–76, clicking on View activates the SOLVIEW command. Once this command is entered, the display screen automatically switches to the paper space environment. Using SOLVIEW will lay out a view based on a series of prompts, depending on the type of view you want to create. Usually, the initial view that serves as the starting point for other orthogonal views is based on the current user coordinate system. This needs to be determined before beginning this command. Once an initial view is created, it is very easy to create Ortho, Section, and even Auxiliary views.

As SOLVIEW is used as a layout tool, the images of these views are taken from the original solid model. In other words, after laying out a view, it does not contain any 2D features such as hidden lines. As shown in Figure 11–76, clicking on Drawing activates the SOLDRAW command used to draw the view once it has been laid out by the SOLVIEW command.

Before using the SOLVIEW command, study the illustration of the solid model in Figure 11–77. In particular, pay close attention to the position of the user coordinate system icon. The current position of the user coordinate system will begin the creation of the first or base view of the 2D drawing.

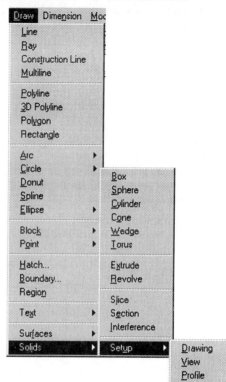

Figure 11–76

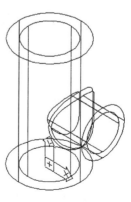

Figure 11–77

Before you continue, remember to load the Hidden linetype. This will automatically assign this linetype to any new layer that requires hidden lines for the drawing mode. If the linetype is not loaded at this point, it must be manually assigned to each layer that contains hidden lines using the Layer and Linetype Properties dialog box.

Activating SOLVIEW automatically switches the display to the paper space environment. Since this is the first view to be laid out, the UCS option will be used to base the view on the current user coordi-nate system. A scale value may be entered for the view. For the View Center, click anywhere on the screen and notice the view being constructed. You can pick numerous times on the screen until the view is in a desired location. When completed, press the ENTER key to place the view. The next series of prompts enable you to clip a corner. This will con-struct a viewport around the view. It is very impor-tant to make this viewport large enough for dimen-sions to fit inside. Once the view is given a name, it is laid out similar to Figure 11−78.

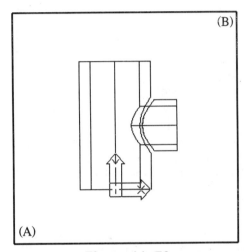

Figure 11−78

 Command:**SOLVIEW**

(The drawing switches to Paper Space)
Ucs/Ortho/Auxiliary/Section/<eXit>:**U** *(For the user coordinate system)*
Named/World/?/<Current>: *(Press* ENTER *to accept the default UCS as the current location)*
Enter view scale<1.0000>: *(Press* ENTER *to accept the default scale value)*
View center: *(Pick a point on the screen to display the view at its center; keep picking until the view is in the desired location)*
View center: *(Press* ENTER *to place the view)*
Clip first corner: *(Pick a point on the screen at "A")*
Clip other corner: *(Pick a point on the screen at "B")*
View name: **FRONT**
Ucs/Ortho/Auxiliary/Section/<eXit>: *(Press* ENTER *to exit this command and check the results)*

Once the view has been laid out using the SOLVIEW command, use SOLDRAW to actually draw the view in two dimensions. If the hidden linetype was loaded prior to using the SOLVIEW command, hidden lines will automatically be assigned to layers that contain hidden line information. The result of using the SOLDRAW command is shown in Figure 11−79.

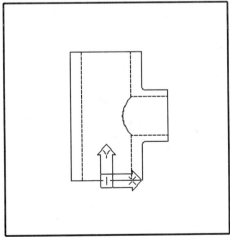

Figure 11−79

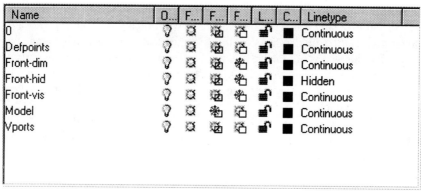

Figure 11–80

 Command:**SOLDRAW**

(The drawing switches to Paper Space)
Select viewports to draw:
Select objects: *(Pick anywhere on the viewport in Figure 11–79)*
Select objects: *(Press* [ENTER] *to perform the soldraw operation)*

The use of layers in 2D view layout is so important that when running the SOLVIEW command, the layers shown in Figure 11–80 are automatically created. With the exception of Model and 0, the layers that begin with "Front" and the Vports layer were all created by the SOLVIEW command. The "Front-dim" layer is designed to hold dimension information for the Front view. "Front-hid" holds all hidden lines information for the Front view; "Front-vis" holds all visible line information for the Front view. All paper space viewports are placed on the "Vports" layer. The Model layer has automatically been frozen in the current viewport to view the visible and hidden lines. To dimension the view shown in Figure 11–81, two items must be set. First, make "Front-dim" the current layer. Second, set the user coordinate system to the current view using the View option. The UCS icon should be similar to the illustration in Figure 11–81. Now, add all dimensions

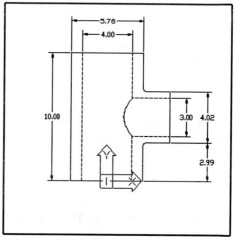

Figure 11–81

to the view using conventional dimensioning commands with the aid of Object snap modes. When
working on adding dimensions to another view, the
same two settings must be made in the new view;
make the layer holding the dimension information
the current layer and update the UCS to the current
view with the View option.

When you draw the views using the SOLDRAW
command and then add the dimensions, switching
back to the model by setting TILEMODE to 1 displays the image shown in Figure 11–82. In addition to the solid model of the object, the constructed
view and dimensions are also displayed. All drawn
views from paper space will display in the model.
Use the LAYER command along with the Freeze
option to freeze all drawing related layers and isolate the solid model.

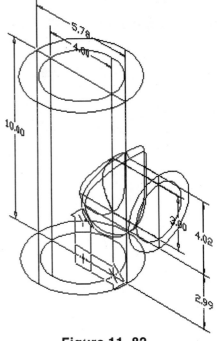

Figure 11–82

Tutorial Exercise
Bplate.Dwg

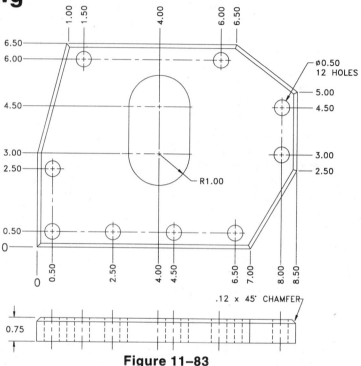

Figure 11–83

Purpose

The purpose of this tutorial is to produce a solid model of the Bplate shown in Figure 11–83.

System Settings

Keep the current limits settings of 0,0 for the lower left corner and 12,9 for the upper right corner. Change the number of decimal places past the zero from four to two using the Units Control dialog box. Turn the grid on and change the Snap value from 1.00 to 0.50. Keep all remaining system settings.

Layers

Special layers do not have to be created for this tutorial exercise.

Suggested Commands

Begin this tutorial by constructing the profile of the Bplate using polylines. Add all circles; use the EXTRUDE command to extrude all objects at a thickness of 0.75 units. Chamfer the top edge of the plate using the CHAMFER command. Use the SUBTRACT command to subtract all cylinders from the Bplate forming the holes in the plate. Perform hidden line removals using the HIDE command and create shaded models using the SHADE command.

Whenever possible, substitute the appropriate command alias in place of the full AutoCAD command in each tutorial step. For example, use "Co" for the COPY command, "L" for the LINE command, and so on. The complete listing of all command aliases is located on pages 1–9 and 1–10.

Step #1

Begin the Bplate by establishing a new coordinate system using the UCS command at the right. Define the origin at 2.00,1.50. Use the UCSICON command to update the user coordinate system icon to the new coordinate system location on the display screen. Use the PLINE command to draw the profile of the Bplate. See Figure 11–84.

 Command: **UCS**

Origin/ZAxis/3point/OBject/View/X/Y/Z/Prev/
    Restore/Save/Del/?/<World>: **O** *(For Origin)*
Origin point: <0,0,0>: **2.00,1.50**

Command: **UCSICON**
ON/OFF/All/Noorigin/ORigin <ON>: **OR** *(For Origin)*

 Command: **PLINE**

From point: **0,0**
Current line-width is 0.00
Arc/Close/Halfwidth/Length/Undo/Width/
    <Endpoint of line>: **@7.00<0**
Arc/Close/Halfwidth/Length/Undo/Width/
    <Endpoint of line>: **@1.50,2.50**
Arc/Close/Halfwidth/Length/Undo/Width/
    <Endpoint of line>: **@2.50<90**
Arc/Close/Halfwidth/Length/Undo/Width/
    <Endpoint of line>: **@-2.00,1.50**
Arc/Close/Halfwidth/Length/Undo/Width/
    <Endpoint of line>: **@5.50<180**
Arc/Close/Halfwidth/Length/Undo/Width/
    <Endpoint of line>: **@-1.00,-3.50**
Arc/Close/Halfwidth/Length/Undo/Width/
    <Endpoint of line>: **C** *(To close the polyline and return to the command prompt)*

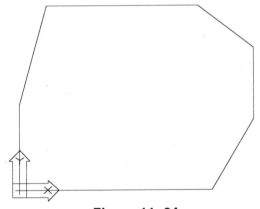

Figure 11–84

Step #2

Draw the twelve circles of 0.50 diameter by placing one circle at "A" and copying the remaining circles to their desired locations (see Figure 11–85). Use the COPY-Multiple command to accomplish this.

 Command: **CIRCLE**

3P/2P/TTR/<Center point>:**0.50,0.50**
Diameter/<Radius>: **D** *(For Diameter)*
Diameter:**0.50**

 Command:**COPY**

Select objects: **L** *(for the last circle)*
Select objects: *(Press* ENTER *to continue with this command)*
<Base point or displacement>/Multiple: **M** *(For Multiple)*
Base point: **@** *(References the center of the 0.50 circle)*
Second point of displacement:**2.50,0.50**
Second point of displacement:**4.50,0.50**
Second point of displacement:**6.50,0.50**
Second point of displacement:**8.00,3.00**
Second point of displacement:**8.00,4.50**
Second point of displacement:**0.50,2.50**
Second point of displacement:**1.50,6.00**
Second point of displacement:**6.00,6.00**
Second point of displacement: *(Press* ENTER *to exit this command)*

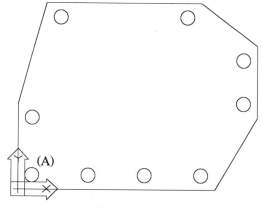

Figure 11–85

Step #3

Form the slot by placing two circles using the CIRCLE command followed by two lines drawn from the quadrants of the circles using the OSNAP-Quadrant mode. See Figure 11–86.

 Command: **CIRCLE**

3P/2P/TTR/<Center point>: **4.00,3.00**
Diameter/<Radius>: **1.00**

 Command: **CIRCLE**

3P/2P/TTR/<Center point>: **4.00,4.50**
Diameter/<Radius>: **1.00**

 Command: **LINE**

From point: **Qua**
of *(Select the quadrant of the circle at "A")*
To point: **Qua**
of *(Select the quadrant of the circle at "B")*
To point: *(Press* ENTER *to exit this command)*

 Command: **LINE**

From point: **Qua**
of *(Select the quadrant of the circle at "C")*
To point: **Qua**
of *(Select the quadrant of the circle at "D")*
To point: *(Press* ENTER *to exit this command)*

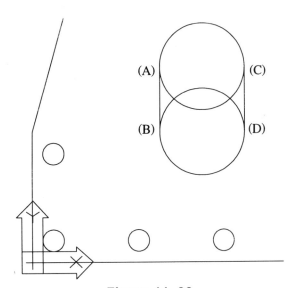

Figure 11–86

Step #4

Use the TRIM command to trim away any unnecessary arcs to form the slot. See Figure 11–87.

Command: **TRIM**

Select cutting edges: *(Projmode = UCS,*
   Edgemode = No extend)
Select objects: *(Select both vertical lines at "A"*
   and "B")
Select objects: *(Press* [ENTER] *to continue with this*
   command)
<Select object to trim>/Project/Edge/Undo:
   (Select the circle at "C")
<Select object to trim>/Project/Edge/Undo:
   (Select the circle at "D")
<Select object to trim>/Project/Edge/Undo:
   (Press [ENTER] *to exit this command).*

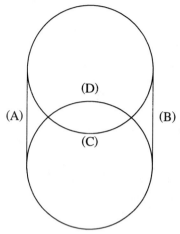

Figure 11–87

Step #5

The next step illustrates the use of the EXTRUDE command to give the plate a thickness of 0.75 units. However, this command only operates on circles and polylines. Currently, all objects may be extruded except for the two arcs and lines representing the slot (see Figure 11–88). Use the PEDIT command to convert these objects into a single polyline. This command is found under Modify > Object > Polyline in the pull-down menu area.

Command: **PEDIT**

Select polyline: *(Select the bottom arc at "A")*
Object selected is not a polyline
Do you want to turn it into one? <Y> *(Press*
   [ENTER] *to accept this default value)*
Close/Join/Width/Edit vertex/Fit/Spline/
   Decurve/Ltype gen/Undo/eXit <X>: **J** *(For*
   Join)
Select objects: *(Select the objects labeled "B,"*
   "C," and "D")
Select objects: *(Press* [ENTER] *to perform the*
   joining operation)
3 segments added to polyline
Open/Join/Width/Edit vertex/Fit/Spline/
   Decurve/Ltype gen/Undo/eXit <X>: *(Press*
   [ENTER] *to exit this command)*

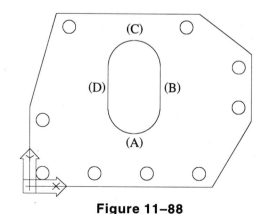

Figure 11–88

Step #6

Use the EXTRUDE command to give Bplate thickness. This command may be entered in from the keyboard or may be selected from the pull-down menu by picking Draw > Solids > Extrude. Use the All option to select all circles and polyline outlines.

 Command: **EXTRUDE**

Select objects: **All**
Select objects: *(Press* ENTER *to continue with this command)*
Path/<Height of Extrusion>: **0.75**
Extrusion taper angle <0>: *(Press* ENTER *to execute the extrusion operation)*

Command: **UCSICON**
ON/OFF/All/Noorigin/ORigin <ON>: **Off**

Step #7

Use the VPOINT command to view the plate in three dimensions using the following prompt sequence and Figure 11–90. Or, pick View from the pull-down menu area and select 3D Viewpoint > SE Isometric to achieve the same results. Next, use the CHAMFER command to place a chamfer of 0.12 units along the top edge of the plate. The top surface may not highlight when prompted for the base surface; use the Next option until the top surface is highlighted. After entering the chamfer distances, use the Loop option to create the chamfer along the entire edge without picking each individual edge along the top.

Command: **VPOINT**
Rotate/<View point> <0.00,0.00,1.00>: **1,-1,1**

 Command: **CHAMFER**

(TRIM mode) Current chamfer Dist1 = 0.50, Dist2 = 0.50
Polyline/Distance/Angle/Trim/Method/<Select first line>: *(Select the line at "A")*
Select base surface:

Enter a value of 0.75 as the height of the extrusion. Keep the default value for the extrusion taper angle. When completed with this operation, turn off the UCS icon with the UCSICON command. See Figure 11–89.

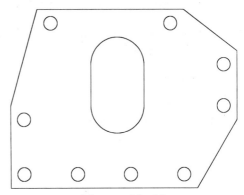

Figure 11–89

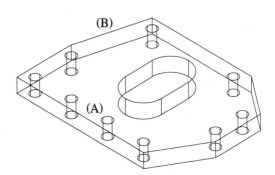

Figure 11–90

Next/<OK>: **N** *(For the next surface which should be the top surface)*
Next/<OK>: *(Press* ⌜ENTER⌝ *to accept the top surface as the base)*
Enter base surface distance <0.50>: **0.12**
Enter other surface distance <0.50>: **0.12**
Loop/<Select edge>: **L** *(For loop)*
Edge/<Select edge loop>: *(Select the edge along the top at "B")*
Edge/<Select edge loop>: *(Press* ⌜ENTER⌝ *to perform the chamfer operation)*

Step #8

Since all of the objects have been extruded a distance of 0.75 units, the holes and slot are considered individual solid objects that do not yet belong to the base. Use the SUBTRACT command to subtract the holes and slot from the base plate (see Figure 11–91). This operation will resemble the drilling of holes and the milling for the slot. This command may be found in the pull-down menu area under Modify > Boolean > Subtract.

 Command: **SUBTRACT**

Select solids and regions to subtract from…
Select objects: *(Select the base of the Bplate along any edge)*
Select objects: *(Press* ⌜ENTER⌝ *to continue with this command)*
Select solids and regions to subtract…
Select objects: *(Select every hole and the slot)*
Select objects: *(Press* ⌜ENTER⌝ *to perform the subtraction operation)*

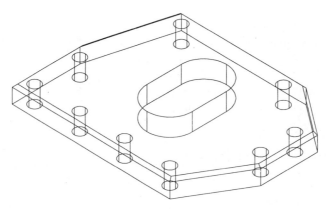

Figure 11–91

Step #9

To see the object with all hidden edges removed, use the HIDE command. This command may be found in the pull-down menu area under View>Hide. See Figure 11–92.

Command: **HIDE**
Regenerating drawing.

Regenerating the display again will return the model back to its wireframe mode.

Command: **REGEN**
Regenerating Drawing.

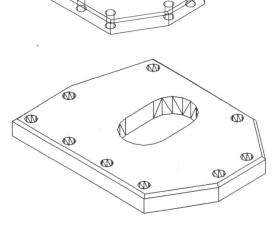

Step #10

From Figure 11–92 in the previous step, the circles and slot look somewhat irregular after performing the Hide operation. To smooth out the circles and arcs representing the slot, use the FACETRES command. This stands for facet resolution and it controls the density circles and arcs are displayed in. Set the FACETRES to a new value of 1, perform a hide, and observe the results. See Figure 11–93.

Command: **FACETRES**
New value for FACETRES <0.5000>: **1**

Command: **HIDE**
Regenerating drawing.

Notice the circles are better defined due the the increased density. Regenerating the display again will return the model back to its wireframe mode.

Command: **REGEN**
Regenerating Drawing.

Figure 11–92

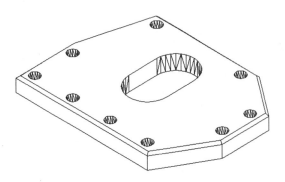

Figure 11–93

Step #11

FACETRES controls the density of circles and arcs when performing hidden line removals. The ISO-LINES command controls the density of lines that represent circles and arcs in the wireframe mode. These lines are called tessellation lines. Change the value of ISOLINES from 4 to a new value of 15, regenerate the screen, and observe the results. See Figure 11–94.

Command: **ISOLINES**
New value for ISOLINES <4>: **15**

Command: **REGEN**
Regenerating drawing.

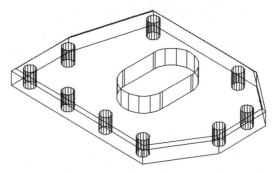

Figure 11–94

Step #12

Use the SHADE command to produce a shaded image consisting of hidden line removal in addition to the shading being performed in the original color of the model. There are no features to control lights, materials, or shadows. This shading mechanism is merely to get a fast shaded image of a model before preparing it to be rendered. See Figure 11–95.

Command: **SHADE**
Regenerating drawing.
Shading complete.

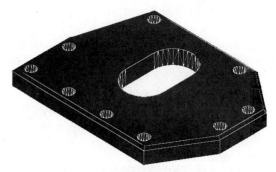

Figure 11–95

Tutorial Exercise
Guide.Dwg

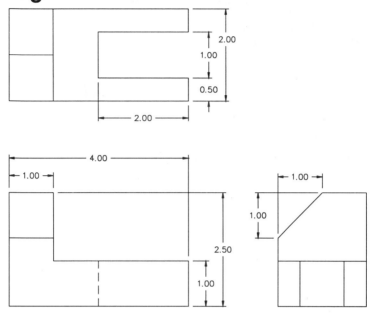

Figure 11–96

Purpose

This tutorial exercise is designed to produce a 3D solid model of the Guide from the information supplied in the orthographic drawing in Figure 11–96.

System Settings

Use the Units Control dialog box, keep the system of measurement set to decimal but change the number of decimal places past the zero from a value of 4 to 2. Set the Snap to a value of 0.50 units. The grid should conform to the current Snap setting. Leave the current limits of 0,0 by 12,9 as the default setting.

Layers

Special layers do not have to be created for this tutorial exercise.

Suggested Commands

Begin this drawing by constructing solid primitives of all components of the Guide using the BOX and WEDGE commands. Move the components into position and begin merging solids using the UNION. To form the rectangular slot, move that solid box into position and use the SUBTRACT command to subtract the rectangle from the solid thus forming the slot. Do the same procedure for the wedge. Perform a hidden line removal and view the solid.

Whenever possible, substitute the appropriate command alias in place of the full AutoCAD command in each tutorial step. For example, use "Co" for the COPY command, "L" for the LINE command, and so on. The complete listing of all command aliases is located on pages 1–9 and 1–10.

Step #1

Begin this tutorial by constructing a solid box 4 units long by 2 units wide and 1 unit in height using the BOX command. See Figure 11–97. Begin this box at absolute coordinate 4.00,5.50. This slab will represent the base of the guide.

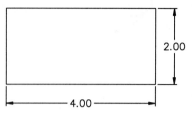

Figure 11–97

 Command:**BOX**

Center/<Corner of box><0,0,0>: **4.00,5.50**
Cube/Length/<other corner>:**L** *(For Length)*
   Length:**4.00**
   Width:**2.00**
   Height:**1.00**

Step #2

Construct a solid box 1 unit long by 2 units wide and 1.5 units in height using the BOX command (see Figure 11–98). Begin this box at absolute coordinate 2.00,1.50. This slab will represent the vertical column of the guide.

 Command:**BOX**

Center/<Corner of box><0,0,0>:**2.00,1.50**
Cube/Length/<other corner>:**L** *(For Length)*
   Length:**1.00**
   Width:**2.00**
   Height:**1.50**

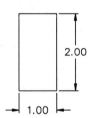

Figure 11–98

Step #3

Construct a solid box 2 units long by 1 unit wide
and 1 unit in height using the BOX command (see
Figure 11–99). Begin this box at absolute coordinate 5.50,1.50. This slab will represent the rectangular slot made into the slab that will be subtracted
at a later time.

 Command: **BOX**

Center/<Corner of box><0,0,0>: **5.50,1.50**
Cube/Length/<other corner>: **L** *(For Length)*
   Length: **2.00**
   Width: **1.00**
   Height: **1.00**

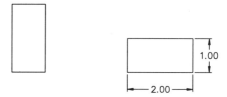

Figure 11–99

Step #4

Use the WEDGE command to draw a wedge 1 unit
in length, 1 unit wide, and 1 unit in height (see Figure 11–100). Begin this primitive at absolute coordinate 9.50,2.00. This wedge will be subtracted from
the vertical column to form the inclined surface.

 Command: **WEDGE**

Center/<Corner of wedge><0,0,0>: **9.50,2.00**
Cube/Length/<other corner>: **L** *(For Length)*
   Length: **1.00**
   Width: **1.00**
   Height: **1.00**

Figure 11–100

Step #5

Use the VPOINT command to view the four solid primitives in three dimensions (see Figure 11–101). Use a new view point of 1,-1,1. A preset viewpoint may be used by selecting the following sequence from the pull-down menu area: View > 3D Viewpoint > SE Isometric. Then use the MOVE command to move the vertical column at "A" to the top of the base at "B".

Command: **VPOINT**
Rotate/<View point> <0,0,1>: **1,-1,1**

 Command: **MOVE**

Select objects: *(Select the solid box at "A")*
Select objects: *(Press* ⏎ *to continue with this command)*
Base point or displacement: **Endp**
of *(Select the endpoint of the solid at "A")*
Second point of displacement: **Endp**
of *(Select the endpoint of the base at "B")*

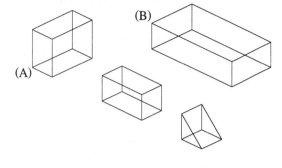

Figure 11–101

Step #6

Use the UNION command to join the base and vertical column into one object. See Figure 11–102.

 Command: **UNION**

Select objects: *(Select the base at "A" and column at "B")*
Select objects: *(Press* ⏎ *to perform the union)*

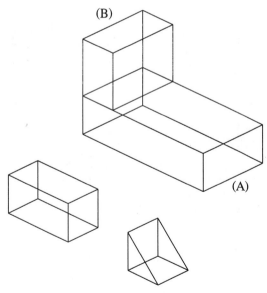

Figure 11–102

Step #7

Use the MOVE command to position the rectangle
from its midpoint at "A" to the midpoint of the base
at "B" in Figure 11–103. In a moment, the small
rectangle will be subtracted forming the rectangular slot in the base.

 Command:**MOVE**

Select objects: *(Select box "A")*
Select objects: *(Press* [ENTER] *to continue with this
    command)*
Base point or displacement: **Mid**
of *(Select the midpoint of the rectangle at "A")*
Second point of displacement: **Mid**
of *(Select the midpoint of the base at "B")*

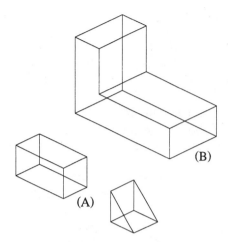

Figure 11–103

Step #8

Use the SUBTRACT command to subtract the
small box from the base of the solid object. See
Figure 11–104.

 Command:**SUBTRACT**

Select solids and regions to subtract from...
Select objects: *(Select solid object "A")*
Select objects: *(Press* [ENTER] *to continue with this
    command)*
Select solids and regions to subtract...
Select objects: *(Select box "B" to subtract)*
Select objects: *(Press* [ENTER] *to perform the
    subtraction operation)*

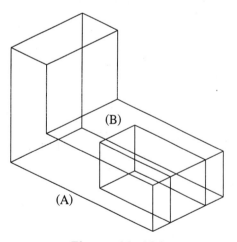

Figure 11–104

Step #9

Use the ALIGN command to match points along a source object (wedge) with points along a destination object (guide). The selection of three sets of points will guide the placement and rotation of the wedge into the guide. See Figure 11–105.

Command: **ALIGN**
Select objects: *(Select the wedge)*
Select objects: *(Press ⏎ to continue with this command)*
Specify 1st source point: **Endp**
of *(Pick the endpoint of the wedge at "A")*
Specify 1st destination point: **Endp**
of *(Pick the endpoint of the guide at "B")*
Specify 2nd source point: **Endp**
of *(Pick the endpoint of the wedge at "C")*
Specify 2nd destination point: **Endp**
of *(Pick the endpoint of the guide at "D")*
Specify 3rd source point or <continue>: **Endp**
of *(Pick the endpoint of the wedge at "E")*
Specify 3rd destination point: **Endp**
of *(Pick the endpoint of the guide at "F")*

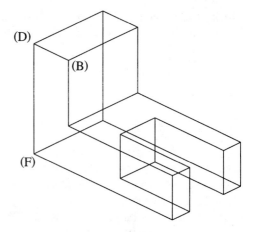

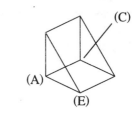

Figure 11–105

Step #10

Use the SUBTRACT command to subtract the wedge from the guide, as shown in Figure 11–106.

 Command: **SUBTRACT**
Select solids and regions to subtract from…
Select objects: *(Select guide "A")*
Select objects: *(Press ⏎ to continue with this command)*
Select solids and regions to subtract…
Select objects: *(Select the wedge at "B")*
Select objects: *(Press ⏎ to perform the subtraction operation)*

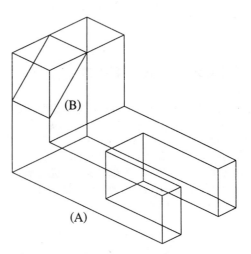

Figure 11–106

Step #11

An alternate method of creating the inclined surface in the guide is to use the CHAMFER command on the vertical column of the guide. See Figure 11–107.

 Command: **CHAMFER**

(TRIM mode) Current chamfer Dist1 = 0.50, Dist2 = 0.50

Polyline/Distance/Angle/Trim/Method/<Select first line>: *(Select the line at "A")*

Select base surface:

Next/<OK>: *(Press* ENTER *to accept the base surface)*

Enter base surface distance <0.50>: **1**

Enter other surface distance <0.50>: **1**

Loop/<Select edge>: *(Select the edge at "A")*

Loop/<Select edge>: *(Press* ENTER *to perform the chamfer operation)*

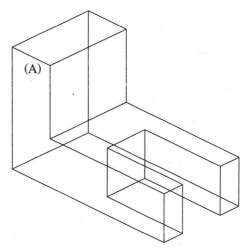

Figure 11–107

Step #12

Using the HIDE command performs a hidden line removal on all surfaces of the object. The results are shown in Figure 11–108.

Command: **HIDE**

Regenerating drawing.

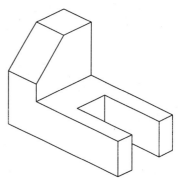

Figure 11–108

Tutorial Exercise
Rotate3D.Dwg

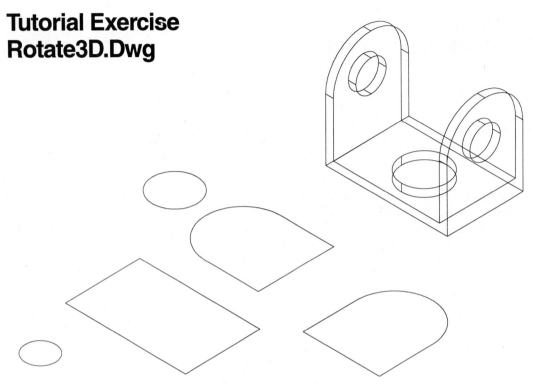

Figure 11-109

Purpose

This tutorial exercise is designed to produce 3D objects and then use the ROTATE3D command to rotate the objects into position before creating unions or performing subtractions. See Figure 11-109.

System Settings

The drawing units, limits, grid, and snap values are already set for this drawing.

Layers

Special layers do not have to be created for this tutorial exercise.

Suggested Commands

Open the drawing file called "ROTATE3D.Dwg." All 2D objects have been converted into polyline objects. The EXTRUDE command is used to create a thickness of 0.50 units for most objects. Then the ROTATE3D command is used to rotate the sides to the proper angle before moving them into place and joining them together. The cylinders are also rotated and moved into place before being subtracted from the base and sides to create holes.

Whenever possible, substitute the appropriate command alias in place of the full AutoCAD command in each tutorial step. For example, use "Co" for the COPY command, "L" for the LINE command, and so on. The complete listing of all command aliases is located on pages 1-9 and 1-10.

Step #1

Extrude the base, two side panels, and large circle to a height of 0.50 units, as shown in Figure 11–110. Then extrude the small circle to a height of 8 units.

This represents the total length of the base and will be used to cut a hole through the two side panels. Use the EXTRUDE command for both operations.

 Command:**EXTRUDE**

Select objects: *(Select the base, two side panels, and large circle)*
Select objects: *(Press* [ENTER] *to continue with this command)*
Path/<Height of Extrusion>:**0.50**
Extrusion taper angle <0>: *(Press* [ENTER] *to execute the extrude operation)*

 Command:**EXTRUDE**

Select objects: *(Select the small circle)*
Select objects: *(Press* [ENTER] *to continue with this command)*
Path/<Height of Extrusion>:**8.00**
Extrusion taper angle <0>: *(Press* [ENTER] *to execute the extrude operation)*

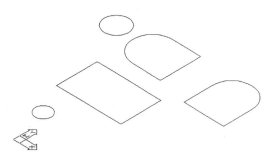

Figure 11–110

Step #2

Use the MOVE command to position the large cylinder in the center of the base. You may also use the

geometry calculator and the MEE function to center the cylinder in the base. Also see Figure 11–111.

 Command:**MOVE**

Select objects: *(Select the large cylinder)*
Select objects: *(Press* [ENTER] *to continue with this command)*
Base point or displacement: **Cen**
of *(Pick the top face of the large cylinder at "A")*
Second point of displacement: **'CAL**
>> Expression:**MEE**
>> Select one endpoint for MEE: *(Pick a point along the top corner of the base at "B")*
>> Select another endpoint for MEE: *(Pick a point along opposite corner of the base at "C")*

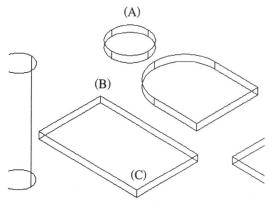

(A)

(B)

(C)

Figure 11–111

Step #3

Begin rotating the remaining shapes into place using the ROTATE3D command (see Figure 11–112). Being careful to observe the current position of the user coordinate system, rotate one of the side panels at an angle of 90 degrees using the Y axis as the axis of rotation.

Command: **ROTATE3D**
Select objects: *(Select the side panel labeled "A")*
Select objects: *(Press ⌷ENTER⌷ to continue with this command)*
Axis by Object/Last/View/Xaxis/Yaxis/Zaxis/
    <2points>: **Y**
Point on Y axis <0,0,0>: **Endp**
of *(Select the endpoint of the side panel at "B")*
<Rotation angle>/Reference: **90**

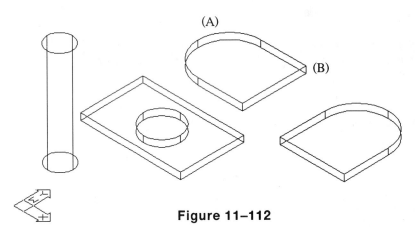

Figure 11–112

Step #4

Position the side panel just rotated onto the top of the base using the MOVE command. See Figure 11–113.

 Command: **MOVE**
Select objects: *(Select the side panel just rotated)*
Select objects: *(Press ⌷ENTER⌷ to continue with this command)*
Base point or displacement: **Endp**
of *(Pick the endpoint of the side panel at "A")*
Second point of displacement: **Endp**
of *(Pick the top endpoint of the base at "B")*

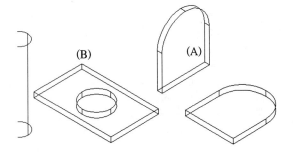

Figure 11–113

Step #5

Now rotate the second side panel into position. This will take two steps to accomplish. First rotate the panel 90 degrees along the X axis. See Figure 11–114.

Command: **ROTATE3D**
Select objects: *(Select the second side panel)*
Select objects: *(Press* ENTER *to continue with this command)*
Axis by Object/Last/View/Xaxis/Yaxis/Zaxis/ <2points>: **X**
Point on X axis <0,0,0>: **Endp**
of *(Pick the endpoint of the side panel at "A")*
<Rotation angle>/Reference: **90**

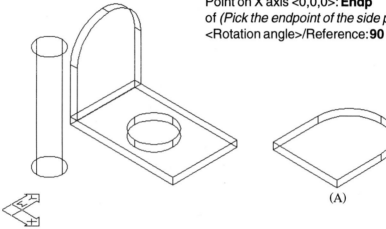

(A)

Figure 11–114

Step #6

Now rotate the same panel 90 degrees along the Z Axis. See Figure 11–115.

Command: **ROTATE3D**
Select objects: *(Select the same side panel)*

Select objects: *(Press* ENTER *to continue with this command)*
Axis by Object/Last/View/Xaxis/Yaxis/Zaxis/ <2points>: **Z**
Point on Z axis <0,0,0>: **Endp**
of *(Pick the endpoint of the side panel at "A")*
<Rotation angle>/Reference: **90**

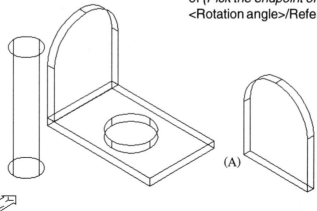

(A)

Figure 11–115

Step #7

Finally position the panel along the top of the base using the MOVE command. See Figure 11–116.

 Command: **MOVE**

Select objects: *(Select the side panel just rotated)*

Select objects: *(Press* ENTER *to continue with this command)*
Base point or displacement: **Endp**
of *(Pick the endpoint of the side panel at "A")*
Second point of displacement: **Endp**
of *(Pick the top of the base at "B")*

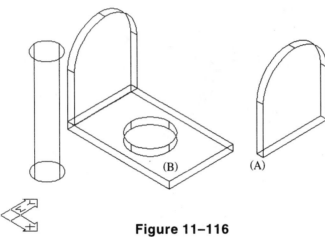

(B)   (A)

Figure 11–116

Step #8

Rotate the long cylinder at 90 degrees using the Y axis as the axis of rotation, as shown in Figure 11–117.

Command: **ROTATE3D**
Select objects: *(Select the long cylinder)*
Select objects: *(Press* ENTER *to continue with this command)*
Axis by Object/Last/View/Xaxis/Yaxis/Zaxis/
<2points>: **Y**
Point on Y axis <0,0,0>: **Cen**
of *(Pick the center of the cylinder at "A")*
<Rotation angle>/Reference: **90**

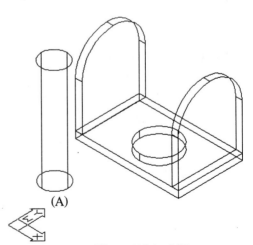

(A)

Figure 11–117

Step #9

Move the long cylinder into position; it should span the complete length of the object and should be lined up with the center of the side panels. See Figure 11–118.

 Command:**MOVE**

Select objects: *(Select the long cylinder)*
Select objects: *(Press* ⏎ *to continue with this command)*
Base point or displacement: **Cen**
of *(Pick the center of the long cylinder at "A")*
Second point of displacement: **Cen**
of *(Pick the outer center of the side panel at "B")*

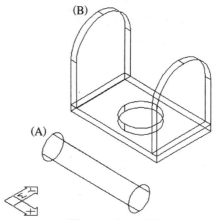

Figure 11–118

Step #10

Finally, use the SUBTRACT command to subtract the two cylinders from the object; the side panels will automatically be unioned to the base when you complete this command. See Figure 11–119.

 Command:**SUBTRACT**

Select solids and regions to subtract from…
Select objects: *(Select the two side panels and the base)*
Select objects: *(Press* ⏎ *to continue with this command)*
Select solids and regions to subtract…
Select objects: *(Select the two cylinders to subtract)*
Select objects: *(Press* ⏎ *to execute this command)*

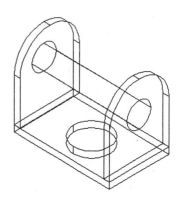

Figure 11–119

Step #11

The completed object is displayed in Figure 11–120.

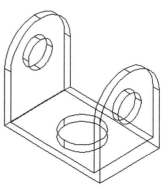

Figure 11–120

Tutorial Exercise
Lever.Dwg

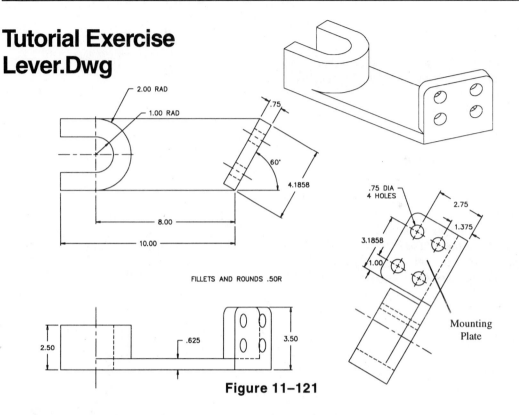

FILLETS AND ROUNDS .50R

Figure 11–121

Purpose

This tutorial is designed to use the UCS command to construct a 3D solid model of the Lever in Figure 11–121.

System Settings

Begin a new drawing called "Lever." Use the Units Control dialog box to change the number of decimal places past the zero from four to two. Keep all default values for the UNITS command. Using the LIMITS command, keep 0,0 for the lower left corner and change the upper right corner from 12,9 to 15.50,9.50. Use the GRID command and change the grid spacing from 1.00 to 0.50 units. Do not turn the Snap or Ortho on.

Layers

Create the following layers with the format:

Name	Color	Linetype
Model	Cyan	Continuous

Suggested Commands

Begin layout of this problem by constructing the plan view of the Lever. Use the EXTRUDE command to create the height of the individual components that make up the lever. Position the UCS to construct the mounting plate.

Whenever possible, substitute the appropriate command alias in place of the full AutoCAD command in each tutorial step. For example, use "Co" for the COPY command, "L" for the LINE command, and so on. The complete listing of all command aliases is located on pages 1–9 and 1–10.

Step #1

Begin this drawing by constructing two circles of radius values 1.00 and 2.00 using the CIRCLE command and 4.00,5.00 as the center of both circles (see Figure 11–122). Use the @ symbol to identify the last known point (4.00,5.00) as the center of the circle.

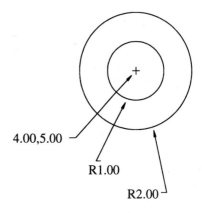

 Command: **CIRCLE**
3P/2P/TTR/<Center point>:**4.00,5.00**
Diameter/<Radius>:**1.00**

 Command: **CIRCLE**
3P/2P/TTR/<Center point>: **@** *(To identify the last point)*
Diameter/<Radius>:**2.00**

Figure 11–122

Step #2

Draw a vertical line 2 units to the left of the center of the circles and 4 units long in the 270 direction, or use the direct distance mode to construct the line in the 270 direction. Use OSNAP-Tracking to accomplish this and be sure Ortho mode is turned off when using tracking mode.

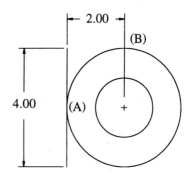

 Command: **LINE**
From point: **TK**
First tracking point: **Qua**
of *(Pick the quadrant at "A")*
Next point *(Press* ENTER *to end tracking):* **Qua**
of *(Pick the quadrant at "B")*
Next point *(Press* ENTER *to end tracking): (Press* ENTER*; tracking is ended)*
To point: **@4<270**
To point: *(Press* ENTER *to exit this command)*

Figure 11–123

Step #3

Draw four horizontal lines from quadrant points on the two circles a distance of 2 units. Polar coordinates or the direct distance mode may be used to accomplish this. These lines should intersect with the vertical line drawn in the previous step. Also see Figure 11–124.

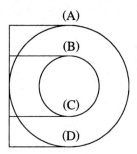

Figure 11–124

 Command: **LINE**

From point: **Qua**
of *(Select the quadrant of the large circle at "A")*
To point: **@2<180**
To point: *(Press* ⌜ENTER⌝ *to exit this command)*

Repeat the above procedure and draw three more lines from points "B," "C," and "D" in Figure 11–124. using the same polar coordinate value for the length, namely, @2<180. An alternate method would be to use the COPY-Multiple command to duplicate the three remaining lines.

Step #4

Use the TRIM command, select the two dashed lines in Figure 11–125 as cutting edges, and trim the left side of the large circle.

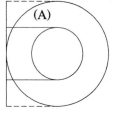
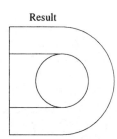

Figure 11–125

 Command: **TRIM**

Select cutting edges: (Projmode = UCS,
   Edgemode = No extend)
Select objects: *(Select the two dashed lines in Figure 11–125)*
Select objects: *(Press* ⌜ENTER⌝ *to continue with this command)*
<Select object to trim>/Project/Edge/Undo:
   (Pick the circle at "A")
<Select object to trim>/Project/Edge/Undo:
   (Press ⌜ENTER⌝ *to exit this command)*

Step #5

Use the TRIM command again and select the two dashed lines in Figure 11–126 as cutting edges. Select the left side of the small circle and the middle of the vertical line as the objects to trim.

 Command: **TRIM**

Select cutting edges: (Projmode = UCS,
   Edgemode = No extend)
Select objects: *(Select the two dashed lines in
   Figure 11–126)*
Select objects: *(Press* ⌷ENTER⌷ *to continue with this
   command)*
<Select object to trim>/Project/Edge/Undo:
   (Select the vertical line at "A")
<Select object to trim>/Project/Edge/Undo:
   (Select the small circle at "B")
<Select object to trim>/Project/Edge/Undo:
   (Press ⌷ENTER⌷ *to exit this command)*

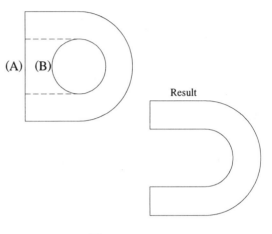

(A) (B)

Result

Figure 11–126

Step #6

Complete a partial plan view by using the LINE command and Figure 11–127 to draw the four lines. Use OSNAP-Endpoint in combination with polar coordinates.

 Command: **LINE**

From point: **Endp**
of *(Select the endpoint of the line or arc labeled
   "Start")*
To point: **@8.00<0**
To point: **@4.1858<60**
To point: **@0.75<150**
To point: **Endp**
of *(Select the endpoint of the line or arc labeled
   "End")*
To point: *(Press* ⌷ENTER⌷ *to exit this command)*

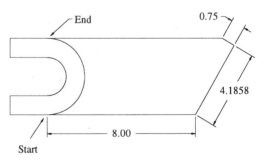

End   0.75

4.1858

8.00

Start

Figure 11–127

Step #7

View the model in 3D using the VPOINT command.
Use a viewing position of 1,-1,1; or pick the VPOINT
command from the following pull-down menu sequence: View > 3D Viewpoint > SE Isometric. See
Figure 11–128.

Command: **VPOINT**
Rotate/<View point> <0.00,0.00,1.00>:**1,-1,1**

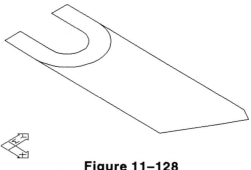

Step #8

Figure 11–128

Move the four lines that represent the base away from
the U-shaped figure as shown in Figure 11–129.
Since both figures are at different heights, they will
be treated as different objects when extruding.

 Command: **MOVE**
Select objects: *(Select all four lines labeled "A,"*
"B," "C," and "D")
Select objects: *(Press [ENTER] to continue with this*
command)
Base point or displacement: **Endp**
of *(Pick the endpoint near "A")*
Second point of displacement: **@2.00<0**

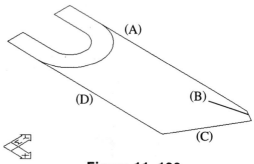

Figure 11–129

Step #9

Copy arc "A" to the endpoint of the line at "C."
This will allow the shape represented by the lines
to be closed through the placement of the arc. See
Figure 11–130.

 Command: **COPY**
Select objects: *(Select arc "A")*
Select objects: *(Press [ENTER] to continue with this*
command)
<Base point or displacement>/Multiple: **Endp**
of *(Pick the endpoint of the arc at "B")*
Second point of displacement: **Endp**
of *(Pick the endpoint of the arc at "C")*

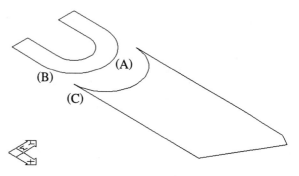

Figure 11–130

Step #10

Convert both shapes into polylines using the PEDIT command. Also refer to Figure 11–131.

 Command:**PEDIT**
Select polyline: *(Select the arc at "A")*
Object selected is not a polyline
Do you want to turn it into one? <Y> *(Press*
    [ENTER] *to accept the default)*
Close/Join/Width/Edit vertex/Fit/Spline/
    Decurve/Ltype gen/Undo/eXit <X>:**J**
Select objects: *(Select all arc and line*
    segments that make up the outline of the U-
    shaped item)
Select objects: *(Press* [ENTER] *to perform the join*
    operation)
7 segments added to polyline
Open/Join/Width/Edit vertex/Fit/Spline/
    Decurve/Ltype gen/Undo/eXit <X>: *(Press*
    [ENTER] *to exit this command)*

 Command:**PEDIT**
Select polyline: *(Select the line at "B")*
Object selected is not a polyline
Do you want to turn it into one? <Y> *(Press*
    [ENTER] *to accept the default)*
Close/Join/Width/Edit vertex/Fit/Spline/
    Decurve/Ltype gen/Undo/eXit <X>:**J**
Select objects: *(Select the other line and arc*
    segments that make up the outline of the
    other shape)
Select objects: *(Press* [ENTER] *to perform the join*
    operation)
4 segments added to polyline
Open/Join/Width/Edit vertex/Fit/Spline/
    Decurve/Ltype gen/Undo/eXit <X>: *(Press*
    [ENTER] *to exit this command)*

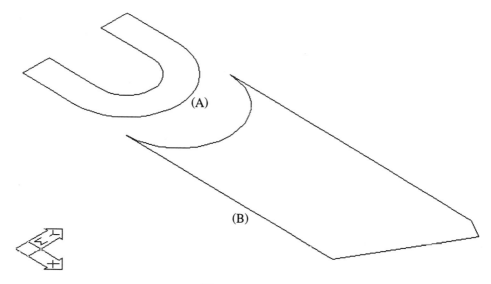

Figure 11–131

Step #11

Extrude the U-shaped figure to a height of 2.50 units. Extrude the other shape to a height of 0.625 units. Perform both operations using the EXTRUDE command and Figure 11–132.

 Command:**EXTRUDE**

Select objects: *(Select the U-shaped figure)*
Select objects: *(Press* ENTER *to continue with this command)*

Path/<Height of Extrusion>:**2.50**
Extrusion taper angle <0>: *(Press* ENTER *to perform the extrusion operation)*

 Command:**EXTRUDE**

Select objects: *(Select the other shape at "A")*
Select objects: *(Press* ENTER *to continue with this command)*
Path/<Height of Extrusion>:**0.625**
Extrusion taper angle <0>: *(Press* ENTER *to perform the extrusion operation)*

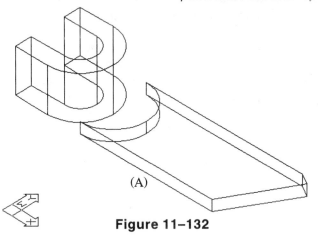

(A)

Figure 11–132

Step #12

With both shapes extruded, connect both shapes together from the quadrant of shape "A" to the quadrant of shape "B" using the MOVE command. See Figure 11–133.

 Command:**MOVE**

Select objects: *(Select shape "A")*
Select objects: *(Press* ENTER *to continue with this command)*
Base point or displacement: **Qua**
of *(Pick the bottom quadrant of the arc at "B")*
Second point of displacement: **Qua**
of *(Pick the bottom quadrant of the arc at "C")*

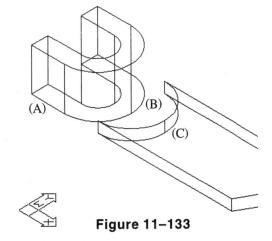

(A)   (B)
                    (C)

Figure 11–133

Step #13

Join both shapes into one using the UNION command, as shown in Figure 11–134.

 Command: **UNION**

Select objects: *(Select shape "A" and "B")*
Select objects: *(Press ⎆ to perform the union operation)*

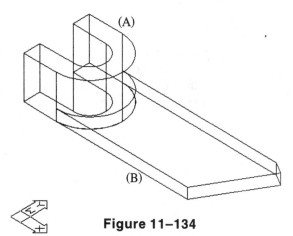

Figure 11–134

Step #14

Define a new origin for the user coordinate system by placing it at the endpoint at "A" in Figure 11–135. If the icon does not move to the new lo-

cation, use the UCSICON command and the Origin option to update the position of the user coordinate system icon.

 Command: **UCS**

Origin/ZAxis/3point/OBject/View/X/Y/Z/Prev/
    Restore/Save/Del/?/<World>: **O** *(For Origin)*
Origin point <0,0,0>: **Endp**
of *(Pick the line at "A" to find the endpoint)*

Command: **UCSICON**
ON/OFF/All/Noorigin/ORigin <ON>: **OR**

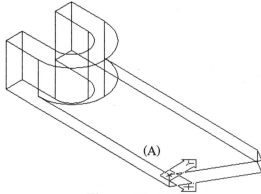

Figure 11–135

Step #15

Before beginning the construction of the mounting plate consisting of the four holes, rotate the UCS

icon 60 degrees about the Z axis and then 90 degrees about the X axis. See Figure 11–136.

 Command: **UCS**

Origin/ZAxis/3point/OBject/View/X/Y/Z/Prev/
    Restore/Save/Del/?/<World>: **Z**
Rotation angle about Z axis <0>: **60**

 Command: **UCS**

Origin/ZAxis/3point/OBject/View/X/Y/Z/Prev/
    Restore/Save/Del/?/<World>: **X**
Rotation angle about X axis <0>: **90**

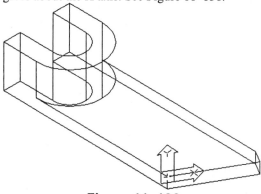

Figure 11–136

Step #16

Construct a polyline representing the front face of the mounting plate (see Figure 11–137). Polar coordinates or the direct distance mode may be used to accomplish this step. Round off the two top corners of the mounting plate using the FILLET command.

 Command: **PLINE**

From point: **0,0,0**
Current line-width is 0.00
Arc/Close/Halfwidth/Length/Undo/Width/
    <Endpoint of line>: **@3.50<90**
Arc/Close/Halfwidth/Length/Undo/Width/
    <Endpoint of line>: **@4.1858<0**
Arc/Close/Halfwidth/Length/Undo/Width/
    <Endpoint of line>: **@3.50<270**
Arc/Close/Halfwidth/Length/Undo/Width/
    <Endpoint of line>: **C** *(To close the polyline and exit the command)*

 Command: **FILLET**

(TRIM mode) Current fillet radius = 0.50
Polyline/Radius/Trim/<Select first object>:
    (Pick the polyline at "A")
Select second object: *(Pick the pline at "B")*

Repeat the FILLET command to round off the other corner.

 Command: **FILLET**

(TRIM mode) Current fillet radius = 0.50
Polyline/Radius/Trim/<Select first object>:
    (Pick the polyline at "B")
Select second object: *(Pick the pline at "C")*

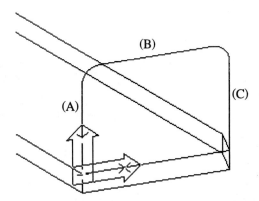

Figure 11–137

Step #17

Extrude the face of the mounting plate back a distance of -0.75 units to give it thickness using the EXTRUDE command. Once the mounting plate has been extruded, join the plate to the Lever using the UNION command. See Figure 11–138.

 Command:**EXTRUDE**

Select objects: *(Select the face of the mounting plate)*
Select objects: *(Press [ENTER] to continue with this command)*
Path/<Height of Extrusion>:**-0.75**
Extrusion taper angle <0>: *(Press [ENTER] to perform the extrusion operation)*

 Command:**UNION**

Select objects: *(Select the mounting plate and the Lever)*
Select objects: *(Press [ENTER] to perform the union operation)*

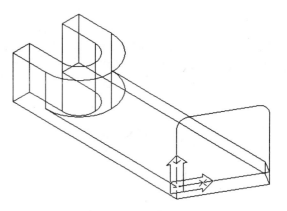

Figure 11–138

Step #18

Lay out one of the holes into the mounting plate. Use the OSNAP-From option to establish a reference point that will allow you to construct the cylinder -1.00 units in the X-axis and 1.375 units along the Y-axis. See Figure 11–139.

 Command:**CYLINDER**

Elliptical/<center point> <0,0,0>:**From**
Base point:**Endp**
of *(Pick the endpoint at "A")*
<Offset>:**@-1.00,1.375**
Diameter/<Radius>:**D** *(For Diameter)*
Diameter:**0.75**
Center of other end/<Height>:**-0.75**

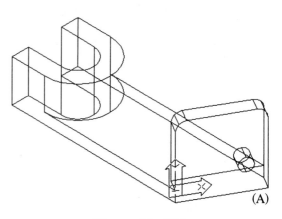

(A)

Figure 11–139

Step #19

Use the ARRAY command to create a rectangular pattern consisting of two rows and two columns. The distance between the rows will be 1.375 units and the distance between columns will be -2.1858 units. Then subtract the four cylinders from the Lever to complete the object using the SUBTRACT command. See Figure 11–140.

 Command: **ARRAY**

Select objects: *(Select the cylinder just created)*
Select objects: *(Press [ENTER] to continue with this command)*
Rectangular or Polar array (<R>/P): **R** *(For rectangular)*
Number of rows (---) <1>: **2**
Number of columns (lll) <1>: **2**
Unit cell or distance between rows (---): **1.375**
Distance between columns (lll): **-2.1858**

 Command: **SUBTRACT**

Select solids and regions to subtract from...
Select objects: *(Select the object at "A")*
Select objects: *(Press [ENTER] to continue with command)*
Select solids and regions to subtract...
Select objects: *(Select the four individual cylinders just created through the ARRAY command)*
Select objects: *(Press [ENTER] to perform the subtraction operation)*

Step #20

Set FACETRES to a value of 2 and perform a hidden line removal to get a better look at the model. See Figure 11–141.

Command: **FACETRES**
New value for FACETRES <0.5000>: **2**

Command: **HIDE**

Perform a REGEN to convert the model back to its wireframe state.

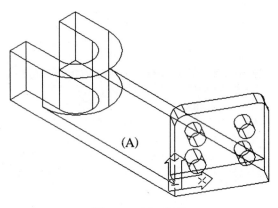

Figure 11–140

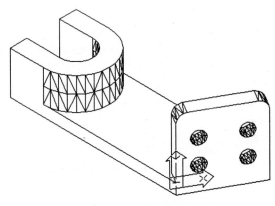

Figure 11–141

Step #21

Return the user coordinate system icon back to its home position by using the UCS-World option. Then use the MASSPROP command to calculate the mass properties of the lever. See Figure 11–142.

 Command: **UCS**

Origin/ZAxis/3point/OBject/View/X/Y/Z/Prev/
Restore/Save/Del/?/<World>: *(Press [ENTER] to accept the default as WORLD)*

 Command: **MASSPROP**

Select objects: *(Select anywhere along the model)*
Select objects: *(Press [ENTER] to perform the mass property calculation)*

_____SOLIDS_____

Mass:	50.22
Volume:	50.22
Bounding box:	X: 4.00–16.09
	Y: 3.00–7.00
	Z: 0.00–4.13
Centroid:	X: 9.71
	Y: 5.03
	Z: 1.12
Moments of inertia:	X: 1461.23
	Y: 5552.24
	Z: 6786.64
Products of inertia:	XY: 2469.71
	YZ: 281.33
	ZX: 571.53
Radii of gyration:	X: 5.39
	Y: 10.51
	Z: 11.62

Principal moments and X-Y-Z directions about centroid:
I: 124.10 along [1.00 0.02 0.04]
J: 756.73 along [-0.02 1.00 -0.08]
K: 782.03 along [-0.04 0.08 1.00]

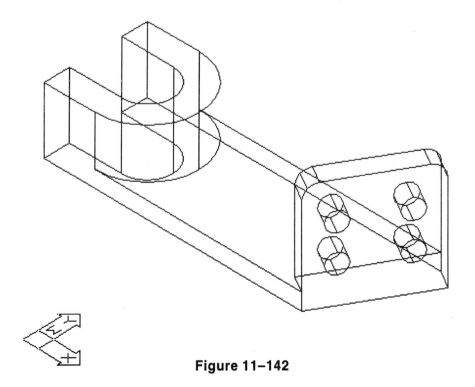

Figure 11–142

Tutorial Exercise
Collar.Dwg

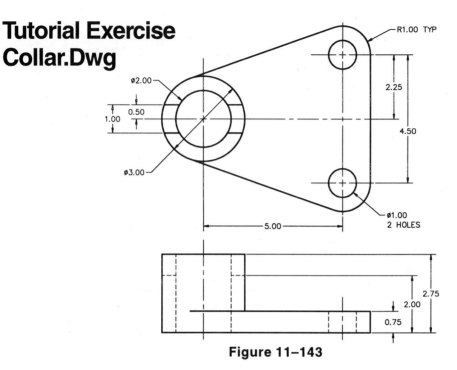

Figure 11–143

Purpose

This tutorial is designed to construct a solid model of the Collar using the dimensions in Figure 11–143.

System Settings

Use the current limits set to 0,0 for the lower left corner and 12,9 for the upper right corner. Change the number of decimal places from four to two using the Units Control dialog box. Snap and Grid values may remain as set by the default although the Snap may be changed to a value of 0.50 thereby affecting the display of the grid.

Layers

Special layers do not have to be created for this tutorial exercise although an object layer may be created using yellow lines:

Name	Color	Linetyp
Model	Cyan	Continuous

Suggested Commands

Begin this tutorial by laying out the Collar in plan view and drawing the basic shape outlined in the top view. Convert the objects into a polyline and extrude the objects to form a solid. Draw a cylinder and combine this object with the base. Add another cylinder and then subtract it to from the large hole through the model. Add two small cylinders and subtract them from the base to form the smaller holes. Construct a solid box, use the MOVE command to move the box into position, and subtract it to form the slot across the large cylinder.

Whenever possible, substitute the appropriate command alias in place of the full AutoCAD command in each tutorial step. For example, use "Co" for the COPY command, "L" for the LINE command, and so on. The complete listing of all command aliases is located on pages 1–9 and 1–10.

Step #1

Begin the Collar by drawing the three circles shown in Figure 11–144 using the CIRCLE command. Place the center of the circle at "A" at 0,0. Perform a ZOOM-All after all three circles have been con- structed. The center marks identifying the centers of the circles are used for illustrative purposes and do not need to be placed in the drawing.

 Command:**CIRCLE**

3P/2P/TTR/<Center point>:**0,0**
Diameter/<Radius>: **D** *(For Diameter)*
Diameter:**3.00**

 Command:**CIRCLE**

3P/2P/TTR/<Center point>:**5.00,2.25**
Diameter/<Radius>:**1.00**

 Command:**CIRCLE**

3P/2P/TTR/<Center point>:**5.00,-2.25**
Diameter/<Radius>:**1.00**

 Command:**ZOOM**

All/Center/Dynamic/Extents/Previous/Scale(X/
    XP)/Window/<Realtime>:**All**

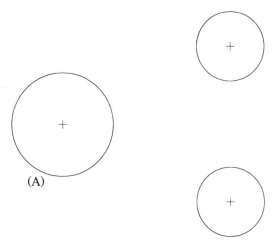

(A)

Figure 11–144

Step #2

Draw lines tangent to the three arcs using the LINE command and the OSNAP-Tangent mode. Also see Figure 11–145.

 Command:**LINE**

From point: **Tan**
to *(Select the circle at "A")*
To point: **Tan**
to *(Select the circle at "B")*
To point: *(Press the* ENTER *key to exit the command)*

Repeat the above procedure to draw lines from "C" to "D" and "E" to "F".

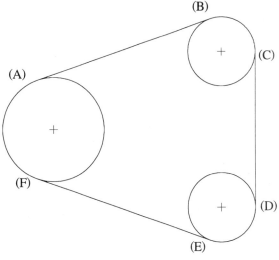

Figure 11–145

Step #3

Use the TRIM command to trim the circles. When prompted to select the cutting edge object, press the ENTER key; this will make cutting edges out of all objects in the drawing. See Figure 11–146.

 Command: **TRIM**

Select cutting edges: *(Projmode = UCS, Edgemode = No extend)*
Select objects: *(Press [ENTER] to create cutting edges out of all objects)*
<Select object to trim>/Project/Edge/Undo: *(Select the circle at "A")*
<Select object to trim>/Project/Edge/Undo: *(Select the circle at "B")*
<Select object to trim>/Project/Edge/Undo: *(Select the circle at "C")*
<Select object to trim>/Project/Edge/Undo: *(Press [ENTER] to exit this command)*

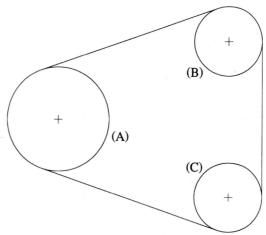

Figure 11–146

Step #4

Prepare to construct the bracket by viewing the object in 3D using the VPOINT command and the coordinates 1,-1,1. You may also select a view-point by picking the following sequence from the pull-down menu area: View > 3D Viewpoint > SE Isometric. See Figure 11–147.

Command: **VPOINT**
Rotate/<Viewpoint><0.00,0.00,1.00>:**1,-1,1**

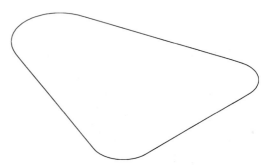

Figure 11–147

Step #5

Convert all objects into a polyline using the Join option of the PEDIT command. See Figure 11–148.

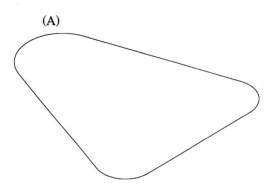

(A)

 Command:**PEDIT**

Select polyline: *(Select the arc at "A")*
Object selected is not a polyline
Do you want to turn it into one? <Y> *(Press*
   ENTER *to accept the default value)*
Close/Join/Width/Edit vertex/Fit/Spline/
   Decurve/Ltype gen/Undo/eXit <X>: **JOIN**
Select objects: *(Select the three lines and*
   remaining two arcs shown in Figure 11–148)
Select objects: *(Press* ENTER *to perform the*
   joining operation)
5 segments added to polyline
Open/Join/Width/Edit vertex/Fit/Spline/
   Decurve/Ltype gen/Undo/eXit <X>: *(Press*
   ENTER *to exit this command)*

Figure 11–148

Step #6

Use the EXTRUDE command to extrude the base to a thickness of 0.75 units. See Figure 11–149.

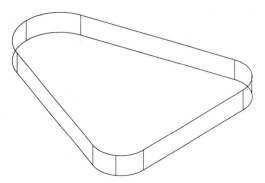

 Command:**EXTRUDE**

Select objects: *(Select the polyline)*
Select objects: *(Press* ENTER *to continue with this*
   command)
Path/<Height of Extrusion>: **0.75**
Extrusion taper angle <0>: *(Press* ENTER *to*
   perform the extrusion operation)

Figure 11–149

Step #7

Create a cylinder using the CYLINDER command.
Begin the center point of the cylinder at 0,0,0 with
a diameter of 3.00 units and a height of 2.75 units.
You may have to perform a ZOOM-All to display
the entire model. See Figure 11–150.

 Command:**CYLINDER**

Elliptical/<center point> <0,0,0>: *(Press* ENTER *to
    accept the default of 0,0,0)*
Diameter/<Radius>: **D** *(For Diameter)*
Diameter:**3**
Center of other end/<Height>:**2.75**

 Command:**ZOOM**

All/Center/Dynamic/Extents/Previous/Scale(X/
    XP)/Window/<Realtime>:**All**

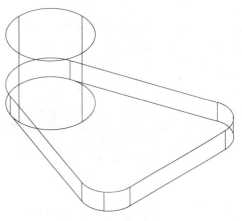

Figure 11–150

Step #8

Merge the cylinder just created with the extruded base
using the UNION command. See Figure 11–151.

 Command:**UNION**

Select objects: *(Select the extruded base and
    cylinder)*
Select objects: *(Press* ENTER *to perform the
    extrusion operation)*

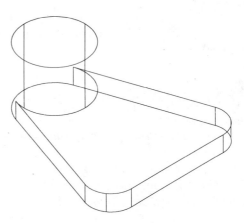

Figure 11–151

Step #9

Use the CYLINDER command to create a 2.00-unit diameter cylinder representing a through hole, as shown in Figure 11–152. The height of the cylinder is 2.75 units with the center point at 0,0,0.

 Command:**CYLINDER**

Elliptical/<center point> <0,0,0>: *(Press* ⟦ENTER⟧ *to accept the default of 0,0,0)*
Diameter/<Radius>:**D** *(For Diameter)*
Diameter:**2.00**
Center of other end/<Height>:**2.75**

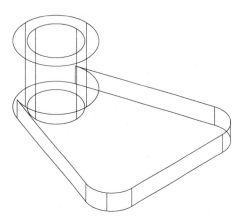

Figure 11–152

Step #10

To cut the hole through the outer cylinder, use the SUBTRACT command. Select the base as the source object; select the inner cylinder as the object to subtract. Use the HIDE command to view the results after performing a hidden line removal. Your display should appear similar to Figure 11–153.

 Command:**SUBTRACT**

Select solids and regions to subtract from...
Select objects: *(Select the base of the Collar)*
Select objects: *(Press* ⟦ENTER⟧ *to continue with this command)*
Select solids and regions to subtract...
Select objects: *(Select the 2.00 diameter cylinder just created)*
Select objects: *(Press* ⟦ENTER⟧ *to perform the subtraction operation)*

Command:**HIDE**

Command:**REGEN**
   (To return to wireframe mode)

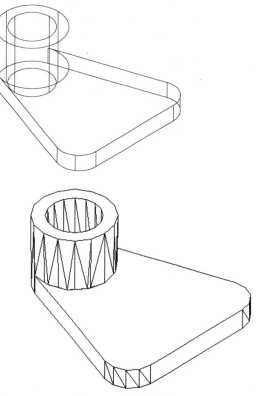

Figure 11–153

Step #11

Begin placing the two small drill holes in the base using the CYLINDER command. Use the OSNAP-Center mode to place each cylinder at the center of arcs "A" and "B" in Figure 11–154.

 Command: **CYLINDER**

Elliptical/<center point> <0,0,0>: **Cen**
of *(Pick the bottom edge of the arc at "A")*
Diameter/<Radius>: **D** *(For Diameter)*
Diameter: **1.00**
Center of other end/<Height>: **0.75**

 Command: **CYLINDER**

Elliptical/<center point> <0,0,0>: **Cen**
of *(Pick the bottom edge of the arc at "B")*
Diameter/<Radius>: **D** *(For Diameter)*
Diameter: **1.00**
Center of other end/<Height>: **0.75**

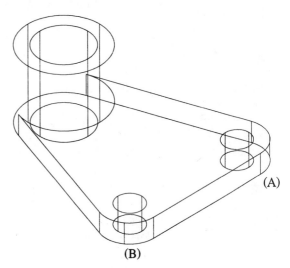

Figure 11–154

Step #12

Subtract both 1.00-diameter cylinders from the base of the model using the SUBTRACT command. Use the HIDE command to view the results after performing a hidden line removal. Your display should appear similar to Figure 11–155.

 Command: **SUBTRACT**

Select solids and regions to subtract from...
Select objects: *(Pick the base of the Collar)*
Select objects: *(Press ⏎ to continue with this command)*
Select solids and regions to subtract...
Select objects: *(Select both 1.00 diameter cylinders at the right)*
Select objects: *(Press ⏎ to perform the subtraction operation)*

Command: **HIDE**

Command: **REGEN**
(To return to wireframe mode)

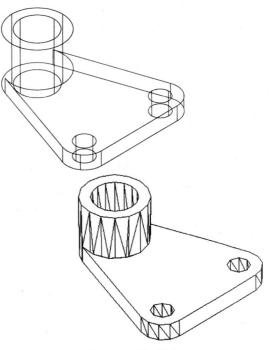

Figure 11–155

Step #13

Begin constructing the rectangular slot that will pass through the two cylinders (see Figure 11–156). Use the BOX command to accomplish this. Locate the center of the box at 0,0,0 and make the box 4 units long, 1 unit wide, and 0.75 units high. Then move the box to the top of the cylinder. Use the geometry calculator to select the base point of displacement at the top of the box with the MEE function. After locating two endpoints, the function will calculate the midpoint.

 Command:**MOVE**

Select objects: *(Select the box just constructed)*
Select objects: *(Press* ⌷ENTER⌷ *to continue with this command)*
Base point or displacement: **'CAL** *(To activate the geometry calculator)*
>> Expression: **MEE**
>> Select one endpoint for MEE: *(Pick a point on one corner of the box at "A")*
>> Select another endpoint for MEE: *(Pick a point on the opposite corner of the box at "B")*
Second point of displacement: **Cen**
of *(Pick the edge of the cylinder at "C")*

Step #14

Use the SUBTRACT command to subtract the rectangular box from the solid model, as in Figure 11–157.

 Command:**SUBTRACT**

Select solids and regions to subtract from...
Select objects: *(Select the model at "A")*
Select objects: *(Press* ⌷ENTER⌷ *to continue with this command)*
Select solids and regions to subtract...
Select objects: *(Select the rectangular box at "B")*
Select objects: *(Press* ⌷ENTER⌷ *to perform the subtraction operation)*

 Command:**BOX**

Center/<Corner of box> <0,0,0>: **C** *(For the center of the box)*
Center of box <0,0,0>: *(Press* ⌷ENTER⌷ *to accept the default center value of 0,0,0).*
Cube/Length/<corner of box>: **L**
  Length:**4**
  Width:**1**
  Height:**0.75**

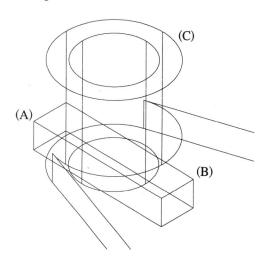

Figure 11–156

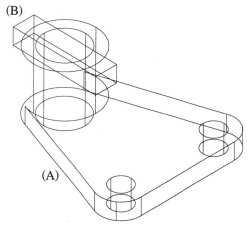

Figure 11–157

Step #15

The completed solid mode should appear similar to
Figure 11–158.

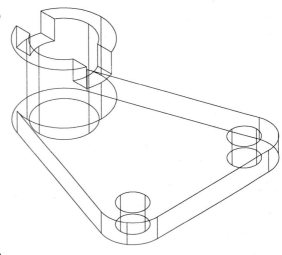

Step #16

Change the facet resolution to a higher value. Then
perform a hidden line removal to see the appearance of the model shown in Figure 11–159.

Figure 11–158

Command:**FACETRES**
New value for FACETRES <0.5000>:**1**

Command:**HIDE**

Command:**REGEN**
   (To return to wireframe mode)

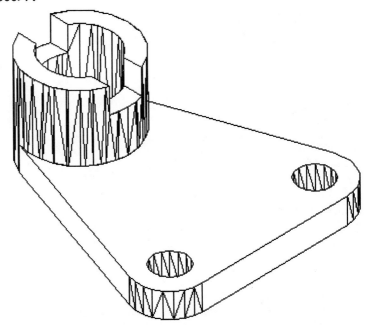

Figure 11–159

Step #17

Important information may be extracted from the solid model to be used for design and analysis. The following calculations are obtained when using the MASSPROP command. The following properties are calculated by this command: Mass, Volume, Bounding Box, Centroid, Moments of Inertia, Products of Inertia, Radii of Gyration, Principal Moments about Centroid.

 Command:**MASSPROP**

Select objects: *(Select the model of the Collar)*
Select objects: *(Press* ENTER *to perform the mass property calculation)*

————————SOLIDS————————

Mass:	29.11
Volume:	29.11
Bounding box:	X: -1.50–6.00
	Y: -3.25–3.25
	Z: 0.00–2.75
Centroid:	X: 2.36
	Y: 0.00
	Z: 0.69
Moments of inertia:	X: 82.92
	Y: 319.72
	Z: 349.84
Products of inertia:	XY: 0.00
	YZ: 0.00
	ZX: 25.72
Radii of gyration:	X: 1.69
	Y: 3.31
	Z: 3.47

Principal moments and X-Y-Z directions about centroid:
I: 65.09 along [0.98 0.00 -0.17]
J: 144.18 along [0.00 1.00 0.00]
K: 192.12 along [0.17 0.00 0.98]

Tutorial Exercise
Column.Dwg

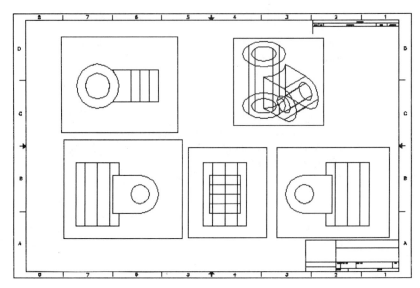

Figure 11–160

Purpose

This tutorial exercise is designed to produce a 3D solid model of the "Column" and then generate orthographic views of the solid using the SOLVIEW and SOLDRAW commands. See Figure 11–160.

System Settings

Begin a new drawing called "Column." Use the Units Control dialog box to change the number of decimal places past the zero from four to two. Keep all default values in the dialog box.

Layers

Create the following layers with the format:

Name	Color	Linetype
Model	Cyan	Continuous

Suggested Commands

Begin this by creating a solid model of the Column. Then, use the SOLVIEW command to lay out the front, top, Right Side, and Isometric views of the column. Next, draw the views in two dimensions using the SOLDRAW command. Dimensioning techniques will be explained. This exercise switches you from Model Space to Paper Space to accomplish the necessary tasks in building the solid model and extracting the views.

Whenever possible, substitute the appropriate command alias in place of the full AutoCAD command in each tutorial step. For example, use "Co" for the COPY command, "L" for the LINE command, and so on. The complete listing of all command aliases is located on pages 1–9 and 1–10.

PHASE I—Creating the Solid Model

Step #1

Choose the Advanced Setup Wizard from the Create New Drawing dialog box shown in Figure 11–161. Click on the Units Tab and change the precision to two decimal places. Next click on the Title Block Tab and for the Title Block Description, pick the ANSI-D(in) title block from the list. Finally click on the Layout Tab; be sure "Yes" is the current value set by the radio button for paper space. You will need advanced paper space capabilities to perform this drawing; Click on the radio button next to "Work on the layout of my drawing." This completes the initial drawing setup for the column. Click on the "Done" button and the display should be similar to Figure 11–162.

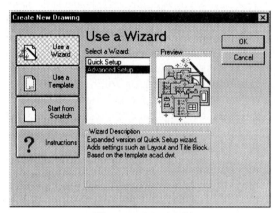

Figure 11–161

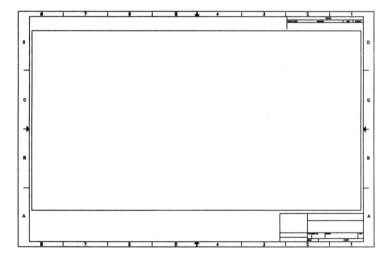

Figure 11–162

Step #2

Figure 11–162 shows the layout or paper space environment that will be used later in this tutorial exercise. Layers have been created to hold all title block and viewport information. If desired, change the "Title_Block" layer the color Magenta. Also notice in addition to the title block, a long rectangle is displayed inside of the title block. This is a viewport where drawing information will be located. However we need to make our own viewports and not rely on the one displayed as a result of using the

Advanced Wizard. Use the ERASE command to delete this viewport from the drawing. Viewports will be created later on in this tutorial.

 Command:**ERASE**

Select objects: *(Select the edge of the viewport at "A" in Figure 11–163)*
Select objects: *(Press* [ENTER] *to execute this command)*

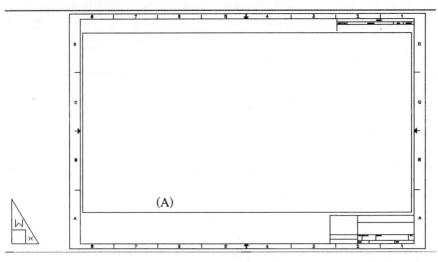

Figure 11–163

Step #3

Leave the paper space environment and enter model space. This area will be used to construct the model of the column before reentering paper space and beginning the layout phase of the drawing. Model space is entered by setting TILEMODE to a value of 1; or in the pull-down menu area, click on View

> Model Space(tiled) to enter the model space environment. Once in model space, create a new layer called MODEL if it has not been already created; make this layer current in the drawing. Then draw a circle 3.50 units in diameter to begin the construction of the large cylinder. See Figure 11–164.

Command:**TILEMODE**
New value for TILEMODE <0>: **1**

 Command:**-LAYER**

?/Make/Set/New/ON/OFF/Color/Ltype/Freeze/
   Thaw/LOck/Unlock: **M** *(To create and make
   the layer current)*
New current layer <0>: **MODEL**
?/Make/Set/New/ON/OFF/Color/Ltype/Freeze/
   Thaw/LOck/Unlock: **C** *(To assign a color to
   the layer MODEL)*
Color: **CYAN**
Layer name(s) for color 4 (cyan) <MODEL>:
   (Press ENTER *to accept this default layer name)*
?/Make/Set/New/ON/OFF/Color/Ltype/Freeze/
   Thaw/LOck/Unlock: *(Press* ENTER *to exit this
   command)*

 Command:**CIRCLE**

3P/2P/TTR/<Center point>:**5,5**
Diameter/<Radius>:**D**
Diameter:**3.5**

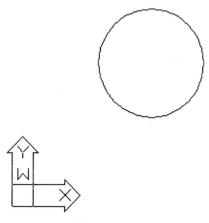

Figure 11–164

Step #4

Click on View in the pull-down menu area and pick 3D Viewpoint > SE Isometric to view the image in the South East viewpoint direction. If the image is too large on the screen, use the ZOOM command

 Command:**ZOOM**

All/Center/Dynamic/Extents/Previous/Scale(X/
XP)/Window/<Realtime>:**0.6**

 Command:**EXTRUDE**

Select objects: *(Select the circle)*
Select objects: *(Press* ENTER *to continue with this command)*
Path/<Height of Extrusion>:**5**
Extrusion taper angle <0>: *(Press* ENTER *to accept the default taper value and perform the extrusion operation)*

Step #5

Begin defining a number of user coodinate systems using the UCS command. Locate the new origin of the UCS at the center of the cylinder base (see Figure 11–166). If the UCS icon does not move to the new origin, use the UCSICON command and acti-

 Command:**UCS**

Origin/ZAxis/3point/OBject/View/X/Y/Z/Prev/
Restore/Save/Del/?/<World>:**O** *(For Origin)*
Origin point <0,0,0>: **Cen**
of *(Pick the edge of the bottom of the cylinder at "A")*

Command:**UCSICON**
ON/OFF/All/Noorigin/ORigin <ON>:**OR**

Command:**UCS**
Origin/ZAxis/3point/OBject/View/X/Y/Z/Prev/
Restore/Save/Del/?/<World>:**S** *(For Save)*
?/Desired UCS name:**BOTTOM**

to demagnify it at a value of 0.6 units (see Figure 11–165). Convert the circle into a cylinder using the EXTRUDE command.

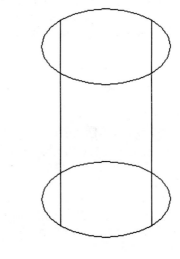

Figure 11–165

vate the Origin option; the UCS icon will now reflect the position of the new user coordinate system. It is also a good practice to save any newly created coordinate systems under a unique name; name this coordinate system "BOTTOM."

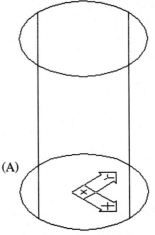

(A)

Figure 11–166

Step #6

Move the UCS icon 2.50 units in the Z direction using the UCS command (see Figure 11–167). Then rotate the UCS icon 90 degrees along the X-axis using the X-axis option of the UCS command. The purpose of this operation is to prepare for the construction of the extension lug coming off of the cylinder. Save this coordinate system as "LUG".

 Command: **UCS**

Origin/ZAxis/3point/OBject/View/X/Y/Z/Prev/
    Restore/Save/Del/?/<World>: **O** *(For Origin)*
Origin point <0,0,0>: **0,0,2.5**

 Command: **UCS**

Origin/ZAxis/3point/OBject/View/X/Y/Z/Prev/
    Restore/Save/Del/?/<World>: **X**
Rotation angle about X axis <0>: **90**

Command: **UCS**
Origin/ZAxis/3point/OBject/View/X/Y/Z/Prev/
    Restore/Save/Del/?/<World>: **S** *(For Save)*
?/Desired UCS name: **LUG**

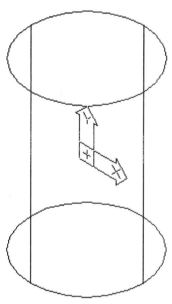

Figure 11–167

Step #7

Create a construction line from 0,0,0 and 3.50 units in the positive X direction. Then offset this line 1.50 units up and down using the OFFSET command. These lines represent the beginning of the extension lug. See Figure 11–168.

 Command: **LINE**

From point: **0,0,0** *(From "A")*
To point: **@3.50<0**
To point: *(Press* ENTER *to exit this command)*

 Command: **OFFSET**

Offset distance or Through <Through>: **1.50**
Select object to offset: *(Select the line at "B")*
Side to offset? *(Pick at "C")*
Select object to offset: *(Select the line at "B")*
Side to offset? *(Pick at "D")*
Select object to offset: *(Press* ENTER *to exit this
    command)*

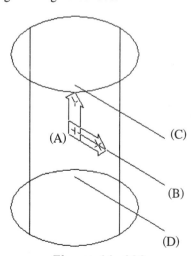

Figure 11–168

Step #8

Use the ERASE command to delete line "B" in Figure 11–168. Then use the FILLET command to cap the endpoints of lines "C" and "D" with an arc, as shown in Figure 11–169. Even if the fillet radius is set to a value, the command will determine the radius based on the distance between the opposite endpoints of two parallel lines. Use the LINE command and connect endpoints "E" and "F".

 Command:**ERASE**

Select objects: *(Select line "B" in Figure 11–168)*
Select objects: *(Press* ⸢ENTER⸣ *to execute this command)*

 Command:**FILLET**

(TRIM mode) Current fillet radius = 0.50
Polyline/Radius/Trim/<Select first object>:
   (Pick the endpoint at "C")
Select second object: *(Pick the endpoint at "D")*

 Command:**LINE**

From point: **Endp**
of *(Pick the endpoint of the line at "E")*
To point: **Endp**
of *(Pick the endpoint of the line at "F")*
To point: *(Press* ⸢ENTER⸣ *to exit this command)*

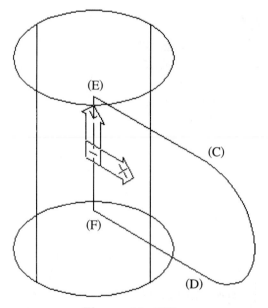

Figure 11–169

Step #9

While in the current user coordinate system (LUG), use the PEDIT command to convert the three line segments and arc segment into a single polyline (see Figure 11–170A). This polyline will be used to create an extrusion representing the lug in a later step. Once the four objects have been converted into a polyline, move the object -1.25 units in the Z direction using the MOVE command. The 1.25 units represents one-half of the total thickness (2.50) of the lug.

 Command:**PEDIT**

Select polyline: *(Pick the line at "A")*
Object selected is not a polyline
Do you want to turn it into one? <Y> *(Press*
  [ENTER] *to turn the line into a polyline)*
Close/Join/Width/Edit vertex/Fit/Spline/
  Decurve/Ltype gen/Undo/eXit <X>:**J**
Select objects: *(Pick the remaining two lines*
  and arc segment)
Select objects: *(Press* [ENTER] *to perform the*
  conversion operation)
3 segments added to polyline
Open/Join/Width/Edit vertex/Fit/Spline/
  Decurve/Ltype gen/Undo/eXit <X>: *(Press*
  [ENTER] *to exit this command)*

 Command:**MOVE**

Select objects: **L** *(for Last; This should*
  highlight the polyline object)
Select objects: *(Press* [ENTER] *to continue with this*
  commmand)
Base point or displacement: **0,0,0**
Second point of displacement: **0,0,-1.25**

Your image should be similar to Fig 11–170B.

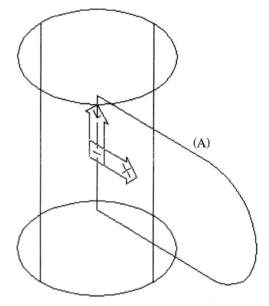

(A)

Figure 11–170A

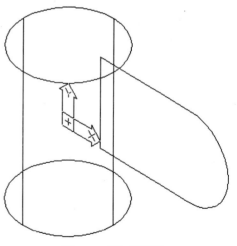

Figure 11–170B

Step #10

Use the EXTRUDE command to extrude the polyline outline representing the lug a distance of 2.50 units. See Figure 11–171.

  Command:**EXTRUDE**

Select objects: *(Pick the polyline outline of the lug at "A")*
Select objects: *(Press* ENTER *to continue with this command)*
Path/<Height of Extrusion>:**2.50**
Extrusion taper angle <0>: *(Press* ENTER *to perform the extrusion operation).*

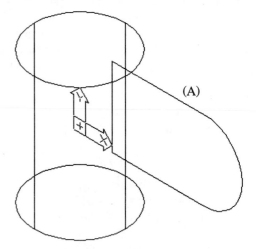

Figure 11–171

Step #11

Use the UNION command to join the lug and cylinder to form a single solid object, as shown in Figure 11–172.

  Command:**UNION**

Select objects: *(Pick the cylinder at "A" and the solid object of the lug at "B")*
Select objects: *(Press* ENTER *to perform the union operation)*

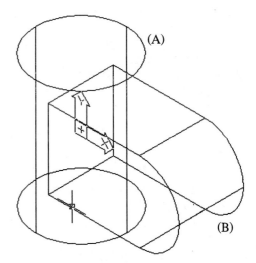

Figure 11–172

Step #12

While in the current user coordinate system, construct a cylinder from the back edge of the lug at a height of 2.50 units (see Figure 11–173). Use the SUBTRACT command to subtract the cylinder from the solid object to create a hole.

 Command:**CYLINDER**

Elliptical/<center point> <0,0,0>:**Cen**
of *(Pick the edge of the arc at "A")*
Diameter/<Radius>:**D**
Diameter:**1.50**
Center of other end/<Height>:**2.50**

 Command:**SUBTRACT**

Select solids and regions to subtract from…
Select objects: *(Select the solid object at "B")*
Select objects: *(Press* ⟨ENTER⟩ *to continue with this command)*
Select solids and regions to subtract…
Select objects: *(Select the small cylinder at "C")*
Select objects: *(Press* ⟨ENTER⟩ *to perform the subtraction operation)*

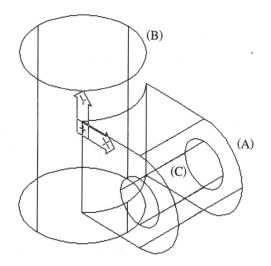

Figure 11–173

Step #13

Restore the previously saved User Coodinate System called "BOTTOM." Construct another cylinder from the bottom of the large cylinder, as shown in Figure 11–174. Create the cylinder with a diameter of 2.00 units and a height of 5.00 units. Sub-

tract the cylinder from the main object. The column is completed.

Continue with this tutorial to create 2D orthographic views from the solid model.

 Command:**UCS**

Origin/ZAxis/3point/OBject/View/X/Y/Z/Prev/
   Restore/Save/Del/?/<World>:**R** *(For
   Restore)*
?/Name of UCS to restore:**BOTTOM**

 Command:**CYLINDER**

Elliptical/<center point> <0,0,0>: *(Press* [ENTER] *to
   accept the default center point location)*
Diameter/<Radius>:**D** *(For Diameter)*
Diameter:**2.00**
Center of other end/<Height>:**5.00**

 Command:**SUBTRACT**

Select solids and regions to subtract from…
Select objects: *(Select the solid object at "A")*
Select objects: *(Press* [ENTER] *to continue with this
   command)*
Select solids and regions to subtract…
Select objects: *(Pick the cylinder at "B")*
Select objects: *(Press* [ENTER] *to perform the
   subtraction operation)*

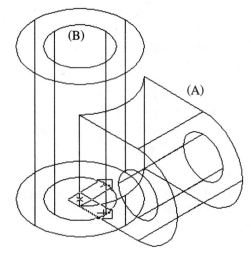

Figure 11–174

Phase II—Extracting the Orthographic Views

With the completed model of the Column, begin the extraction process which will create the orthographic views of Front, Top, and Right Side in addition to a Section view and Isometric View. After all views are extracted, drawn, sectioned, and dimensioned, the completed drawing will appear similar to Figure 11–175.

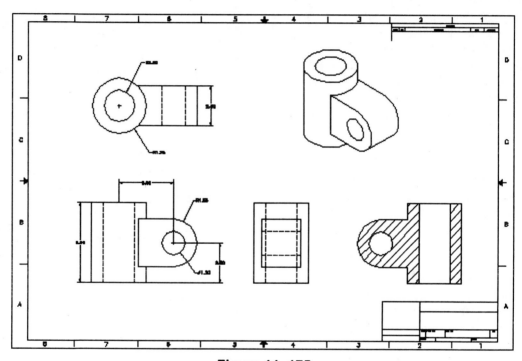

Figure 11–175

Step #1

From Figure 11–174, note the position of the user coordinate system. This location will begin the extraction process to create the initial orthographic view. All other views will be based on the initial view. To begin, enter the paper space environment by setting TILEMODE to 0. Once inside of paper space, bring up the Layer and Linetype Properties dialog box. Delete the layer called VIEWPORT; this layer was created when beginning the drawing in Step #1. A different layer will be created shortly to hold all viewport information. Next, to further prepare for the extraction and drawing of the orthographic views, load the HIDDEN linetype. You may also make color assignments to various layers.

Command: **TILEMODE**
New value for TILEMODE <1>: **0**

In the Layer and Linetype Properties dialog box, delete the VIEWPORT layer.

Load the HIDDEN linetype.

Click OK to save the changes to the Layer and Linetype Properties dialog box.

Your layer display should appear similar to Figure 11–176.

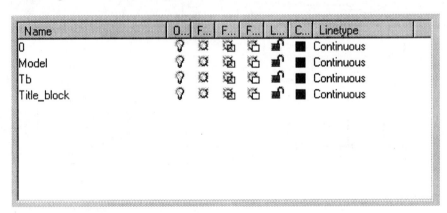

Figure 11–176

Step #2

While inside of paper space, use the SOLVIEW command to begin the extraction process. This command will allow you to build an initial view based on the current position of the user coordinate system. Once you locate the view on the drawing, you may change the scale of the view, give the view a name, and construct a viewport around the view. When constructing the viewport, take care to make it large enough to accommodate the dimensions that need to be added later on to the view. See Figure 11–177 and the following prompts.

 Command:**SOLVIEW**

Ucs/Ortho/Auxiliary/Section/<eXit>:**U** *(To base the view on the current user coordinate system)*

Named/World/?/<Current>: *(Press* ENTER *to accept the <Current> default)*
Enter view scale<1.00>: *(Press* ENTER *to accept the default scale value)*
View center: *(Locate the image approximately in the upper left corner of the border area)*
View center: *(Press* ENTER *to place the view and continue with the command)*
Clip first corner: *(Pick a point approximately at "A")*
Clip other corner: *(Pick a point approximately at "B")*
View name: **TOP**
Ucs/Ortho/Auxiliary/Section/<eXit>:*(Press* ENTER *to exit this command)*

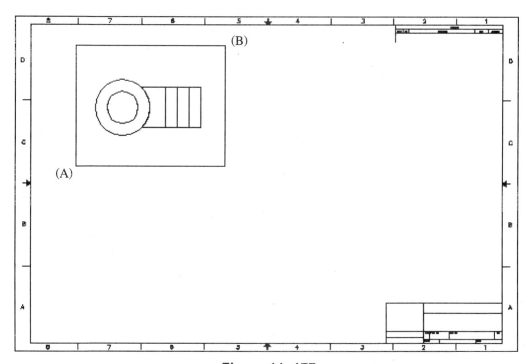

Figure 11–177

Step #3

Before continuing with the SOLVIEW command, view the layers at this point through the Layer and Linetype Properties dialog box in Figure 11–178. Three layers were automatically created as a result of laying out the first view using SOLVIEW; namely, TOP-DIM, TOP-HID, and TOP-VIS. TOP-DIM will hold all dimension information located in the top view. TOP-HID will hold all hidden line information for the top view, and TOP-VIS will hold all visible line information in the top view. These three layers are automatically visible in the current viewport and automatically frozen in other viewports; the presence of the snowflake symbol states this.

Now return to the SOLVIEW command and lay out the remaining orthographic views.

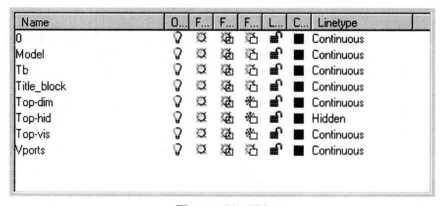

Figure 11–178

Step #4

Use SOLVIEW and the Ortho option to create the front view based on the initial view already placed (see Figure 11–179). As the new view is being located, Ortho mode is automatically turned on to keep both views lined up with each other in the same orientation. Also, the OSNAP-Midpoint mode is activated allowing you to select the midpoint of the viewports edge and project the new view down. Create a viewport by following the prompts to clip a first corner and other corner. Again, you will add dimensions to this view later on so make the viewport a good size. Name the view FRONT and continue on by laying out the R_SIDE view in using the same prompt sequence. Create a small viewport since no dimensions will be placed on this view.

 Command:**SOLVIEW**

Ucs/Ortho/Auxiliary/Section/<eXit>:**O** *(To create an orthogonal view)*

Pick side of viewport to project: *(Pick the viewport at "A")*

View center: *(Pick a point below to place the view)*

View center: *(Press* ENTER *to place the view and continue with the command)*

Clip first corner: *(Pick a point near "B")*

Clip other corner: *(Pick a point near "C")*

View name:**FRONT**

Ucs/Ortho/Auxiliary/Section/<eXit>:**O** *(For ortho mode)*

Pick side of viewport to project: *(Pick the viewport at "D")*

View center: *(Pick a point at the right to place the view)*

View center: *(Press* ENTER *to place the view and continue with the command)*

Clip first corner: *(Pick a point near "E")*

Clip other corner: *(Pick a point near "F")*

View name:**R_SIDE**

Ucs/Ortho/Auxiliary/Section/<eXit>:*(Press* ENTER *to exit this command)*

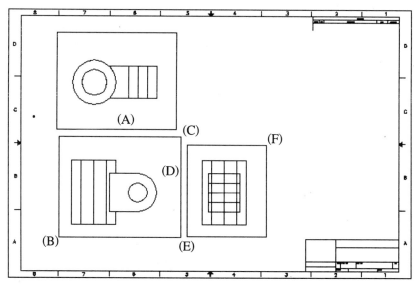

Figure 11–179

Step #5

To demonstrate the flexibility of the SOLVIEW command, a section view will be created by constructing a cutting plane line in the R_SIDE view (see Figure 11–180). Be sure the R_SIDE view is the active viewport when creating this section.

 Command:**SOLVIEW**

Ucs/Ortho/Auxiliary/Section/<eXit>:**S** *(For creating a section view)*
Cutting Plane's 1st point: **Mid**
of *(Pick the midpoint at "A" in the R_SIDE view)*
Cutting Plane's 2nd point: **Mid**
of *(Pick the midpoint at "B" in the R_SIDE view)*

Side to view from: *(Pick a point near "C" in the R_SIDE view)*
Enter view scale<1.00>: *(Press* ENTER *to accept this default value)*
View center: *(Pick a point at a convenient location to the right of the R_SIDE view)*
View center: *(Press* ENTER *to place the view and continue with the command)*
Clip first corner: *(Pick a point near "D")*
Clip other corner: *(Pick a point near "E")*
View name: **SECTION**
Ucs/Ortho/Auxiliary/Section/<eXit>:*(Press* ENTER *to exit this command)*

The drawing should appear similar to Figure 11–180.

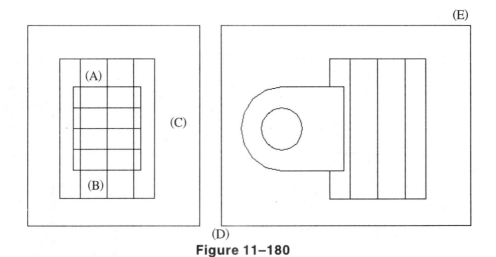

Figure 11–180

Step #6

To create a projection of an isometric view, return to the solid model; this is accomplished back in model space by setting TILEMODE to 1. Then use the UCS command to create a new user coorinate system based on the current view. See Figure 11–181.

Command:**TILEMODE**
New value for TILEMODE <0>: **1**

 Command: **UCS**
Origin/ZAxis/3point/OBject/View/X/Y/Z/Prev/
    Restore/Save/Del/?/<World>: **V** *(For View)*

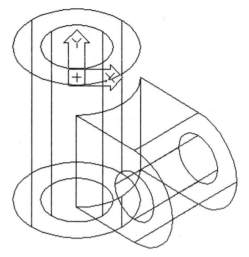

Figure 11–181

Step #7

Return back to paper space by setting TILEMODE back to 0. Issue the SOLVIEW command and use the UCS option to layout the isometric view of the column based on the current user coordinate system. See Figure 11–182.

Command:**TILEMODE**
New value for TILEMODE <1>: **0**

 Command:**SOLVIEW**
Ucs/Ortho/Auxiliary/Section/<eXit>:**U**

Named/World/?/<Current>: *(Press* ENTER *to accept the current position of the user coordinate system)*
Enter view scale<1.00>: *(Press* ENTER *to accept this default value)*
View center: *(Locate the isometric view in the upper right corner of the title block area)*
View center: *(Press* ENTER *to place the view and continue with the command)*
Clip first corner: *(Pick a point near "A")*
Clip other corner: *(Pick a point near "B")*
View name: **ISO**
Ucs/Ortho/Auxiliary/Section/<eXit>:*(Press* ENTER *to exit this command)*

All views have been successfully laid out. However none of the orthographic views show hidden features defined by hidden lines. Also, the section view does not show section lines to illustrate which surfaces were cut by the imaginary cutting plane line.

The SOLVIEW command is used only as a lay out tool. To actually draw the view complete with visible, object, and section lines, use the SOLDRAW command. This will be covered in Phase III of this project.

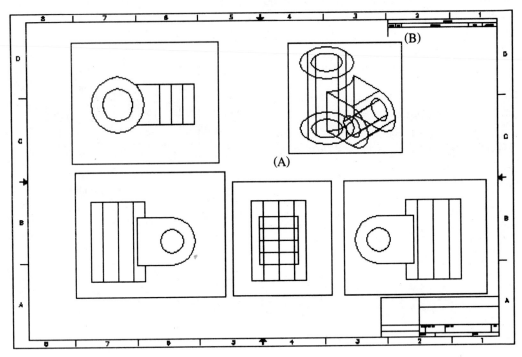

Figure 11–182

PHASE III—Using SOLDRAW to Construct the 2D Views

Step #1

The SOLDRAW command is designed to follow the SOLVIEW command. SOLVIEW is used as a layout tool; it allows you to position the Top, Front, R_side, Section, and Isometric views in the paper space environment. See Figure 11–183. SOLDRAW is used to draw object, hidden, and section lines. Dimensions will be placed in the drawing during another phase.

 Command:**SOLDRAW**
Select viewports to draw:
Select objects: *(Select the edges of the five viewports to draw the orthogonal views)*
Select objects: *(Press* [ENTER] *to perform the SOLDRAW operation)*

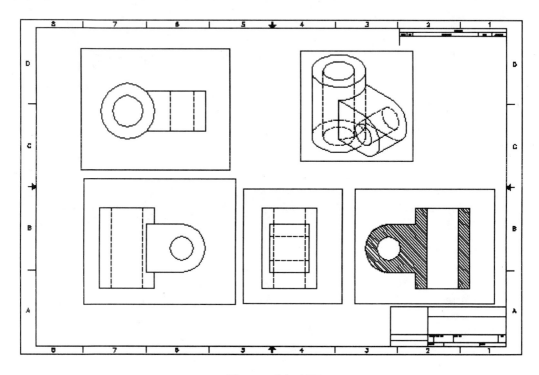

Figure 11–183

Step #2

Notice the isometric view in Figure 11–184A has been drawn complete with hidden lines; these lines need to be made invisible in the isometric view. To accomplish this, first enter floating model space. Click inside of the viewport that holds the drawing of the isometric view in Figure 11–184A; this makes the viewport current. Then issue the LAYER command and freeze the layer containing the hidden lines just in the Isometric view in Figure 11–184B

Command: **MS** *(To enter model space)*

Activate the viewport holding the Isometric view.

 Command:**LAYER**

While inside of the Layer and Linetype Properties dialog box, locate the layer called "ISO-HID" and freeze this layer in the current viewport in the illustration provided.

Figure 11–184A

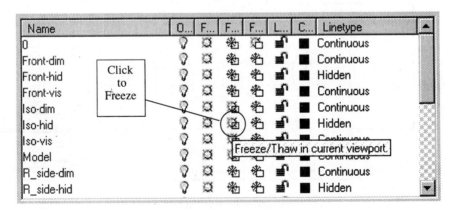

Figure 11–184B

Step #3

While still in Model Space, activate the viewport holding the section view information. The cross-hatching looks too dense and needs to be scaled up in size using the Hatchedit dialog box in Figure 11–185.

(Click in the viewport that contains the section view)

Command:**HATCHEDIT**
Select hatch object: *(Select the hatch object located in the viewport in Figure 11–186)*

In the Scale: area of the Hatchedit dialog box, change the hatch pattern scale to a new value of 3 units. Click the Apply button to change the scale of the hatch pattern.

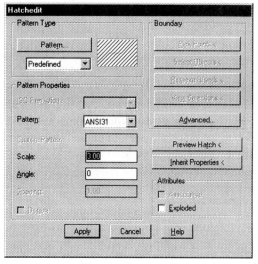

Figure 11–185

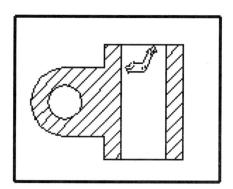

Figure 11–186

PHASE IV—Adding Dimensions to the Views

Step #1

Begin adding dimensions to the Front View; do this while still in model space. First, click in the viewport containing the front view information. Then, issue the UCS command along with the View option. This will line up the user coordinate system icon parallel to the display screen. This will enable the dimensions to be placed as in a normal 2D drawing. Once the icon has been changed, make the FRONT-DIM layer current and begin placing the dimensions in the front viewport. As you place the dimensions, notice they do not appear in other viewports. The SOLVIEW command automatically creates dimension layers and freezes the layers in the viewports that they do not apply in.

(Click in the viewport at "A" in Figure 11–187 that contains the front view)

 Command: **UCS**
Origin/ZAxis/3point/OBject/View/X/Y/Z/Prev/
   Restore/Save/Del/?/<World>: **V** *(For View)*

 Command: **-LAYER**
?/Make/Set/New/ON/OFF/Color/Ltype/Freeze/
   Thaw/LOck/Unlock: **S** *(To set a layer
   current)*
New current layer <MODEL>: **FRONT-DIM**
?/Make/Set/New/ON/OFF/Color/Ltype/Freeze/
   Thaw/LOck/Unlock: *(Press ENTER to make the
   FRONT-DIM layer current)*

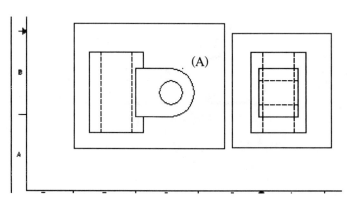

Figure 11-187

Step #2

Begin adding dimensions to the front view. Before performing this operation, lock yourself into the OSNAP-Intersection mode since numerous intersections will be required in dimensioning. Enter the Dimension Styles dialog box (DDIM), click on the Annotation button, click on the Units button, and change the decimal precision to two places. With the front viewport active and the user coordinate system set to the current view, begin placing the linear, radius, and diameter dimensions. See Figure 11–188.

 Command:**DIMLIN**

First extension line origin or press ENTER to
   select: *(Select the intersection at "A")*
Second extension line origin: *(Select the*
   intersection at "B")
Dimension line location (Mtext/Text/Angle/
   Horizontal/Vertical/Rotated): *(Pick at "C")*
Dimension text = 5.00

 Command:**DIMLIN**

First extension line origin or press ENTER to
   select: *(Select the intersection at "D")*
Second extension line origin: **Cen**
of *(Select the center at "E")*

Dimension line location (Mtext/Text/Angle/
   Horizontal/Vertical/Rotated): *(Pick at "F")*
Dimension text = 2.50

 Command:**DIMLIN**

First extension line origin or press ENTER to
   select: **Mid**
of *(Select the midpoint at "G")*
Second extension line origin: **Cen**
of *(Select the center at "E")*
Dimension line location (Mtext/Text/Angle/
   Horizontal/Vertical/Rotated): *(Pick at "H")*
Dimension text = 3.50

 Command:**DIMRAD**

Select arc or circle: *(Select at "E")*
Dimension text = 1.50
Dimension line location (Mtext/Text/Angle):
   (Select at "I")

 Command:**DIMDIA**

Select arc or circle: *(Select at "J")*
Dimension text = 1.50
Dimension line location (Mtext/Text/Angle):
(Select at "K")

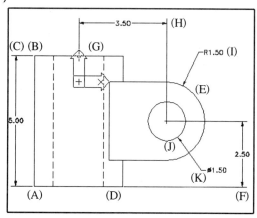

Figure 11–188

Step #3

Begin adding dimensions to the Top View; do this while still in model space. First, click in the viewport containing the top view information. Then, issue the UCS command along with the View option. This will position the user coordinate system icon parallel to the display screen. This will also enable the dimensions to be placed as in a normal 2D drawing. Once the icon has been aligned to the view, make the TOP-DIM layer current and begin placing the dimensions in the top viewport. As the dimensions are being placed, notice they do not appear in other viewports. The SOLVIEW command automatically creates dimension layers and freezes the layers in the viewports that they do not apply in.

(Click in the viewport at "A" in Figure 11–189 that contains the top view)

 Command: **UCS**
Origin/ZAxis/3point/OBject/View/X/Y/Z/Prev/
   Restore/Save/Del/?/<World>: **V** *(For View)*

 Command: **-LAYER**
?/Make/Set/New/ON/OFF/Color/Ltype/Freeze/
   Thaw/LOck/Unlock: **S** *(To set a layer*
   current)
New current layer <MODEL>: **TOP-DIM**
?/Make/Set/New/ON/OFF/Color/Ltype/Freeze/
   Thaw/LOck/Unlock: *(Press* ENTER *to make the*
   TOP-DIM layer current)

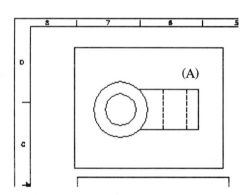

Figure 11–189

Step #4

Begin adding linear, radius, and diameter dimensions to the top view. The OSNAP-Intersection mode should still be active or running. See Figure 11–190.

 Command: **DIMLIN**

First extension line origin or press ENTER to select: *(Select the intersection at "A")*
Second extension line origin: *(Select the intersection at "B")*
Dimension line location (Mtext/Text/Angle/ Horizontal/Vertical/Rotated): *(Pick at "C")*
Dimension text = 2.50

 Command: **DIMRAD**

Select arc or circle: *(Select at "D")*
Dimension text = 1.75
Dimension line location (Mtext/Text/Angle):
  (Locate the radius dimension at "E")

 Command: **DIMDIA**

Select arc or circle: *(Select at "F")*
Dimension text = 2.00
Dimension line location (Mtext/Text/Angle):
  (Locate the diameter dimension at "G")

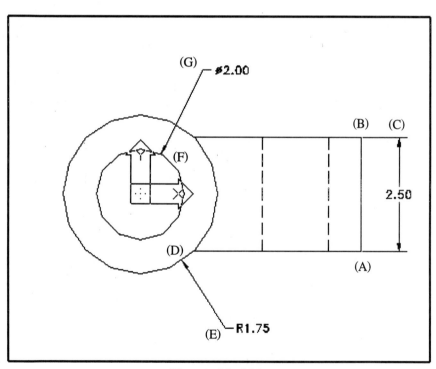

Figure 11–190

Step #5

Switch back to paper space. Turn off viewports and engineering drawing is complete (see Figure 11–191). Centerlines could be added to all views. As this exercise seemed to take a while to complete, use the steps as guides for extracting the views of other solid models created in AutoCAD.

Command: **PS** *(For Paper Space)*

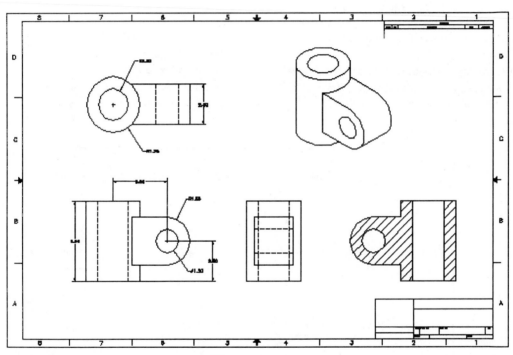

Figure 11–191

Problems for Unit 11

Directions for Problems 11-1
> *1. Create a 3D solid model of each object on layer "Model."*
> *2. When completed, calculate the volume of the solid model using the MASSPROP command.*
> *3. Using the Tutorial Exercise of Column.Dwg on page 11–92 as a guide, create an engineering drawing of each model consisting of front, top, right side, and isometric views.*
> *4. Properly dimension the engineering drawing.*

Problem 11–1

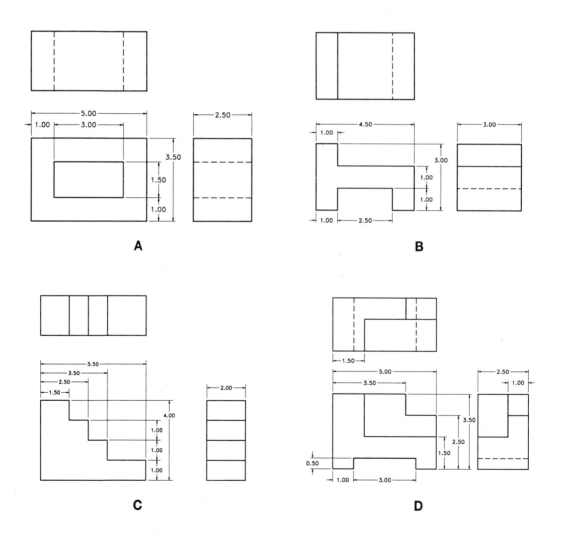

Problem 11–1 *(continued)*

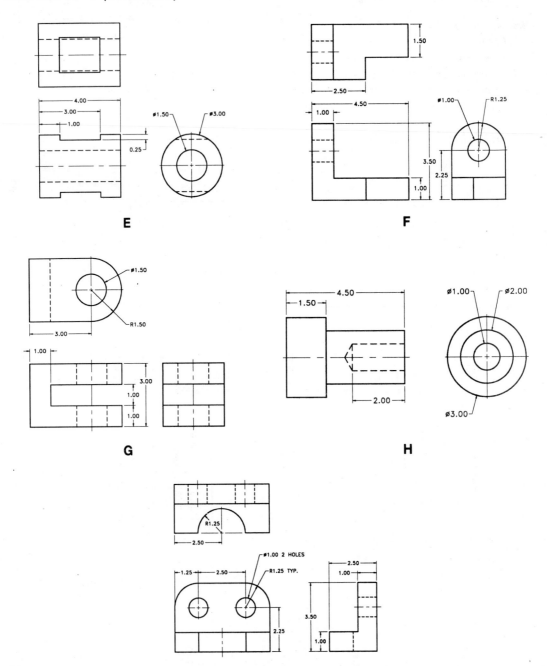

Directions for Problems 11–2 through 11–30
1. *Create a 3D solid model of each object on Layer "Model."*
2. *When completed, calculate the volume of the solid model using the MASSPROP command.*
3. *Using the Tutorial Exercise of Column.Dwg on page 11–92 as a guide, create an engineering drawing of each model consisting of front, top, right side, and isometric views.*
4. *Properly dimension the engineering drawing.*

Problem 11–2

Plate Thickness of 0.50

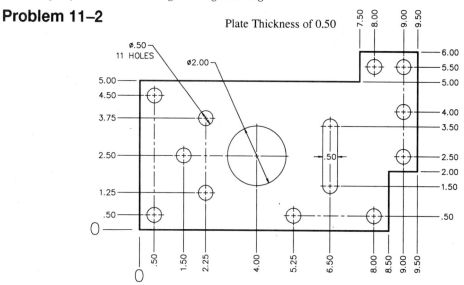

Problem 11–3

Plate Thickness of 0.75

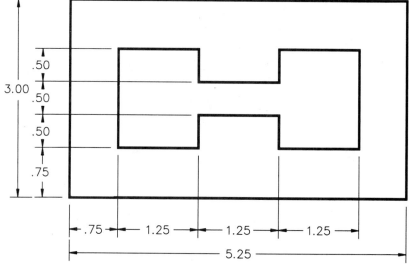

Problem 11–4

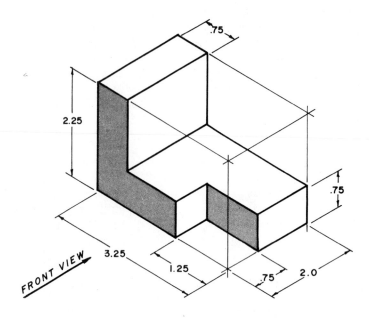

.75

2.25

.75

FRONT VIEW   3.25

1.25

.75   2.0

Problem 11–5

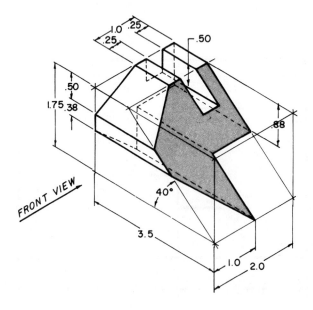

1.0 .25

.25

.50

.50

1.75 .38

.88

FRONT VIEW

40°

3.5

1.0   2.0

Problem 11–6

Problem 11–7

Problem 11–8

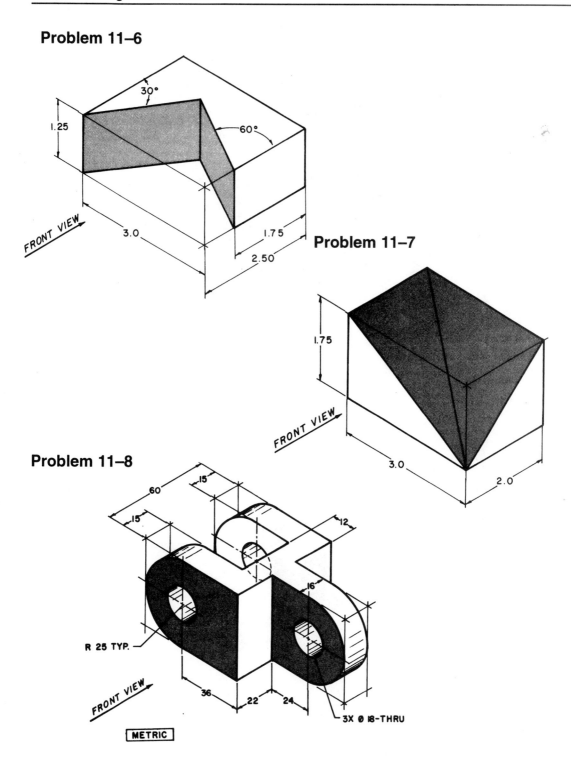

Problem 11–9

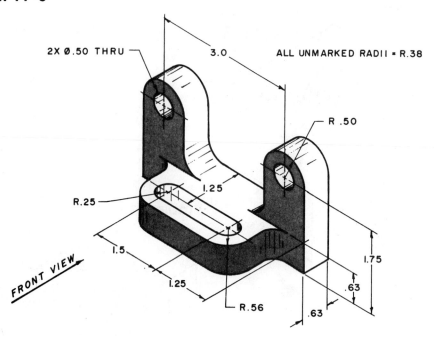

2X Ø.50 THRU

3.0

ALL UNMARKED RADII = R.38

R .50

1.25

R.25

1.5

1.25

R.56

.63

.63

1.75

FRONT VIEW

Problem 11–10

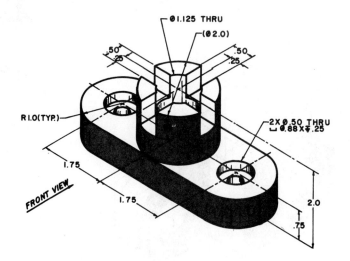

Ø1.125 THRU

(Ø 2.0)

.50
.25

.50
.25

R 1.0 (TYP.)

2X Ø.50 THRU
⌴ Ø.88 X ⌵ .25

1.75

1.75

2.0

.75

FRONT VIEW

Problem 11–11

FRONT VIEW

Problem 11–12

FRONT VIEW

R .25 (TYP)

Problem 11–13

Ø 25 - THRU

FRONT VIEW

METRIC

Problem 11–14

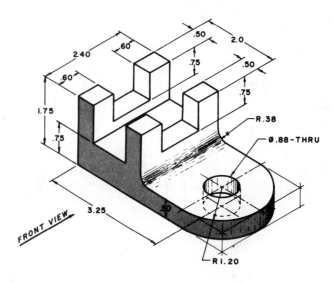

Problem 11–15

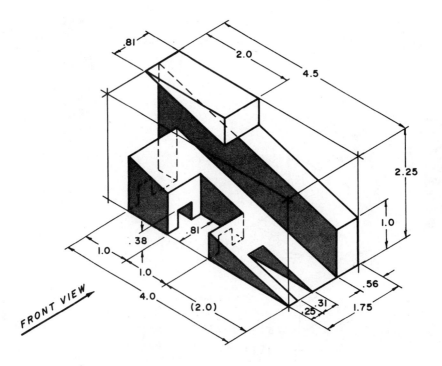

Problem 11–16

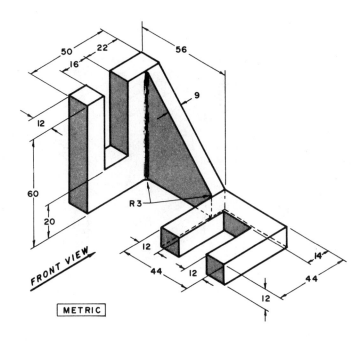

FRONT VIEW

METRIC

Problem 11–17

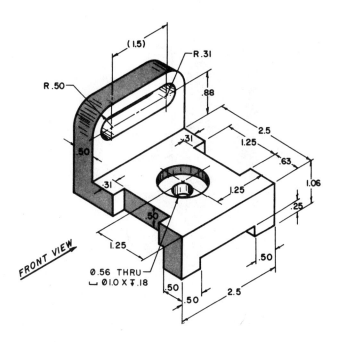

FRONT VIEW

Problem 11–18

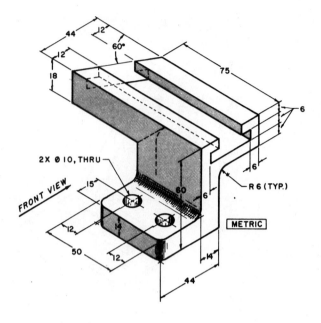

Problem 11–19

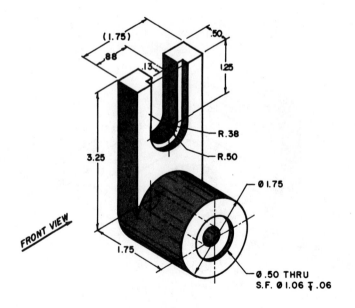

Problem 11–20

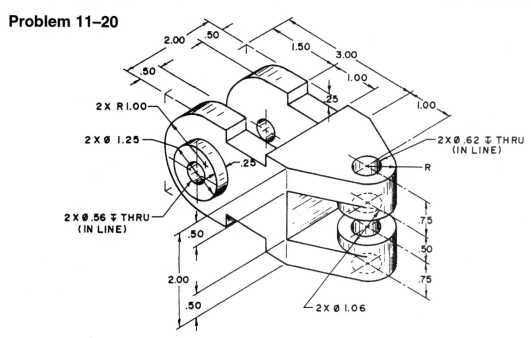

Problem 11–21

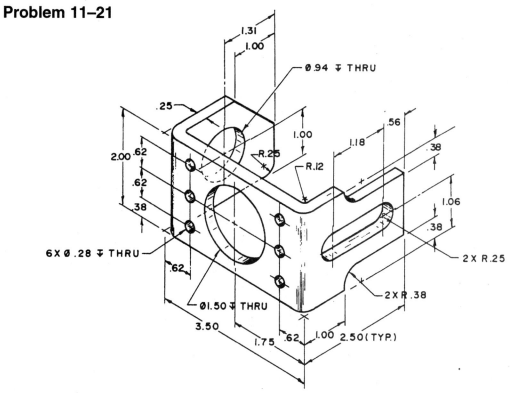

Fillet Radius Values at "A" and "B" = 0.25

Problem 11–22

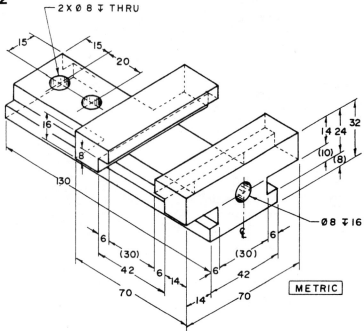

Problem 11–23

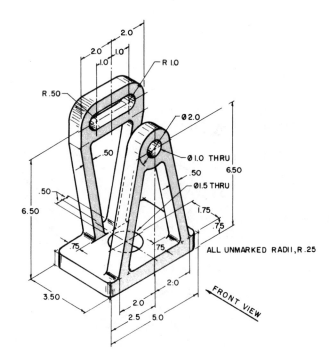

Problem 11–24

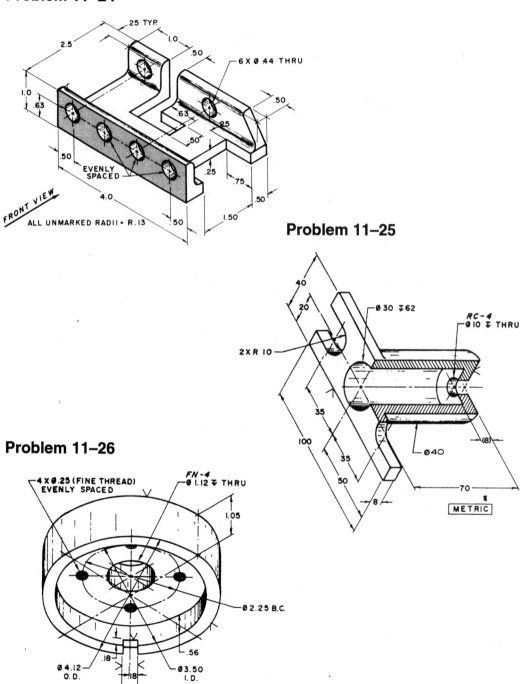

.25 TYP.

2.5

1.0

.50

6 X Ø .44 THRU

1.0

.63

.50

.63

.25

.50

.50

EVENLY
SPACED

.25

.75

.50

4.0

FRONT VIEW

ALL UNMARKED RADII = R.13

.50

.50

1.50

Problem 11–25

40

20

Ø 30 ∓.62

RC-4
Ø10 ∓ THRU

2 X R 10

35

100

35

50

Ø40

(8)

8

70

METRIC

Problem 11–26

4 X Ø.25 (FINE THREAD)
EVENLY SPACED

FN-4
Ø 1.12 ∓ THRU

1.05

Ø 2.25 B.C.

.18

.56

Ø 4.12
O.D.

.18

Ø 3.50
I.D.

.36

Problem 11–27

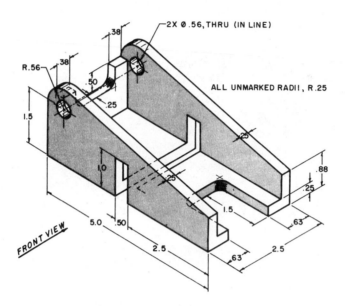

2X Ø.56,THRU (IN LINE)

R.56

38

38

.50

.25

R.56

1.5

ALL UNMARKED RADII, R.25

.25

1.0

.25

.88

.25

1.5

FRONT VIEW

5.0   .50

2.5

.63

.63   2.5

Problem 11–28

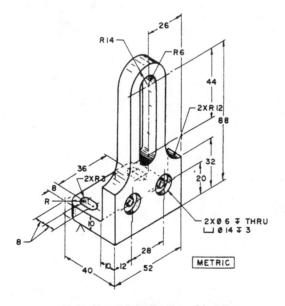

26

R14

R6

44

2XR12

88

32

20

36

2XR3

8

R

10

2XØ6 ⫪ THRU
⊔ Ø14 ⫪ 3

8

28

40   10   12   52

METRIC

Note: Use a Fillet Radius of 2 units
for the inside unmarked radius.

Problem 11–29

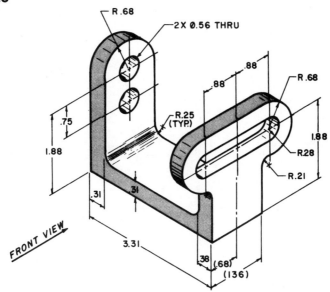

Problem 11–30

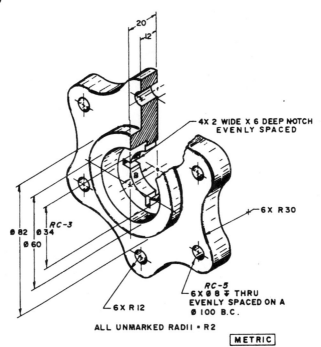

4X 2 WIDE X 6 DEEP NOTCH
EVENLY SPACED

6X R30

6X R12

RC-3

RC-5
6X Ø 8 ⊽ THRU
EVENLY SPACED ON A
Ø 100 B.C.

Ø 82
Ø 34
Ø 60

ALL UNMARKED RADII = R2

METRIC

A

Figure A–1

Figure A–2

Bonus Commands

Figure A–3

Performing a Full installation of AutoCAD Release 14 displays three additional toolbars designed to speed up your drafting and design tasks. The Bonus Text Tools toolbar is illustrated in Figure A–1. A second toolbar is the Bonus Standard Toolbar in Figure A–2. The third toolbar consists of Bonus Layer Tools in Figure A–3. These tools are designed to act as productivity tools to provide extra functionality to the existing AutoCAD commands already provided in Release 14. The bonus routines consist mainly of AutoLISP programs; other powerful applications are ARX programs. Since you can select most commands from the three toolbars, additional tools are available in the Bonus pull-down menu area. A few bonus routines are accessible only from the keyboard.

Follow the next series of pages for a detailed description of most bonus routines provided in AutoCAD Release 14.

The Bonus Standard Toolbar

The Bonus Standard Toolbar consists mainly of a series of Modify and Drawing programs. Modification programs include the ability to trim or extend to a block, creating individual multiple polylines using a multiple polyline edit command, an extended change properties command, a command that has move, copy, and rotate built into a single command, an extended or "cookie cutter" trim command, and a command that allows you to stretch a multiple number of objects using a single command. You can find these commands in the Bonus Standard Toolbar as shown in Figure A–4, and also in the Bonus pull-down menu area under "Modify".

Drawing programs include the ability to create a revision cloud and a very powerful quick leader command. A few utility commands are available to list block objects in the current drawing and you can list all files associated with a drawing file to share the drawing and support files with another individual. You can also locate the drawing programs in the Bonus pull-down menu area under "Draw."

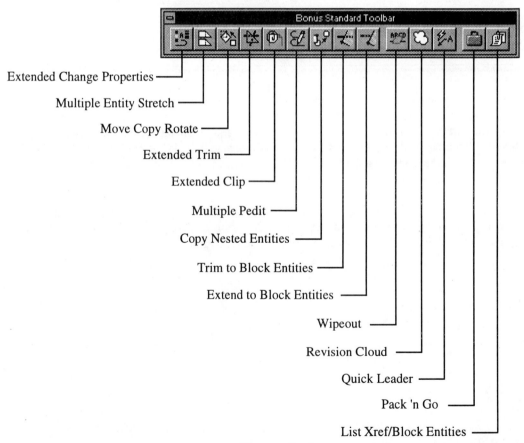

Extended Change Properties

Multiple Entity Stretch

Move Copy Rotate

Extended Trim

Extended Clip

Multiple Pedit

Copy Nested Entities

Trim to Block Entities

Extend to Block Entities

Wipeout

Revision Cloud

Quick Leader

Pack 'n Go

List Xref/Block Entities

Figure A–4

Extended Change Properties

The Extended Change Properties command allows for added controls not normally found in the DDCHPROP or DDMODIFY commands. Executing the EXCHPROP command and selecting the polyline rectangle at "A" in Figure A–5 activates the dialog box shown in Figure A–6. In addition to the regular Color, Layer, Linetype, Linetype Scale, and Thickness controls found in other dialog boxes, the Change Properties dialog box also allows you to control the Width and Elevation of the polyline. Picking the mtext object at "B" in Figure A–5 activates the dialog box in Figure A–7. Notice how you are able to change the text height and style of selected text, mtext, and attribute objects.

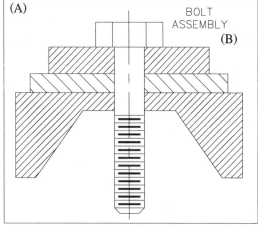

Figure A–5

 Command:**EXCHPROP**

Select object: *(Select an object to change)*

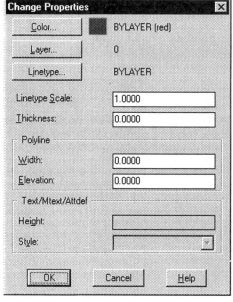

Figure A–6

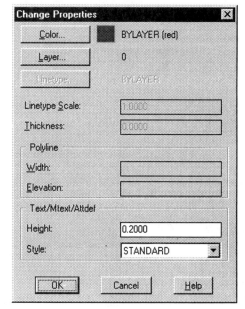

Figure A–7

Multiple Entity Stretch

Normally, the STRETCH command allows you to create only one crossing box to stretch a group of objects. Use this new bonus routine to create multiple crossing boxes to stretch multiple objects while inside the same stretch operation. The MSTRETCH command automatically is activated in a crossing box mode. It does not matter where you pick from; the MSTRETCH command is always in the crossing box mode.

Use Figure A–8 and the prompt sequence below as guides for using the MSTRETCH command.

 Command: **MSTRETCH**

Define crossing windows or crossing
  polygons…
CP(crossing polygon)/<Crossing First point>:
  (Pick at "A")
Other corner: (Pick at "B")
CP/Undo/<Crossing First point>: *(Pick at "C")*
Other corner: (Pick at "D")
CP/Undo/<Crossing First point>: *(Pick at "E")*
Other corner: (Pick at "F")
CP/Undo/<Crossing First point>: *(Pick at "G")*
Other corner: (Pick at "H")
CP/Undo/<Crossing First point>: *(Press* ENTER *to
  continue with this command)*
Done defining windows for stretch…
Remove objects/<Base point>: *(Pick a point
  anywhere on the screen)*
Second base point: **@1<0** *(Or use the Direct
  Distance Mode)*

The result of the multiple stretch operation is illustrated in Figure A–9.

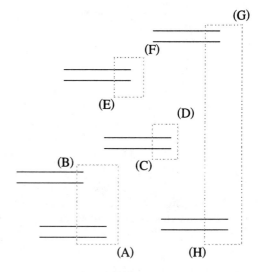

Figure A–8

Figure A–9

Move Copy Rotate

MOCORO, or Move/Copy/Rotate, is a single command that contains options to either move, copy, and/or rotate an object. The object in Figure A–10A will be used to illustrated the use of this command. When you first activate this command and select objects such as the square and hexagon shapes in Figure A–10B, the object highlights along with a marker identifying the current base point. Follow this prompt sequence and Figures A–10A though A–11B to understand how to use this command.

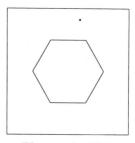

Figure A–10A

 Command: **MOCORO**

Select objects: *(Select the square and hexagon in Figure A–10A)*
Select objects: *(Press ENTER to continue)*
Base point: **Endp**
of *(Select the Endpoint of the square at "A" in Figure A–10B)*
Move/Copy/Rotate/Scale/Base pt/Undo/ <eXit>: **C** *(To Copy)*
Second point of displacement/Undo/<eXit>: *(Pick a new location to copy)*
Second point of displacement/Undo/<eXit>: *(Press ENTER to exit)*
Move/Copy/Rotate/Scale/Base pt/Undo/ <eXit>: **R** *(To Rotate)*
Second point or rotation angle: **45** *(Figure A–11A)*
Move/Copy/Rotate/Scale/Base pt/Undo/ <eXit>: **S** *(To Scale)*
Second Point or Scale factor: **0.75** *(Figure A–11B)*
Move/Copy/Rotate/Scale/Base pt/Undo/ <eXit>: *(Press ENTER to exit the command)*

(A)

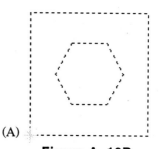

Figure A–10B

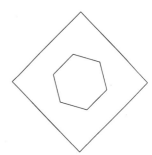

Figure A–11A

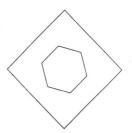

Figure A–11B

Extended Trim

Use this bonus routine to select a cutting edge, pick a side to trim, and have all objects in contact with the cutting edge trim to the selected side. Use the prompt sequence below and Figure A–12 to understand how to use this command.

 Command: **EXTRIM**

Pick a POLYLINE, LINE, CIRCLE, or ARC for cutting edge..
Select objects: *(Pick the edge of the circle at "A" in Figure A–12)*
Pick the side to trim on: *(Pick a point outside of the circle at "B")*

The result is shown in Figure A–13 with all lines outside of the cutting edge trimmed. If a point was selected inside of the cutting edge, all interior lines would be trimmed to the cutting edge.

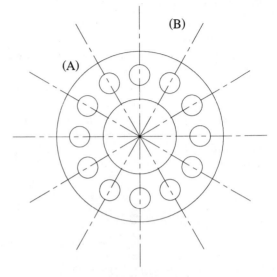

Figure A–12

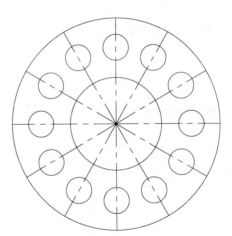

Figure A–13

Multiple Pedit

Multiple polyline edit, or MPEDIT allows you to convert multiple objects into polylines. When you use the PEDIT command, you can convert only one object into a polyline. Of course, you can convert other adjacent objects into polylines with the Join option. However for separate objects, PEDIT only converts one object at a time into a polyline. Use the following prompt sequence and Figure A–14 to understand how this command is used.

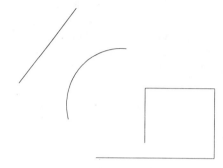

Figure A–14

 Command:**MPEDIT**

Select objects: *(Select the arc and line*
   segments in Figure A–14)
Select objects: *(Press* ⟦ENTER⟧ *to continue)*
Convert Lines and Arcs to polylines? <Yes>:
   (Press ⟦ENTER⟧ *to accept the default)*
Open/Close/Width/Fit/Spline/Decurve/Ltype
   gen/eXit <X>: **W** *(For width)*
Enter new width for all segments: **0.05**
Open/Close/Width/Fit/Spline/Decurve/Ltype
   gen/eXit <X>: *(Press* ⟦ENTER⟧ *to exit this*
   command)

The results of using the MPEDIT command are shown in Figure A–15 with all objects converted into polylines and the width assigned to 0.05 units.

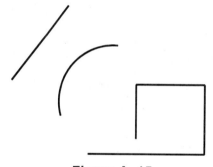

Figure A–15

Trim to Block Entities

The ability to trim to a block object is now possible using the BTRIM command. When you select cutting edges to trim to, the block object temporarily is converted into individual objects. This allows you to pick individual cutting edges from the block. In Figure A–16, the outside edges of the bolt act as cutting edges.

 Command:**BTRIM**

Select cutting edges: *(Pick the edge of the bolt at "A")*

Select cutting edges: *(Pick the edge of the bolt at "B")*

Select cutting edges: *(Press* ENTER *to continue with this command)*

Select objects:

<Select object to trim>/Project/Edge/Undo: *(Pick the line at "C")*

<Select object to trim>/Project/Edge/Undo: *(Pick the line at "D")*

<Select object to trim>/Project/Edge/Undo: *(Pick the line at "E")*

<Select object to trim>/Project/Edge/Undo: *(Press* ENTER *to exit this command)*

The result of using the BTRIM command is shown in Figure A–17. Once this command has been completed, the block returns back to its original state.

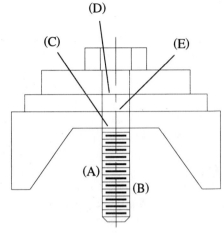

Figure A–16

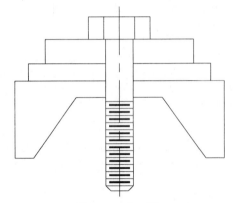

Figure A–17

Extend to Block Entities

The ability to extend to a block object is also possible using the BEXTEND command. When you select boundary edges to extend to, the block object temporarily is converted into individual objects. This allows you to pick individual boundary edges from the block. In Figure A–18, the outside edges of the bolt act as cutting edges.

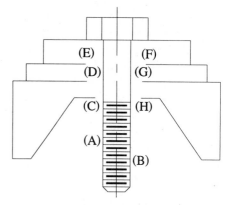

Figure A–18

 Command:**BEXTEND**

Select edges for extend: *(Pick the edge of the bolt at "A")*
Select edges for extend: *(Pick the edge of the bolt at "B")*
Select edges for extend: *(Press* ⌷ENTER⌷ *to continue with this command)*
Select objects:
<Select object to extend>/Project/Edge/Undo: *(Pick the line at "C")*
<Select object to extend>/Project/Edge/Undo: *(Pick the line at "D")*
<Select object to extend>/Project/Edge/Undo: *(Pick the line at "E")*
<Select object to extend>/Project/Edge/Undo: *(Pick the line at "F")*
<Select object to extend>/Project/Edge/Undo: *(Pick the line at "G")*
<Select object to extend>/Project/Edge/Undo: *(Pick the line at "H")*
<Select object to extend>/Project/Edge/Undo: *(Press* ⌷ENTER⌷ *to exit this command)*

The results of using this command are shown in Figure A–19.

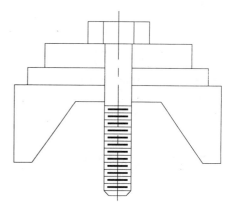

Figure A–19

Wipeout

This bonus routine covers a selected area with a mask of the current background color. Before you use the command, you must first construct a closed polyline around the object or group of objects to mask out. You may use any multisided polyline as the frame for the mask. When you issue the WIPEOUT command, it prompts you to identify a new frame. After you select the polyline, an additional prompt displays and offers you the choice to erase or save the polyline.

 Command:**WIPEOUT**

Frame/New <New>: *(Press* ENTER *to accept default)*
Select a polyline: *(Pick the polyline at "A" in Figure A–20)*
Erase polyline? Yes/No <No>:**YES**
Wipeout created.

The results of using WIPEOUT are illustrated in Figure A–21 with the island in the center of the kitchen being wiped out or masked over. In addition to the island being masked, also notice the centerline struck from the window and the hidden line at the right side of the countertip are also masked.

If it appears the polyline is not deleted, it is because the frame is still turned on. Reissue the WIPEOUT command, use the Frame option, and turn it off.

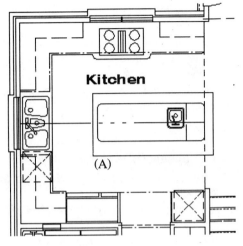

Figure A–20

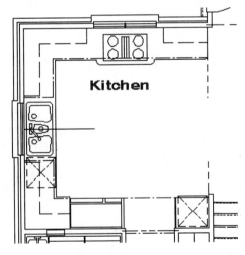

Figure A–21

Revision Cloud

Use this routine to create a revision cloud. The components of the cloud consist of a series of arcs in sequence that form a closed polyline object. You can adjust an arc length to reflect a cloud that has larger or smaller arcs. First pick a starting point for the arc; then move the crosshairs in the counterclockwise direction. This automatically creates the cloud in the current direction until the cloud is finished. Figure A–22 shows a 3D assembly model that has a revision cloud drawn around one of the components.

It is recommended that the REVCLOUD command be used with Snap turned off. If you get caught in an endless loop because snap was turned on, press the ESC key to cancel the command.

 Command:**REVCLOUD**

Arc length set at 0.375
Arc length/<Pick cloud starting point>: *(Pick the start of the cloud)*
Guide crosshairs along cloud path…
(Move the crosshairs in a counterclockwise direction until the cloud begins to form)
Cloud finished.

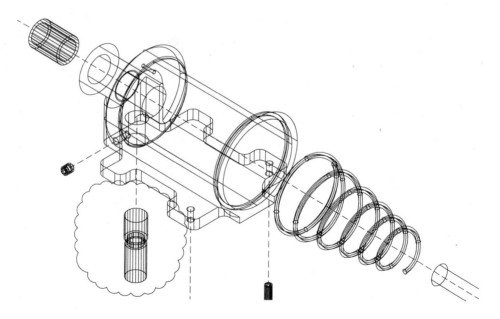

Figure A–22

Quick Leader

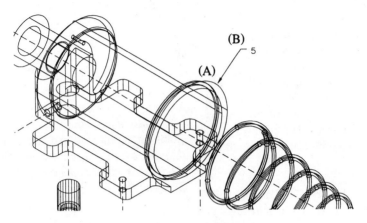

Figure A–23

The QLEADER bonus routine allows you to draw a quick leader (see Figure A–23). Respond to the prompts by identifying a first point, and then a series of next points until the leader is drawn. Pressing the ENTER key in response to the first leader point activates the dialog box shown in Figure A–24; use this to specify what kind of leader you want to create.

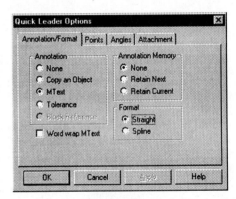

Figure A–24

 Command:**QLEADER**

First Leader point or press [ENTER] to set Options:
    (Pick at "A" in Figure A–23)
Next Leader point: *(Pick at "B")*
Next Leader point: *(Press [ENTER] to continue)*
Enter Leader text:**5**
Enter Leader text: *(Press [ENTER] to draw the
    leader and place the text)*

Other commands associated with the Quick Leader Options dialog box include:

QLATTACH—Use this command to attach a leader line to an mtext, tolerance, or block object.

QLDETACHSET—Use this command to detach a leader line from an mtext, tolerance, or block object.

QLATTACHSET—This command globally attaches leader lines to mtext, tolerance, or block objects.

Pack 'n Go

Use this bonus command to copy all files associated with a drawing along with all associated support files to a designated location. In Figure A–25, three drawing files are listed in the dialog box. Three drawing files are associated with this file along with the standard AutoCAD font map file (acad.fmp) and numerous ttf (true type font) and shx (AutoCAD font) files. This routine is primarily designed to make it easier to share information with another individual or company by providing all support files associated with the drawing. In this way, files are not accidentally forgotten when they are being shared.

 Command: **PACK**

(The list of files is automatically generated in Figure A–25. Providing a destination in the "Copy to:" edit box copies all files in the list to the designated area.)

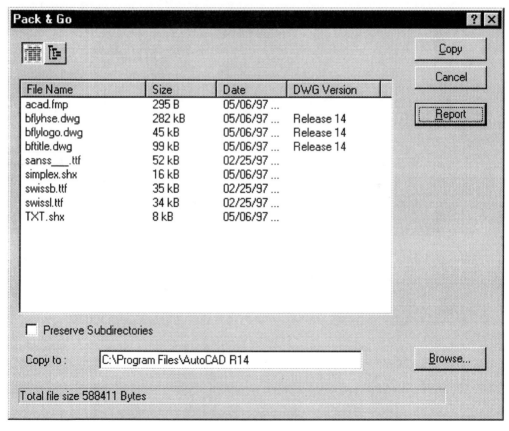

Figure A–25

The Bonus Text Toolbar

The Bonus Text Tools toolbar consists of programs designed to manipulate text and attributes.

Commands included in the Bonus Text Tools toolbar in Figure A–26 include the ability to: expand or compress text using a text fit command; mask objects that may cross on top of text objects; globally change text style, height, color, layer and other properties; explode text into individual objects to create text that is placed along an arc and edit this text; perform search and replace operations on text objects and two items dealing with attributes; explode attributes into individual text objects; and globally edit attributes.

In addition to locating these commands in the toolbar in Figure A–26, you can also find these programs in the Bonus pull-down menu area under "Text".

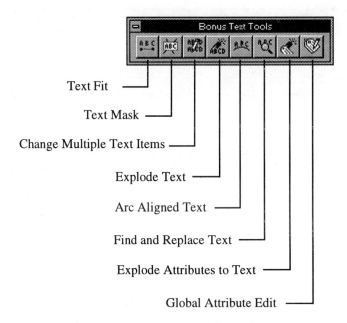

Text Fit

Text Mask

Change Multiple Text Items

Explode Text

Arc Aligned Text

Find and Replace Text

Explode Attributes to Text

Global Attribute Edit

Figure A–26

Text Fit

Use this bonus text routine to compress or stretch text objects by picking new start points and end-points. At this time, the command will only operate on text objects that you made using the DTEXT command. When selecting the text to stretch or compress, the operation begins at the original insertion point of the text object. Moving the cursor to the right of the insertion point compresses the text; moving to the left of the insertion point expands the text.

To compress text, use the following prompt sequence and Figure A–27.

 Command:**TEXTFIT**

Select Text to stretch/shrink: *(Pick the text "INDUSTRIAL TECHNOLOGY")*
Starting Point/<Pick new ending point>: **S** *(for starting point)*
Pick new starting point: *(Pick at "A" in Figure A–27)*
ending point: *(Pick at "B")*

To expand text, use the following prompt sequence Figure A–28.

 Command:**TEXTFIT**

Select Text to stretch/shrink: *(Pick the text "INDUSTRIAL")*
Starting Point/<Pick new ending point>: **S** *(for starting point)*
Pick new starting point: *(Pick at "A" in Figure A–28)*
ending point: *(Pick at "B")*

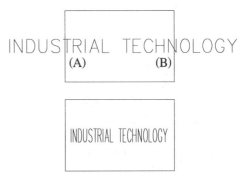

Figure A–27

Figure A–28

Text Mask

Use the TEXTMASK command to hide all objects that come in contact with a selected text object. This command works on both text and mtext objects. It uses a rectangular frame and an offset distance from the text. You may specify this distance at the command prompt. TEXTMASK actually works in conjunction with the WIPEOUT command to hide objects crossing over the top of text. Use the following prompt sequence below and Figure A–29 for this command.

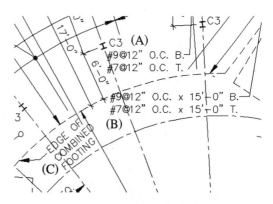

Figure A–29

 Command:**TEXTMASK**

Enter offset factor relative to text height
    <0.35>: *(Press ⌜ENTER⌝ to accept this default)*
Select Text to MASK…
Select objects: *(Select the text objects at "A",*
    "B", and "C" in Figure A–29)
Select objects: *(Press ⌜ENTER⌝ to perform the text*
    mask operation)
Wipeout created.
Wipeout created.
Wipeout created.

The result of using the TEXTMASK command is shown in Figure A–30. Previously, center and object lines crossed through the text objects. Now, the text has a mask underneath that prevents the lines from crossing through the text object. The TEXTMASK command does not operate on associative dimension text at this time.

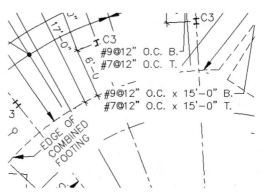

Figure A–30

Change Multiple Text Items

This bonus text routine is sometimes compared to a text processor inside of AutoCAD. It can be used to edit text strings individually or globally. A few of the command options include editing the text height, justification, insertion point (location), rotation angle, style, and width factor. You can also edit the actual text string. Figure A–31 consists of the letters A through F; this figure was originally drawn using the DTEXT command in the STANDARD text style. The identifying letters AB through EF along with the distances consist of a single mtext object. The CHT command works on both mtext as well as objects created with the DTEXT command. Use the following prompt sequence and Figure A–31 to change all text to a new height of 0.15 units and to change the text style from STANDARD to ROMAND. (The ROMAND text style must already be created.)

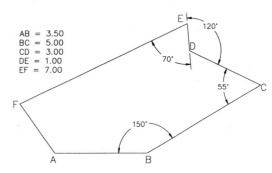

Figure A–31

 Command: **CHT**

Select annotation objects to change.
Select objects: *(Select all text and mtext objects in Figure A–31; do not select the dimension text)*
Select objects: *(Press [ENTER] to continue)*
7 annotation objects found.
Height/Justification/Location/Rotation/Style/
Text/Undo/Width: **H** *(For height)*
Individual/List/<New height for all text objects>:
0.15
Height/Justification/Location/Rotation/Style/
Text/Undo/Width: **S** *(For style)*
Individual/List/Select style/<New style name for all text objects>: *(Press [ENTER] to accept default)*
New style name for all text objects: **ROMAND**
Height/Justification/Location/Rotation/Style/
Text/Undo/Width: *(Press [ENTER] to exit this command)*

The completed object is shown in Figure A–32.

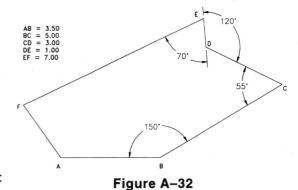

Figure A–32

Explode Text

Use this bonus text routine to explode text or mtext objects into geometry consisting of individual lines and/or arcs.

The illustration at "A" in Figure A–33 shows the word "NORMAL" which consists of a regular mtext object. When selecting this object, all lines and arcs that define the text highlight.

The illustration at "B" in Figure A–33 shows the word "EXPLODED." When selecting this text object at "C," only the line highlights. This is the result of exploding a text or mtext object; it is broken down into the individual segments that define the text.

In some cases, the lines and arcs of an exploded text object may be used to machine text into an object using a CAM (computer-aided manufacturing) package. The text lines and arcs act as contours that guide the milling cutter.

Use the following prompt sequence to explode text.

 Command: **TXTEXP**
Select text to be EXPLODED:
Select objects: *(Select the word "EXPLODED" in Figure A–33)*
Select objects: *(Press* ENTER *to execute the exploding of text)*
1 found.
1 text object(s) have been exploded to lines.
The line objects have been placed on layer 0.

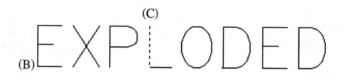

Figure A–33

Arc Aligned Text

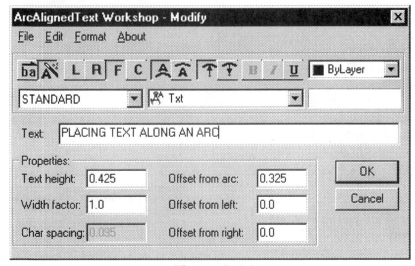

Figure A–34

The ARCTEXT command is used to place text along an arc. Activating the command and selecting an arc displays the dialog box in Figure A–34. Numerous settings such as text height, width factor, offset distance from the arc, and text font and style information can be changed through this dialog box. Also, a series of controls at the top of the dialog box allow you to place text above or below the arc, to have the text left, right, or center justified, or bold and/or underlined. Arctext is placed using the following command sequence.

 Command: **ARCTEXT**

Select an Arc or an ArcAlignedText: *(Select an arc or an existing arc aligned text object)*

To edit arc text, use the same command (ARCTEXT) and select the arctext already placed. The dialog box will again appear similar to Figure A–34.

Figure A–35A displays arc text placed above an arc; Figure A–35B displays arc text placed below an arc; Figure A–35C displays arc text that has its font changed to ROMAND.

Figure A–35A

Figure A–35B

Figure A–35C

Find and Replace Text

(A)

Use this bonus text routine to locate and replace text or text strings. This command operates on text created with the DTEXT command and is not yet available to find and replace text consisting of attribute or mtext objects.

Part of an entity can be removed using the Break command. You can break lines, circles, arcs, polylines, ellipses, splines, xlines, and rays. When breaking an entity, you can either select the entity at the first break point and then specify a second break point, or you can select the entire entity and then specify the two break points.

Figure A–36

 Command: **FIND**

(After the command is entered, the Find and Replace dialog box appears. Make the necessary changes in Figure A–37 and click the OK button)

Select objects: *(Pick the text at "A" in Figure A–36)*

Select objects: *(Press* ⎡ENTER⎤ *to continue)*

(When the Find and Replace dialog box appears similar to Figure A–38, click on the Replace button to replace the text)

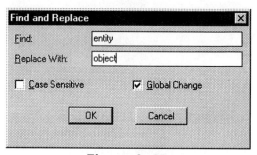

Figure A–37

In Figure A–37, the text to Find is "entity"; the replacement text is "object". Clicking the OK button and selecting a line of text at "A" displays the dialog box in Figure A–38. "1 of 1" means Find and Replace found one line of text out of one line of text. Clicking on the Replace button replaces the text.

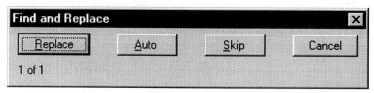

Figure A–38

Numerous lines of text may be selected at the command prompt to perform a find and replace operation using a single command. The results of using the FIND command to replace "entity" with "object" is shown in Figure A–39.

Part of an object can be removed using the Break command. You can break lines, circles, arcs, polylines, ellipses, splines, xlines, and rays. When breaking an object, you can either select the object at the first break point and then specify a second break point, or you can select the entire object and then specify the two break points.

Figure A–39

Explode Attributes to Text

The BURST text command works on blocks that contain attribute information. This command explodes a block object and converts its attribute values into text objects. Figure A–40 shows a block symbol. Two attribute values are present; namely, DIFFUSER and a catalog number SC249EA.

Whenever you explode a block containing attribute values, the results are similar to Figure A– 41. In this example, the block is converted into individual objects; however the attribute text lose their values and convert to their tags. The two tags that define the values in Figure A–40 are NAME and CATALOG_NO.

In addition to exploding the symbol into individual lines such as "A" in Figure A–42, the BURST command maintains the attribute values by not exploding the text back into the original attribute tag. Study the following command prompt and Figure A–42.

 Command:**BURST**

Select objects: *(Select the image of the diffuser in Figure A–42)*
Select objects: *(Press* ENTER *to execute the command)*

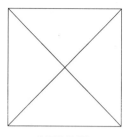

DIFFUSER
SC249EA

Figure A–40

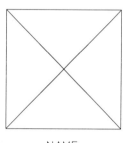

NAME
CATALOG_NO.

Figure A–41

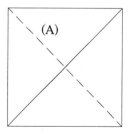

(A)

DIFFUSER
SC249EA

Figure A–42

The Bonus Layer Toolbar

The Bonus Layer Tools toolbar shown in Figure A–43 contains eight programs to assist in controlling layers. The layer manager command (LMAN) allows you to name, save, and restore the settings made in the Layer & Linetype Properties dialog box. These layer settings may be saved to a .LAY extension which may be imported into other drawings. Other layer commands include the ability to: match an objects layer with that of another; pick an object to make the layer of that object current; and pick a layer and isolate it where all other layers are turned off. Still other commands allow you to freeze, turn off, lock or unlock an object's layer by just picking an object.

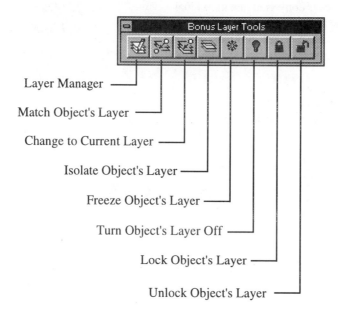

Figure A–43

Layer Manager

Use this powerful bonus tool to manage the settings made in the Layer and Linetype properties dialog box. The Layer Manager or LMAN command allows you to save and restore layer configurations into "layer states." Certain layers can be turned on or off and saved in the Layer Manager dialog box in Figure A–44. Layer states are saved in the drawing and can also be exported to or imported from a LAY file.

Figure A–45 is an office plan drawing. The following layer states have been saved in Figure A– 44:
   Furniture
   Total_Drawing
   Walls

Since the layer state called "Walls" has all layers but the building walls frozen, clicking on WALLS in the Layer Manager dialog box in Figure A–44 displays the image in Figure A–46.

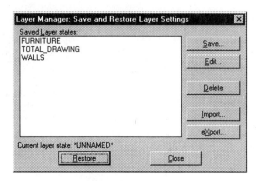

Figure A–44

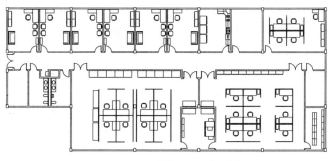

Figure A–45

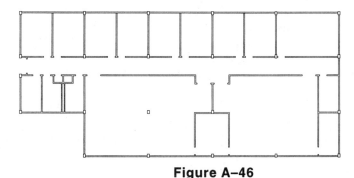

Figure A–46

Match Object's Layer

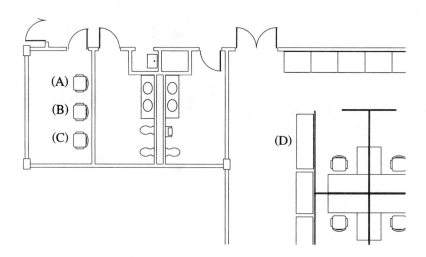

Figure A–47

This layer bonus routine allows you to change the layers of selected objects to match the layer of a selected destination object. The command is LAYMCH. First select the objects to be changed such as the three chairs illustrated in Figure A–47. Next select the object on the destination layer as in the shelf at "D." The three chairs now take on the same layer as the shelf.

 Command:**LAYMCH**
Select objects to be changed:
Select objects: *(Select the three chairs labeled "A", "B", and "C")*
Select objects: *(Press* ENTER *to continue)*
3 found.
Type name/Select entity on destination layer:
  (Pick the shelf at "D")
3 objects changed to layer FURNITURE.

Change to Current Layer

This layer bonus routine is used to change the layer of one or more objects to the current layer. This is particularly helpful if objects were constructed on the wrong layer. In the past, the DDCHPROP or DDMODIFY commands were used to move objects to a new layer. The LAYCUR command operates in a more efficient manner than previously used commands. In Figure A–48, the three chairs labeled "A," "B," and "C" were drawn on the wrong layer. First, make the desired layer current; in Figure A–48, the current layer is FURNITURE. Selecting the three chairs after issuing the LAYCUR command changes the three chairs to the FURNITURE layer.

 Command: **LAYCUR**

Select objects to be CHANGED to the current layer:

Select objects: *(Select the three chairs "A", "B", and "C")*

Select objects: *(Press [ENTER] to continue)*

3 found.

3 objects changed to layer FURNITURE *(the current layer).*

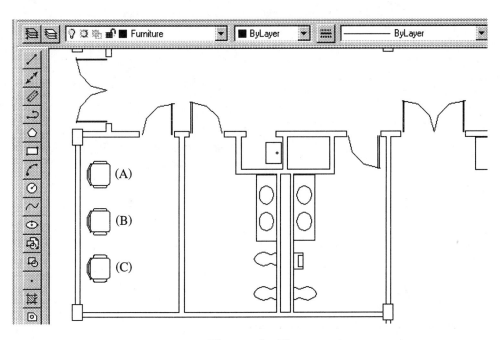

Figure A–48

Isolate Object's Layer

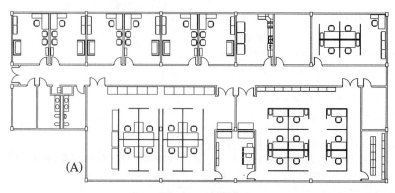

(A)

Figure A–49

This layer bonus routine isolates the layer or layers of one or more selected objects by turning all other layers off. The effects of this command are the same as using the Layer & Linetype Properties dialog box and picking the layer names to turn off. The LAYISO command is yet another bonus tool that operates in a more efficient manner than conventional AutoCAD commands. After issuing the command, you are prompted to select the object or objects on the layer or layers to be isolated. In Figure A–49, pick any outside wall and press the ENTER key to execute the command. A message in the command prompt area alerts you that layer 1-WALL was iso-

lated. All objects on 1-WALL are shown in Figure A–50; all other layers are turned off.

 Command: **LAYISO**

Select object(s) on the layer(s) to be ISOLATED:

Select objects: *(Pick the wall at "A" in Figure A–49)*

Select objects: *(Press ENTER to isolate the layers based on the objects selected)*

Layer 1-WALL has been isolated.

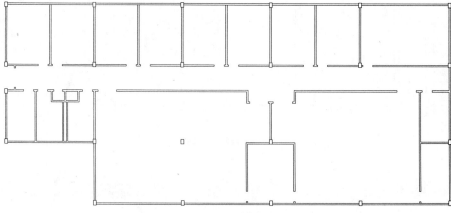

Figure A–50

Freeze Object's Layer

This layer bonus routine freezes layers by picking objects to control the layers to be frozen. Again, the Layer & Linetype Properties dialog box is normally used to freeze and thaw layers. As with other layer concepts, the current layer can never be frozen.

 Command: **LAYFRZ**

Options/Undo/<Pick an object on the layer to be FROZEN>: *(Pick one of the chairs in Figure A–49)*
Layer CHAIRS has been frozen.
Options/Undo/<Pick an object on the layer to be FROZEN>: *(Pick one of the shelves in Figure A–49)*
Layer FURNITURE has been frozen.

Options/Undo/<Pick an object on the layer to be FROZEN>: *(Pick one of the bathroom sinks in Figure A–49)*
Layer FIXTURES has been frozen.
Options/Undo/<Pick an object on the layer to be FROZEN>: *(Pick one of the kitchen refrigerators in Figure A–49)*
Layer KITCHEN has been frozen.
Options/Undo/<Pick an object on the layer to be FROZEN>: *(Pick one of the doors in Figure A–49)*
Layer DOOR has been frozen.
Options/Undo/<Pick an object on the layer to be FROZEN>: *(Press ENTER to exit this command. The results should be similar to Figure A–51 below)*

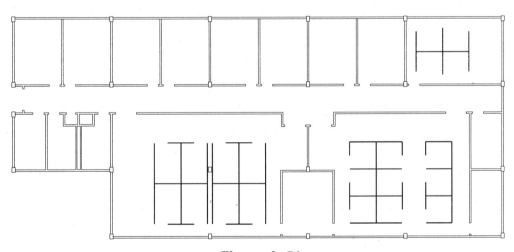

Figure A–51

Turn Object's Layer Off

This layer bonus routine is similar to the LAYFRZ command except instead of freezing layers, you can turn off a layer or group of layers by selecting an object or group of objects. The LAYOFF command follows and references Figure A–49.

 Command:**LAYOFF**

Options/Undo/<Pick an object on the layer to be turned OFF>: *(Pick one of the shelves in Figure A–49)*
Layer FURNITURE has been turned off.

Options/Undo/<Pick an object on the layer to be turned OFF>: *(Pick one of the office partitions in Figure A–49)*
Layer 1-PART has been turned off.
Options/Undo/<Pick an object on the layer to be turned OFF>: *(Pick one of the chairs in Figure A–49)*
Layer CHAIRS has been turned off.
Options/Undo/<Pick an object on the layer to be turned OFF>: *(Press* ENTER *to exit this command. The results should be similar to Figure A–52)*

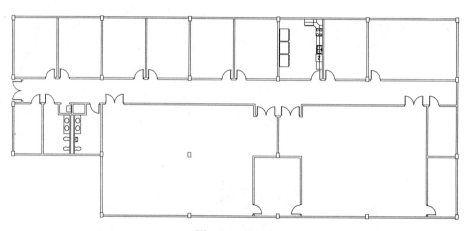

Figure A–52

Lock Object's Layer

This layer bonus routine locks the layer of a selected object. A locked layer is visible on the display screen; however, any object associated with a locked layer is nonselectable. The command to lock a layer is LAYLCK.

 Command: **LAYLCK**

Pick an object on the layer to be LOCKED: *(Pick one of the exterior walls in Figure A–49)*

Layer 1-WALL has been locked.

 # Unlock Object's Layer

This layer bonus routine unlocks the layer of a selected object. Objects on an unlocked layer can now be selected whenever the prompt "Select objects:" appears. The command to lock a layer is LAYULK.

 Command: **LAYULK**

Pick an object on the layer to be UNLOCKED: *(Pick one of the exterior walls in Figure A–49)*

Layer 1-WALL has been unlocked.

Activating the Bonus Popup Menu

A special feature of the bonus routines can be found in the Tools area of the Bonus pull-down menu area. Picking on Popup Menu in Figure A–53 activates the bonus pop up menu. Pressing CTRL-Right Mouse Click displays the bonus popup; pressing ALT-Right Mouse Click allows for the configuration of a different bonus popup menu. The bonus popup menu capability may also be entered in from the keyboard using the command BONUSPOPUP.

By default, pressing CTRL-Right Mouse Click displays the bonus popup menu illustrated in Figure A–54.

Pressing ALT-Right Mouse Click displays the "Pick a Popup Menu: dialog box shown in Figure A–55. Clicking on one of the existing pull-down menu categories assigns a new alternate popup menu.

To display this popup menu, press CTRL-Right Mouse Click.

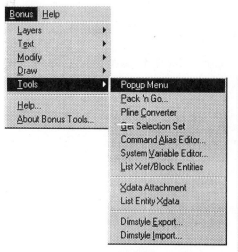

Figure A–53

Figure A–54

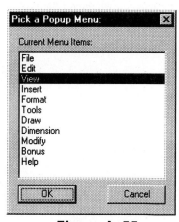

Figure A–55

Exporting and Importing Dimension Styles

Other useful utilities found by clicking Tools located in the Bonus pull-down menu area are Dimstyle Export and Dimstyle Import shown in Figure A–56. Dimstyle Export, or DIMEX, is designed to create a file containing selected dimension styles along with the dimension settings associated with this file. A special extension of .DIM is added to this file. Entering DIMEX displays the Dimension Style Export dialog box shown in Figure A–57. A list of all available dimension styles is displayed; click on the desired dimension style or styles to export in the .DIM file format.

Once you have created a .DIM file, you can import it into any drawing file by using the Dimstyle Import or DIMIM command. Entering in this command displays the Dimension Style Import dialog box shown in Figure A–58. Use this dialog box to browse through all .DIM files until the desired file is found and loaded into the current drawing file.

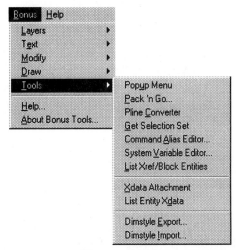

Figure A–56

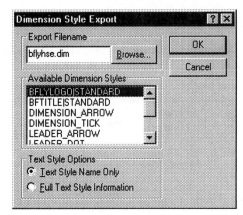

Figure A–57

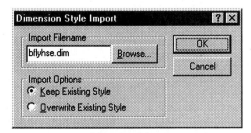

Figure A–58

Index